ISBN 3-540-96346-4 Springer-Verlag Berlin Heidelberg New York Tokyo
ISBN 0-387-96346-4 Springer-Verlag New York Heidelberg Berlin Tokyo

PREFACE

This book represents an outgrowth of an interdisciplinary session
held at the Seventh International Estuarine Research Federation
Conference held at Virginia Beach, Virginia, October 1983. At that
meeting, the participants agreed to contribute to and develop a
monograph entitled "Tidal Mixing and Plankton Dynamics" by inviting
an expanded group of authors to contribute chapters on this theme.
The emphasis would be to review and summarize the considerable body
of knowledge that has accumulated over the last decade or so on the
fundamental role tidal mixing plays in energetic shallow seas and
estuaries in stimulating and controlling biological production.

We have attempted to provide a mix of contributions, composed of
reviews of the state-of-the-art, reports on current research activities, summaries of the design and testing of a new generation of
innovative instruments for biological and chemical sampling and
sorting, and some imaginative ideas for future experiments on
stimulated mixing in continental shelf seas.

We encouraged the contributors to present critical and thought-
provoking assessments of current wisdom specifying the sorts of
techniques and observational strategies needed to validate the
various hypotheses linking physical structure, mixing and circulation
to plankton biomass and production.

We hope this volume will appeal to incoming research students and
established scholars alike. We certainly have enjoyed working with
all the authors in compiling this book. We thank the numerous
scientists who have served as reviewers, P. Boisvert for typing the
manuscripts and W. Bellows for proofreading.

MALCOLM J. BOWMAN
CLARICE M. YENTSCH
WILLIAM T. PETERSON

Lecture Notes on Coastal and Estuarine Studies

Lecture Notes on
Coastal and Estuarine
Studies

Managing Editors:
Malcolm J. Bowman
Richard T. Barber
Christopher N. K. Mooers

17

Tidal Mixing
and Plankton Dynamics

Edited by Malcolm J. Bowman, Clarice M. Yentsch
and William T. Peterson

Springer-Verlag
Berlin Heidelberg New York Tokyo

Stac Lee and Boreray Islands in the St. Kilda group of islands off the northwest coast of Scotland. Stac Lee is greater than 200 m high and is host to the largest gannet colony in Europe. For a discussion of island mixing effects, see Simpson and Tett (this volume).

Photo credit - J.H. Simpson

TABLE OF CONTENTS

LIST OF AUTHORS AND CONTRIBUTORS

W.M. BALCH : Institute of Marine Resources A018
Scripps Institute of Oceanography
University of California, San Diego
La Jolla, CA 92093
U.S.A.

J.D. BOON, III : Virginia Institute of Marine Sciences
Gloucester Point, VA 23062
U.S.A.

M.J. BOWMAN : Marine Sciences Research Center
State University of New York
Stony Brook, NY 11794
U.S.A.

J.W. CAMPBELL : Bigelow Laboratory for Ocean Sciences
McKown Point
W. Boothbay Harbor, ME 04575
U.S.A.

L.A. CODISPOTI : Bigelow Laboratory for Ocean Sciences
McKown Point
W. Boothbay Harbor, ME 04575
U.S.A.

T.L. CUCCI : Bigelow Laboratory for Ocean Sciences
McKown Point
W. Boothbay Harbor, ME 04575
U.S.A.

G.R. DABORN : Department of Biology
Acadia University
Wolfville, Nova Scotia BOP 1X0
CANADA

S. DEMERS : Centre Champlain des Sciences de la Mer
Ministère des Pêches et des Océans
C.P. 15 500, 901 Cap Diamant
Québec, G1K 7Y7
CANADA

B.R. DILKE : P.O. Box 2085
Sidney, B.C.
CANADA

W.E. ESAIAS : Code 671
NASA
Goddard Space Flight Center
Greenbelt, MD 20771
U.S.A.

D.A. EVANS : Virginia Institute of Marine Science
School of Marine Science
College of William and Mary
Gloucester Point, VA 23062
U.S.A.

G.C. FELDMAN : Marine Sciences Research Center
State University of New York
Stony Brook, NY 11794
U.S.A.

G.E. FRIEDERICH : Bigelow Laboratory for Ocean Sciences
McKown Point
W. Boothbay Harbor, ME 04575
U.S.A.

K.D. FRIEDLAND : Virginia Institute of Marine Sciences
Gloucester Point, VA 23062
U.S.A.

J.J. GRAHAM : Marine Department of Marine Resources
W. Boothbay Harbor, ME 04575
U.S.A.

L.W. HAAS : Virginia Institute of Marine Sciences
Gloucester Point, VA 23062
U.S.A.

D. HAYWARD : Virginia Institute of Marine Sciences
Gloucester Point, VA 23062
U.S.A.

P.M. HOLLIGAN : Marine Biological Association
Citadel Hill, Plymouth PL2 2PB
UNITED KINGDOM

A.W.G. JOHN : Institute for Marine Environmental Research
Prospect Place, The Hoe PL1 3DH
UNITED KINGDOM

P.J. KELLY : Bigelow Laboratory for Ocean Sciences
McKown Point
W. Boothbay Harbor, ME 04575
U.S.A.

L. LEGENDRE : GIROQ, Départment de Biologie
Université Laval
Québec, G1K 7P4
CANADA

G.T. MARDELL : Marine Biological Association of the U.K.
Citadel Hill, Plymouth PL1 2PB
UNITED KINGDOM

A. OKUBO : Marine Sciences Research Center
State University of New York
Stony Brook, NY 11794
U.S.A.

R.I. PERRY : Department of Fisheries and Oceans
Marine Fish Division
Biological Station
St. Andrews, N.B. EOG 2X0
CANADA

W.T. PETERSON : Marine Sciences Research Center
State University of New York
Stony Brook, NY 11794
U.S.A.

D.A. PHINNEY : Bigelow Laboratory for Ocean Sciences
McKown Point
W. Boothbay Harbor, ME 04575
U.S.A.

R.D. PINGREE : Marine Biological Association of the U.K.
Citadel Hill, Plymouth PL1 2PB
UNITED KINGDOM

P.C. REID : Institute for Marine Environmental Research
Prospect Place, The Hoe
Plymouth PL1 3DH
UNITED KINGDOM

E.P. RUZECKI : Virginia Institute of Marine Science
School of Marine Science
College of William and Mary
Gloucester Point, VA 23062
U.S.A.

J.H. SIMPSON : Marine Science Laboratories
Menai Bridge
Gwynedd, LL59 5EY
UNITED KINGDOM

D.K. STEVENSON : Department of Zoology
University of Maine, Orono, ME 04469 and
Marine Department of Marine Resources
W. Boothbay Harbor, ME 04575
U.S.A.

P.B. TETT : Scottish Marine Biological Association
Dunstaffnage
Oban, Argyll PA34 4AD
UNITED KINGDOM

J.-C. THERRIAULT : Centre Champlain des Sciences de la Mer
Ministère des Pêches et des Océans
C.P. 15 500, 901 Cap Diamant
Québec, G1K 7Y7
CANADA

J.A. TOPINKA : Bigelow Laboratory for Ocean Sciences
McKown Point
W. Boothbay Harbor, ME 04575
U.S.A.

D.W. TOWNSEND : Bigelow Laboratory for Ocean Sciences
McKown Point
W. Boothbay Harbor, ME 04575
U.S.A.

A. TVIRBUTAS : Charles S. Draper Laboratory
555 Technology Square
Cambridge, MA 02138
U.S.A.

T.E. WHITLEDGE : Oceanographic Sciences Division
                   Brookhaven National Laboratory
                   Upton, NY 11973
                   U.S.A.

R.E. WILSON   : Marine Sciences Research Center
                   State University of New York
                   Stony Brook, NY 11794
                   U.S.A.

C.D. WIRICK   : Department of Marine Science
                   University of South Florida
                   St. Petersburg, FL 33701
                   U.S.A.

C.M. YENTSCH   : Bigelow Laboratory for Ocean Sciences
                   McKown Point
                   W. Boothbay Harbor, ME 04575
                   U.S.A.

C.S. YENTSCH   : Bigelow Laboratory for Ocean Sciences
                   McKown Point
                   W. Boothbay Harbor, ME 04575
                   U.S.A.

PHYTOPLANKTON RESPONSES TO VERTICAL TIDAL MIXING*

S. Demers
Centre Champlain des Sciences de la Mer
Ministère des Pêches et des Océans
C.P. 15 500, 901 Cap Diamant
Québec, Québec, Canada G1K 7Y7

L. Legendre
GIROQ, Département de biologie
Université Laval
Québec, Québec, Canada G1K 7P4

J.-C. Therriault
Centre Champlain des Sciences de la Mer
Ministère des Pêches et des Océans
C.P. 15 500, 901 Cap Diamant
Québec, Québec, Canada G1K 7Y7

INTRODUCTION

Vertical mixing of the water column in estuaries and other coastal environments requires an input of mechanical energy that is mainly provided by the tides, the wind stress on the water surface and the fresh water runoff. Variations in these three sources are known to have a marked influence on the phytoplankton. At the seasonal scale, river runoff has been identified as an important driving force of phytoplankton dynamics. For example, Gilmartin (1964) has shown that the increased river runoff during the winter in Indian Arm (a fjord of Western Canada) destabilizes the water column which favors the replenishment of the surface mixed layer in nutrients. These nutrients are subsequently used for the initiation of the phytoplankton spring bloom, when the water column stabilizes following the decrease in runoff. At the time scale of a few days, physical transient phenomena (wind storms, periodic upwelling, etc.) associated with the passage of frontal disturbances (Heath, 1973; Walsh et al., 1977) are also strong destabilizing agents of the water

*Authors listed alphabetically

Lecture Notes on Coastal and Estuarine Studies, Vol. 17
Tidal Mixing and Plankton Dynamics. Edited by J. Bowman, M. Yentsch and W.T. Peterson
© Springer-Verlag Berlin Heidelberg 1986

column. Iverson et al. (1974), Takahashi et al. (1977), Walsh et al. (1978), Walsh (1981) and Legendre et al. (1982) have reported intermittent phytoplankton blooms following stabilization of a water column previously destabilized by strong winds. Côté and Platt (1983) have found that the normal progression of the phytoplankton community structure in a small coastal inlet was periodically perturbed by transient phenomena. Therriault et al. (1978) and Therriault and Platt (1981) have also demonstrated for another coastal inlet that, in conditions of low turbulent mixing (wind <5 m·s^{-1}), phytoplankton patchiness was induced by local differences in growth rates and that, in conditions of high wind (≥ 5 m·s^{-1}), spatial variations were damped out by the horizontal turbulent processes. Similarly, George and Heaney (1978) showed that wind velocities ≥ 1.2 m·s^{-1} were sufficient to break down patches of phytoplankton in a small lake. It seems, however, that stronger winds would be needed to homogenize an estuarine environment subject to strong tidal currents as compared to more stable coastal areas (Levasseur et al., 1983). In an estuarine environment such as Long Island Sound, the tide is the principal source of energy that is used in the mixing processes (Bokuniewicz and Gordon, 1980), since a tidal stream does an amount of mixing equivalent to that of a wind with a velocity 25 times higher (Sinclair et al., 1981). The influence of vertical tidal mixing processes on the dynamics of phytoplankton is the specific topic that will be reviewed in this paper. Horizontal mixing and dispersion by the tides will not be discussed, as horizontal effects on phytoplankton are more mechanical than physiological (Legendre and Demers, 1984).

Since tidal mixing can be considered as a special case of vertical mixing, we shall examine, in a first section, the general implications of vertical mixing for phytoplankton. Then, in subsequent sections, we shall consider the observed and predicted responses of phytoplankton to tidal changes in vertical stability, before dealing with internal tides and tidal fronts. Considering that several phytoplankton responses to either internal tides or tidal fronts are special cases of the response to tidal changes in vertical stability, many cross references will be made. Finally, we shall summarize the information in the last section by presenting a general conceptual framework showing how tidal mixing can affect phytoplankton on the M_2 and MS_f spatio-temporal scales.

Effects of Vertical Mixing on Phytoplankton

When discussing the effects of vertical mixing on phytoplankton, one must bear in mind that these effects are generally not direct, but through the proximal agency of light and/or nutrient fluctuations. White (1976) has reported growth inhibition caused by agitation in the marine dinoflagellate Gonyaulax. Similarly, Savidge (1981) has shown that turbulence can affect the growth kinetics, the nutrient uptake and the photosynthesis of marine phytoplankton cultures. Such small-scale effects of agitation have unfortunately not been demonstrated on natural phytoplankton, but its larger-scale effects through the agency of light and nutrients are well documented. The depth of the layer for optimal phytoplankton growth is determined by the interrelationship between the downward light gradient and the upward nutrient gradient. The downward light flux cannot be vertically mixed, but the phytoplankton cells can be vertically redistributed relative to the light gradient, so as to escape prolonged exposure to near-surface inhibiting intensities or sinking below the photic layer. On the contrary, the upward flux of nutrients can be greatly increased by the vertical mixing. If the phytoplankton cells are taken as the fixed reference, rather than the actual physical space, vertical mixing can result in homogenizing both the nutrient and light responses of the cells on the vertical. The range of time and space scales of vertical cycling of phytoplankton by turbulent eddy diffusion, internal wave motions, Langmuir circulation, double diffusive processes and sinking have been estimated by Denman and Gargett (1983).

Increased photosynthetic activity has been demonstrated, both in the laboratory and in natural conditions, for phytoplankton exposed to cyclical changes in light intensity (Harris and Lott, 1973; Jewson and Wood, 1975; Harris and Piccinin, 1977; Harris, 1978; Marra, 1978a,b). This might be due to a shortened time of exposure of phytoplankton cells to bright light, with the result that high rates of photosynthesis could be maintained before photoinhibition and photorespiration become prevalent (Harris, 1980b). Gallegos et al. (1983) have indicated that Arctic populations from subsurface chlorophyll maxima are less susceptible to surface photoinhibition than those from below the thermocline. Vertical mixing can also favor the maintenance in adequate light conditions of cells that would have otherwise sunk below the photic zone (Malone et al., 1983). It is also well known that light and shade adaptation can modify the pool of photosynthetic pigments (Steemann-Nielsen, 1975; Beardall and Morris, 1976; Prézelin, 1976), the proportion of light-harvesting pigments to P_{700} (reaction-center chlorophyll) (Thornber et al., 1977), the characteristics of the photosynthesis-irradiance curves as shown in Fig. 1 (Steemann-Nielsen and Jorgensen, 1968; Dubinsky, 1980) and the size and number of photosynthetic units (Prézelin, 1981; Perry et al., 1981). Therefore, it appears that the interrelationship between the rate of vertical mixing and the rate of light and shade adaptation by phytoplankton cells (Falkowski, 1980, 1983) is of prime importance for the water column productivity, especially in the case of non-uniform vertical distributions of cells (Falkowski and Wirick, 1981).

Both Falkowski (1980) and Harris (1980a) have suggested that physiological changes resulting from vertical mixing may influence the succession of species and other properties at the community

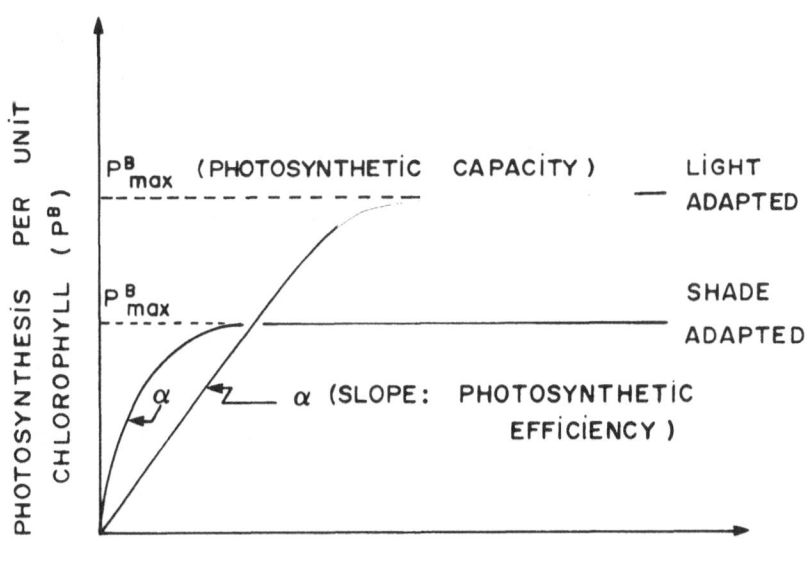

Figure 1. Photosynthesis versus irradiance curves for phytoplankton adapted to high (light adapted) or low (shade adapted) light intensities. Light adapted cells have higher photosynthetic capacity (maximum photosynthetic rate) and lower photosynthetic efficiency (photosynthetic rate per unit light, in the light limited part of the curve). Adapted from Dubinsky (1980).

level. Using a model, Kemp and Mitsch (1979) have shown a relationship between species diversity and vertical mixing. The coexistence
of three species in the same niche was only possible if the model
included a random turbulent component with a frequency approaching
the turnover rate of the phytoplankton. Margalef (1978), Margalef et
al. (1979) and Harris and Piccinin (1980) have also argued that
species succession can result from changes in vertical mixing
conditions.

The effects of increased nutrients on phytoplankton photosynthesis are well documented. Numerous observations of a positive phytoplankton response to nutrient increases, on various time and space
scales, were summarized by Legendre (1981) in a simple conceptual
model (Fig. 2). According to this model, a phytoplankton burst
occurs upon stabilization of previously destabilized (and thus
nutrient replenished) waters, the transition between the two states
being either spatial (e.g. tidal front) or temporal (e.g. spring
bloom). Since neither stabilization (which may lead to nutrient
limitation) nor destabilization (which may cause light limitation)
favors phytoplankton production, the phytoplankton burst on the
stable side of the front is short, with the result that the phytoplankton production potential, when considering the nutrients,
depends mainly on the frequency of stabilization-destabilization of
the water column. In addition, recent studies by Turpin and Harrison
(1979, 1980) and Quarmby et al. (1982) have suggested a strong
response of phytoplankton to pulsed limiting nutrients (Fig. 3).

Phytoplankton Responses to Tidal Changes in Vertical Stability

As mentioned before, the tide is the principal source of power
for vertical mixing in many coastal environments. The dissipation of
tidal energy causes changes in the vertical stability of the water
column. The frequency of changes in vertical stability due to these
tidal effects depends upon the relative importance of various har-
monics of the local tide. In practice, most phytoplankton studies in
tidally active environments have dealt with either the semi-diurnal
(M_2) or the fortnightly (MS_f) tidal components, so that the following
discussion will mainly refer to these two tidal periodicities.

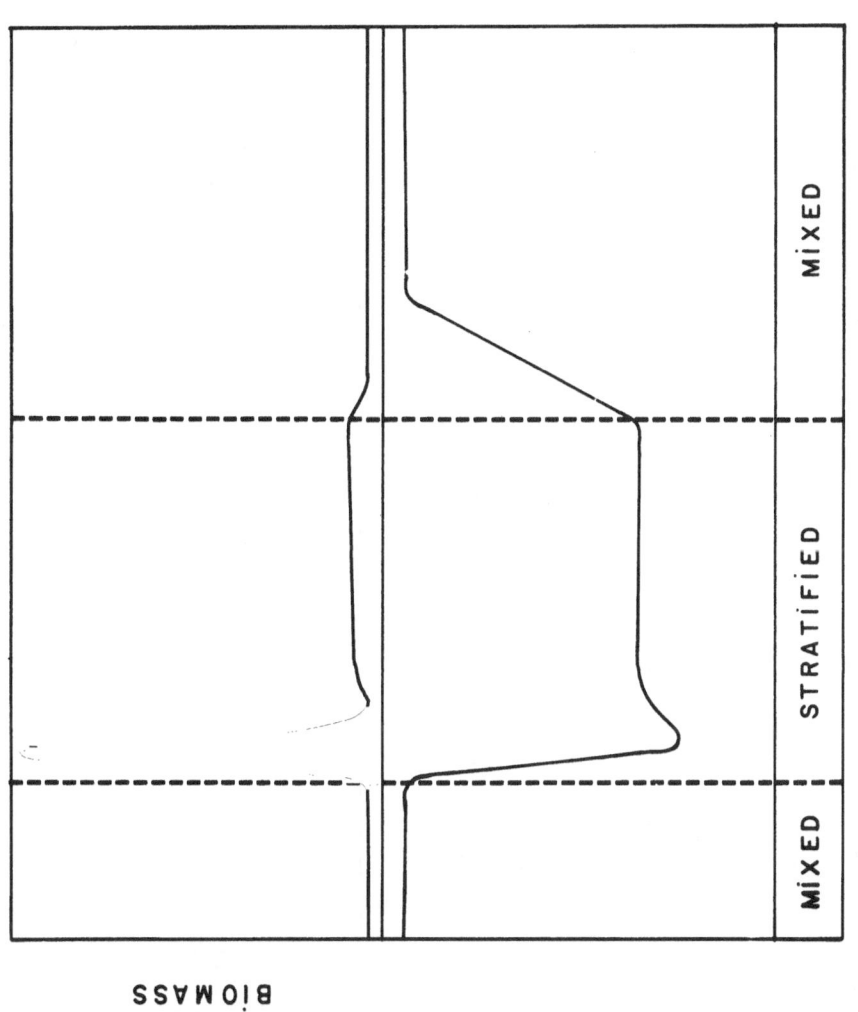

Figure 2. Simple conceptual model of phytoplankton dynamics, after Legendre (1981). The boundary between well-mixed and the stratified waters may be spatial (front) or temporal (stabilization of a wind-mixed water column, M_2 transition, etc.), with the phytoplankton maximum always on the stable side of the boundary.

7

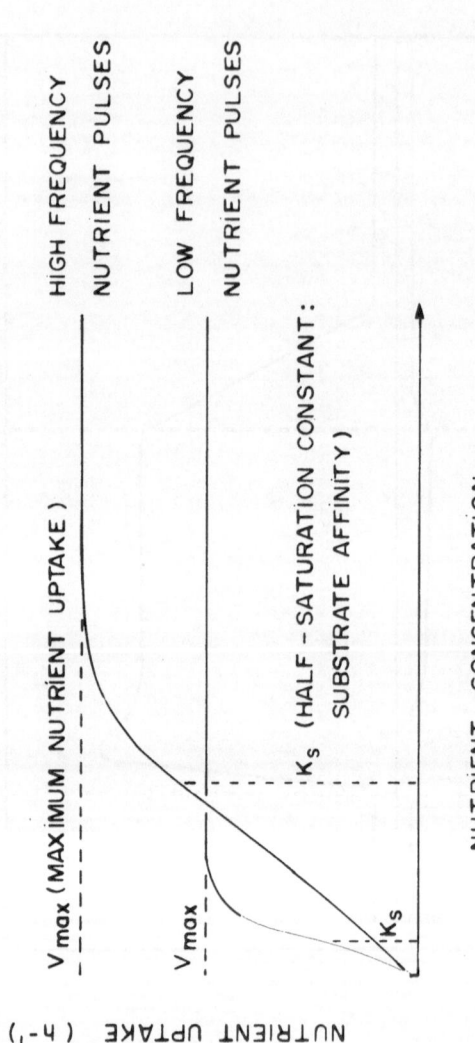

Figure 3. Nutrient uptake by phytoplankton according to nutrient patchiness. The maximum uptake rate and the substrate affinity increase with the frequency of nutrient pulses. Schematized from Turpin and Harrison (1979, 1980) and Quarmby et al. (1982).

Vertical stability and light

Phytoplankton responses to vertical mixing depend on its intensity and persistence, since the light history of the cells is known to influence phytoplankton photosynthesis (Steemann-Nielsen and Hansen, 1959; Ryther and Menzel, 1959; Beardall and Morris, 1976; Marra, 1978a; Falkowski, 1980). On the semidiurnal time scale (M_2), the phytoplankton can be subjected to two different conditions of mixing. When the vertical mixing is moderate, the environmental conditions change at a rate slower than the physiological adaptation time of the phytoplankton, so that the cells can continuously adjust their metabolic activities to the new conditions (Vincent, 1980). This can lead to variations of some phytoplankton characteristics with a M_2 tidal periodicity. Fortier and Legendre (1979) and Fréchette and Legendre (1982) have reported 12 to 13-h cycles in the photosynthetic capacity of natural populations, which were ascribed to periodic variations of the light conditions due to tidal changes in vertical stability. Furthermore, responses of the phytoplankton phased on the M_2 signal, possibly of endogenous nature, have been evidenced by Auclair et al. (1982) for chlorophyll synthesis. This physiological strategy is important since it can serve as a regulatory mechanism for maximizing the efficiency of utilization of the available light (Auclair et al., 1982).

In contrast, when the vertical mixing is persistent and intense, changes in environmental conditions are faster than the physiological adjustment time of the phytoplankton, so that the cells can only adjust to the mean environmental conditions (Savidge, 1979; Falkowski, 1980). In a tidally well-mixed area of the St. Lawrence Estuary, Demers and Legendre (1979, 1981) have observed endogenous circadian periodicities in the photosynthetic activity of the phytoplankton, whereas M_2 tidal periodicities in chlorophyll and species composition had been ascribed to advective processes (Fortier et al., 1978; Lafleur et al., 1979).

Neap-spring (MS_f) tidal changes in the vertical stability of the water column have also been shown to influence the phytoplankton. Sinclair (1978) evidenced neap-spring tidal variations of phytoplankton production in the St. Lawrence Estuary and in Puget Sound (reinterpreted data from Winter et al., 1975) which were well correlated with the MS_f cycle of stratification-destratification. Demers et al.

(1979) and Demers and Legendre (1981) also found, in the St. Lawrence Estuary, lower photosynthetic capacity and cellular chlorophyll content in the spring tide. This was explained by a poorer physiological state of the phytoplankton population as the increased vertical mixing of the spring tide was reducing the residence time of the cells in the photic layer. A decrease in the amplitude of the circadian variations of the photosynthetic capacity was also noticed by Demers and Legendre (1981). Fortier and Legendre (1979) have explained a drastic decrease of biomass at a fixed station in the lower St. Lawrence Estuary by the periodic advection of turbid waters following the fortnightly tidal cycle (MS_f). Lafleur et al. (1979) also ascribed to fortnightly advection the occurrence of fresh water taxa at a station in the upper St. Lawrence Estuary.

Bowman et al. (1981) have linked the spatial distribution of phytoplankton morphological groups (i.e. motility) to variations in vertical mixing and optical depth of the water column (s-kh diagrams: see Pingree, 1978a and Pingree et al., 1978b). In Long Island and Block Island Sounds, highest diatom concentrations were found in marginally stratified, well illuminated, shallow waters, while μ-flagellates dominated in the surface layers of deeper, darker, stratified waters.

Vertical stability and nutrients

As mentioned before, the availability of nutrients in the mixed layer is strongly influenced by the tidal mixing, or equivalently by changes in the vertical stability of the water column. Therefore, in a tidal environment, the phytoplankton is usually subjected to pulses of nutrients, the frequency and the dilution rates of which are controlled by the magnitude of the tidal energy.

Turpin and Harrison (1979, 1980) have demonstrated that pulsed limiting nutrients can be a critical factor for the competition between species in natural phytoplankton populations. According to the frequency at which limiting nutrients are pulsed, certain species can take advantage of this fluctuating environment while others cannot. The intermittent addition of limiting nutrients can affect various characteristics of the phytoplankton cells. In cultures, Turpin and Harrison (1980) have shown an increase in cell size with low frequency nutrient pulses (Fig. 4). Relative abundances of

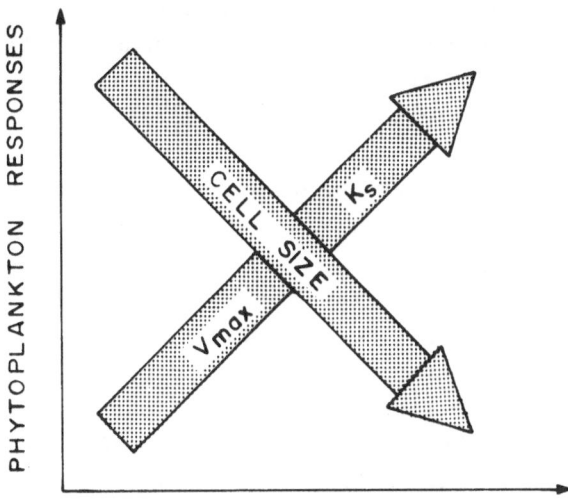

Figure 4. Phytoplankton responses to the frequency of nutrient pulses. As the frequency increases, V_{max} and K_s (Fig. 3) increase and cell size decrease. Schematized from Turpin and Harrison (1979, 1980) and Quarmby et al. (1982).

species and diversity can also be modified by changing the frequency of the nutrient pulses (Turpin and Harrison, 1979). Optimization of a patchy limiting nutrient environment would be the result of enhanced uptake rates (V_{max}), while optimization of an homogeneous limiting nutrient system appears to be more a function of substrate affinity (K_s). Quarmby et al. (1982) have demonstrated the induction of a diel periodicity in the fluorescence yield and in the photosynthetic rate by a once-per-day addition of ammonium in cultures growing under continuous light. Federov and Sermin (1970), Paasche (1971) and Falkowski and Stone (1975) have shown a suppression of inorganic carbon fixation by enrichment with NO_3^- and NH_4^+, which was interpreted as a competition for adenosine triphosphate (ATP) between the processes of inorganic carbon fixation and nitrogen assimilation. The results of Quarmby et al. (1982) could be explained by this competition mechanism.

In areas where nutrients are limited, the tidal variations of vertical mixing may act as a mechanism to pulse nutrients in the

mixed layer. On the semidiurnal time scale, the effects of such
pulsed nutrient additions on phytoplankton dynamics have been demonstrated in association with internal tides (see below). In the St.
Lawrence Estuary, Fortier and Legendre (1979) have shown an inverse
semi-diurnal relationship between phytoplankton biomass and nitrate,
linking M_2 phytoplankton bursts to variations in the vertical stability of the water column and in nutrient concentrations.

As mentioned before, nutrient concentrations in the mixed layer
can vary on a neap-spring tidal periodicity and phytoplankton appear
to respond rapidly to these changes (Legendre, 1981). Several studies
have ascribed the occurrence of intermittent summer phytoplankton
blooms to the stabilization of the water column during the neap
tides, following the introduction of new nutrients in the mixed layer
during the destabilized spring tidal period (e.g. Takahashi et al.,
1977, as reinterpreted by Webb and D'Elia, 1980; Legendre et al.,
1982).

Very few studies have considered the combined effects of the
tidal variations of both nutrients and light on phytoplankton
dynamics. Levasseur et al. (1984) have recently proposed a model
showing how phytoplankton succession is hierarchically controlled by
physical and chemical factors in the St. Lawrence Estuary. In their
model, the selection of growth rates is exerted by the frequency of
destabilization of the water column which determines nutrient limitation, the presence or absence of diatoms is determined by the mean
light intensity in the mixed layer, which is dependent upon solar
radiation and changes in density stratification, and finally conditions for optical metabolic activities are determined by temperature
conditions.

Internal Tides and Phytoplankton Dynamics

When a long barotropic wave (the surface tide) passes over a
continental shelf, a coastal bank or any other step-like topographic
feature, internal waves of the same frequency can be generated in the
upper part of the water column between the layers of varying densities to compensate for the mismatch in horizontal and vertical
velocity fields (LeBlond and Mysak, 1978). This phenomenon has
received wide attention and is a mechanism that is now well described
(Zeilon, 1913, 1934; Rattray, 1960; Rattray et al., 1969; Baines,

1973, 1974). Internal tides usually take the form of a progressive internal wave travelling in vertically non-homogeneous waters, and especially at the interface between two water masses of different densities (Proudman, 1953; Defant, 1961; LaFond, 1962; LeBlond and Mysak, 1978). LaFond (1962) has described the theoretical water motions associated with such waves and has indicated that, in addition to simple vertical oscillations, strong lateral and shear motions were associated with an internal wave. Also, because of the short wavelength of these waves, relatively strong zones of convergence and divergence between the crests and troughs can generally be observed (Fig. 5). Sea-surface slicks often seen as streaks or patches of relatively calm surface water surrounded by rippled water are visual evidences of these zones (c.f LaFond, 1962; Ingram, 1978; Haury et al., 1979). Forrester (1974) also explained by this phenomenon the build-up and relaxation of ice pressure frequently reported by ships operating in heavy ice in the St. Lawrence Estuary. Recent observations have shown that internal tides of semi-diurnal frequency can have very large amplitude in the vicinity of continental slopes (Wunsch and Hendry, 1972; Gould and McKee, 1973; Wunsch, 1975; Barbee et al., 1975) and in some particular coastal environments such as the St. Lawrence Estuary (Forrester, 1974; Ingram, 1975; Therriault and Lacroix, 1976), and can contribute a significant fraction of the horizontal and vertical velocity fields at tidal frequencies. The physical characteristics of such internal tides have been particularly well studied for regions such as the Southern California coast (Cairns, 1966, 1967, 1968; Cairns and LaFond, 1966; Cairns and Nelson, 1968), the Celtic and Armorican shelf break (Pingree and Griffiths, 1978; Pingree and Mardell, 1981; Pingree et al., 1982) and the St. Lawrence Estuary (Forrester, 1974; Muir, 1979).

Internal tides affect almost all the characteristics that are influenced by the temperature and/or density changes in the water column, including the living organisms and the chemical and physical properties of the water (LaFond, 1962). If the internal tide is located near the sea surface, it is expected that the vertical motions in the water column will exert strong effects on the phytoplankton, primarily by modifying the conditions of irradiance and of nutrient supply. The effects of these modifications on the phytoplankton are the object of some speculation in this section.

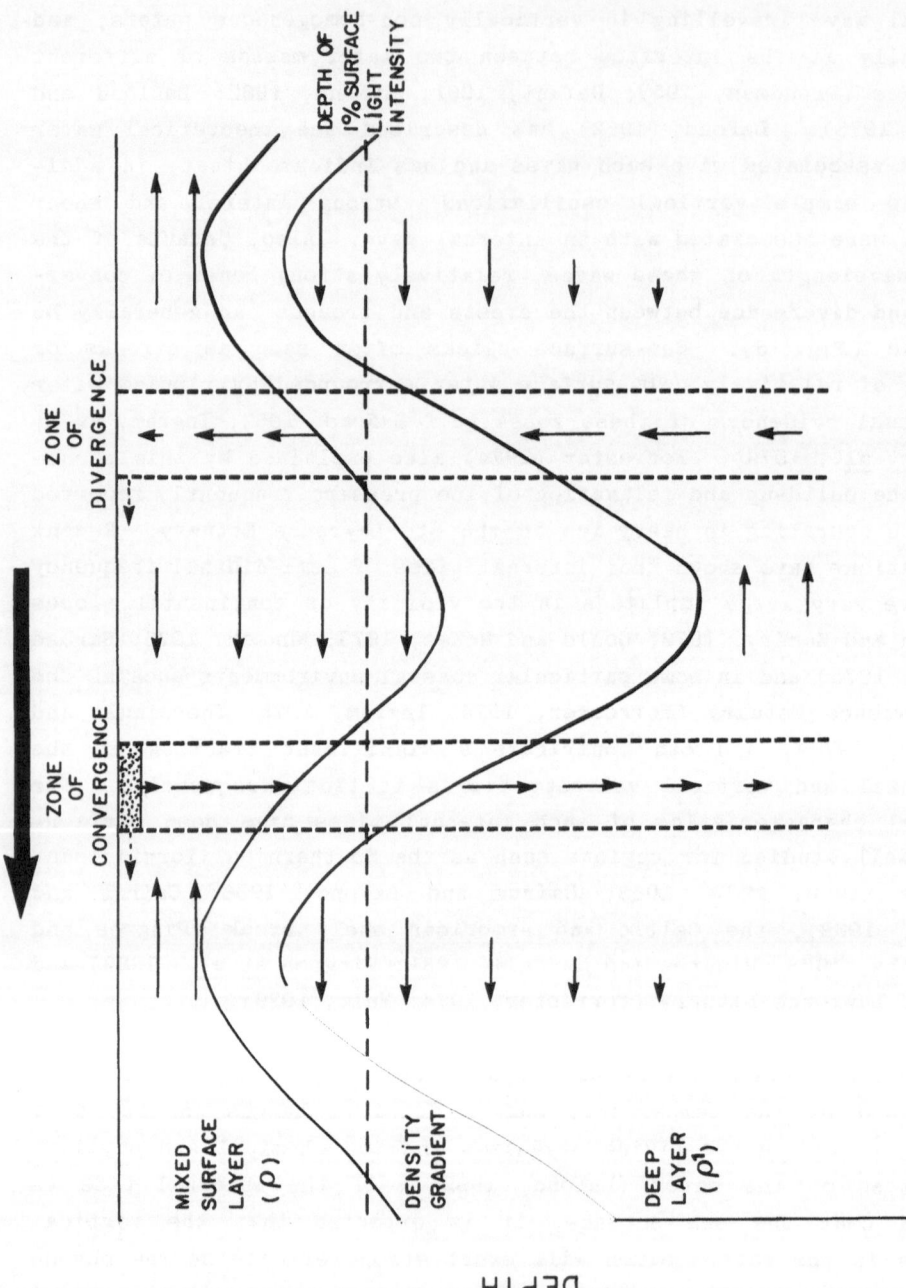

Figure 5. Theoretical motion of the water particles associated with the propagation of an internal tide between water masses of different density. The large arrow indicates the direction of propagation of the internal wave, while the small arrows indicate the directions and relative velocities of the currents.

A number of studies have indicated a relationship between internal waves and phytoplankton. Internal tides have been invoked to explain horizontal heterogeneities observed in the distribution of chlorophyll (Denman and Platt, 1975; Denman, 1976, 1977; Fasham and Pugh, 1976; Therriault and Lacroix, 1976; Haury et al., 1979; Sinclair et al., 1980; Pingree and Mardell, 1981; Pingree et al., 1982; Cullen et al., 1983), cell numbers (Venrick, 1972; Lafleur et al., 1979) and nutrients (Armstrong and LaFond, 1968; Therriault and Lacroix, 1976; Greisman and Ingram, 1977; Ingram, 1979; Cullen et al., 1983). Fréchette and Legendre (1982) have demonstrated an influence of internal tides on the phytoplankton photosynthetic response to light. Kamykoswki (1974, 1976, 1979), considering different strategic behaviors of the organisms, has theoretically examined on several temporal scales the influence of internal tides on the spatial distribution, primary production and physiology of phytoplankton. He has concluded that internal tides can contribute significantly to the biological variability observed at a geographical location. Therefore, there is already a good deal of information on the interrelationships between internal tides and phytoplankton. Let us now speculate on these interrelationships by systematically examining the changes on the light and nutrient regimes brought about by the propagation of a shallow internal tide with a semidiurnal frequency.

Internal tides and light

If we assume first that there is no turbulent mixing across the sharp density gradient separating two water masses, the organisms will experience no changes in temperature or nutrients during the passage of the internal tide, but the radiation field experienced by the cells, however, will be significantly affected by the vertical movements in relation to the diurnal cycle. In practice this means rapidly changing the depth of the mixed layer (Z_m) relative to that of the euphotic (Z_{eu}) zone (Fig. 5). Changes and the rate of change in the ratio Z_{eu}/Z_m are of fundamental importance for the phytoplankton community in marine as well as in fresh water environments (e.g. Riley, 1957; Hitchcock and Smayda, 1977; Haffner et al., 1980; Harris and Piccinin, 1980; Sinclair et al., 1981; Levasseur et al., 1984). Kamykowski (1974) has demonstrated that the range of radiation intensities, to which various organisms located near the density gradient are exposed every day, varies with their relative position along the

internal wave at sunrise. It is obvious that organisms which are at
the crest of the wave at noon will receive much more light than those
which are in the crest at sunrise and sunset. The effect this may
have on phytoplankton photosynthesis depends on the ability of the
different species to adapt to rapidly changing light intensities as
well as spectral qualities (Jerlov, 1968). Depending on the seasonal
characteristics of the wave and of the light cycle, an organism that
is located at the crest of the wave at noon may even experience
strong photoinhibition (Harris and Lott, 1973). Semidiurnal internal
tides and daylight do not maintain a constant phase relationship. It
is therefore expected that the area along the wave that has low phytoplankton growth due to lack of light will exhibit increased growth on
successive days as its horizontal position at the crest is shifted
towards noon (Kamykowski, 1974, 1979). The end result will be low
frequency cycles of light availability (of the order of 15 days for a
12:12-h day-night cycle and a 12.4-h semidiurnal tidal cycle) which
can result in photosynthetic as well as community changes (Kamykowski, 1974, 1976). In estuarine environments where advection of
surface water is important, the relationship between the phases of
the semidiurnal internal tide and of the daylight cycle is not as
straightforward as described above, and the period of the long term
cycle of light availability may be considerably reduced, therefore
leaving less time for community structure changes.

    Our discussion until now was based solely on light considera-
tions. Let us also include the effects of the water motions associated with the internal tide (Fig. 5). Over the crest of the wave,
stratification increases and produces a sharper density gradient
(e.g. Therriault and Lacroix, 1976; Fréchette and Legendre, 1982), so
that the depth of the mixed layer is greatly reduced. The lateral or
horizontal currents are increased because of the funnelling effect
(LaFond, 1962). This does not imply, however that the theoretical
internal wave of Fig. 5 will stay together for a long time, since the
squeezing and thickening of the pycnocline might not travel at the
same speed as the wave propagating along the density interface. The
propagation of the internal tide can result in a shallower mixed
surface layer above the density gradient in which phytoplankton cells
can be strongly mixed vertically. As explained previously, depending
on the intensity of the vertical mixing, these cells will respond to
the new conditions by adjusting their photosynthetic activity (Demers
and Legendre, 1981, 1982). Reduced or no photoinhibition may also

occur, depending on the ratio Z_{eu}/Z_m. Over the trough of the wave, the density gradient is usually much less sharply defined and the lateral current velocity is also reduced. This situation usually gives rise to a more stable surface layer in which vertical mixing may be reduced. Again, phytoplankton organisms will have to adjust to these new conditions, as described above. The observed succession of high and low values of the Richardson Number (Ri) with a M_2 frequency, in the St. Lawrence Estuary (Fortier and Legendre, 1979) and at the Celtic Sea shelf-break (Pingree and Mardell, 1981), are examples of variability in the stability of the water column related to internal tides.

As generally observed (e.g. Muir, 1979; Cairns, 1968; Forrester, 1974), the amplitude and phase structure of the internal wave field is quite different in spring tides and neap tides, the maximum amplitudes of the internal tides occurring at the spring tide. Neap tide stratification is usually much stronger than that of the spring tide, since there is much more energy available at spring tides for mixing (Muir, 1979). Therefore one would expect to see phytoplankton characteristics to exhibit MS_f as well as M_2 periodicities in response to the internal tide. In the St. Lawrence estuary, where large amplitude internal waves (50-100 m) are present (Forrester, 1974; Ingram, 1975; Therriault and Lacroix, 1976), 12 to 13-h (M_2) cycles in chlorophyll (Therriault and Lacroix, 1976; Fortier and Legendre, 1979) and in photosynthetic capacity (Fortier and Legendre, 1979; Fréchette and Legendre, 1982) are observed as well as neap-spring (MS_f) cycles in chlorophyll concentrations (Sinclair, 1978), cellular chlorophyll content (Demers et al., 1979) and amplitude of the circadian variations of the photosynthetic capacity (Demers and Legendre, 1981). Since the amplitude of the diurnal inequality may be as large as the spring to neap inequality (Muir, 1979), one could assume that the internal tide characteristics could be quite different from one tide to the next. Also since the wavelength of the internal tide is relatively short, often of the size range of the topographic features in estuaries, high frequency internal waves (solitons, internal wave packets) are generated as the internal tide is forced by the barotropic tide (Armstrong and LaFond, 1968; Ingram, 1978; Haury et al., 1979; Muir, 1979; Sinclair et al., 1980). It seems therefore more appropriate to consider that the phytoplankton have to respond to a wide spectrum of light variations due to tidally induced internal waves. Another degree of complexity can be added when we consider that the characteristics of the internal tides change with the season. Denman and Gargett (1983) have indicated the general characteristic scales in time and space we can expect for vertical displacement of phytoplankton by internal waves in different environments.

Water motions associated with the internal tide can also be conducive to patchiness. For example, zones of convergence and divergence (Fig. 5) may cause patchiness (LaFond, 1962; Forrester, 1974; Kamykowski, 1974). When the organisms have some degree of control on their relative positions in the water column around the density gradient, spatial heterogeneity invariably results (Kamykowski, 1974, 1976, 1979; Kamykowski and Zentera, 1977; Cullen et al., 1983). Interactions between the internal tide and the zooplankton migration and feeding behavior can also cause phytoplankton patchiness (Kamykowski, 1974).

Internal tides and nutrients

Until now we have considered a simple situation where there was no mixing of nutrients across the density gradient separating two water layers with different nutrient characteristics. In practice, it is known that shear instabilities which are often associated with internal waves (cf. Woods and Wiley, 1977; McGowan and Hayward, 1978; Haury et al., 1979) and breaking of internal waves may be conditions conducive to mixing nutrients from the bottom layer into the nutrient depleted surface layer (cf. Winant and Olson, 1976). This results in spatially localized nutrient pulses in the surface layer to which phytoplankton, especially when nutrient-limited, might respond quickly (cf. Turpin and Harrison, 1979, 1980). The end result will be spatial variability in phytoplankton characteristics. Evidence of such a nutrient pulse event is given by Cullen et al., (1983) for the Southern California shelf.

Effects of internal tides in canyon like features may also result in mixing or upwelling events. Thompson and Golding (1981) have described a tidally induced upwelling over the Great Barrier Reef under certain conditions of interaction between the internal tide and the tidal flow in a reef-edge channel. In the St. Lawrence Estuary, it is apparent that part of the nutrient rich deeper layer is upwelled with a semidiurnal periodicity and mixed in the surface layer (Therriault and Lacroix, 1976; Greisman and Ingram, 1977; Ingram, 1979). This nutrient input results from isopycnal shoaling, shear instabilities and tidal mixing due to the generation of large amplitude internal tides (50 to 100 m) near the end of the Laurentian Channel (Greisman and Ingram, 1977). Steven (1975), Therriault and Lacroix (1976) and Greisman and Ingram (1977) have attributed the relatively high productivity of the St. Lawrence Estuary to this periodic introduction of nutrients in the surface layer. Since there is a neap-spring relationship in the internal tide amplitude, higher concentrations of nutrient are introduced in the surface layer around the spring tides, so that spatial variability in phytoplankton productivity may result from the fortnightly oscillations of the nutrient source (Greisman and Ingram, 1977).

Tidal Fronts and Phytoplankton Dynamics

The geographical distribution of water column stability on the continental shelf may be characterized by contouring the Simpson and Hunter (1974) stratification parameter

$$\underline{S} = \log_{10} \frac{h}{\underline{C}_D \, \underline{u}^3} \tag{1}$$

where $\underline{h}$ is the water depth, $\underline{u}$ is the mean tidal stream velocity, and $\underline{C}_D$ is the drag coefficient (~ 0.0025). For example, this was done for the Bay of Fundy and Gulf of Maine (Garrett et al., 1978), the northwest European shelf (Pingree and Griffiths, 1978), the Gulf of St. Lawrence (Pingree and Griffiths, 1980), Greater Cook Strait (Bowman et al., 1980) and for Hudson Bay (Griffiths et al., 1981). The same approach was successfully used by Bowman and Esaias (1981) for a moderately stratified estuary. The boundary between vertically mixed and stratified waters is often delineated by a well defined front (at $\underline{S} \simeq 1.5$, if c.g.s. units are used) caused by the inter-action between the tide and the water depth. According to the mechanisms involved, tidal fronts have been described as shallow sea fronts (Simpson and Pingree, 1978), headland fronts (Pingree et al., 1978a) or estuarine fronts (Bowman and Iverson, 1978).

The hydrodynamic mechanisms responsible for the shallow sea fronts have been reviewed in several papers (Pingree, 1978a; Simpson and Pingree, 1978; Simpson, 1981). Basically, the energy required to vertically mix the water column in shallow seas is derived from the dissipation of the M_2 tide and from the wind stress on the water surface. Since the effects of wind mixing are relatively uniform, shallow sea fronts are mainly caused by the distribution of tidal energy dissipation, relative to water depth. In regions where $\underline{S} < 1$, the water column is continuously vertically mixed and, in those were $\underline{S} > 2$, the water column can stabilize under the influence of the summer heat flux. In shallow seas where there is a transition zone between well-mixed ($\underline{S} < 1$) and well-stratified ($\underline{S} > 2$) waters, a summer front can often be found associated with the $\underline{S} \simeq 1.5$ contour. A significant feature of shallow sea tidal fronts is that the frontal region is expected to slightly invade the more stratified areas during the spring tide and to recede away from it in the neap tide, thus enriching the surface layer on the restabilized side of the front with recently mixed water.

The MS_f frontal excursions described above are a mechanism conducive to enhance phytoplankton production, since the surface waters on the stable side of the front are periodically replenished in nutrients. A number of studies report high phytoplankton biomass associated with shallow sea tidal fronts, as a probable response to neap-spring frontal movements (Pingree et al., 1975, 1976, 1977, 1978b, 1982, 1983; Pingree, 1978a; Simpson and Pingree, 1978; Parsons et al., 1983). However, recent satellite observations and numerical models (Simpson, 1981) suggest that the oscillations in frontal positions over the MS_f cycle are probably restricted to only a few kilometers, so that additional mechanisms are required to account for the observed enhanced phytoplankton biomass resulting from cross-frontal mixing. Cyclonic eddies along the front are such a candidate mechanism. According to Pingree (1978a,b), cyclonic eddies are likely to be important in the transfer of heat, salt and nutrients across the frontal zone, in the summer time. In support of this view, Pingree et al. (1979) have reported a good relationship between the surface distribution of chlorophyll $\underline{a}$ and cyclonic eddies. As another possible mechanism of cross-frontal mixing, Garrett and Loder (1981) have shown that the flux associated with the frictionally induced mean flow appears to be more important than the cross-frontal transfer by baroclinic eddies, for typical shallow sea fronts.

These mechanisms of frontal nutrient enrichment can only be directly applied, however, to the relatively deep areas of the continental shelf, where the whole water column does not become

nutrient-depleted in the summer. In shallower environments, high
mixing may not lead to light limitation of phytoplankton (Legendre,
1981), while phytoplankton on the well-mixed side of the front may
become nutrient-limited (Perry et al., 1983). Under such conditions,
both sides of the front are nutrient-limited, so that high phytoplankton biomass in the frontal region must be explained by other
mechanisms such as shelf-break upwelling or advection of nutrient-
rich water (Perry et al., 1983).

A second type of front on the continental shelf is called a
headland front, as it is formed in association with flow around
headlands. These fronts develop in response to periodic tidal flow
past a headland, so that they are expected to go through a generation
and dissipation phase within a tidal period. This short characteristic time scale (M_2) distinguishes them from the shallow sea fronts.
According to Pingree et al. (1978a), the strong tidal streaming in
the neighborhood of the headland will cause a local minimum in the
value of the stratification parameter $\underline{S}$ (well-mixed waters), with an
abrupt transition (front) to stratified conditions, if these exist
offshore. As a result, nutrient-rich bottom water will tend to move
towards the inshore region, thus replenishing the nutrients used by
the phytoplankton (Pingree, 1978a). Bowman and Esaias (1977) have
shown the co-occurrence of nutrient and chlorophyll maxima associated
with a headland in Long Island Sound.

A third type of tidal front can be observed parallel to the main
axis of estuaries. These estuarine fronts form in shoaling areas,
where tidally (M_2) generated bottom turbulent stirring is strong
enough to break down the vertical stratification. The front, which
can be located by plotting the stratification parameter $\underline{S}$, is found
between the inshore well-mixed water and the offshore stratified
water. The stratification in the deeper offshore waters is due to
the buoyancy of the surface brackish layer (Bowman and Iverson,
1978). Since lateral shear is important in maintaining these fronts,
they often strengthen on the ebb tide, and weaken or vanish on the
flood tide. Due to strong convergence velocities, estuarine fronts
can accumulate floating organic matter and detritus. Their effect on
phytoplankton biomass is not well documented.

Fronts do not develop in areas where the mean tidal velocity $\underline{u}$
is small, since the dissipation of a weak M_2 tide cannot prevent the

development of vertical stratification. However, when an island is located in such an area, it can act as a stirring rod by oscillating at the tidal frequency relative to the water, and thus produce zones of intensified vertical mixing. The mixing process around the island combines top and bottom waters from the stratified regime into a nutrient-rich mixture of intermediate density. In a non-rotating system, this water would intrude into the pycnocline, around the well-mixed zone surrounding the island. If the surface layer away from the island is nutrient-limited, the injection of the nutrient-rich mixed water will result in enhanced phytoplankton production. Such a situation was studied by Simpson et al. (1982), who found enhanced standing crop (6-fold) over an area that is at least 20 times the island area, with correspondingly enhanced primary production.

High phytoplankton biomass has been reported, for the continental shelf, in association with other fronts than those tidally induced. This is the case for the shelf-break fronts of the Nova Scotian shelf (Fournier, 1978; Fournier et al., 1977, 1979; Herman and Denman, 1979), the Bering Sea (Iverson et al., 1979) and the Celtic Sea (Pingree and Mardell, 1981; Pingree et al., 1982), and for the coastal fronts (Holligan, 1981) or the fronts related to freshwater advection (Norwegian coast: Steele, 1957; Chesapeake Bay: Seliger et al., 1981). Since tidal mixing is not involved as such in those fronts, they will not be discussed here.

Most phytoplankton measurements in fronts concern the photosynthetic biomass (chlorophyll a), which was often estimated from in vivo fluorescence (Lorenzen, 1966). However, there is no clear-cut answer to the question "how are the large biomasses often found in association with tidal fronts produced?" This is due to a lack of information on in situ rates of photosynthesis comparable to chlorophyll distributions. The mechanisms described above, that explain the enhanced standing crops in frontal regions in terms of vertical mixing regimes are, however, supported by both indirect and direct evidences.

Among the indirect supports to the vertical mixing hypothesis is the demonstration by Savidge (1976), from enrichment experiments, that phytoplankton growth is nitrate-limited in fronts, so that the flux of nitrate by cross-frontal and vertical mixing would determine the overall net production (Holligan, 1981). In addition, Tett (1981) was able to closely predict the distribution of chlorophyll across a frontal region, using a model based on the vertical mixing hypothesis. Direct measurements of primary production led Parsons et al. (1983) to conclude that chlorophyll maxima in a frontal zone were caused by increased production. As in the model above, nutrients entering the system from the well-mixed waters, during maximum exchange (spring tides) were the driving force for new production. They also concluded to a rapid uptake of nitrate in the frontal zone, which is consistent with the nitrate limitation found by Savidge (1976). If the mixing hypothesis is correct, then phytoplankton must be able to respond to brief periods of stabilization, which led Pingree (1978a) to hypothesize a fast division rate of cells in frontal regions, which would allow them to take advantage of conditions when the water column stabilizes for only a few days in the spring to neap cycle of tidal mixing. The frontal regime thus restricts phytoplankton growth to a narrow band on the stable side of two discontinuities in stability: (1) the frontal transition, located between well-mixed and well-stabilized water, and (2) the temporal transition, that occur from spring to neap tides. This is in agreement with the general conceptual model of Legendre (1981), reported above (Fig. 2).

The presence of fronts also influences the composition of the phytoplankton assemblages. According to Wangersky (1977), diatom growth responds to stabilization of nutrient rich water, while dinoflagellates and μ-flagellates become dominant when the major source of nutrients is in situ regeneration, from the degradation of dissolved organic matter by bacteria. On the other hand, Levasseur et al. (1984) argued that the most important factor for phytoplankton succession is the frequency of destabilization of the water column, which determines nutrient limitation and consequently selects a range of growth rates. When the frequency of destabilization is high, small cells with high division rates are expected; when the frequency of destabilization is low, a complete summer succession from fast growing small-celled diatoms to larger diatoms with lower growth rates and to large dinoflagellates with still lower growth rates should result. These two hypotheses are not mutually exclusive.

Figure 6 summarizes the information on phytoplankton distributions in frontal areas reported by Pingree et al. (1978b, 1979) and

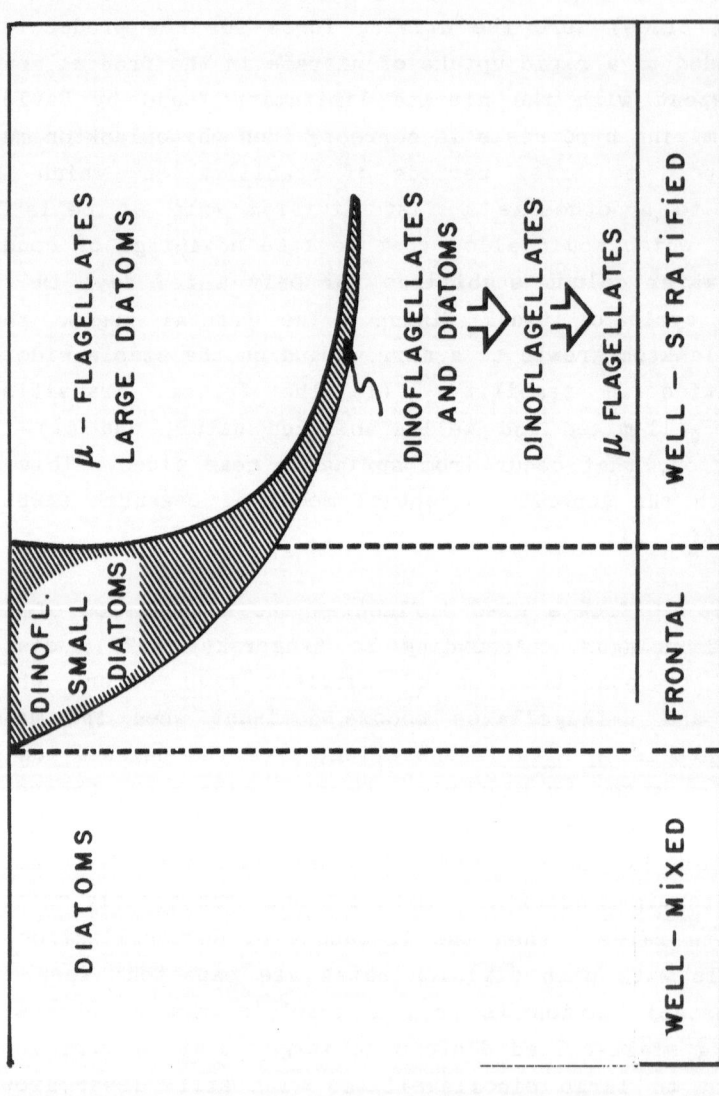

Figure 6. Composite generalized distribution of phytoplankton assemblages in frontal regions, in relation with the often observed surface and sub-surface chlorophyll maxima (shaded), on the stable side of the front. The open arrows refer to a seasonal succession within the subsurface chlorophyll maximum (see text). After Pingree et al. (1978b, 1979) and Holligan (1978, 1981).

Holligan (1978, 1981). It schematically shows the surface chlorophyll maximum, on the stable side of the front, and the associated
subsurface chlorophyll maximum, in the pycnocline. These authors
report that the well-mixed waters, in the summer time, are dominated
by diatoms, including benthic species. In the well stratified surface
waters, on the other hand, μ-flagellates and some large diatoms are
usually the dominant taxa. Three groups of species can be found in
the thermocline, a mixture of dinoflagellates and diatoms, or dinoflagellates, or flagellates; these represent a successional series
that occurs after the establishment of the thermocline (Holligan and
Harbour, 1977). Finally, mixed populations of dinoflagellates and
small diatoms form the high chlorophyll surface concentrations on the
stabilized side of the front. The hypothesis (previous paragraph) of
Wangersky (1977) can account for the dominance of the well-stratified
surface waters by μ-flagellates, and that of Levasseur et al. (1984)
for the dominance by large diatoms. The mixture of dinoflagellates
(Wangersky: in situ nutrient regeneration) and of small diatoms
(Levasseur et al.: recently stabilized water) in the surface frontal
regions agrees with the model of alternating stratification (regeneration) and vertical mixing on the neap-spring tidal cycle, described
above.

In the frontal region of shallow sea fronts, nutrient-rich water
is fortnightly stabilized for a longer and longer period of the tidal
cycle, in the summer (Pingree, 1978a). This leads to nutrient pulses
with a fortnightly periodicity. In headland and estuarine fronts, on
the other hand, this periodicity may be as short as 12.4 h. By
reference to the experimental studies reported previously on the
responses of phytoplankton to pulsed nutrients (Turpin and Harrison,
1979, 1980; Quarmby et al., 1982), it can be expected that frontal
populations will be influenced by the frequency of frontal stabilization. For example, the short pulses of stability in headland and
estuarine fronts, and also in shallow sea fronts at the beginning of
the summer, should select for small diatoms with high V_{max}. However,
due to lack of published information on species succession in fronts
and on the responses of phytoplankton to low frequency nutrient
pulses, it is difficult to check this hypothesis.

Phytoplankton Responses to Tidal Mixing

It is obvious from the above discussion that the effects of light and/or nutrient changes on phytoplankton are quite different depending on the time scale of tidal mixing considered. Table 1 synthesizes the information discussed above in a general framework, where specific phytoplankton responses are linked to tidal mixing conditions through identifiable changes in the two proximal agents light and nutrients.

In an environment characterized by no tidal mixing (condition 1 in Table 1), there are no M_2 or MS_f periodicities in the light field, and phytoplankton photosynthesis may become inhibited near the surface unless vertical mixing is generated otherwise (wind, waves, Langmuir circulation, etc.). Because of the lack of mixing in this condition, the mixed layer can rapidly become strongly nutrient depleted, so that phytoplankton growth rapidly depends on nutrient regeneration within the mixed layer. Only circadian cycles are then observed in the photosynthetic parameters. Closely related to the photosynthesis, phytoplankton biomass shows a heterogeneous vertical distribution with either subsurface or deep chlorophyll maxima, the latter being probably related to the availability of higher nutrient concentrations in or below the pycnocline. In summer time, well-stratified and nutrient limited waters are successively dominated by large diatoms and by dinoflagellates.

Depending mainly on the topography, the dissipation of the M_2 tidal energy can result in either continuously intense turbulent mixing (condition 3) or in alternating periods of stability and moderate vertical mixing (condition 2) of M_2 frequency. For the sake of simplicity, intermediate mixing conditions between 2 and 3 are not considered here. Condition 2 is characterized by M_2 changes in the velocity of the vertical mixing, that is an alternation (12.4-h cycle) of null or weak vertical mixing and of relatively strong mixing conditions. Such an alternation occurs along the temporal axis, but it can be extended to the spatial domain when considering headland and estuarine fronts. For the phytoplankton, these M_2 changes in vertical mixing correspond to alternating periods of stable and variable light intensity on a 12.4-h cycle, to which they respond by changing their P^B_{max} (Fortier and Legendre, 1979; Fréchette and Legendre, 1982), μ_{pot} (potential growth rate: Fortier

Table 1. Phytoplankton responses to tidal mixing through the proximal agency of light or nutrient changes. All the phytoplankton responses are documented in the literature.

TIDAL MIXING	PROXIMAL AGENTS	PHYTOPLANKTON RESPONSES
1. No tidal mixing	Light stable on M_2 and MS_f	Photoinhibition near the surface
	Nutrients available only through regeneration; nutrients can become limiting	Biomass vertically stratified (subsurface or deep chlorophyll maximum)
		Dominance of large diatoms and dinoflagellates
2. Dissipation of the M_2 tide: alternation of stability and moderate vertical mixing		
· M_2 changes in the velocity of the vertical mixing	Alternating periods of stable and varying light intensity on a 12.4-h cycle	M_2 cycles in P_{max}^B, μ_{pot}, and chlorophyll/cell
· M_2 changes in the turbulent diffusion across the pycnocline	Pulses of nutrients on a 12.4-h	M_2 cycles in P_{max}^B and α High new production Selection for small-sized diatoms, with high V_{max}
3. Dissipation of the M_2 tide: continuous intense vertical mixing	Rapidly changing light environment (no M_2 cycles)	P_{max} responds to the average light intensity: circadian periodicity only
	Nutrients continuously replenished	Selection for small-sized diatoms, with low K_s

Table 1. Continued.

4. MS$_f$ modulation of (2): MS$_f$ changes in the relative importance (duration and intensity) of stable vs. mixed conditions	Spring tides: light and nutrients Neap tides: light and nutrients	Spring tides: phytoplankton responses as in (3) Neap tides: phytoplankton responses as in (2)
5. MS$_f$ modulation of (3): MS$_f$ changes in the depth of the mixed layer	MS$_f$ variations of the average light intensity in the mixed layer MS$_f$ variations of nutrient levels in the mixed layer	MS$_f$ variations in P$_{max}^{B}$, chlorophyll/cell, and primary production MS$_f$ variations in the amplitude of the circadian cycles MS$_f$ cycles in chlorophyll, number of cells, and new primary production

and Legendre, 1979) and intracellular chlorophyll (Auclair et al., 1982). Condition 2 is also characterized by M_2 changes in the turbulent diffusion across the thermocline, which can result in pulses of limiting nutrients such as NO_3 (Fortier and Legendre, 1979), thus increasing the new production (and not only the regenerated production: Eppley and Peterson, 1979; Eppley et al., 1979). From laboratory evidences, these nutrient pulses could lead to M_2 cycles in P^B_{max} and α (Falkowski and Stone, 1975; Quarmby et al., 1982), and to the selection of small-sized diatoms (Turpin and Harrison, 1979, 1980) with high V_{max} (Turpin and Harrison, 1979; Quarmby et al., 1982). The ecological approach (Levasseur et al., 1984) also predicts the dominance of small-sized diatoms in frequently mixed environments. This represents an interesting convergence between experimental and field evidence. In environments with diurnal inequality in tidal amplitude, it is even possible that the 12.4-h cycles in phytoplankton responses become 24.8-h cycles, being then practically undistinguishable from circadian cycles, except through very long time series. It must be noted that the M_2 transition from well-mixed to stabilized waters is the temporal analogue of a transverse crossing of a shallow-sea front, as shown in Fig. 2. Similar to the M_2 tidal boundary, the stratified side of the front is characterized by high biomass (Fig. 6) and photosynthesis.

Condition 3 is characterized by continuous intense vertical mixing. Hydrodynamically, condition 3 is at the opposite end of the mixing range from condition 1. Intense mixing can produce changes in the environmental light conditions that are faster than the physiological adjustment time of the cells, so that the phytoplankton can only adjust its responses to the average light conditions (Demers and Legendre, 1981, 1982). It is known from the ecological approach that these well-mixed waters are dominated by small-sized diatoms (Holligan, 1978; Pingree et al., 1978b, 1979; Levasseur et al., 1984). Similarly, experimental evidence suggest that the high nutrient flux associated with high mixing should result in the dominance of diatoms (Turpin and Harrison, 1979), which would have a low K_s (increased substrate affinity) in the absence of limiting-nutrient pulses. High new production is generally not expected, since intense vertical mixing normally subjects the cells to an average low light intensity.

The MS_f tide modulates the amplitude of the M_2 tide, so that it influences the phytoplankton on a fortnightly periodicity. The MS_f

modulation of an intensely-mixed water column (condition 5 in Table 1) will cause fortnightly variations in the depth of the mixed layer (if the whole water column is not mixed from surface to bottom). Phytoplankton cells mixed at a greater depth will be exposed to a lower average light intensity, with corresponding MS_f variations in P^B_{max} and the amount of intracellular chlorophyll (Demers et al., 1979), and also probably water-column primary production. MS_f variations in the amplitude of the circadian cycles were also observed under such conditions (Demers and Legendre, 1981). If the deep waters are nutrient-rich, MS_f variations of nutrient levels will occur in the mixed layer, eventually leading to MS_f cycles in the levels of new production (Eppley and Peterson, 1979), and therefore in phytoplankton biomass (chlorophyll and number of cells).

Condition 4 in Table 1 results in rather different light and nutrient regimes, in spring vs. neap tides. As indicated in the table, spring tides will push the system towards intense vertical mixing (condition 3), with shorter periods of stability and deeper mixing (as in the York River estuary, Webb and D'Elia, 1980). On the other hand, the water column in the neap tides will offer longer periods of stability and will be mixed to shallower depths, being more similar to condition 2. Under such a regime, MS_f variations were observed for phytoplankton biomass (chlorophyll: Winter et al., 1975, as reinterpreted by Sinclair, 1978; Takahashi et al., 1977, as reinterpreted by Webb and D'Elia, 1980; Sinclair, 1978; chlorophyll and number of cells: Levasseur et al., 1984) and for water-column production (Winter et al., 1975; Sinclair, 1978; Levasseur et al., 1984).

It follows from the discussion of Table 1 that both the M_2 and MS_f tidal mixing have significant effects on almost all the phytoplankton characteristics. Sampling the phytoplankton in tidally mixed environments must therefore take into account the physical scales of variability that dominate the phytoplankton environment. This is often considered as a major source of difficulties in designing proper sampling schemes for tidally dominated environment. On the other hand, these environments offer unique opportunities to develop and test hypotheses about the general effects of vertical mixing on phytoplankton, since (1) the tides are a very predictable hydrodynamic signal and (2) different conditions of vertical mixing

can often be encountered just a few kilometers or a few days apart,
in the same area.

REFERENCES

Armstrong, F.A.J. and E.C. LaFond. 1968. Chemical nutrient
concentrations and their relationship to internal waves and
turbidity off Southern California. Limnol. Oceanogr. 11:
538-547.

Auclair, J.-C., S. Demers, M. Fréchette, L. Legendre and C.L. Trump.
1982. High frequency endogenous periodicities of chlorophyll
synthesis in estuarine phytoplankton. Limnol. Oceanogr. 27:
348-352.

Baines, P.G. 1973. The generation of internal tides by flat-bump
topography. Deep-Sea Res. 20: 179-205.

Baines, P.G. 1974. The generation of internal tides over steep
continental slopes. Phil. Trans. R. Soc. Lond. A277: 27-58.

Barbee, W.B., J.G. Dworski, J.D. Irish, L.H. Larsen and M. Rattray,
Jr. 1975. Measurement of internal waves of tidal frequency
near a continental boundary. J. Geophys. Res. 80: 1965-1974.

Beardall, J. and I. Morris. 1976. The concept of light intensity
adaptation in marine phytoplankton: some experiments with
Phaeodactylum tricornutum. Mar. Biol. 37: 377-387.

Bokuniewicz, H.J. and R.B. Gordon. 1980. Storm and tidal energy in
Long Island Sound. Adv. Geophysics 22: 41-67.

Bowman, M.J. and W.E. Esaias. 1977. Fronts, jets, and phytoplankton
patchiness. In: Bottom Turbulence, pp. 255-268. J.C.J. Nihoul
(ed.). Elsevier, Amsterdam.

Bowman, M.J. and W.E. Esaias. 1981. Fronts, stratification, and
mixing in Long Island and Block Island Sounds. J. Geophys. Res.
86: 4260-4264.

Bowman, M.J., W.E. Esaias and M.B. Schnitzer. 1981. Tidal stirring
and the distribution of phytoplankton in Long Island and Block
Island Sounds. J. Mar. Res. 39: 587-603.

Bowman, M.J. and R.L. Iverson. 1978. Estuarine and plume fronts.
In: Oceanic Fronts in Coastal Processes, pp. 87-104. M.J.
Bowman and W.E. Esaias (eds.). Springer-Verlag, Berlin
Heidelberg.

Bowman, M.J., A.C. Kibblewhite and D.E. Ash. 1980. M_2 tidal effects
in Greater Cook Strait, New Zealand. J. Geophys. Res. 85:
2728-2742.

Cairns, J.L. 1966. Depth and strength of the seasonal thermocline
in shallow water off southern California. U.S.N. Symp. Military
Oceanogr. 3rd V. 2: 27-28.

Cairns, J.L. 1967. Asymmetry of internal tidal waves in shallow
coastal waters. J. Geophys. Res. 72: 3563-3565.

Cairns, J.L. 1968. Thermocline strength fluctuations in coastal waters. J. Geophys. Res. 73: 2591-2595.

Cairns, J.L. and E.C. LaFond. 1966. Periodic motions of the seasonal thermocline along the southern California coast. J. Geophys. Res. 71: 3903-3915.

Cairns, J.L. and K.W. Nelson. 1968. Seasonal characteristics of the seasonal thermocline in shallow coastal water. U.S.N. Symp. Military Oceanogr. 5th, V. 1: 38-55.

Côté, B. and T. Platt. 1983. Day-to-day variations in the spring-summer photosynthetic parameters of coastal marine phytoplankton. Limnol. Oceanogr. 28: 320-344.

Cullen, J.J., E. Stewart, E. Renger, R.W. Eppley and C.D. Winant. 1983. Vertical motion of the thermocline, nitracline and chlorophyll maximum layers in relation to currents on the Southern California Shelf. J. Mar. Res. 41: 239-262.

Defant, A. 1961. Physical Oceanography. Vol. II. Pergamon Press Inc., New York, NY 598 p.

Demers, S., P.E. Lafleur, L. Legendre and C.L. Trump. 1979. Short-term covariability of chlorophyll and temperature in the St. Lawrence Estuary. J. Fish. Res. Bd. Can. 36: 568-573.

Demers, S. and L. Legendre. 1979. Effets des marées sur la variation circadienne de la capacité photosynthétique du phytoplancton de l'estuaire du Saint-Laurent. J. Exp. Mar. Biol. Ecol. 39: 87-99.

Demers, S. and L. Legendre. 1981. Mélange vertical et capacité photosynthétique du phytoplancton estuarien (estuaire du Saint-Laurent). Mar. Biol. 64: 243-250.

Demers, S. and L. Legendre. 1982. Water column stability and photosynthetic capacity of estuarine phytoplankton: long-term relationships. Mar. Ecol. Prog. Ser. 7: 337-340.

Denman, K.L. 1976. Covariability of chlorophyll and temperature in the sea. Deep-Sea Res. 23: 539-550.

Denman, K.L. 1977. Short term variability in vertical chlorophyll structure. Limnol. Oceanogr. 22: 434-441.

Denman, K.L. and A.E. Gargett. 1983. Time and space scales of vertical mixing and advection of phytoplankton in the upper ocean. Limnol. Oceanogr. 28: 801-815.

Denman, K.L. and T. Platt. 1975. Coherences in the horizontal distributions of phytoplankton and temperature in the upper ocean. Mém. Soc. R. Sci. Liège, 6th series, 7: 19-30.

Dubinsky, Z. 1980. Light utilization efficiency in natural phytoplankton communities. In: Primary Productivity in the Sea, pp. 83-97. P.G. Falkowski (ed.). Plenum Press, New York.

Eppley, R.W. and B.J. Peterson. 1979. Particulate organic matter flux and planktonic new production in the deep ocean. Nature (London) 282: 677-680.

Eppley, R.W., E.H. Renger and W.G. Harrison. 1979. Nitrate and phytoplankton production in southern California coastal waters. Limnol. Oceanogr. 24: 483-494.

Falkowski, P.G. 1980. Light and shade adaptation in marine phytoplankton. In: Primary Productivity in the Sea, pp. 99-119. P.G. Falkowski (ed.). Plenum Press, New York.

Falkowski, P.G. 1983. Light-shade adaptation and vertical mixing of marine phytoplankton: A comparative field study. J. Mar. Res. 41: 215-237.

Falkowski, P.G. and D.P. Stone. 1975. Nitrate uptake in marine phytoplankton: Energy sources and the interaction with carbon fixation. Mar. Biol. 32: 77-84.

Falkowski, P.G. and C.D. Wirick. 1981. A simulation model of the effects of vertical mixing on primary productivity. Mar. Biol. 65: 69-75.

Fasham, M.J. and P.R. Pugh. 1976. Observations on the horizontal coherence of chlorophyll a and temperature. Deep-Sea Res. 23: 539-550.

Federov, V.D. and V.A. Sermin. 1970. Primary production in relation to the hydrological regime of a sea basin (exemplified by the White Sea). Oceanology 10: 242-253.

Forrester, W.D. 1974. Internal tides in the St. Lawrence estuary. J. Mar. Res. 32: 55-66.

Fortier, L. and L. Legendre. 1979. Le contrôle de la variabilité à court terme du phytoplancton estuarien: stabilité verticale et profondeur critique. J. Fish. Res. Bd. Can. 36: 1325-1335.

Fortier, L., L. Legendre, A. Cardinal and C.L. Trump. 1978. Variabilité à court terme du phytoplancton de l'estuarire du Saint-Laurent. Mar. Biol. 46: 349-354.

Fournier, R.O. 1978. Biological aspects of the Nova Scotian shelf break fronts. In: Oceanic Fronts in Coastal Processes, pp. 69-77. M.J. Bowman and W.E. Esaias (eds.). Springer-Verlag, Berlin Heidelberg.

Fournier, R.O., J. Marra, R. Bohrer and M. Van Det. 1977. Plankton dynamics and nutrient enrichment of the Scotian shelf. J. Fish. Res. Bd. Can. 34: 1004-1018.

Fournier, R.O., M. Van Det, J.S. Wilson and N.B. Hargreaves. 1979. Influence of the shelf-break front off Nova Scotia on phytoplankton standing stock. J. Fish. Res. Bd. Can. 36: 1228-1237.

Fréchette, M. and L. Legendre. 1982. Phytoplankton photosynthetic response to light in an internal tide dominated environment. Estuaries 5: 287-293.

Gallegos, C.L., T. Platt, W.G. Harrison and B. Irwing. 1983. Photosynthetic parameters of Arctic marine phytoplankton: Vertical variation and time scales of adaptation. Limnol. Oceanogr. 28: 698-708.

Garrett, C.J.R., J.R. Keeley and D.A. Greenberg. 1978. Tidal mixing versus thermal stratification in the Bay of Fundy and Gulf of Maine. Atm. Ocean 16: 403-423.

Garrett, C.J.R. and J.W. Loder. 1981. Dynamical aspects of shallow sea fronts. Phil. Trans. R. Soc. Lond. A302: 563-581.

George, D.G. and S.I. Heaney. 1978. Factors influencing the spatial distribution of phytoplankton in a small productive lake. J. Ecol. 66: 133-155.

Gilmartin, M. 1964. The primary production of a British Columbia fjord. J. Fish. Res. Bd. Can. 21: 505-538.

Gould, J. and W. McKee. 1973. Observation of the vertical structure of semi-diurnal tidal currents in the Bay of Biscay. Nature (London) 244: 88.

Greisman, P. and R.G. Ingram. 1977. Nutrient distribution in the St. Lawrence Estuary. J. Fish. Res. Bd. Can. 34: 2117-2123.

Griffiths, D.K., R.D. Pingree and M. Sinclair. 1981. Summer tidal fronts in the near-Arctic regions of Foxe Basin and Hudson Bay. Deep-Sea Res. 28: 865-873.

Haffner, G.D., G.P. Harris and M.K. Jarai. 1980. Physical variability and phytoplankton communities. III. Vertical structure in phytoplankton populations. Arch. Hydrobiol. 89: 363-381.

Harris, G.P. 1978. Photosynthesis, productivity and growth: The physiological ecology of phytoplankton. Arch. Hydrobiol. Beih. Ergebn. Limnol. 10: 1-171.

Harris, G.P. 1980a. Temporal and spatial scales in phytoplankton ecology. Mechanisms, methods, models and management. Can. J. Fish. Aquat. Sci. 37: 877-900.

Harris, G.P. 1980b. The measurement of photosynthesis in natural populations of phytoplankton. In: The Physiological Ecology of Phytoplankton, pp. 129-187. I. Morris (ed.). Blackwell, Oxford.

Harris, G.P. and J.W.A. Lott. 1973. Light intensity and photosynthetic rates in phytoplankton. J. Fish. Res. Bd. Can. 30: 1771-1778.

Harris, G.P. and B.B. Piccinin. 1977. Photosynthesis by natural phytoplankton populations. Arch. Hydrobiol. 80: 405-457.

Harris, G.P. and B.B. Piccinin. 1980. Physical variability and phytoplankton communities. IV. Temporal changes in the phytoplankton community of a physically variable lake. Arch. Hydrobiol. 89: 447-473.

Haury, L.R., M.G. Briscoe and M.H. Orr. 1979. Tidally generated internal wave packets in Massachusetts Bay. Nature (London) 278: 312-317.

Heath, R.A. 1973. Flushing of coastal embayments by changes in atmospheric conditions. Limnol. Oceanogr. 18: 849-862.

Herman, A.W. and K.L. Denman. 1979. Intrusions and vertical mixing at the shelf/slope water front south of Nova Scotia. J. Fish. Res. Bd. Can. 36: 1445-1453.

Hitchcock, G.L. and T.J. Smayda. 1977. The importance of light in the initiation of the 1972-1973 winter-spring bloom in Narragansett Bay. Limnol. Oceanogr. 22: 126-131.

Holligan, P.M. 1978. Patchiness in subsurface phytoplankton populations on the Northwest European continental shelf. In: Spatial Pattern in Plankton Communities, pp. 221-238. J.H. Steele (ed.). Plenum Press, New York.

Holligan, P.M. 1981. Biological implications of fronts on the northwest European continental shelf. Phil. Trans. R. Soc. Lond. A302: 547-562.

Holligan, P.M. and D.S. Harbour. 1977. The vertical distribution and succession of phytoplankton in the western English Channel in 1975 and 1976. J. Mar. Biol. Ass. U.K. 57: 1075-1093.

Ingram, R.G. 1975. Influence of tidal induced mixing on primary productivity in the St. Lawrence estuary. Mém. Soc. R. Sci. Liège 7: 59-74.

Ingram, R.G. 1978. Internal wave observations off Isle Verte. J. Mar. Res. 36: 715-724.

Ingram, R.G. 1979. Water mass modification in the St. Lawrence estuary. Naturaliste Can. 106: 45-54.

Iverson, R.L., H.C. Curl, Jr., H.B. O'Connors, Jr., D. Kirk and K. Zakar. 1974. Summer phytoplankton blooms in Auke Bay, Alaska, driven by wind mixing of the water column. Limnol. Oceanogr. 19: 271-278.

Iverson, R.L., T.E. Whitledge and J.J. Goering. 1979. Chlorophyll and nitrate fine structure in the southeastern Bering Sea shelf break front. Nature (London) 281: 664-666.

Jerlov, N.G. 1968. Optical Oceanography. Elsevier, New York. 194 p.

Jewson, D.H. and R.B. Wood. 1975. Some effect on integral photosynthesis of artificial circulation of phytoplankton through light gradients. Vehr. Internat. Verein. Limnol. 19: 1037-1044.

Kamykowski, D. 1974. Possible interactions between phytoplankton and semi-diurnal tides. J. Mar. Res. 32: 67-89.

Kamykowski, D. 1976. Possible interactions between plankton and semi-diurnal internal tides. II. Deep thermoclines and trophic effects. J. Mar. Res. 34: 499-509.

Kamykowski, D. 1979. The growth response of a model Gymnodinium splendens in stationary and wavy water columns. Mar. Biol. 50: 289-303.

Kamykowski, D. and S.-J. Zentera. 1977. The diurnal vertical migration of motile phytoplankton through temperature gradients. Limnol. Oceanogr. 22: 148-151.

Kemp, W.M. and W.J. Mitsch. 1979. Turbulence and phytoplankton diversity: A general model of the "Paradox of Plankton." Ecological Modelling 7: 202-222.

Lafleur, P.E., L. Legendre and A. Cardinal. 1979. Dynamique d'une population estuarienne de diatomées planctoniques: Effet de l'alternance des marées de morte-eau et de vive-eau. Oceanol. Acta 2: 307-315.

LaFond, E.C. 1962. Internal waves. Part. I. In: The Sea, Vol. 1, pp. 731-751. M.N. Hill (ed.). Intersciences Publishers, Inc. New York, NY.

LeBlond, P.H. and L.A. Mysak. 1978. Waves in the Ocean. Elsevier, Amsterdam. 602 p.

Legendre, L. 1981. Hydrodynamic control of marine phytoplankton production: The paradox of stability. In: Ecohydrodynamics, pp. 191-207. J.C.J. Nihoul (ed.). Elsevier, Amsterdam.

Legendre, L. and S. Demers. 1984. Towards dynamic biological oceanography and limnology. Can. J. Fish. Aquat. Sci. 41: 2-19.

Legendre, L., R.G. Ingram and Y. Simard. 1982. Aperiodic change of water column stability and phytoplankton in an Arctic coastal embayment, Manitounuk Sound, Hudson Bay. Naturaliste Can. 109: 775-786.

Levasseur, M., J.-C. Therriault and L. Legendre. 1983. Tidal currents, wind and the morphology of phytoplankton spatial structures. J. Mar. Res. 41: 655-672.

Levasseur, M., J.-C. Therriault and L. Legendre. 1984. Hierarchical control of phytoplankton succession by physical factors. Mar. Ecol. Prog. Ser. 19: 211-222.

Lorenzen, C.J. 1966. A method for the continuous measurement of in vivo chlorophyll concentration. Deep-Sea Res. 13: 223-227.

Malone, C.M., P.G. Falkowski, T.S. Hopkins, G.T. Rowe and T.E. Whitledge. 1983. Mesoscale response of diatom populations to a wind event in the plume of the Hudson River. Deep-Sea Res. 30: 149-170.

Margalef, R. 1978. Life-forms of phytoplankton as survival alternatives in an unstable environment. Oceanol. Acta 1: 493-509.

Margalef, R., M. Estrada and D. Blasco. 1979. Functional morphology of organisms involved in red tides, as adapted to decaying turbulence. In: Toxic Dinoflagellate Blooms, Vol. I, pp. 89-94. L. Taylor and H.H. Seliger (eds.). Elsevier, Amsterdam.

Marra, J. 1978a. Effect of short-term variations in light intensity on photosynthesis of a marine phytoplankter. A laboratory simulation study. Mar. Biol. 46: 191-202.

Marra, J. 1978b. Phytoplankton photosynthetic response to vertical movement in a mixed layer. Mar. Biol. 46: 203-208.

McGowan, J.A. and T.L. Hayward. 1978. Mixing and oceanic productivity. Deep-Sea Res. 25: 771-793.

Muir, L.A. 1979. Internal tides in the Middle Estuary of the St. Lawrence. Naturaliste Can. 106: 27-36.

Paasche, E. 1971. Effect of ammonia and nitrate on growth, photosynthesis and ribulose diphosphate carboxylase content of Dunaliella tertiolecta. Physiol. Plant. 25: 294-299.

Parsons, T.R., R.I. Perry, E.D. Nutbrown, W. Hsieh and C.M. Lalli. 1983. Frontal zone analysis at the mouth of Saanich Inlet, British Columbia, Canada. Mar. Biol. 73: 1-5.

Perry, R.I., B.R. Dilke and T.R. Parsons. 1983. Tidal mixing and summer plankton distributions in Hecate Strait, British Columbia. Can. J. Fish. Aquat. Sci. 40: 871-887.

Perry, M.J., M.C. Talbot and R.S. Alberte. 1981. Photoadaptation in marine phytoplankton: response of the photosynthetic unit. Mar. Biol. 62: 91-101.

Pingree, R.D. 1978a. Mixing and stabilization of phytoplankton distributions on the Northwest European continental shelf. In: Spatial Pattern in Plankton Communities, pp. 181-220. J.H. Steele (ed.). Plenum Press, New York.

Pingree, R.D. 1978b. Cyclonic eddies and cross-frontal mixing. J. Mar. Biol. Ass. U.K. 58: 955-963.

Pingree, R.D., M.J. Bowman and W.E. Esaias. 1978a. Headland fronts. In: Oceanic Fronts in Coastal Processes, pp. 78-86. M.J. Bowman and W.E. Esaias (eds.). Springer-Verlag, Berlin Heidelberg.

Pingree, R.D. and D.K. Griffiths. 1978. Tidal fronts on the shelf seas around the British Isles. J. Geophys. Res. 83: 4615-4622.

Pingree, R.D. and D.K. Griffiths. 1980. A numerical model of the M_2 tide in the Gulf of St. Lawrence. Oceanol. Acta 3: 221-225.

Pingree, R.D., P.M. Holligan and R.N. Head. 1977. Survival of dinoflagellate blooms in the western English Channel. Nature (London) 265: 266-269.

Pingree, R.D., P.M. Holligan and B.T. Mardell. 1978b. The effects of vertical stability on phytoplankton distributions in the summer on the northwest European shelf. Deep-Sea Res. 25: 1011-1028.

Pingree, R.D., P.M. Holligan and G.T. Mardell. 1979. Phytoplankton growth and cyclonic eddies. Nature (London) 278: 245-247.

Pingree, R.D., P.M. Holligan, G.T. Mardell and R.N. Head. 1976. The influence of physical stability on spring, summer and autumn phytoplankton blooms in the Celtic Sea. J. Mar. Biol. Ass. U.K. 56: 845-873.

Pingree, R.D. and G.T. Mardell. 1981. Slope turbulence, internal waves and phytoplankton growth at the Celtic Sea shelf-break. Phil. Trans. R. Soc. London A302: 663-682.

Pingree, R.D., G.T. Mardell, P.M. Holligan, D.K. Griffiths and J. Smithers. 1982. Celtic Sea and Armorican current structure and the vertical distributions of temperature and chlorophyll. Cont. Shelf Res. 1: 99-116.

Pingree, R.D., G.T. Mardell and L. Maddock. 1983. A marginal front in Lyme Bay. J. Mar. Biol. Ass. U.K. 63: 9-15.

Pingree, R.D., P.R. Pugh, P.M. Holligan and G.R. Forster. 1975. Summer phytoplankton blooms and red tides along tidal fronts in the approaches to the English Channel. Nature (London) 258: 672-677.

Prézelin, B.B. 1976. The role of peridinin-chlorophyll a-proteins in the photosynthetic light adaptation of the marine dinoflagellate, Glenodinium sp. Planta 130: 225-233.

Prézelin, B.B. 1981. Light reactions in photosynthesis. In: Physiological Bases of Phytoplankton Ecology, pp. 1-43. T. Platt (ed.). Can. Bull. Fish. Aquat. Sci. (210).

Proudman, J. 1953. Dynamical Oceanography. Methuen and Co., Ltd., London. 409 p.

Quarmby, L.M., D.H. Turpin and P.J. Harrison. 1982. Physiological responses of two marine diatoms to pulsed additions of ammonium. J. Exp. Mar. Biol. Ecol. 63: 173-181.

Rattray, M., Jr. 1960. On the coastal generation of internal tides. Tellus 12: 54-62.

Rattray, M., Jr., J.G. Dworski and P.E. Kovala. 1969. Generation of long internal waves at the continental slope. Deep-Sea Res., Suppl. to Vol. 16: 179-195.

Riley, G.A. 1957. The plankton of estuaries. In: Estuaries, pp. 316-326. G.J. Lauff (ed.). AAAS Pub. H88.

Ryther, J.H. and D.W. Menzel. 1959. Light adaptation by marine phytoplankton. Limnol. Oceanogr. 4: 492-497.

Savidge, G. 1976. A preliminary study of the distribution of chlorophyll a in the vicinity of fronts in the Celtic and Western Irish Seas. Estuar. Coast. Mar. Sci. 4: 617-625.

Savidge, G. 1979. Photosynthetic characteristics of marine phytoplankton from contrasting physical environments. Mar. Biol. 53: 1-12.

Savidge, G. 1981. Studies on the effects of small-scale turbulence on phytoplankton. J. Mar. Biol. Ass. U.K. 61: 477-488.

Seliger, H.H., K.R. McKinley, W.H. Biggley, R.B. Rivkin and K.R.H. Aspden. 1981. Phytoplankton patchiness and frontal regions. Mar. Biol. 61: 119-131.

Simpson, J.H. 1981. The shelf-sea fronts: Implications of their existence and behaviour. Phil. Trans. R. Soc. London A302: 531-546.

Simpson, J.H. and J.R. Hunter. 1974. Fronts in the Irish Sea. Nature (London) 250: 404-406.

Simpson, J.H. and R.D. Pingree. 1978. Shallow sea fronts produced by tidal stirring. In: Oceanic Fronts in Coastal Processes, pp. 29-42. M.J. Bowman and W.E. Esaias (eds.). Springer-Verlag, Berlin Heidelberg.

Simpson, J.H., P.B. Tett, M.L. Argote-Espinoza, A. Edwards, K.J. Jones and G. Savidge. 1982. Mixing and phytoplankton growth around an island in a stratified sea. Cont. Shelf Res. 1: 15-31.

Sinclair, M. 1978. Summer phytoplankton variability in the lower St. Lawrence estuary. J. Fish. Res. Bd. Can. 35: 1171-1185.

Sinclair, M., J.P. Chanut and M. El-Sabh. 1980. Phytoplankton distributions observed during a 3½ days fixed-station in the lower St. Lawrence estuary. Hydrobiology 75: 129-147.

Sinclair, M., D.V. Subba Rao and R. Couture. 1981. Phytoplankton temporal distributions in estuaries. Oceanol. Acta 4: 239-246.

Steele, J.H. 1957. The role of lateral eddy diffusion in the northern North Sea. J. Cons. Int. Explor Mer. 22: 152-162.

Steemann-Nielsen, E. 1975. Marine Photosynthesis. Elsevier, Amsterdam. 141 p.

Steemann-Nielsen, E. and V.Kr. Hansen. 1959. Light adaptation in marine phytoplankton population and its interrelation with temperature. Physiol. Plant. 12: 353-370.

Steemann-Nielsen, E. and E.G. Jorgensen. 1968. The adaptation of plankton algae. I. General part. Physiol. Plant. 21: 401-413.

Steven, D.M. 1975. Biological production in the Gulf of St. Lawrence. In: Energy Flow - Its Biological Dimensions, A. Summary of the IBP in Canada, pp. 229-248. W.M. Cameron and L.W. Billingsley (eds.). Royal Society of Canada, Ottawa.

Takahashi, M., D.L. Siebert and W.H. Thomas. 1977. Occasional blooms of phytoplankton during summer in Saanich Inlet, B.C., Canada. Deep-Sea Res. 24: 775-780.

Tett, P. 1981. Modelling phytoplankton production at shelf-sea fronts. Phil. Trans. R. Soc. London A302: 605-615.

Therriault, J.-C. and G. Lacroix. 1976. Nutrients, chlorophyll and internal tides in the St. Lawrence Estuary. J. Fish. Res. Bd. Can. 33: 2747-2757.

Therriault, J.-C., D.J. Lawrence and T. Platt. 1978. Spatial variability of phytoplankton turnover in relation to physical processes in a coastal environment. Limnol. Oceanogr. 23: 900-911.

Therriault, J.-C. and T. Platt. 1981. Environmental control of phytoplankton patchiness. Can. J. Fish. Aquat. Sci. 38: 638-641.

Thompson, R.O.R.Y. and T.J. Golding. 1981. Tidally induced "upwelling" by the Great Barrier Reef. J. Geophys Res. 86: 6517-6521.

Thornber, P.J., R.S. Alberte, F.A. Hunter, J.A. Shiozawa and K.-S. Kan. 1977. The organization of chlorophyll in the plant photosynthetic unit. Brookhaven Symp. Biol. 28: 132-148.

Turpin, D.H. and P.J. Harrison. 1979. Limiting nutrient patchiness and its role in phytoplankton ecology. J. Exp. Mar. Biol. Ecol. 39: 151-166.

Turpin, D.H. and P.J. Harrison. 1980. Cell size manipulation in natural marine, planktonic, diatom communities. Can. J. Fish. Aquat. Sci. 37: 1193-1195.

Venrick, E.L. 1972. Small-scale distributions of oceanic diatoms. Fish. Bull. 70: 363-372.

Vincent, W.F. 1980. Mechanisms of rapid photosynthetic adaptation in natural phytoplankton communities. II. Changes in photochemical capacity as measured by DCMU-induced chlorophyll fluorescence. J. Phycol. 16: 568-577.

Walsh, J.J. 1981. Shelf-sea ecosystems. In: Analysis of Marine Ecosystems, pp. 159-196. A.R. Longhurst (ed.). Academic Press, New York.

Walsh, J.J., T.E. Whitledge, F.W. Barvenik, C.D. Wirick and S.O. Howe. 1978. Wind events and food chain dynamics within the New York Bight. Limnol. Oceanogr. 23: 659-683.

Walsh, J.J., T.E. Whitledge, J.C. Kelley, S.A. Huntsman and R.D. Pillsbury. 1977. Further transition states of the Baja California upwelling ecosystem. Limnol. Oceanogr. 22: 264-280.

Wangersky, P.J. 1977. The role of particulate matter in the productivity of surface waters. Helgol. Wiss. Meeresunters. 30: 546-564.

Webb, K.L. and C.F. D'Elia. 1980. Nutrient and oxygen redistribution during a spring neap tidal cycle in a temperature estuary. Science 2078: 983-985.

White, A.W. 1976. Growth inhibition caused by turbulence in the toxic marine dinoflagellate Gonyaulax excavata. J. Fish. Res. Bd. Can. 33: 2598-2602.

Winant, C.D. and R.J. Olson. 1976. The vertical structure of coastal currents. Deep-Sea Res. 23: 925-936.

Winter, D.F., K. Banse and G.C. Anderson. 1975. The dynamics of phytoplankton blooms in Puget Sound, a fjord in the northwestern United States. Mar. Biol. 29: 139-176.

Woods, J.D. and R.L. Wiley. 1977. Billow turbulence and ocean microstructure. Deep-Sea Res. 19: 87-121.

Wunsch, C. 1975. Internal tides in the ocean. Rev. Geophys. Space Phys. 13: 17-182.

Wunsch, C. and R. Hendry. 1972. Array measurements of the bottom boundary layer and the internal wave field on the continental slope. Geophys. Fluid. Dyn. 4: 101-145.

Zeilon, N. 1913. On the seiches of the Gullman Fjord. Svenska Hydrog. Biolog. Komm. Skrifter 5.

Zeilon, N. 1934. Experiments on boundary rides. Gateboy Veten-sksamh. Handl. Folj. 5: B3, No. 10.

ISLAND STIRRING EFFECTS ON PHYTOPLANKTON GROWTH

J.H. Simpson
Marine Science Laboratories
Menai Bridge
Gwynedd, LL59 5EY
UNITED KINGDOM

P.B. Tett
Scottish Marine Biological Association
Dunstaffnage
Oban, Argyll PA34 4AD
UNITED KINGDOM

INTRODUCTION

It has long been suspected that the disturbance of the flow caused by the presence of an oceanic island may be responsible for increased biomass and production near the island. Doty and Oguri (1956), reporting observations near Oahu, Hawaii, coined the phrase "island mass effect" to describe such enhancement, which they explained with the suggestion that benthic algae or corals on the island's flanks might concentrate inorganic nutrients from the passing waters, grazing eventually making these nutrients available to phytoplankton. They drew on older literature (e.g. Gran, 1912) reporting an increase in oceanic plankton near continental margins. Dandonneau and Charpy (submitted), in a detailed statistical analysis of data from the S.W. Pacific, could find only occasional and weak relationships between sea-surface photosynthetic pigment concentration and distance from the nearest island. Remote sensing of pigments has not yet been used extensively to observe biomass distributions around islands (but see Simpson and Tett, this volume, for an example of the exciting possibilities in this area). Other accounts of island mass effects are anecdotal and often unsupported by measurements of the physical structure and flow field. It thus appears to remain true, as Gilmartin and Revelante (1974) remarked in a further study of the Hawaiian islands, that 'except for the work of Doty and Oguri (1956), the so-called 'island mass' effect is neither well documented nor are the causes for the increases clear.'

Lecture Notes on Coastal and Estuarine Studies, Vol. 17
Tidal Mixing and Plankton Dynamics. Edited by J. Bowman, M. Yentsch and W.T. Peterson
© Springer-Verlag Berlin Heidelberg 1986

More recently, however, attention has been directed to the effects of islands in shallow stratified seas, where the influence of tidal flow is dominant and the occurrence of stratification is determined by the water depth h and tidal amplitude u in the form of the parameter h/u^3 (Simpson and Hunter, 1974). In such circumstances, a partial breakdown of seasonal stratification may occur around the island with the release of some nutrient-rich bottom water into the surface layers and a consequent enhancement of phytoplankton biomass both around the island and in any wake region associated with the non-tidal mean flow.

These tidal stirring effects of islands are better defined and more accessible than those associated with 'steady' flow past oceanic islands, and form the natural focus for a review at this stage. They should, however, be seen as part of the general problem of flow disturbance by islands and we set them in this context with a brief review of the basic effects induced by idealized and real islands in steady and oscillating flows.

Flow Around Islands

i) Steady current

Many of the important features of the flow around real islands can be identified by reference to the case of steady non-rotating flow of a homogeneous fluid past a circular cylinder. This case has been much studied in laboratory experiments. Whatever the imposed flow condition, the presence of the cylinder causes an increase of the flow speed on the "flanks" of the body with stagnation points along the central flow axis (Fig. 1a). For the case of potential flow (Lamb, 1932), the maximum velocity is greater by a factor of exactly 2 than that in the far field flow. This augmented flow is required to transfer fluid from ahead of the cylinder to fill up the void that the cylinder would otherwise leave behind in advancing relative to the fluid.

In practice, the flow around the cylinder does not usually conform to the symmetrical streamlines indicated for potential flow, the form of the wake depending on the value of the Reynolds number:

43

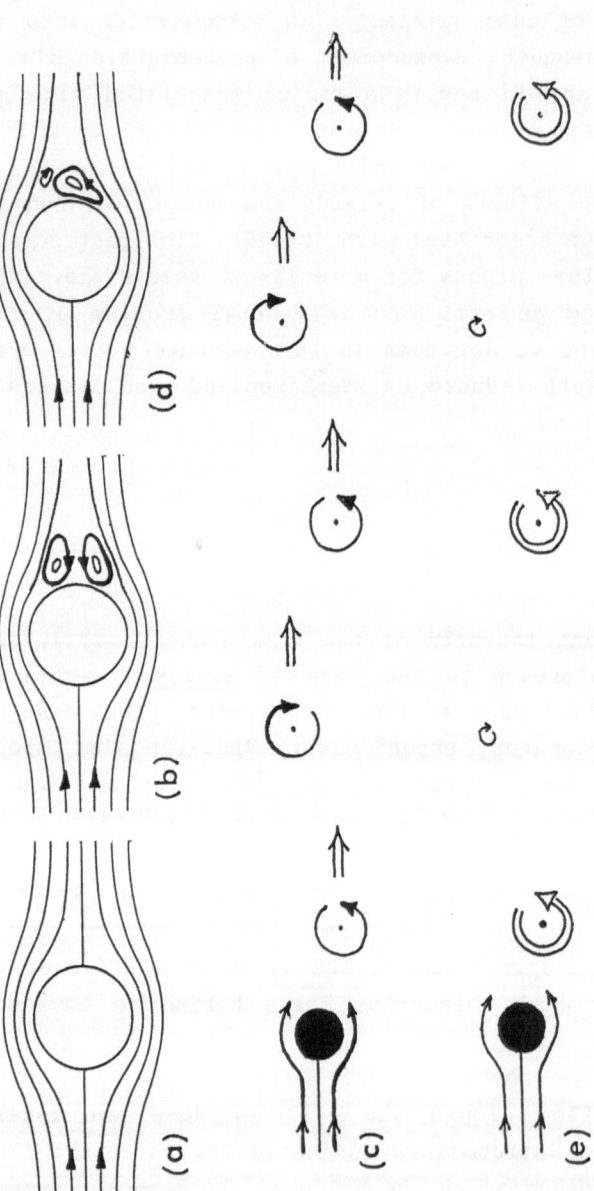

Figure 1. Flow past a circular cylinder. (a) Re ~ 1. No rotation. Fully attached flow. (b) 1 < Re < 70. No rotation. Steady separation region in the wake. (c) 70 < Re < 2500. No rotation. Regular shedding of vortices at a frequency ~ 0.105 U/R. (d) Re ~ 50 with rotation. Asymmetric steady separation region. (e) Re ~ 500 with rotation. Regular shedding of vortices alternately strong cyclonic and weaker anticyclonic.

The transition from steady flow to vortex shedding for flow in a rotating system is not determined by Re alone, but depends on Ro and Ek whose ratio Ro/Ek = Re. Increasing the rotation rate decreases Ek and Ro and inhibits the onset of eddy shedding.

$$Re = \frac{2UR}{\nu}$$

where U = free stream velocity, R = cylinder radius and ν = kinematic vicosity. Only at low Reynolds numbers ($\sim$1), where viscous effects are important, does the flow exhibit upstream-downstream symmetry. As the Reynolds number increases from unity to $\sim$70, an increasingly large separation region forms in the wake with two contra-rotating stationary eddies apparent (Fig. 1b). At higher Reynolds number, separation effects dominate and detached eddies are observed in the wake of the cylinder (see Batchelor (1970) p. 352). For the range 70 < Re < 2500, the wake takes the form of a regular array of eddies termed a "vortex street" (Fig. 1c). Eddies with alternating positive and negative vorticity are shed by the cylinder to form two rows in the wake. The frequency n at which eddies are shed from the cylinder is closely related to the scale frequency U/2R so that the Strouhal number

$$St = \frac{2nR}{U}$$

has an approximately constant value of 0.21 for Re > 300, implying that the eddies have a spacing of $\sim$ 10R if they move downstream with the mean flow. In fact they move somewhat slower than the mean flow ($\sim$0.85U) and, with increasing distance downstream, there is a slow increase of the distances between the two rows of vortices and between neighbouring vortices. Far downstream beyond the vortex street, the wake is usually turbulent. For Re > 2500, turbulence extends to the whole wake region and eddy shedding is no longer periodic.

Recently, the study of wakes in cylindrical flow has been extended to include the effects of constant rotation (f plane) and rotation varying with latitude (β plane) both of which are likely to be important in the interpretation of large scale geophysical flows. In the f plane studies, dimensional considerations dictate that the controlling parameters are the Ekman and Rossby numbers which, for a rotation rate, ω, are defined as

$$Ek = \frac{\nu}{2\omega R^2} \qquad\qquad Ro = \frac{U}{2\omega R}$$

whose ratio   Ro/Ek = Re.

Experimental studies by Bowyer and Kmetz (1983) and Boyer and Davies (1982) have demonstrated the relevance of Ek and Ro to controlling the onset of instability in the wake. For a given value of Re, the occurrence of eddy shedding is inhibited by decreasing Ek as the rotation rate is increased.

There is also a general tendency for the wakes to exhibit an asymmetry between left and right hand sides (looking downstream from the cylinder - see Fig. 1d,e). With increasing flow speed, eddies are shed first from the left hand side of the cylinder with an eddy remaining attached on the right side (for anti-clockwise rotation). At higher speeds, eddies are shed from both sides in the non-rotating case, but the right side (cyclonic) eddies are substantially stronger than those on the left (Fig. 1e). Such asymmetry has been convincingly observed in a number of satellite images of cloud patterns in the atmospheric wakes of oceanic islands (Pitts et al., 1977; Scorer, 1978).

(ii) Oscillatory flow

The disturbance of the flow described above for a steady current encountering an island is further complicated when we consider the reciprocating current of amplitude U_o associated with tidal flow at frequency ω' for islands which are small enough that

$$St' = \frac{2\omega'R}{U_o} \ll 1$$

the tidal excursion is large in relation to the scale of the island so that an extended wake region occurs. Flow in this region may be regarded as "quasi-steady" and should exhibit a range of eddy-shedding behavior as the Reynolds number varies with the tidal velocity. Such effects have been little studied in the ocean though Wolanski et al. (1984) have recently reported observations of the tidal flow past a small island (1.3 km x 0.3 km) in shallow water ($\sim$20 m) at relatively low St'. In this case the wake takes the form of a persistent clockwise eddy corresponding approximately to the steady 2D flow past an inclined plate at Re $\sim$ 10. The effects of bottom friction are apparently responsible for this low effective Re and the suppression of any tendency towards the development of a vortex street. On the other hand, observations by one of us (JHS) using time-lapse photography have clearly revealed the development of a vortex street in the wake of a small island in the Menai Straits, where the ratio of depth to island radius is greater.

Generally, however, for the range of tidal currents present in the shelf seas, St' << 1 only for very small features (<< 1 km) and for most islands of interest St' is of order unity or greater so that the tidal cycle is too short to allow the development of significant organized wakes. A 2D numerical model of such an oscillating flow without eddy shedding effects has been given by Pingree and Maddock (1979) for the case of a conical island with R = 4.5 km at the surface and bottom depth increasing uniformly to 35 m. The model, which includes the Coriolis and bottom frictional forces, is driven by a far-field rectilinear flow of 50 cm s^{-1}. The resulting distribution of h/u^3 (Fig. 2a) exhibits a number of similarities to the steady potential flow model. High dissipation regions occur on the flanks of the island with correspondingly low energy on the axis of the flow (in the vicinity of the stagnation points of the potential flow model). The effect of rotation is, however, discernible in the slight clockwise displacement of the whole pattern relative to the non-rotating case.

Two other features of this model flow are of interest in the present context. A rectification of the tidal flow results in four residual eddies (Fig. 2b) which combine to produce offshore streaming on the flanks of the island. In the rotating case, the Coriolis forces introduce a net clockwise component of circulation (in the northern hemisphere). Longuet-Higgins (1969) has shown analytically that this type of flow is a Stokes drift associated with the induced wave which propagates around the island.

Close to the island where the tidal current is obliged to follow a path of high curvature, centrifugal forces must be balanced by a pressure gradient which implies a lowering of sea level on the flanks of the island (Fig. 2c). In the near-bottom flow, friction reduces the current velocity and hence the centrifugal forces with the consequence that there is an unbalanced component of the pressure gradient which drives a flow towards the island. The resulting

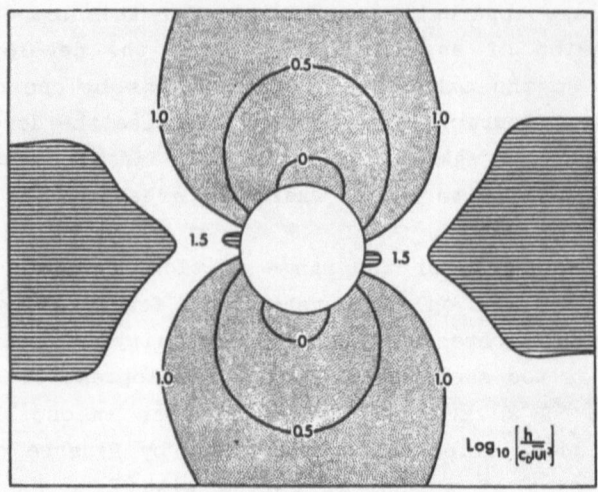

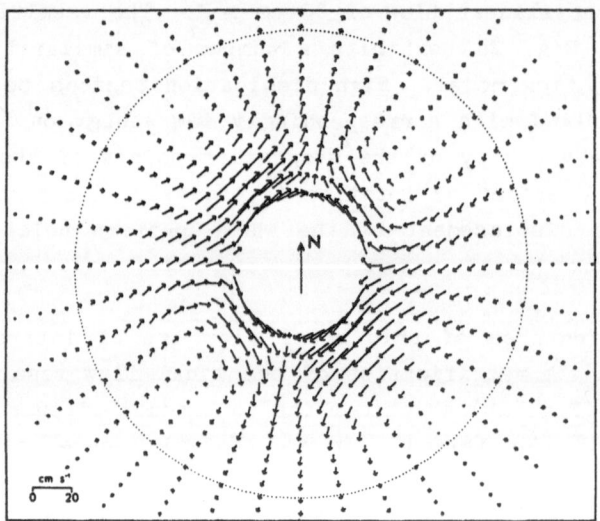

Figure 2. Tidal flow around an idealized conical island of surface radius R = 4.5 km. Away from the island (R > 15 km) depth h = 35 m the far field flow is rectilinear in the east-west direction with an amplitude of 50 cm s^{-1}.

(a) Distribution of h/u^3. Plotted parameter is in c.g.s. units with C_D = 0.0025. This form of log h/u^3 is 1.4 less than that used in Fig. 3.

(b) Residual flow forced by the tidal motion.

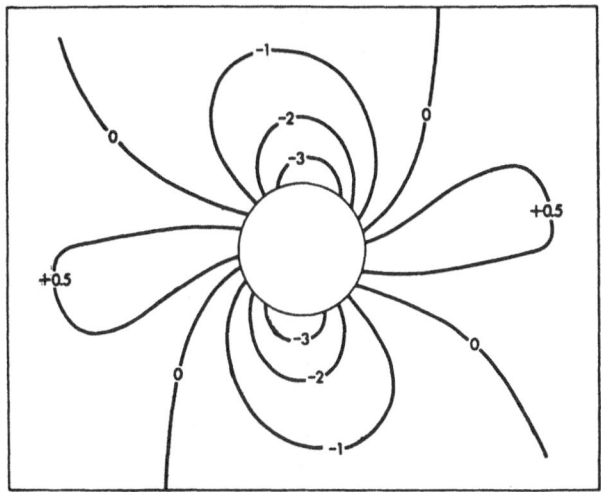

Figure 2. (c) Depression of sea level (in cm) due to centrifugal
            forces acting on the flow around the flanks of the island.

upwelling (Garrett and Loucks, 1976) may be locally important in
enhancing the flux of bottom water properties into the surface
layers.

Numerical models have also been used to determine the details of
tidal flow and dissipation around real islands. Argote (1983)
describes a detailed model of the Scilly Isles region based on a 2 km
grid. The boundary conditions specify a rotary M_2 tidal flow derived
from a larger scale model of the European shelf (Pingree and
Griffiths, 1978). Details of the modification of the tidal ellipse
by the island are illustrated in Figure 3 which shows maximum velocities close to the island to be approximately double those in the far
field. The resulting h/u^3 distribution (Fig. 3d) retains many of the
features of the idealized island model but includes important effects
due to the real topography. The prominent mixing zones ($\log h/u^3 < 2.5$)
frequently coincide closely with regions of minimum temperature
observed in infrared images, as can be seen by comparison of Fig. 4a
and Fig. 3d. The mixed water near the island may be up to 3° cooler
than the surface layer of the stratified region.

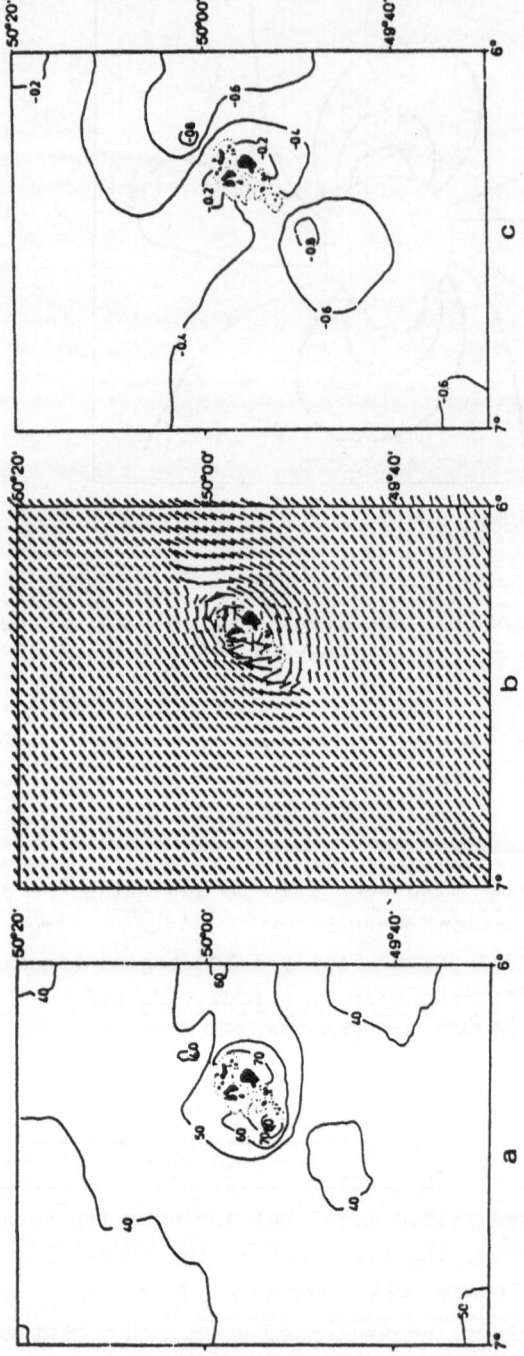

Figure 3. Model of tidal flow over real topography around the Scilly Isles.

(a) Amplitude of the major axis tidal stream amplitude for M_2 in cm s^{-1}.

(b) Major axis of the M_2 tidal stream.

(c) Ratio b/a of the minor and major axes of the tidal ellipse. Negative values indicate rotation in a clockwise sense.

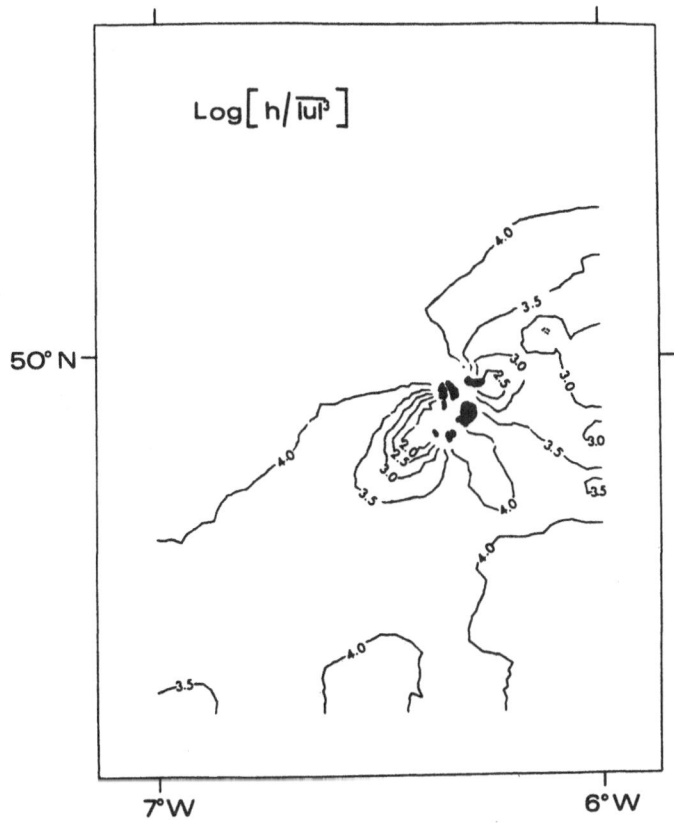

Figure 3. (d) Distribution of h/u³. Note the strong mixing zones
(log h/u³ < 2.7) to the northeast and southwest of the
islands. (from Argote, 1983).

This drop in surface temperature in shallower water is a clear
diagnostic of the enhanced vertical mixing which, in terms of impact
on the biology, is perhaps the most important aspect of the flow
disturbance by an island in a shelf regime. The continuous stirring
action of the island bringing cold water to the surface also supplies
a vertical flux of nutrients into the photic zone which, as we shall
discuss in section 3, can provide for substantial extra production of
biomass.

Before considering these biological effects we should note one
final physical aspect of island flow, that involving the interaction
of a mean flow with the tidal mixing zone around an island. A well-

Figure 4. Satellite infrared images of island mixing zones.

(a) Scilly Isles. TIROS-N AVHRR iamge. 145 GMT. 1/6/79.
The tip of Cornwall peninsula is visible at the right hand
margin of this image. Image size approximately 120 x 120
km^2.

defined residual flow will tend to sweep the mixed water away from
the island into an identifiable wake with lower stability and lower
surface temperature than the ambient water.

A model of this process based on depth-uniform residual flow has
been studied by Argote (1983). She included advective terms in the
equation for the potential energy anomaly ϕ (Simpson, 1981) and
integrated forward in time from the onset of stratification in the
spring. Using u^3 estimates from the tidal model of the area, predictions were made of the structure around the Scilly Isles for various
flow fields. In the example shown in Figure 5, a low stability wake
is seen to develop in response to a sustained steady flow of only 1.

Figure 4. Satellite infrared images of island mixing zones.

(b) St. Kilda. NOAA-7. AVHRR. 1352 GMT. 5/6/82.

In reality, the flow is likely to be much more variable than this but
the important general result illustrated here is the sensitivity of
the structure to low level mean flows. Evidence of wake formation
can be seen in some I-R images of Scilly Isles in which the observed
displacement is much bigger than can be accounted for by the tidal
movement.

The interaction of periodic tidal mixing with the mean flow is
manifest in the generation of 'fringes' in the vertically mixed
waters surrounding the Channel Islands (Pingree et al., 1985). While
the exact mechanism producing these fringes is not clear, their
spacing may be readily determined from I-R imagery and, since one
fringe is produced each semi-diurnal tidal cycle, used to give an

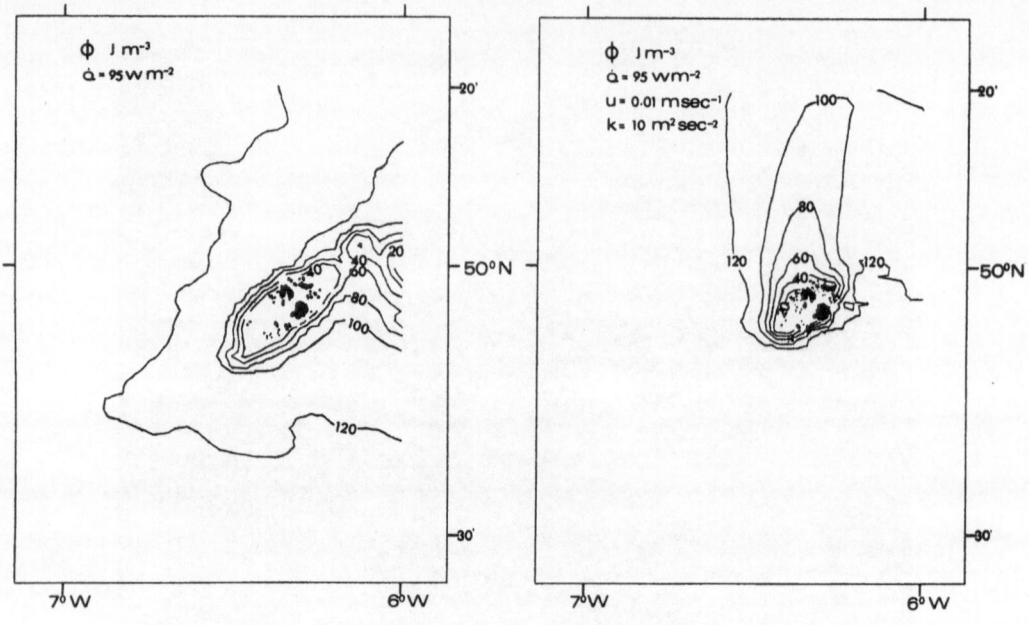

Figure 5. Predictions of potential energy anomaly φ (J m⁻³) for the
Scilly Isles region at midsummer.

(a) with no advection or diffusion.

(b) with a mean far-field flow of 1 cm s⁻¹ to the north and
diffusivity K = 10 m² s⁻¹.

Note the development of a wake of lower stability in
response to a weak mean flow.

estimate of the mean flow.   The decay rate of the fringe in the wake
is largely controlled by lateral diffusion and may be used to set a
bound to the diffusion coefficient.

Constraints on phytoplankton growth

The growth of planktonic plants requires energy, obtained from
the photosynthetic fixation of inorganic carbon, and supplies of
inorganic compounds of other elements, in particular those of
nitrogen, phosphorus, and silicon.   Nitrogen is generally considered
the nutrient most likely to limit phytoplankton production in the sea
(e.g. Burton in Raymont, 1980), and Dugdale (1967) made a useful
distinction between 'new' and 'regenerated' nitrogen.   The former is

mainly nitrate transported by physical processes to the superficial layers of the sea from deeper regions where organic matter is remineralized. The latter includes reduced compounds of nitrogen excreted locally by zooplankton as ammonium or dissolved organic nitrogen (DON). Such recycling can greatly augment primary production (e.g. Harrison et al., 1983; Holligan et al., 1984), but the supply of new nitrogen provides the ultimate control on phytoplankton biomass and productivity. In the most general terms, primary production and phytoplankton standing crop are usually low in well-stratified water columns in summer because a large proportion of the potentially available nitrogen is lying inactive in deep water instead of providing a basis for phytoplankton biomass and activity in the well-illuminated zone above the pycnocline.

Vertical mixing is a primary mechanism for transport of new nitrogen from a deep water source to the photic zone, where uptake by phytoplankton creates a sink. Since deep water acts as a sink for phytoplankton biomass, vertical mixing also causes a net transfer of phytoplankton out of the photic zone. Pingree et al. (1975) brought together the three components of a light-nutrient-mixing (LNM) model, pointing out that growth of a phytoplankton population can only take place where there is adequate light and where mixing provides an adequate nutrient flux yet is not so strong as to remove phytoplankton from the photic zone before they can exploit these nutrients. Tett and Edwards (1984) review the history of the LNM theory, which provides the simplest basis for assessing the biological effects of island stirring. Indeed, any form of transport of nutrient rich water into a nutrient-depleted photic zone in summer should be associated with enhanced phytoplankton growth in summer, so long as the new water remains near the sea surface for sufficiently long. Given typical values of phytoplankton specific growth rate (μ = $(dX/dt) \cdot (1/X)$ where X is biomass) of 0.1 to 1.0 day^{-1} in summer, the relevant time scale (given by $1/\mu$) is 1 to 10 days.

Prediction of island-enhanced biomass relies on the premises that particulate organic nitrogen in the euphotic zone is intimately associated with phytoplankton biomass (which is most easily measured as chlorophyll concentration), and that the photosynthetic formation of organic carbon is always eventually linked with uptake of dissolved nitrogen compounds. Although 'Cell-Quota' theory (Droop,

1983; see also Tett and Droop, in press) shows how, in the short run, uptake of various nutrients may be uncoupled from each other and from photosynthesis, we can nevertheless assume that in the medium term phytoplankton carbon:nitrogen atomic intake ratios (net of carbon losses in plant respiration) fall into the range 4:1 to 20:1, with a likely modal value close to the Redfield ratio (Redfield, 1958) of about 7:1 (atom:atom). We also assume that the chlorophyll:nitrogen ratio of healthy phytoplankton falls within the range 1 to 5 mg chlorophyll (mM N)$^{-1}$ with a typical value of 2.

A steady-state nitrate enhancement of 0.05 mM N m^{-3} could thus support in the photic zone an extra phytoplankton biomass of 0.1 (0.05-0.25) mg chl m^{-3}, and a flux into the photic zone of 0.5 mM N m^{-2} day^{-1} might be associated with a 'new' primary production of about 40 (20-120) mg C m^{-2} day^{-1}. Larger fluxes should give rise to greater primary production, although it is clear that production and chlorophyll enhancement need not increase linearly with nitrate flux. Thus a strongly mixed but deep water column might receive a large but useless injection of nitrate into its superficial waters if phytoplankton growth in these is light-limited.

Several possible fates await the nitrogen associated with newly formed plant biomass. It may be grazed and passed up the food chain to larger animals such as fish or sea-birds; it may be grazed and excreted as ammonium or dissolved organic nitrogen (DON) capable of sustaining additional phytoplankton growth in the euphotic zone; or it may pass out of the euphotic zone through mixing, downwelling, or sinking of phytoplankton, zooplankton, or detrital material derived from those. These latter losses eventually contribute, through deep water or sediment remineralization, to deep water nitrate. The processes that may be involved in nitrogen cycling near islands are summarized in Fig. 6; they are particular instances of those responsible for the cycling of nitrogen in the oceans as a whole (see Fogg, 1982).

The direct assessment of island mixing effects on biological production thus involves the questions of nutrient flux into the euphotic zone, the extent of nitrogen limitation therein, and the uptake, cycling and fate of additional nitrogen and the organic carbon fixation associated with it. 'New' nitrogen flux could, in

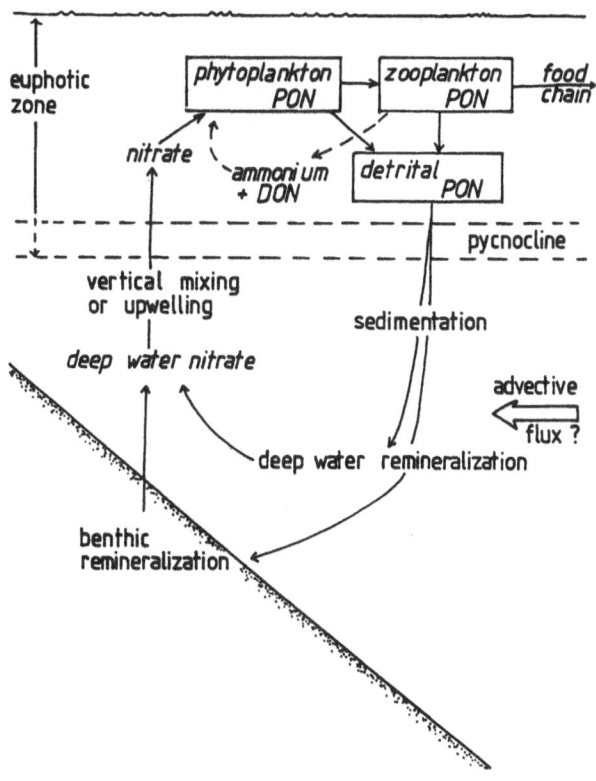

Figure 6. Main pathways of nitrogen cycling near an island in the sea.

principle, be estimated from (the inverse of) heat flux, since the new nitrogen is almost always associated with colder water. It is, however, not usually possible to measure heat or nitrogen fluxes directly, nor is it easy to make accurate and direct measurements of primary production over large areas at sea. The effects of island mixing are thus best investigated through relationships between the distributions of temperature, nitrate (or other nutrients acting as tracers of deep water), and phytoplankton biomass, which are easily measured and which may be made the state variables of theoretical or numerical models driven by the physical processes. This is the method mainly employed in the next session.

Practical Examples in Shelf Seas

We shall concentrate on recent studies of two small island
groups on the European shelf: the Scilly Isles in the southwestern
approaches to the British Isles and the St. Kilda group on the
western Scottish shelf (Fig. 7). Both island groups are of 5 km
radius and rise steeply out of the surrounding shelf seas from depths
of 90 m (Scilly Isles) and 125 m (St. Kilda). Although there are
several islands in the Scillys group, the topography is such that
they may be regarded as a single island block. This is almost the
case for St. Kilda except that the island of Boreray is separated
from the rest of the group by relatively deep water (see Fig. 11b).

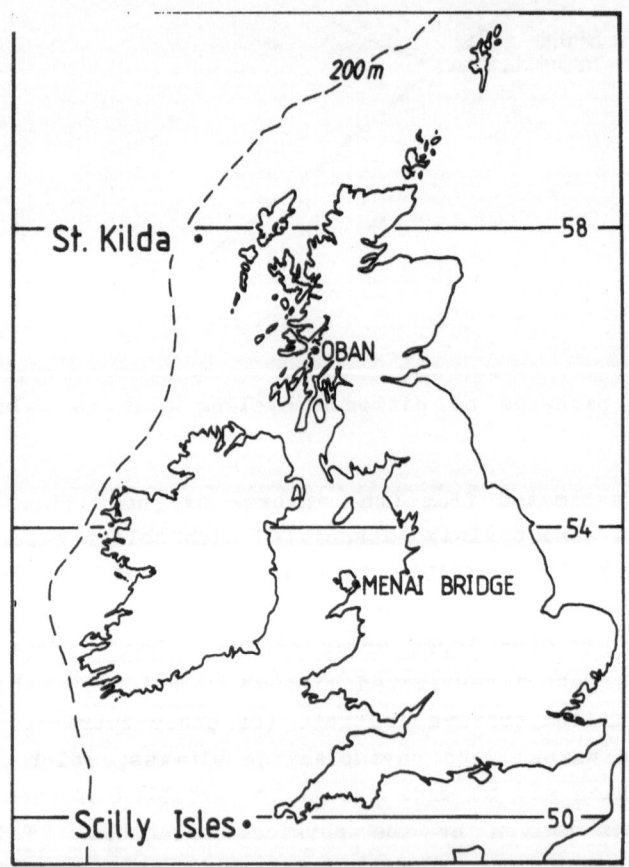

Figure 7. The British Isles and surrounding continental shelf.

Both the Scilly Isles and St. Kilda are subject to significant tidal flows. In the case of the Scilly Isles, details of the flow regime based on Argote (1983) were discussed in section 2. The equivalent fine-scale model of St. Kilda predicts a similar pattern of flow with augmented streaming around the east and west flanks of the island. Tidal stream amplitudes for M_2 increase from far field values of 0.15 m s^{-1} to local maxima of $\sim$0.5 m s^{-1} near the islands, values which are markedly lower than those for the Scilly Isles. Evidence of the resulting enhanced mixing around both island groups comes from satellite I-R imagery like that of Fig. 4b in which cold water is clearly manifest around both island groups.

Direct observations of the physical structure and biomass distributions in the vicinity of the Scilly Isles were conducted from RRS Challenger in July 1979 (Simpson et al., 1982). The observed pattern of stratification around the islands (Fig. 8a) is similar to that predicted by the h/u^3 contours (Figs. 3d and 5a) with pronounced minima on the west and east sides of the islands where ϕ fell to 20 J m^{-3}. There is, however, some indication of northward displacement of the structure relative to the h/u^3 contours and this would be consistent with advection by a weak current residual in the manner discussed in section 2. Evidence of a northwards residual flow $\sim$4 cm s^{-1} in the far field is apparent in the mean current vectors from moorings to the north, west and south of the islands. The surface temperature field (Fig. 8b) also exhibits strong minima on the flanks of the island with temperatures 3-4°C less than in the open Celtic Sea to the west.

The impact of the physical structure on the biomass field is illustrated in Fig. 8c,d which depicts the surface and column-integrated chlorophyll data, respectively. Both these quantities are markedly increased in areas close to the mixing zones with surface values up to 10 times the level in the nutrient-depleted surface water to the west. ^{14}C uptake data, also shown in Fig. 8d, indicate that production as well as standing crop is increased by the island mixing effect.

The scale of island influence may usefully be assessed from Fig. 9 which is a plot of the radial variation of surface chlorophyll for the semi-circle to the west of the islands (distributions in the equivalent eastern semi-circle are complicated by the presence of the

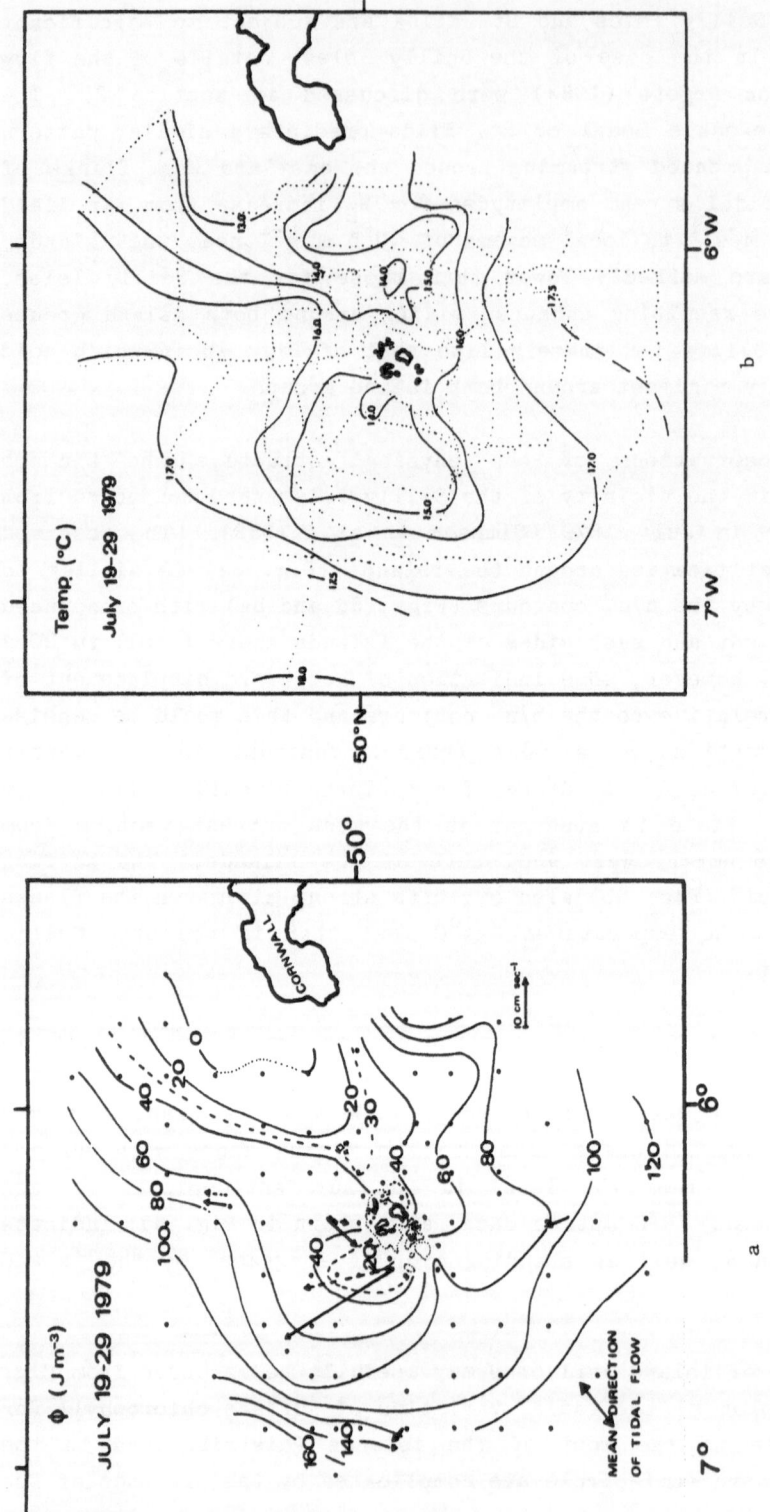

Figure 8. Maps of various properties near the Scilly Isles, July 19-29, 1979. For details of methods see Simpson et al., 1982.

(a) Station grid, stratification (as measured by φ, energy required to mix) and residual surface and bottom currents: the latter shown by a broken line.

(b) Ship's track and surface temperature.

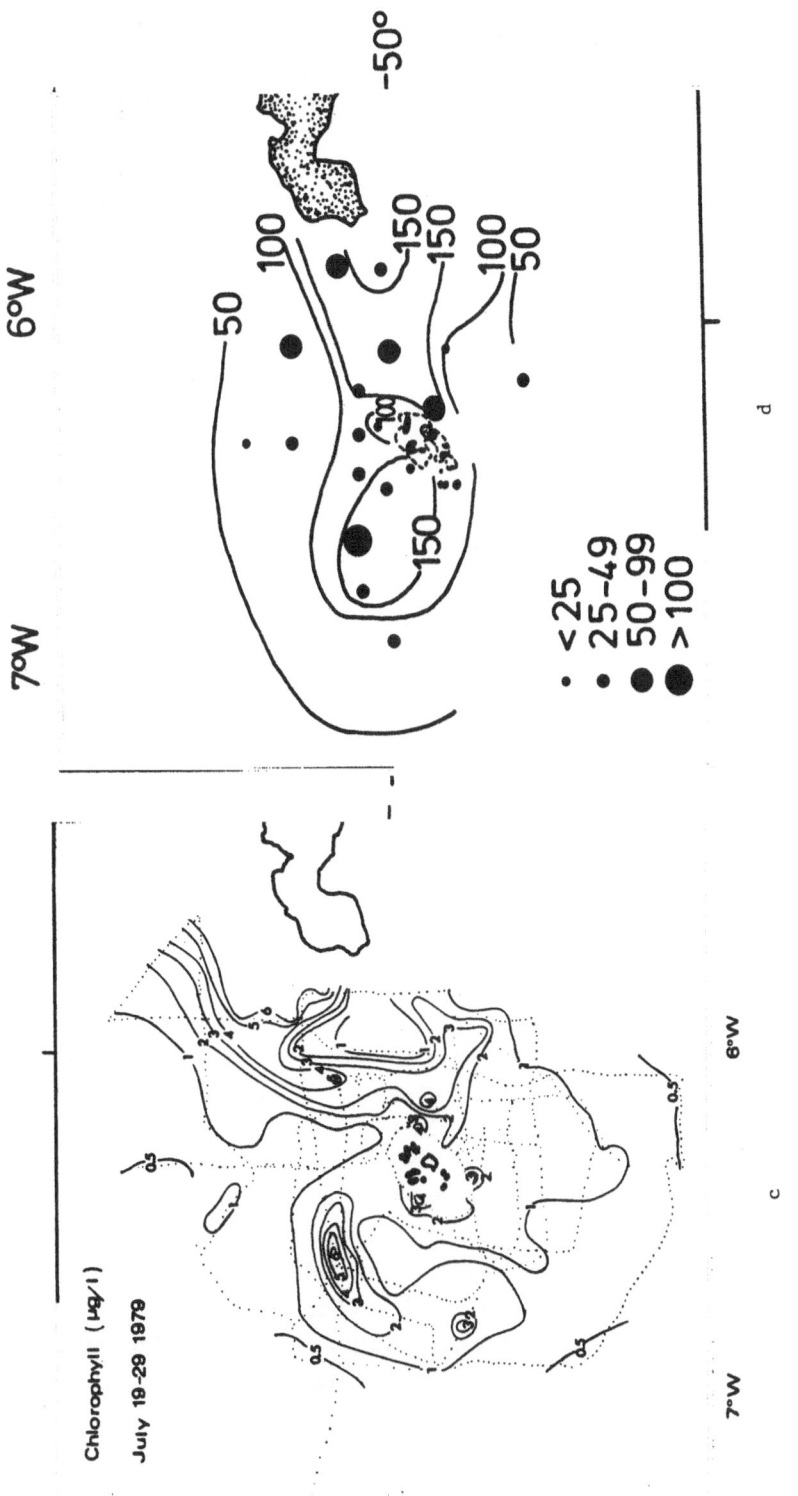

Figure 8. Maps of various properties near the Scilly Isles, July 19-29, 1979. For details of methods see Simpson et al., 1982.

(c) Ship's track and surface chlorophyll (from fluorescence in a flow-through system).

(d) Column production (circles, mg C m⁻² d⁻¹) and isopleths of column chlorophyll (mg m⁻²)

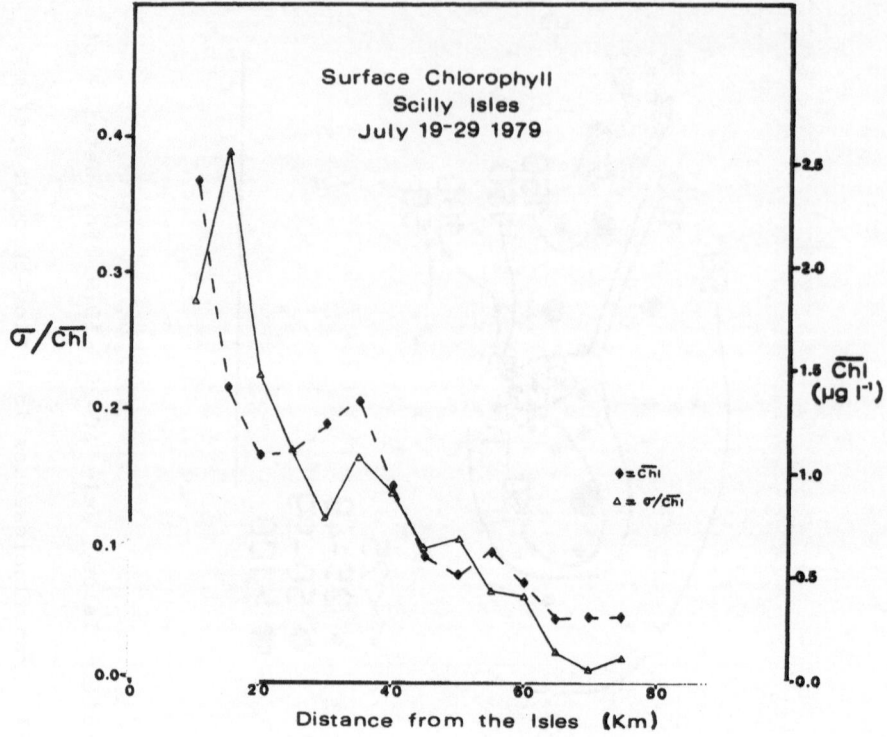

Figure 9. Scilly Isles, July 1979: mean and coefficient of variation
of (fluorescence-derived) chlorophyll concentration as a
function of radial distance from the islands. Values are
based on averages over ∿2 km of ship's track and only data
from the 180° sector to the west of the islands are
included.

Cornwall peninsula). Both mean chlorophyll and its variability
declined with distance from the island with a 1/e scale ∿30 km which
is 3 times the island group diameter.

The vertical distribution of chlorophyll on a section extending
to the west (Fig. 10d) contained a pronounced maximum extending
outwards from the mixing zone along the pycnocline. This section
also illustrates the effect of mixing in increasing surface levels of
inorganic nitrogen close to the islands with a corresponding reduction in the near bottom water (Fig. 10b). Total nitrogen in the
water column (Fig. 10c) was also notably increased.

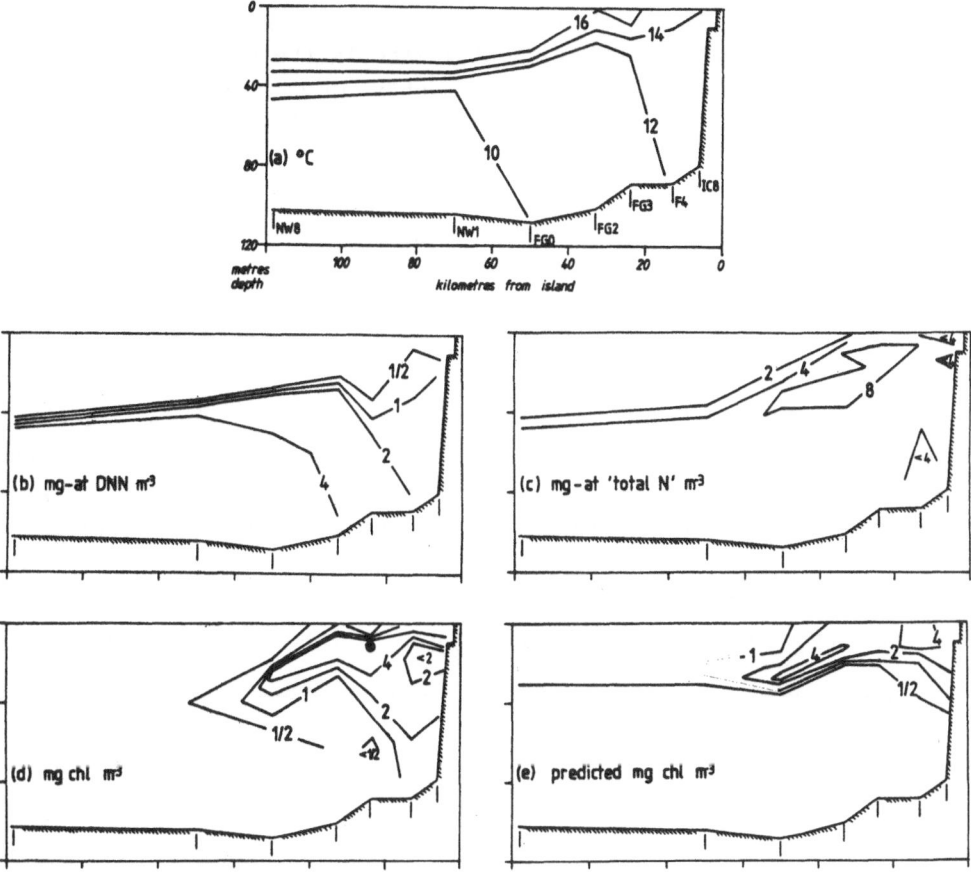

Figure 10. Western radial sections, Scilly Isles, July 1979. Data reworked from Simpson et al., 1982, and unpublished results of K. Jones. Interpolation of contours based on discrete samples has been guided by detailed density structure.

(a) Temperature: from CTD profiles at each station.

(b) Nitrate + nitrate concentration: about 6 water bottle samples at each station.

(c) 'Total N' concentration: the sum of nitrate, nitrite and particulate organic nitrogen concentrations.

(d) (Extracted, phaeopigment-free) chlorophyll a concentration: about 6 water bottle samples at each station; the filled circle at station FG3 represents a single sample with 31 mg chlorophyll a mg^{-3}.

The St. Kilda island group represents perhaps the nearest approach to a topographically ideal island in the stratified areas of the European shelf. It is considerably further from the nearest landmass than the Scilly Isles is from Cornwall and, apart from some shoaling to the east, is surrounded by water of relatively uniform depth (Fig. 11b). Interdisciplinary surveys of the island region were undertaken in July'83 and July'84 to ascertain the extent of mixing activity and biomass response. The results will be presented in full elsewhere but we include here a summary of the principal physical and biological data in Figs. 11, 12 and 13.

Although the tidal stirring is significantly less, the main features of the physical effects identified in the Scilly Isles can be seen around St. Kilda. Data from both years show that the highly stratified conditions in the surrounding waters give way to values of $\phi < 40$ J m^{-3} near the islands (Fig. 11c,d). Relatively very cold water is apparent close to the islands in both surveys (Fig. 12a,b) without any consistent spatial pattern. It seems probable, however,

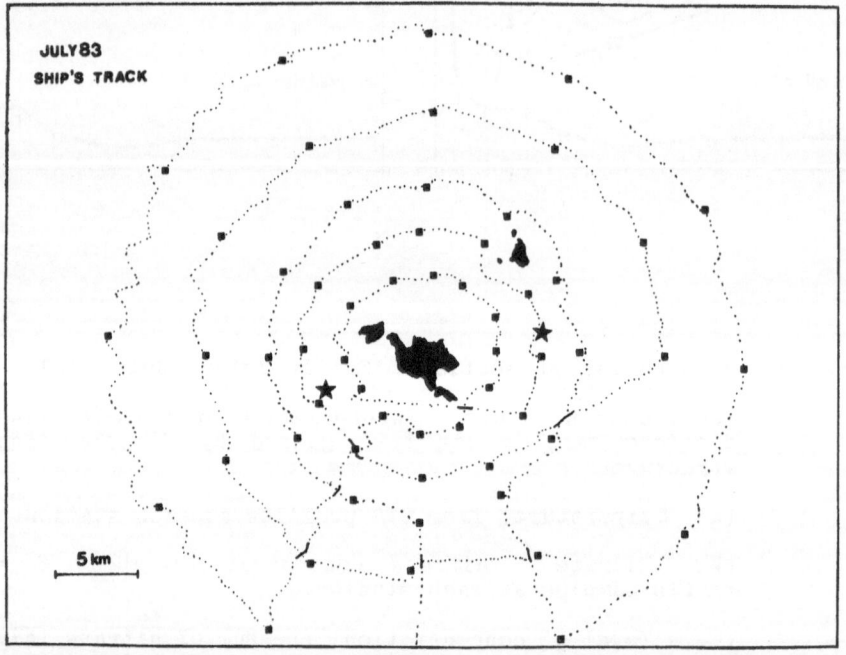

Figure 11.   St. Kilda survey July 1983 and June-July 1984.

(a) Ship's track and station position for 83 surveys. Stars mark the two current meter mooring positions.

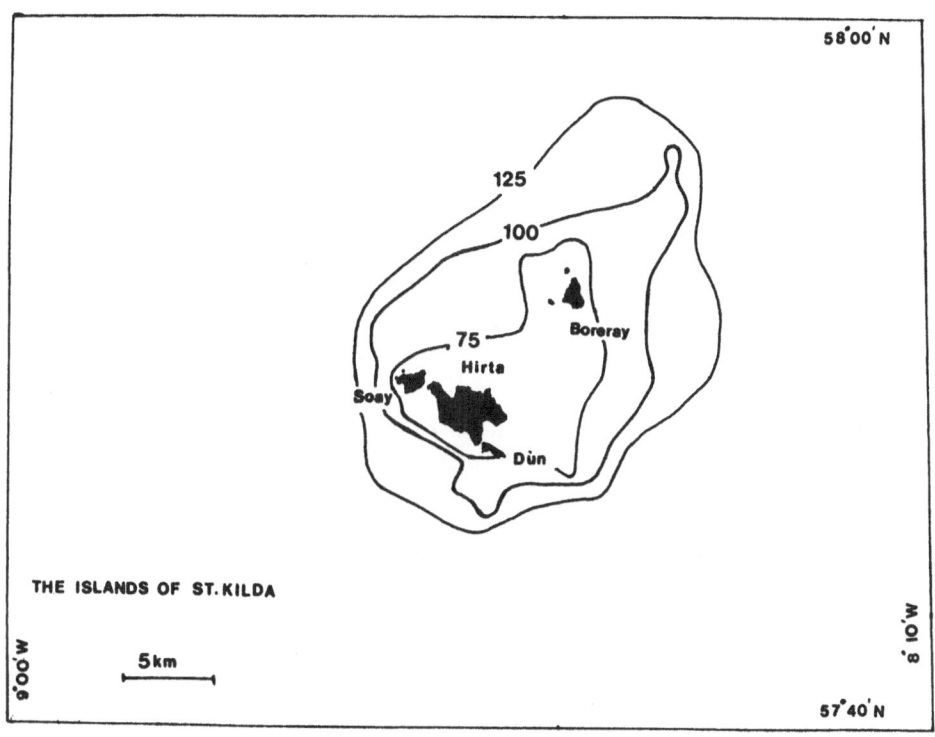

Figure 11.   St. Kilda survey July 1983 and June–July 1984.

(b)   Bottom topography.

that the centrifugal upwelling process described in section 2 may be responsible for local upwelling of cold water.   The observation of a persistent cold water patch close to the western point of Soay on both ebb and flood phases of the tide during the 1984 survey is strongly suggestive of such an effect.

Although topographically isolated, St. Kilda is unfortunately not in a region of complete horizontal uniformity.   This is because of the presence of the Scottish coastal current, part of which flows up the western margin of the Hebrides and, as can be seen from the isohalines (Fig. 12c,d), reaches out to influence the St. Kilda area. An assessment of biomass enhancement is therefore perhaps best made by comparison of N–S sections through the island center and in the coastal current water.   Figure 12 represents such a comparison. Surface layer chlorophyll was significantly increased in the vicinity of the island whereas there as no comparable signal at the latitude

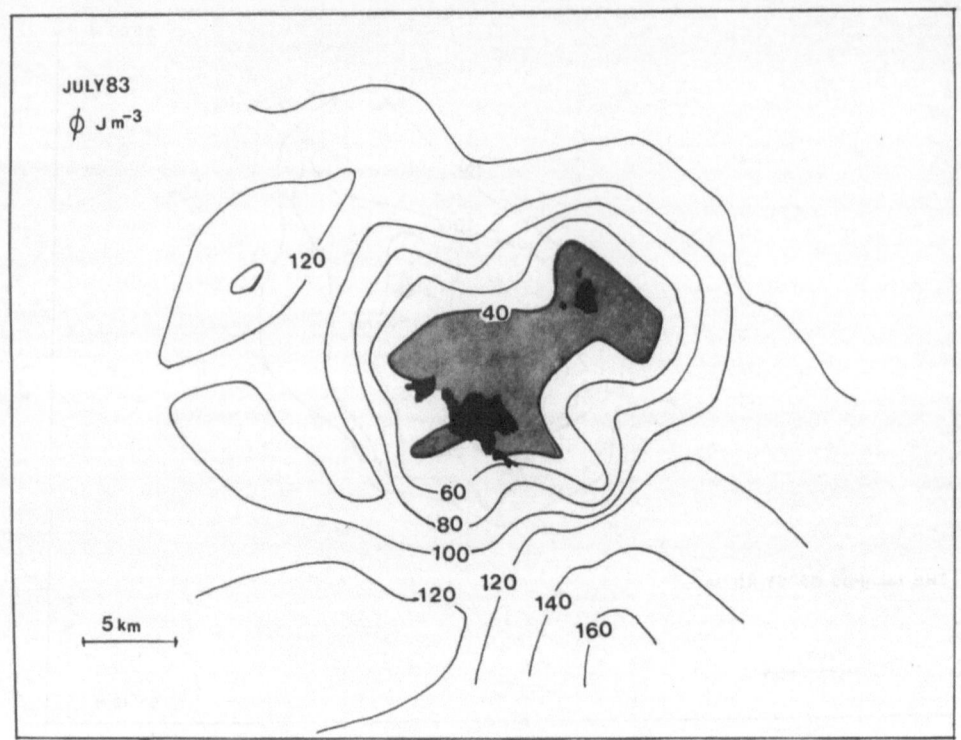

Figure 11. St. Kilda survey July 1983 and June–July 1984.

(c) Potential energy anomaly ϕ (J m^{-3}) for 1983 survey.

of St. Kilda on the control section. Strong enhancement further south on this section was associated with another and stronger tidal mixing zone off Barra Head. As Table 1 indicates, mixing due to the St. Kilda group appeared in 1984 to increase column phytoplankton production by ∿40% over an area ∿5000 km^2 or about 50 times that of the smallest circle surrounding the islands.

This island-induced increase is less marked than in the case of the Scilly Isles. It may be, however, that standing crop is not the best index of the islands' impact on the production: direct measurements of productivity are clearly needed to assess the real contribution of the well-defined mixing effect to the local food chain. At the top of this food chain are the numerous sea-birds that nest on the islands; Quine (1982) estimates the bird stock as about 10^6 of which gannets and puffins alone might consume up to 250 tonnes of fish a day during a 120 day breeding session.

Table 1. Enhancement of phytoplankton biomass by St. Kilda: July 1984.

Zone	Distance from St. Kilda (km)	Area (km^2)	Mean column chlorophyll (mg m^{-2})[2]	(Column chlorophyll s.d., x/÷)	Number of stations	t value for difference[3]
Near	0-10	314	95	(1.20)	(6)	1.2
Intermediate	10-20	943	83	(1.41)	(4)	1.2
Distant	20-40	3770	69	(1.51)	(9)	
Total island-enhanced	0-40	5027	73	-	(19)	2.0
Far field	>60[1]	-	51	(1.26)	(4)	

(t values between zones: Near–Intermediate 1.2; Intermediate–Distant 1.2; 2.6*; 3.3*; 7.5*)

[1] Ignoring other enhanced regions, such as Barra Head.

[2] Sample chlorophyll concentrations log-transferred to interpolate missing depths at stations; station values for column chlorophyll log-transferred for calculation of statistics.

[3] Based on logarithmic means and standard errors; starred values significant at $p < 0.05$.

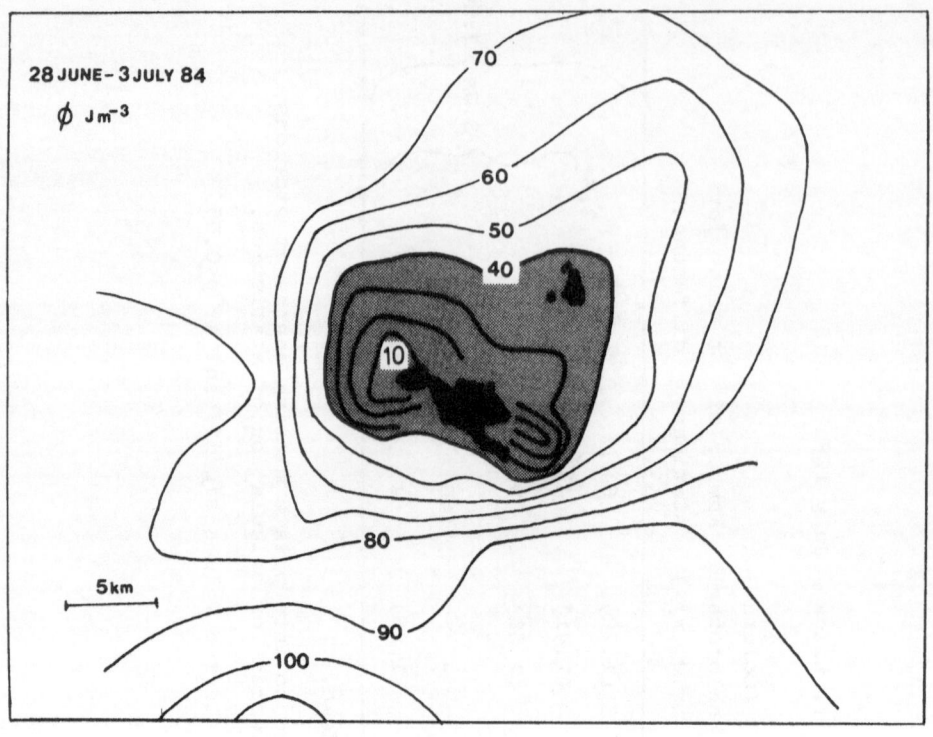

Figure 11. St. Kilda survey July 1983 and June-July 1984.

(d) ϕ (J m^{-3}) for 1984 survey.

Likely errors in such estimates are discussed by Furness (1984); we can nevertheless use the figures to make an order-of-magnitude estimate of the plant production necessary to support this harvest. Assuming the shortest possible food chain implies that the birds may each year be taking fish equivalent to a primary production of about 100 g C m^{-2} over the 5000 km^2 influenced by St. Kilda. Total annual primary production around the islands might be expected to be considerably greater than this, and thus perhaps substantially in excess of the productivity of 100 g C m^{-2} yr^{-1} given by Steele (1974) for the northern North Sea, the adjacent and most closely comparable shelf sea.

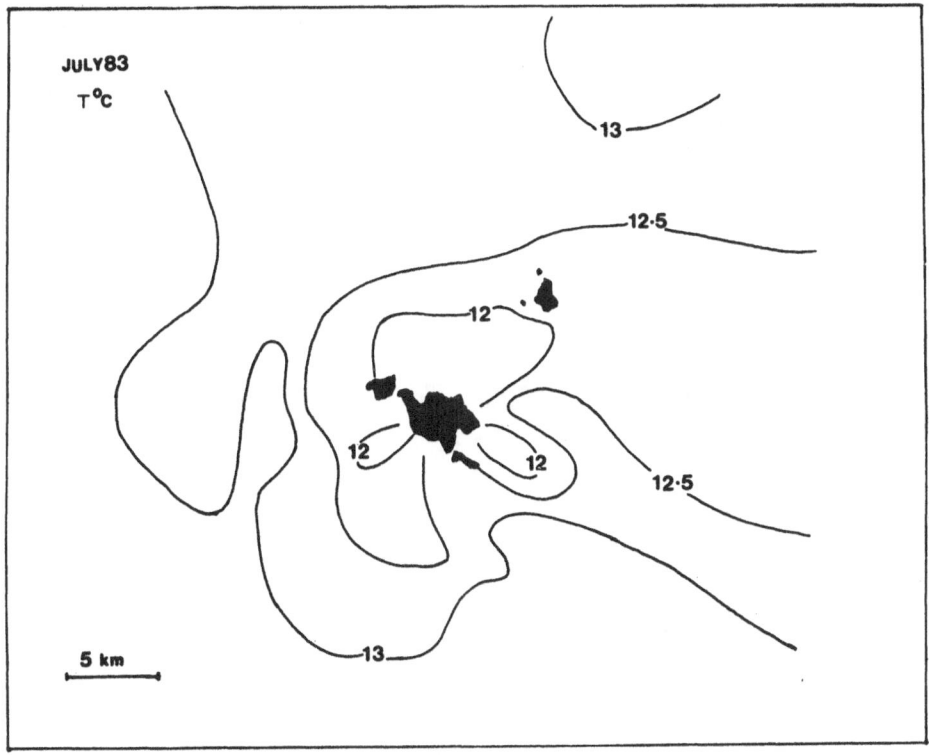

Figure 12.   St. Kilda survey July 1983 and June–July 1984.

(a)   Surface temperature July 1983.

DISCUSSION

The case studies discussed above strongly suggest that island mixing zones can have a major biological impact, probably by increasing the proportion of water column nitrogen available in summer to the planktonic food chain. Further support is lent to this hypothesis by Kinder et al. (1983), who demonstrated large summer concentrations of certain sea birds near tidal fronts surrounding the Pribilof Islands on the Alaskan shelf. Townsend et al. (1983) suggest that vertical mixing and upwelling associated with Monhegan Island causes enhanced midwater chlorophyll concentrations on flanks of the island. The hydrography of the Gulf of Maine is, however, complex, and our view is that their data do not unequivocably demonstrate a relationship between island mixing and enhanced biomass in this case.

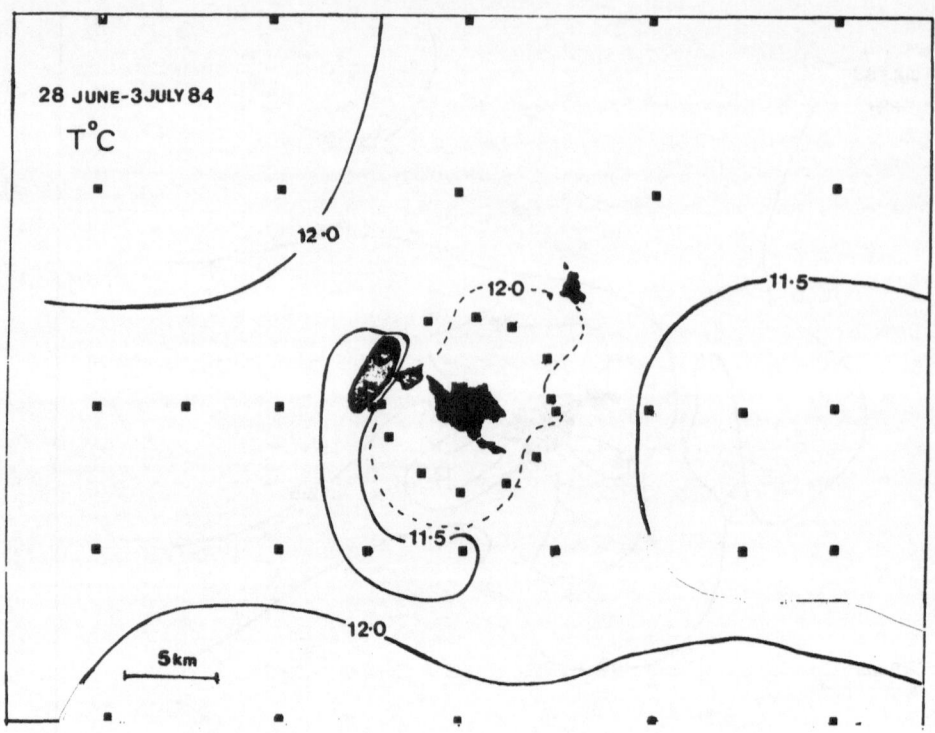

Figure 12. St. Kilda survey July 1983 and June–July 1984.

(b) Surface temperature June–July 1984.

There have been several suggestions for the specific mechanism involved in increased nutrient flux into the euphotic zone. Tett (1981) used a numerical model to examine the simplest hypothesis, that of enhanced vertical mixing without any horizontal transport; the chlorophyll distribution predicted for the Scilly Isles radial section by a slightly revised version of this model (Tett, Edwards and Jones, ms) is given in Figure 10e.

In addition to such vertical transport, we might expect the mixed, nutrient-rich water produced in the mixing zone to spread radially along the pycnocline, stimulating enhanced phytoplankton growth and biomass so long as the pycnocline remains within the photic zone. Regarding the island as a stirring engine, the flux of mixed water may be estimated from the total turbulent energy production around the island assuming that the efficiency of mixing has its

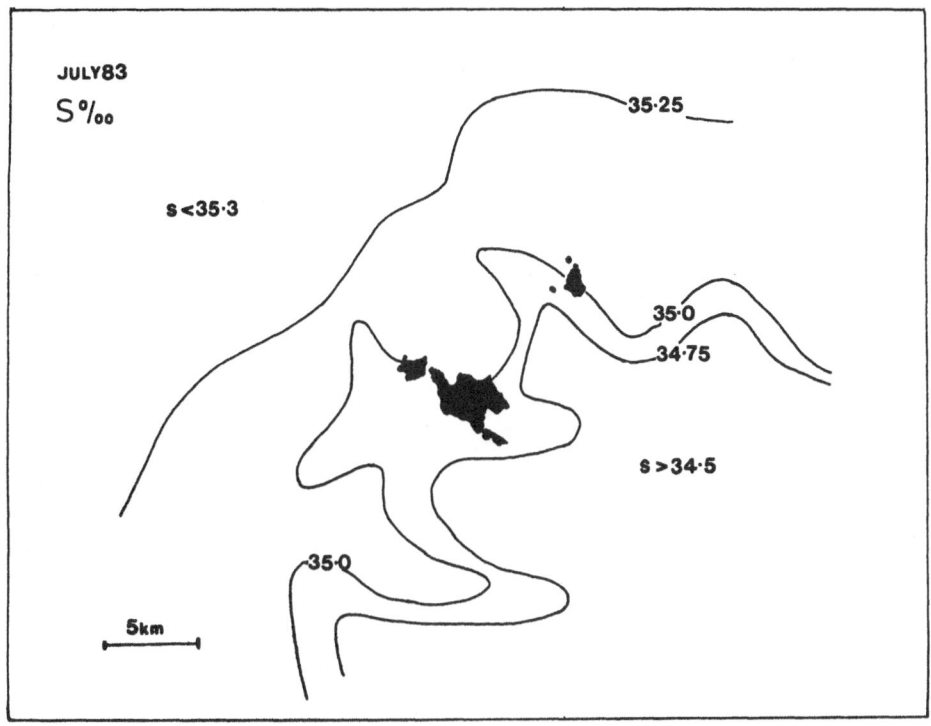

Figure 12.   St. Kilda survey July 1983 and June–July 1984.

(c)   Surface salinity July 1983.

frontal zone value of ∿0.004. Nutrient fluxes into the pycnocline around the Scilly Isles predicted in this way (Simpson et al., 1982) are of the same order as that required to sustain the observed enhanced production around the islands.

Advection of mixed water away from the island by the mean flow may also be important biologically as, for example, in the case discussed by Dooley (1981). He proposed that nutrient-rich water from the mixing zone between the Scottish Orkney and Shetland islands is swept into the North Sea to enhance phytoplankton growth and biomass in the photic zone as the water mass re-stratifies.

Finally, it is also possible that, as Doty and Oguri (1956) suggested, total available nitrogen may be higher in water columns

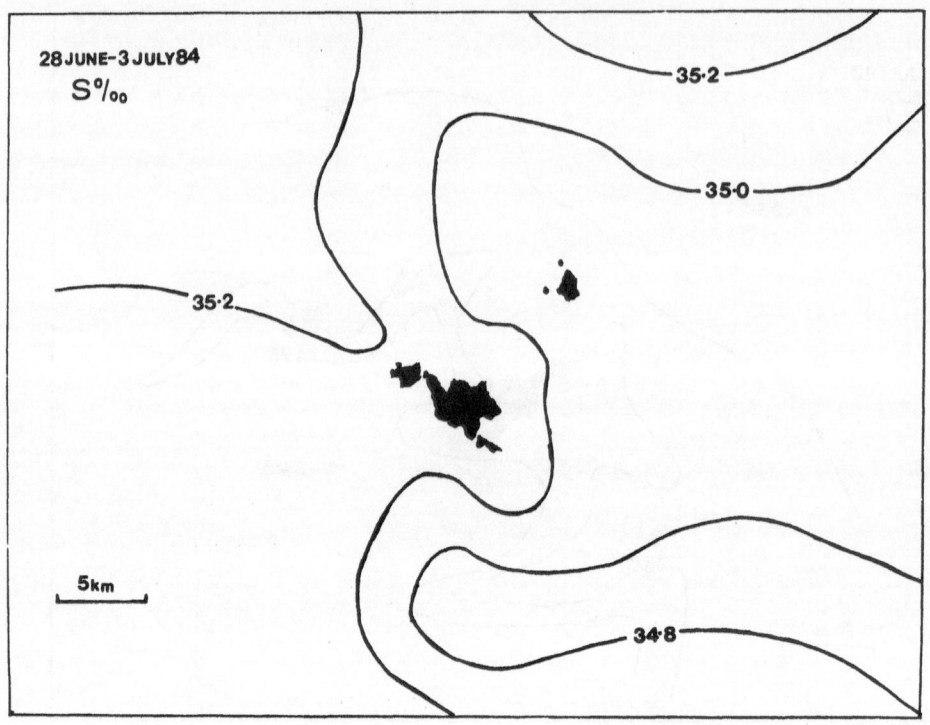

Figure 12.   St. Kilda survey July 1983 and June–July 1984.

(d)   Surface salinity June–July 1984.

near islands as a result of 'trapping' of nitrogen from passing
waters by the enhanced cycling of nitrogen, through water column and
benthos, perhaps engendered by higher water column production and
organic sedimentation.   Some support is provided for this idea by our
Scilly Isles data, which showed column total nitrogen to be greater
at stations close to the islands.   For stations FG2 to I8 in the
section in Figure 10, mean column total nitrogen was 0.53 (se 0.08) M
m^{-2}, whereas for the 'far field' stations NW8 and NW1, it was 0.37
(se 0.03) M m^{-2}.   These data, however, exclude water column DON and
any potentially available sediment nitrogen.

These different mechanisms are not, of course, incompatible and
may all contribute to a full explanation of the biological effects of
the Scilly Isles and St. Kilda.   We are, however, faced with formid-

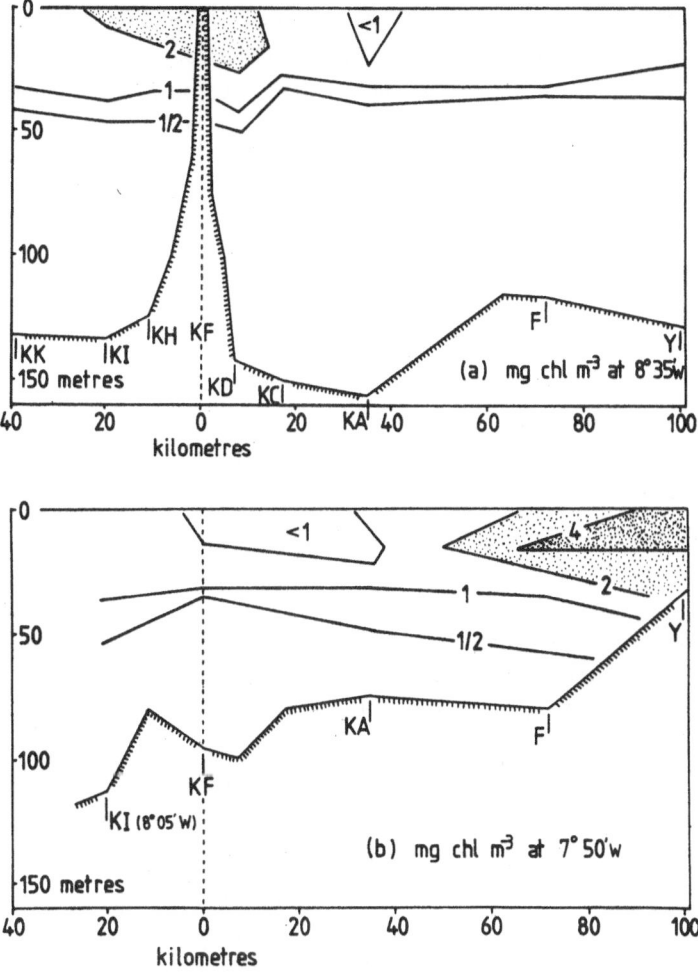

Figure 13. North-south sections on the western Scottish shelf, 28 June – 2 July 1984, contoured for (extracted) chlorophyll a concentration. About 6 samples at each station. The letters KK to Y mark east-west lines of stations. The KF line, at 57°49'N, passes through Hirta in the St. Kilda group. The Y line, at 56°55'N, passes through the Barra Head group of islands at the southern end of the main Hebridean chain.

(a) Section at 8°35'W passing through, and showing enhanced near-surface chlorophyll around, the St. Kilda group.

(b) Section at 7°50'W, closer to Hebridean coast, showing enhanced near-surface chlorophyll probably resulting from increased mixing around Barra Head.

able problems in trying to separate and quantify their respective
contributions. It seems doubtful, at this stage, that such understanding will emerge from further observational survey work of the
kind already undertaken. The high level of variability in time and
space of the biomass field, together with the topographic complexity
of real islands and the sampling problems associated with tidal
advection, will probably frustrate progress in detailed studies so
that a total flux approach may remain unavoidable in modelling island
effects.

At the same time, the difficulties associated with real islands,
illustrated by the examples considered here, may lead to the consideration of "artificial mixing experiments" which would provide a
more satisfactory experimental basis. An assessment of the possible
form of such a shelf sea experiment is given elsewhere in this
volume.

For the case of deep ocean islands, however, there is still a
need for more exploratory studies to ascertain the nature and importance of induced flow perturbations and mixing. On the basis of
theory and model experiments presented in 2(i), it should be possible
to predict the range of island scales and steady flow speeds for
which flow separation will occur. In practice, this is difficult
because of uncertainties about the effective value of the viscosity
in such flows. There is a clear requirement, therefore, for a series
of eddy-resolving measurements around islands in a variety of steady
flow regimes. Since eddies will probably not be detectable by a
strong surface signature in temperature, remote-sensing methods are
unlikely to provide indications of wake structure like those
available to meteorologists from cloud patterns in atmospheric flow.
We shall need, therefore, to undertake direct observations of island
flow with the fast-sampling techniques of undulating CTD and acoustic
doppler current measurement to resolve the evolution of the flow.

Such measurements will provide the basis for a better understanding of the role of islands in stirring the deep ocean where they may
be a major source of diapycnal mixing. At the same time, a study of
the physics of wakes will lay the foundation for studies of their
biology. Rapid sampling of nutrient and phytoplankton biomass in
parallel with with physical variables is required to provide data

sets equivalent to those already available for the shelf sea islands. Whenever possible, measurement programs should also fully exploit the remote sensing of pigments (Simpson and Tett, this volume) which seems to offer a powerful new technique for monitoring the long term variability of biomass and generalizing the results of limited ship observations of phytoplankton distributions. Only from such integrated efforts are we likely to get a better understanding of "island mass effects" and their contribution to the productivity of island fisheries.

ACKNOWLEDGEMENTS

We are grateful to K. Jones for the nitrogen data in Fig. 10 and to C. Hudson for processing the physical results of the St. Kilda surveys and drawing Figures 11 and 12. The Scottish Marine Biological Association is grant-aided by the U.K. Natural Environment Research Council. P. Tett was in receipt of a grant from the Royal Society Browne Research Fund/Marshall and Orr bequest during the preparation of this paper.

REFERENCES

Argote, M.L. 1983. Perturbation of the density field by an island in a stratified sea. Ph.D. Thesis, University of Wales.

Batchelor, G.K. 1970. An Introduction to Fluid Dynamics. Cambridge University Press.

Boyer, D.L. and P.A. Davies. 1982. Flow past a circular cylinder on a B-plane. Phil. Trans. R. Soc. London A306: 533-556.

Boyer, D.L. and M.L. Kmetz. 1983. Vortex shedding in rotating flows. Geophys. Astrophys. Fluid Dynamics 26: 51-83.

Dandonneau, Y. and L. Charpy. An empirical approach to the island mass effect in the south tropical Pacific based on sea surface chlorophyll concentrations. Deep-Sea Res., submitted.

Dooley, H.D. 1981. The role of axially varying mixing along the path of a current in generating phytoplankton production. Phil. Trans. R. Soc. London, A302: 649-660.

Doty, M.S. and M. Oguri. 1956. The island mass effect. J. Conseil 22: 33-37.

Droop, M.R. 1983. 25 years of algal growth kinetics. Bot. Marina 26: 99-112.

Dugdale, R.C. 1967. Nutrient limitation in the sea: dynamics, identification and significance. Limnol. Oceanogr. 12: 685-695.

Fogg, G.E. 1982. Nitrogen cycling in sea waters. Phil. Trans. R. Soc. London, B296: 511-520.

Furness, R.W. 1984. Seabird biomass and food consumption in the North Sea. Mar. Poll. Bull. 15: 244-248.

Garrett, C.J.R. and R.H. Loucks. 1976. Upwelling along the Yarmouth shore of Nova Scotia. J. Fish. Bd. Can. 33: 116-117.

Gran, H.H. 1912. Pelagic plant life. In: The Depths of the Ocean, pp. 307-386. Murray and Hjort (eds.). MacMillan and Co., Ltd., London.

Gilmartin, M. and N. Revelante. 1974. The 'island mass' effect on the phytoplankton and primary production of the Hawaiian islands. J. Exp. Mar. Biol. Ecol. 16: 181-204.

Harrison, W.G., D. Douglas, P. Falkowski, G. Rowe and J. Vidal. 1983. Summer nutrient dynamics of the Middle Atlantic Bight: nitrogen uptake and regeneration. J. Plank. Res. 5: 539-556.

Holligan, P.M., P.J.leB. Williams, D. Purdie and R.P. Harris. 1984. Photosynthesis, respiration and nitrogen supply of plankton populations in stratified, frontal and tidally mixed waters. Mar. Ecol. Prog. Ser. 17: 201-213.

Kinder, T.H., G.L. Hunt, Jr., D. Schneider and J.D. Schumacher. 1983. Correlations between seabirds and oceanic fronts around the Pribilof Islands, Alaska. Estuar. Coast. Shelf Science 16: 309-319.

Lamb, H. 1932. Hydrodynamics. Cambridge University Press.

Longuet-Higgins, M.S. 1969. Steady currents induced by oscillations around islands. J. Fluid Mechanics 42(4): 701-720.

Pingree, R.D., P.R. Pugh, P.M. Holligan and G.R. Forster. 1975. Summer phytoplankton blooms and red tides along tidal fronts in the approaches to the English Channel. Nature (London) 258: 672-677.

Pingree, R.D. and D.K. Griffiths. 1978. Tidal fronts on the shelf seas around the British Isles. J. Geophys. Res. 83(C9): 4615-4622.

Pingree, R.D. and L. Maddock. 1979. Tidal flow around an island with regularly sloping bottom topography. J. Mar. Biol. Assoc. U.K. 59(3): 699-710.

Pingree, R.D., G.T. Mardell and L. Maddock. 1985. Tidal mixing in the Channel Islands region devised from the results of remote sensing and measurement at sea. Estuar. Coast. Shelf Sci. 20: 1-18.

Pitts, D.E., J.T. Lees, J. Fein, T. Sasaki, K. Wagner and R. Johnson. 1977. Mesoscale cloud features observed from Skylab. In: Skylab Explores the Earth, pp. 479-501. NASA Special Publication 380.

Quine, D.A. 1982. St. Kilda Revisited. Dowland Press, Frome, England.

Raymont, J.E.G. 1980. Plankton and Productivity in the Oceans, 2nd Edition, Volume 1 - Phytoplankton. Pergamon Press, Oxford.

Redfield, A.C. 1958. The biological control of chemical factors in the environment. American Scientist 46: 205-221.

Scorer, R.S. 1978. Environmental aerodynamics. Ellis Horwood, Chichester.

Simpson, J.H. 1981. The shelf sea fronts: implications of their existence and behavior. Phil. Trans. R. Soc. London A302: 531-546.

Simpson, J.H. and J.R. Hunter. 1974. Fronts in the Irish Sea. Nature (London) 250: 404-406.

Simpson, J.H., P.B. Tett, M.L. Argote-Espinoza, A. Edwards, K.J. Jones and G. Savidge. 1982. Mixing and phytoplankton growth around an island in a stratified sea. Continental Shelf Science 1(1): 15-31.

Steele, J.H. 1974. The Structure of Marine Ecosystems. Blackwell, Oxford.

Tett, P. 1981. Modelling phytoplankton production at shelf-sea fronts. Phil. Trans. R. Soc. London A303: 605-615.

Tett, P. and A. Edwards. 1984. Mixing and plankton: an interdisciplinary theme in oceanography. Oceanography and Marine Biology, An Annual Review. M. Barnes (ed.). 22: 99-123.

Tett, P. and M.R. Droop. In press. Cell quota models and planktonic primary production. In: Handbook of Laboratory Model Systems for Microbial Ecosystem Research. J.W.T. Wimpenny (ed.) CRC Press.

Tett, P., A. Edwards and K. Jones. ms. A model for the growth of shelf-sea phytoplankton in summer.

Townsend, D.W., C.M. Yentsch, C.E. Parker, W.M. Balch and E.D. True. 1983. An island mixing effect in the coastal Gulf of Maine. Helgol. Meeresunters. 36: 347-356.

Wolanski, E., J. Imberger and M.L. Heron. 1984. Island wake in shallow coastal waters. J. Geophys. Res. 89(C6): 10553-10569.

PATTERNS OF PHYTOPLANKTON PRODUCTION AROUND THE GALAPAGOS ISLANDS

G. C. Feldman[1]
Marine Sciences Research Center
State University of New York
Stony Brook, NY 11794

INTRODUCTION

"The currents about these islands are very remarkable...."

Fitz-Roy (1839)

For the highly productive regions of the world's oceans, it is the pattern of physical processes that to a large extent determines the character and richness of the ecosystem. It has long been recognized that the ocean waters around islands are inherently more productive than waters far removed from land (Gilmartin and Revelante, 1974; Barber and Chavez, 1983). This increased productivity, which is often referred to in the literature as the Island Mass Effect (Doty and Oguri, 1956) depends primarily on the supply of nutrients to the euphotic zone brought about by enhanced vertical mixing around islands. Localized upwelling on the downstream sides of islands (LaFond and LaFond, 1971), topographically induced upwelling (Houvenaghel, 1978), the formation of island wakes (White, 1973), wind driven coastal or equatorial upwelling and tidal mixing (Kogelschatz et al., 1985) are some of the mechanisms that can produce vertical mixing around islands. Although the mechanisms may differ, the end result is often the same; an increase in the vertical transport of nutrients to the surface waters supporting enhanced levels of phytoplankton biomass and primary production.

Straddling the equator approximately 900 km to the west of the South American mainland, the Galapagos Islands (Figures 1 and 2) lie within the heart of the equatorial current system. Rising from the sea floor, the volcanic islands of the Galapagos are set on top of a distinctive platform. The main portion of the Galapagos Platform is relatively flat and less than 1000 m in depth. The steepest slopes are found along the western and southern flanks of the platform with a gradual slope toward the east.

---

[1]Current address: NASA, Goddard Space Flight Center, Code 636, Greenbelt, MD 20771

Lecture Notes on Coastal and Estuarine Studies, Vol. 17
Tidal Mixing and Plankton Dynamics. Edited by J. Bowman, M. Yentsch and W. T. Peterson
© Springer-Verlag Berlin Heidelberg 1986

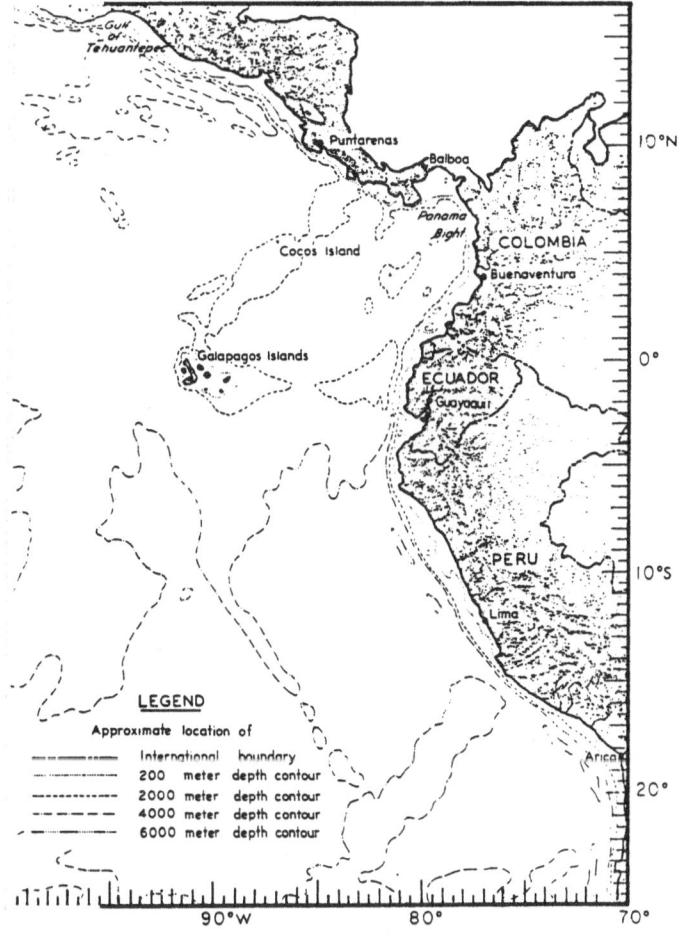

Figure 1. Map of the eastern equatorial Pacific Ocean showing the major features of the submarine bathymetry.

Numerous submarine banks and topographic features interrupt the flow of two of the major ocean currents of this region. The surface circulation around the Galapagos Islands is dominated by the generally westward flows of the South Equatorial Current (SEC). The strength and direction of flows associated with the SEC normally have a strong annual cycle which is related to the intensity of the South-east Trade Winds (Halpern, 1983), westward surface flows associated with maximum westward winds (July through December) and eastward near-surface currents correlated with minimum westward wind or weak easterlies (March and April). The thickness of the SEC is minimal at the equator (20 to 50 m) deepening to the north and south. Below

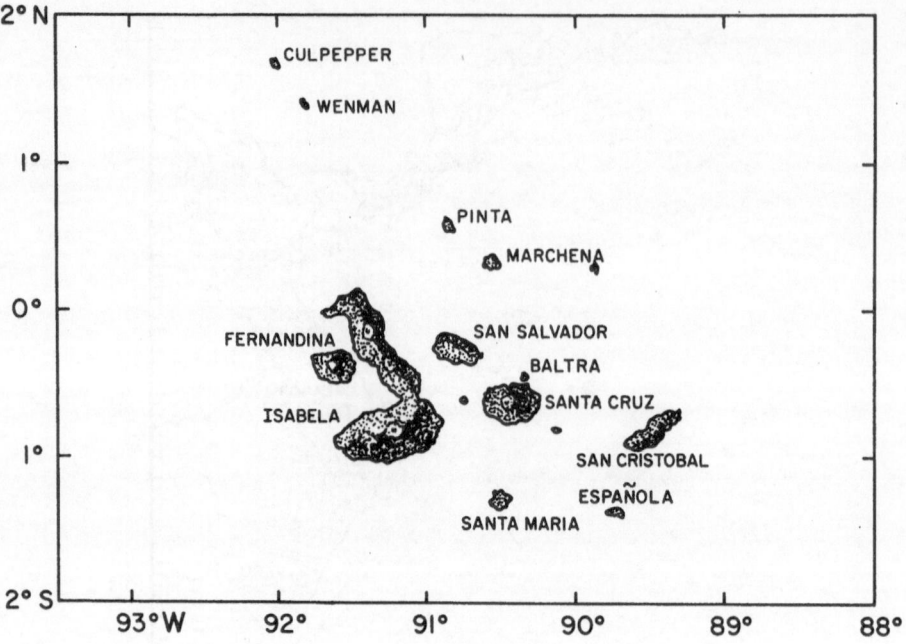

Figure 2. Chart of the Galapagos Islands in which the major islands of the archipelago are shown.

this shallow westward-flowing surface layer are the waters of the Equatorial Undercurrent (EUC), also called the Cromwell Current, which is a subsurface, eastward flowing current approximately 200 m thick with its high velocity core normally found at ∿75 m. The influence of these two opposing flow regimes makes the circulation patterns around these islands remarkably complex. The logs written by captains of 18th and 19th century sailing vessels are filled with references to the bewildering currents and winds they encountered as they explored, provisioned their vessels, or hunted the whales that were abundant in the waters around the Galapagos Islands.

Since the title of this volume is 'Tidal Mixing and Plankton Dynamics', it is appropriate at this point to discuss the influence of tides and tidal mixing in this region. Unfortunately, little is known and even less is written about the tides around the Galapagos Islands. What is known, however, is that the tides are predominantly semidiurnal, having two nearly equal high and low waters each day. The tidal range (approximately 1.5-2.5 m) varies from one location to another throughout the archipelago, and since the tidal currents in a particular area are often proportional to the range of the tide, it can be assumed that tidal currents also vary throughout the islands. Beebe (1926) wrote that

> "The captain discovered the fact that at the surface at least there is a tidal current. On the lowering tide the current sets strongly north, and on the rising tide it turns and sets as strongly southward along the coast, at least ten miles out."

Houvenaghel (1978) concluded that tidal mixing played a role in reducing water column stratification, while Kogelschatz et al. (1985) believed that the strongly periodic surface enrichment of nutrients they observed at their Academy Bay time series site may have been caused by tidal destratification. Clearly, more work needs to be done in order to fully assess the role of tidal mixing around the Galapagos.

What all the descriptive and scientific evidence points to is the fact that this is an exceptionally dynamic region and that a great number of mixing processes are at work, any one of which could provide the vertical transport of nutrients necessary to support enhanced phytoplankton production. Not only do phytoplankton represent the first link in the oceanic food chain, but their patterns of distribution in time and space may indicate changes in the physical environment (Steele, 1978) and provide clues as to how oceanographic processes regulate primary production (Yentsch, 1983). Until recently, the limitations imposed by traditional sampling techniques and strategies have made it difficult to assess the variability of the biological response to changes in the physical environment. The development of satellite sensing systems (Hovis, 1981), notably the Nimbus-7 Costal Zone Color Scanner (CZCS), has provided the synoptic perspective, the quantitative areal data and temporal coverage required for an accurate description of this region.

It is not the purpose of this paper to describe the mixing processes associated with these islands, but rather to show how mixing processes affect the distribution and abundance of phytoplankton in the surrounding waters. Satellite ocean color observations are used to describe the patterns of distribution and the degree of temporal and spatial variability of phytoplankton biomass around the Galapagos Islands. The relationship between temperature (where colder temperatures at the surface often indicate the injection of newly upwelled, or recently mixed waters rich in nutrients) and the distributions and abundances of phytoplankton will be described using sea surface temperature and ocean color imagery.

In addition it is shown that changes in the speed and direction of flows (winds and currents) past the islands produce significant variations in the patterns of phytoplankton distribution in the region and that these patterns are correlated with the seasonal cycle evident in the meteorological and oceanographic data. Finally, the satellite data are used to determine the spatial extent of the area of enhanced biological production, referred to here as the productive habitat, associated with the Galapagos Islands.

MATERIAL AND METHODS

Twenty Coastal Zone Color Scanner (CZCS) images of the Galapagos Islands covering the period from December 1978 through March 1983 were processed according to procedures described by Gordon et al. (1983). The subtle changes in ocean color detected by the CZCS provide a quantitative measure of near-surface phytoplankton pigment concentrations. These concentrations, which for remote sensing applications represent the sum of chlorophyll-a and phaeophytin-a, are an index of phytoplankton biomass and may be empirically related to primary production (Smith et al., 1982; Platt and Herman, 1983; Eppley, 1984; Feldman, 1985). The depth to which these measurements apply is inversely related to the concentration of phytoplankton and suspended material in the water column. The CZCS-derived values represent the average pigment concentration to a depth of 1 optical attenuation length (approximately the top 22% of the euphotic zone). Direct comparison between ship-measured and satellite-derived pigment concentrations have shown that over the 0.08-1.5 mg/m^3 concentration range, the algorithms used to relate the retrieved spectral radiances to phytoplankton pigment concentrations are accurate to within 30-40% (Gordon et al., 1983). The principal advantage to be gained by the use of satellites is the vastly increased spatial and temporal coverage possible as compared with that available from ships. The errors imposed by the inherent limitations of satellite measurements appear to be comparable to the errors caused by the spatial inhomogeneities

in the biomass fields which are not accurately assessed by the ship
surveys (Platt and Herman, 1983).

Five representative color-encoded maps of the phytoplankton
pigment distributions around the Galapagos Islands are presented. In
each case the islands are black and clouds white. The color scale
representing specific pigment concentration ranges (e.g. dark green
covers the concentration range from 0.41 to 0.50 mg/m^3) that was
applied to each of the computer-processes CZCS images presented in
this paper is included with the images. The CZCS images that were
used in the statistical analyses were remapped to uniform spatial
coordinates so that any given pixel (the smallest element resolved by
the sensor) represents the same geographical location on the earth's
surface in each image.

RESULTS AND DISCUSSION

CZCS data of the Galapagos Islands from Nimbus-8 orbit 739 taken
on 16 December 1978 were processed to produce simultaneous, co-regis-
tered images of phytoplankton pigment concentrations (Plate 1a) and
sea surface temperature (Plate 1b). Although direct observations of
the oceanographic conditions around the islands are not available for
December 1978, data gathered during the November 1978 cruise in the
vicinity of the Galapagos by the research vessel Orion of the Oceanographic Institute of Ecuador (Arcos, 1981; Jimenez, 1981) provide
surface information useful in the interpretation of the satellite
images. The time difference between the ship survey and the
satellite overpass makes direct comparisons difficult, however, a
general agreement of the large-scale features does exist.

The satellite-derived sea surface temperature distributions
(Plate 1b) show that warm waters (yellow) were found predominantly in
the northern and southern regions of the archipelago. The warm,
tropical surface waters to the north are separated from the generally
cooler waters (blue) of the south by the Equatorial Front which
extends zonally, and generally crosses the equator in the vicinity of
the Galapagos Islands (Hayes, 1985). The Equatorial Front, sometimes
referred to as the Galapagos Front in the vicinity of the islands,
was very strong on the western side of the archipelago in November
1978 and was located just off the northwestern tip of Isabela.

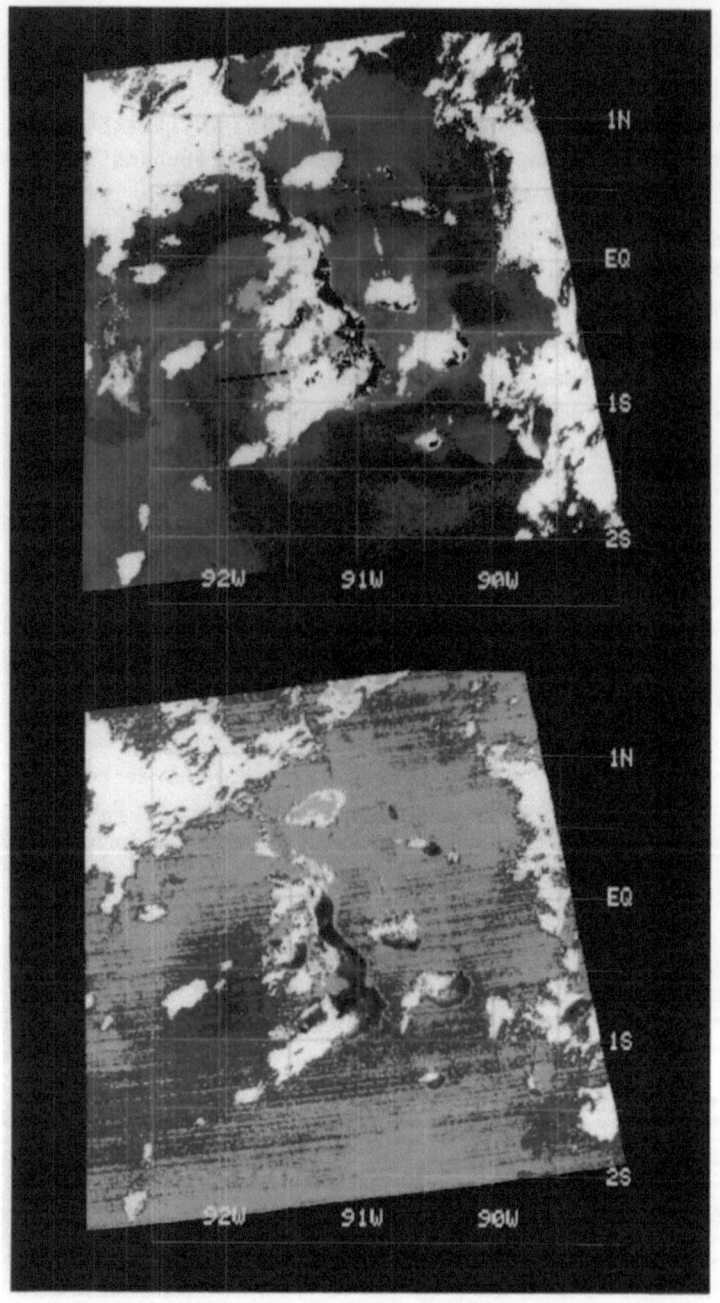

Plate 1a. Satellite ocean color image showing the distribution of phytoplankton pigments around the Galapagos Islands acquired on 16 December 1978 (Nimbus-7 orbit 739). Major islands are black and clouds white. The dotted line across the sharp color front located to the west of Isabela Island represents the trackline from which the data presented in Fig. 3 were derived.

Plate 1b. Satellite-derived sea surface temperature distributions around the Galapagos Islands acquired on 16 December 1978 (Nimbus-7 orbit 739). Regions of warm waters (above 21°C) are yellow and red; intermediate temperatures (20 to 21°C) are light blue; coolest temperatures ($<20^\circ$C) associated with the upwelling of the Equatorial Undercurrent are deep blue. Major islands are gray and clouds white.

One striking feature evident in the temperature image is the patch of relatively cold water (~19°C) located to the west of Isabela Island between the equator and 1.5°S. Jimenez (1981) reported that during November 1978 the axis of the eastward flowing subsurface Equatorial Undercurrent (EUC) was displaced slightly to the south of the equator; other studies (Bubnov and Egorikhin, 1981; Lukas, 1981) have given some indication of the behavior and the extent and period of the meandering of the axis of the EUC. Measurements have shown that subsurface waters belonging to the EUC upwell in the Galapagos area and influence the hydrology and productivity of the entire archipelago (Houvenaghel, 1978). Both Houvenaghel and Jimenez report the existence of a large patch of cold (19°) water to the west of Isabela Island. This patch, representing the topographically induced surfacing of the EUC, is clearly seen in Plate 1b. The satellite-derived temperature distributions seem to indicate that at the time of the satellite overpass, the EUC flowed primarily through the southern portion of the archipelago although its signature is not nearly as distinct as it is to the west. These data support the observations reported by others (Stevenson and Taft, 1971; Lukas, 1981; Hayes, 1985) that the EUC flows southward through the islands.

The CZCS-derived pigment concentrations for this period (Plate 1a) show that phytoplankton biomass was by no means uniformly distributed around the archipelago. Pigment concentrations were greatest on the western side of Isabela, particularly in Elizabeth Bay (located between Isabela and Fernandina Islands), while a small patch (~25 km) of relatively phytoplankton-rich water was found just to the north of Fernandina. The lowest concentrations are associated with the warm waters to the south of the islands and with the regions to the north of the reported position of the Equatorial Front. The region of relatively low pigment water located to the west of Isabela corresponds with the center of the EUC upwelling water as seen in the sea surface temperature image. It has long been recognized that newly upwelled water, although generally high in nutrients is usually low in phytoplankton biomass (Barber and Ryther, 1969). Jimenez's (1981) findings and the satellite data support this idea. Both show a sharp decrease in pigment concentrations where the EUC core upwells; however, the waters surrounding the upwelling center have some of the highest pigment concentrations in the region.

A particularly sharp and nearly coincident temperature and color front can be seen on the eastern edge of the EUC upwelling center. The bathymetry of the region (Figure 1) indicates that this front is located in waters with depths greater than 1000 m. Sea surface temperatures and pigment concentrations across this front (Figure 3) extracted from CZCS imagery along the dotted trackline shown in Plate 1a, show the sharpest gradient across the color front (0.4-6.1 mg/m^3) occurs between pixels 22 to 26 (a distance of approximately 3 km). Mean pigment concentrations on the east side of the front (0.95 mg/m^3) extending into Elizabeth Bay are significantly higher than those to the west (0.28 mg/m^3), which are generally within the newly upwelled EUC waters. The temperature data shows a marked, although less dramatic difference across the front. Temperatures were cooler (~18.8^{o}C) in the western region, increasing across the front and averaging 19.6^{o}C on the eastern side.

Pigment distributions on 11 June 1981 (Orbit 13288, Plate 2) show that although different in some respects to the patterns observed during December 1978, several features are clearly similar. Once again, the highest pigment concentrations are found on the

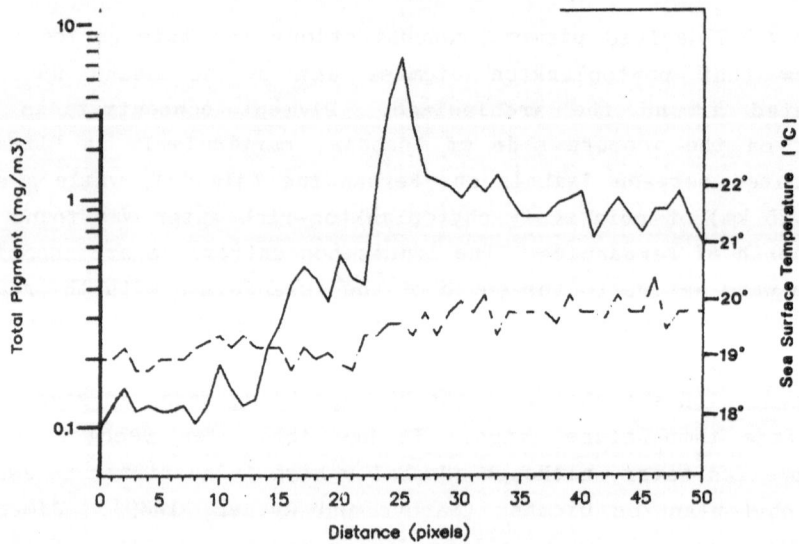

Figure 3. Satellite-derived sea surface temperatures (^{o}C, dashed line) and phytoplankton pigment concentrations (mg/m^3, solid line) versus distance (1 pixel = 825 meters) along the trackline depicted in Plate 1b extracted from the CZCS data acquired on 6 December 1978 (Nimbus-7 orbit 739).

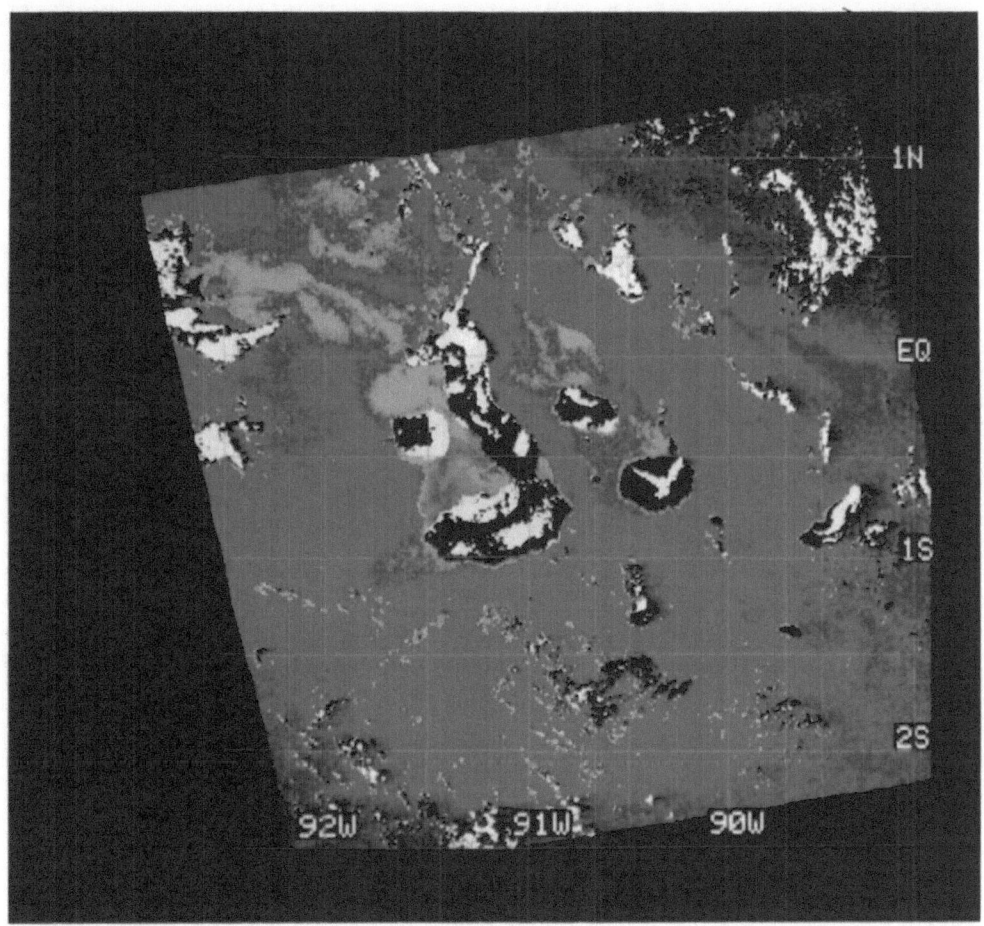

Plate 2. Satellite ocean color image showing the distribution of phytoplankton pigments around the Galapagos Islands acquired on 11 June 1981 (Nimbus-7 orbit 13288).

western side of Isabela in Elizabeth Bay. The sharp color front seen in December 1978 also appears in the June 1981 image, although its position is displaced approximately 20 km to the east. A small patch of pigment-rich water is again found off the northern tip of Fernandina Island. The most notable differences between the two scenes are (1) the appearance of phytoplankton-rich plumes extending to the northwest from most of the islands in June 1981 and (2) an overall increase in pigment concentrations in the waters around the archipelago. The overall mean pigment concentration for the entire region covered by the CZCS image increased from 0.26 mg/m^3 in December 1978 to 0.42 mg/m^3 in June 1981. This increase is particularly evident in the offshore waters, indicating the possibility of either a large-scale seasonal or interannual effect. The seasonal influence will be addressed later in this paper, whereas Feldman (1985) has discussed the interannual variability in phytoplankton biomass and production experienced by this region.

Oceanographic measurements of the large-scale circulation patterns which are available for the June 1981 period, proved useful in interpreting the patterns seen in the satellite image. The trajectories of four satellite-tracked drifting buoys deployed to the east of the Galapagos Islands (Pazos and Paul, 1984) during the latter half of June 1981 are presented in Figure 4; the tic-marks along the trajectories represent successive ten day intervals. The horizontal displacements of the buoys have been interpreted as providing a measure of the velocity and direction of water movement in the upper layer (~50 m) of the ocean (Hansen and Paul, 1984). The trajectories show that the surface waters to the north of the equator around the Galapagos Islands were moving in a generally west-northwest direction, and in a west-southwest direction for the waters to the south. The plumes of pigment-rich water seen in Plate 2 therefore, are forming on the downstream sides of the islands.

Simpson et al. (1982) and Simpson and Tett (this volume) described the mixing processes associated with flows around islands. It was noted that at times when significant non-tidal residual flows were present (4-20 cm/sec was reported by Simpson et al. for the waters around the Scilly Isles), mixed water would be swept away from the island and produce a plume downstream. Houvenaghel (1978) reported that in the Galapagos, newly upwelled waters of the EUC were mixed with and carried along by the flows of the South Equatorial Current. The appearance of distinctive island waters in many of the ocean color images of the Galapagos clearly indicates the presence and significance of the mean surface flows in determining the patterns of phytoplankton distribution around these islands.

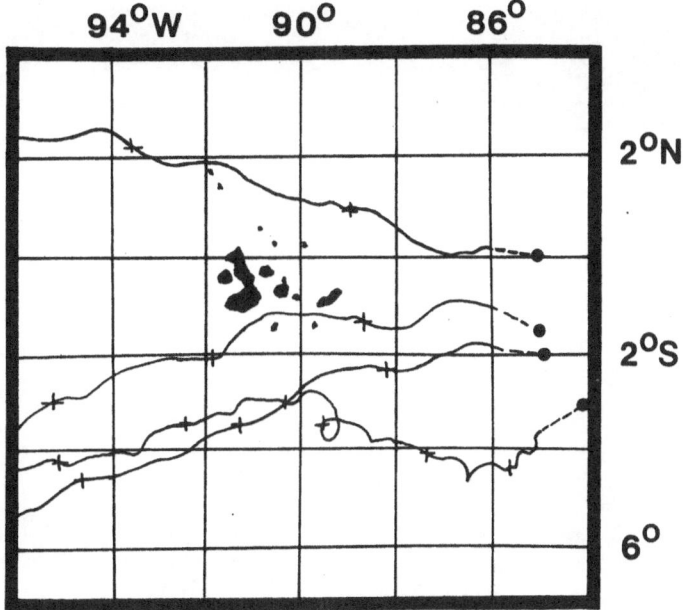

Figure 4. The trajectories of four satellite-tracked drifting buoys deployed to the east of the Galapagos Islands during the third week of June 1981. The deployment locations are indicated (closed circles) and the tic-marks along the trajectories represent subsequent ten-day intervals (after Pazos and Paul, 1984).

The trajectories given in Figure 4 indicate that the buoy to the north of the islands traveled further during the same period of time than did the buoy to the south. For the second ten-day interval, the period when the buoys were closest to the islands, the northern buoy covered approximately 54 km/day (~63 cm/sec) while the southern buoy averaged only 35 km/day (~41 cm/sec). The slower westward movement of this southern buoy may possibly be due to the influence of the eastward flows of the EUC, particularly if, as seen in Plate 1a, the major portion of the EUC flows were through the southern part of the archipelago.

Buoy displacements can also be used to help interpret the patterns of phytoplankton distribution observed on 20 August 1980 (Orbit 9211, Plate 3). In this scene, the large-scale spatial resolution of the CZCS is utilized to show the extent to which the Galapagos Islands enhance the production of the surrounding waters. Although the islands themselves are obscured by clouds (Isabela and Fernandina are masked in black), the plumes of phytoplankton-rich

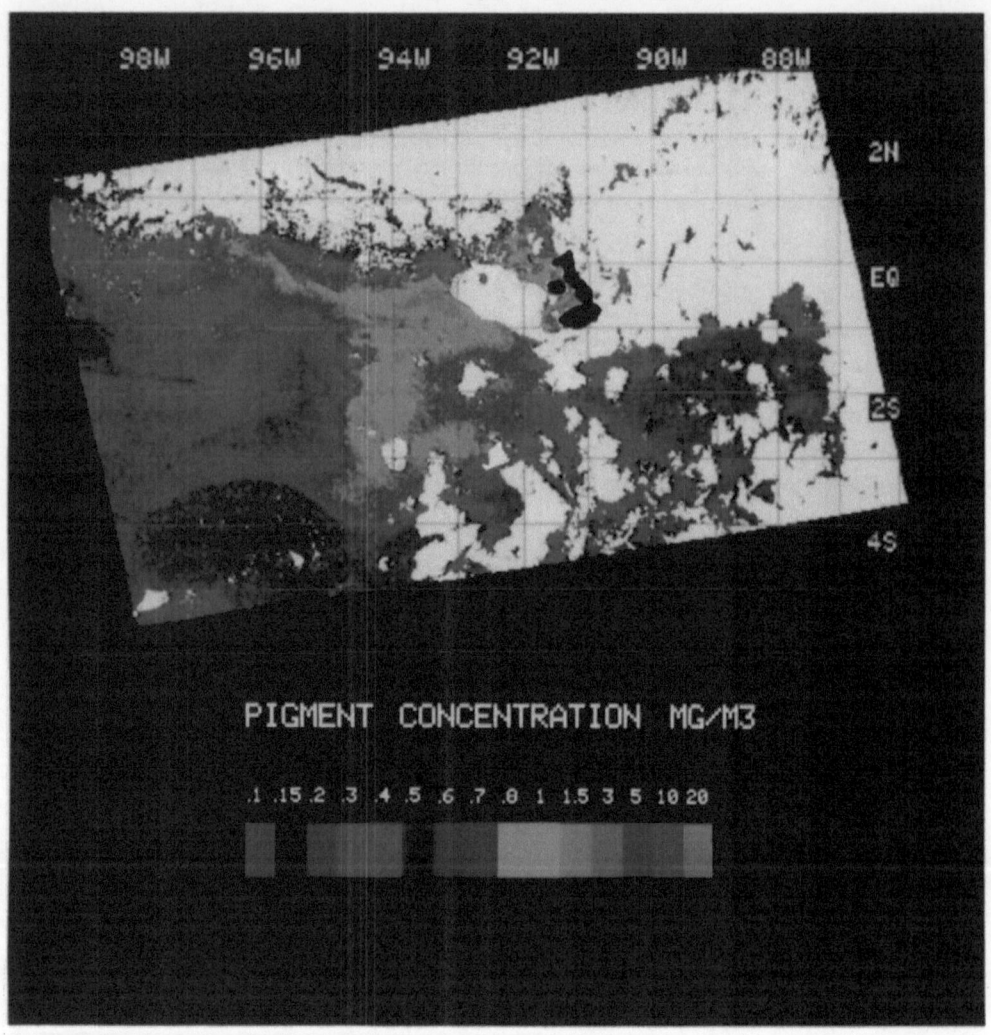

Plate 3. Satellite ocean color image showing the distribution of
phytoplankton pigments around the Galapagos Islands
acquired on 20 August 1980 (Nimbus-7 orbit 9211). This
image is presented at full swath resolution (~1500 km wide)
so that the spatial extent of the area of enhanced
phytoplankton production associated with the Galapagos
Islands can be assessed. Although the islands are obscured
by cloud cover, the locations of Isabela and Fernandina
have been masked in black.

waters can be seen extending nearly 600 km towards the west. The twin plumes observed in the satellite image appear to reflect the drifting buoy trajectories for this period as described by Pazos and Paul (1984) which were primarily westward to the north of the equator and in a generally southwesterly direction to the south. This image will be discussed in more detail in a later section of this paper where the spatial extent of the area of enhanced biological production (the productive habitat) associated with islands is estimated.

The distributions of phytoplankton pigments observed on 26 April 1980 (Orbit 7608, Plate 4) are dramatically different from those found in the previously described scenes. Although the mean pigment concentration for the entire region was only slightly lower in April 1980 than during June 1981 (0.37 vs. 0.42 mg/m^3), the greatest difference is seen in the patterns of phytoplankton distribution around the islands. The regions of highest phytoplankton biomass are no longer located on the western side of Isabela, but rather are found to the south and east. Mean pigment concentrations for the waters around Elizabeth Bay were only 0.35 mg/m^3 while those between Isabela, San Salvador and Santa Cruz were 0.74 mg/m^3.

A question that can be asked is what differing oceanographic conditions existed during these periods that could possibly explain the contrasting patterns of phytoplankton distributions observed in the satellite image. One possibility is the altered circulation patterns observed during April 1980. Near-surface (15 m) current meter measurements taken at 0°, 110^oW for the period February 1980 through September 1981 are presented in Figure 5a (Halpern, pers. comm.). Although these observations were made nearly 2100 km to the west of the Galapagos Islands, they document the dominant oceanic circulation features of the region during the times of three of the satellite overpasses presented in this paper. Coincident sea level records from the Galapagos Islands are presented in Figure 5b (Hayes, pers. comm.). A recent study (Hayes and Halpern, 1984) has shown that for the period March 1980 to July 1981 sea level at the Galapagos was highly correlated with the currents measured at 0°, 110^oW. Specifically, they found that eastward flows, expressed as vertically averaged zonal velocities, resulted in rising sea level at the Galapagos.

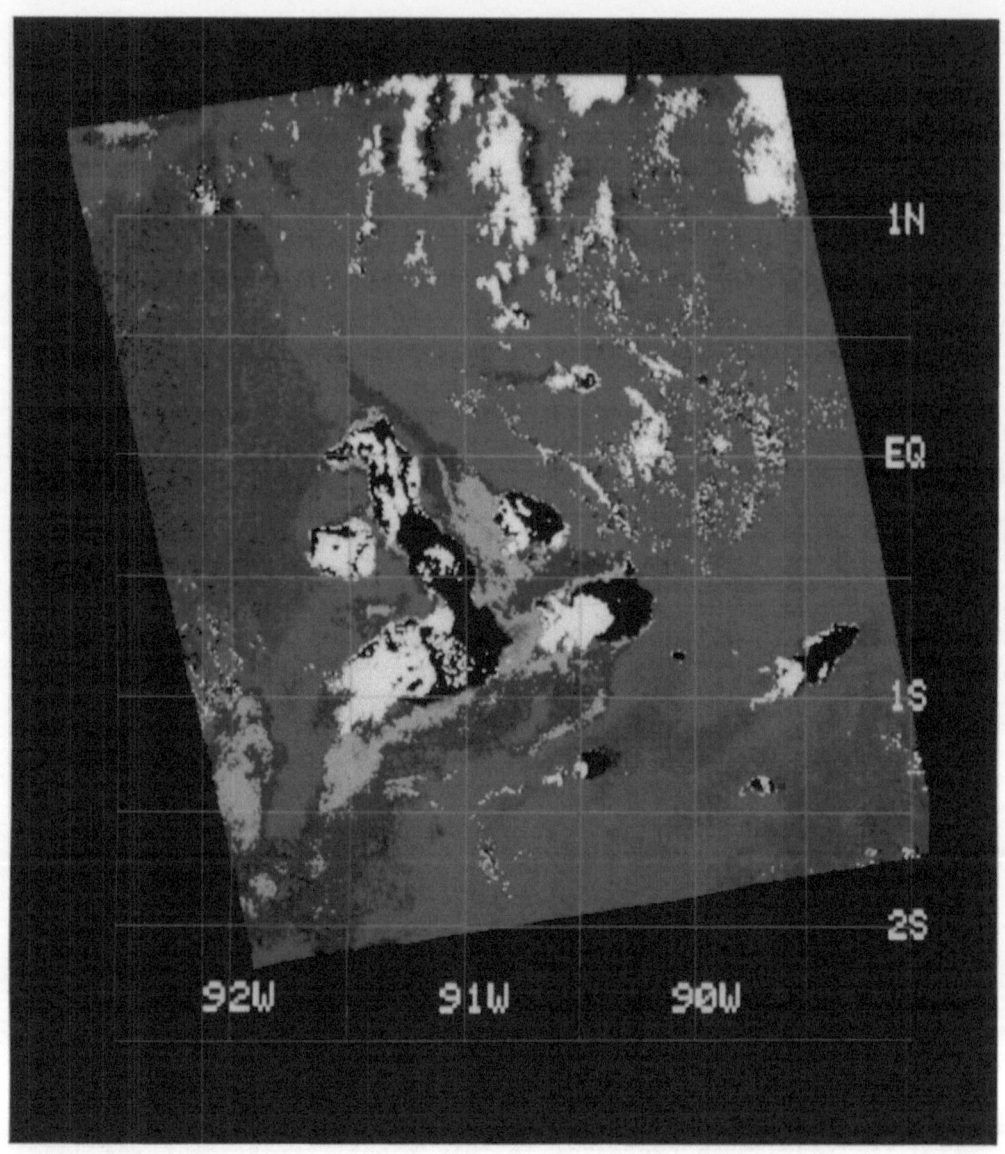

Plate 4.   Satellite ocean color image showing the distribution of
           phytoplankton pigments around the Galapagos Islands
           acquired on 26 April 1980 (Nimbus 7 orbit 7608). Note that
           phytoplankton pigment concentrations are greatest in the
           waters to the east of Isabela Island and that relatively
           low phytoplankton abundances (pigment concentrations less
           than 0.4 mg/m^3) are observed in Elizabeth Bay.

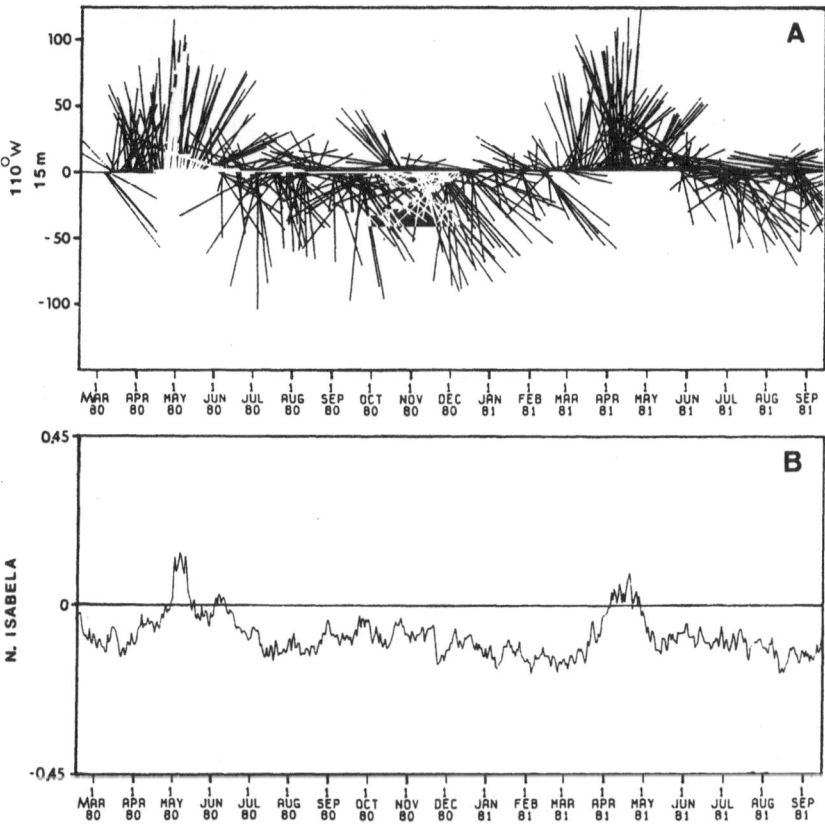

Figure 5. A) Daily vector-averaged current meter records showing speed and direction at a depth of 15 m measured at $0°$, $100°W$ (Halpern, pers. comm., 1984). Eastward direction is upward. B) Low pass filtered and detided sea level record from the Galapagos Islands (recorded at the northern tip of Isabela Island) for the same period covered by the current meter data (Hayes, pers. comm., 1984). In this presentation, pressure changes can be related to sea level fluctuations assuming 1 cm = 0.01 db.

Enhanced eastward surface flows and elevated sea levels are clearly seen during the March through May periods of each year. By the latter part of April 1980, sea level at the Galapagos had been rising rapidly and strong eastward flows had been recorded throughout the preceding month. This is in sharp contrast to the conditions that were observed during August 1980 and June 1981 during which time sea levels were low, and surface flows although variable, were in a generally westward direction.

The position of the Galapagos Islands near the eastern end of the equatorial waveguide, subjects them not only to perturbations in local forcing, but also to disturbances generated in the central and western Pacific (Kogelschatz et al., 1985). These propagating disturbances, the most dramatic of which are associated with El Niño, have been shown to have profound effects on the biota of these islands. A recent investigation utilizing satellite ocean color observations and complemented with coincident oceanographic measurements demonstrated the tight coupling that exists between the distribution of phytoplankton biomass around the Galapagos Islands and the oceanographic and atmospheric conditions observed during the 1982-83 El Niño (Feldman et al., 1984).

The CZCS data presented in Plate 5 was acquired on 8 November 1982 (Orbit 20406) during the onset phase of El Niño. When the overall mean pigment concentration for the entire archipelago was computed in the same manner as the other CZCS scenes presented in this paper, an interesting fact emerged; the mean pigment concentration calculated for the November 1982 scene (0.23 mg/m^3) was not significantly lower than was calculated for December 1978 (0.26 mg/m^3), which supposedly represents the non-Niño conditions. In fact, the sharp color front and highest pigment concentrations are once again found outside of Elizabeth Bay. What is particularly striking about this scene, however, is the large region to the north of the islands where waters with exceptionally low pigment concentrations were observed. In the satellite-derived sea surface temperature distributions of the eastern equatorial Pacific on 8 November 1982 presented in Philander et al. (1985), two features appear which are also observable in the CZCS image of that day. The Equatorial Front was in a position very close to that observed in the CZCS image as the boundary between the low pigment waters to the north and the richer waters to the south. The front is displaced to the south during El Niño (Barber and Chavez, 1983; Hayes, 1985) and it appears as if the 8 November CZCS image caught the front just as it reached the Galapagos. The sea surface temperature data also indicates a plume of relatively cool water extending to the south of Isabela. A similar plume of pigment-rich water can be seen in the CZCS image. It is not known whether the plumes were generated locally or swept around the island from the west as a result of the changing oceanographic and meteorological conditions at the time. Although locally high levels of phytoplankton biomass were observed around the Gala-

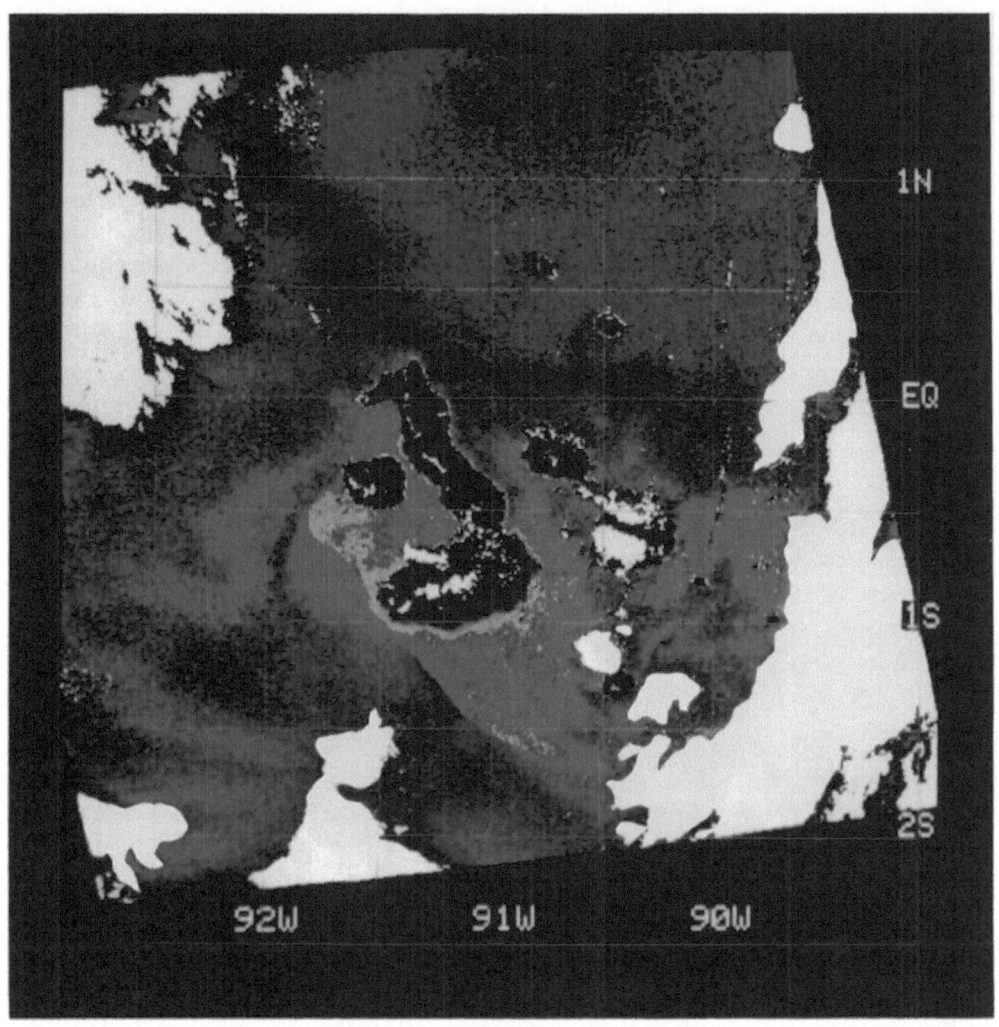

Plate 5. Satellite ocean color image showing the distribution of
phytoplankton pigments around the Galapagos Islands
acquired on 8 November 1982 (Nimbus-7 orbit 20406) during
the onset of the 1982-83 El Niño. Note the sharp boundary
between waters very low in phytoplankton abundances
(pigment concentrations less than 0.1 mg/m^3) in the
northern portion of the image and the generally richer
waters to the south.

pagos Islands during El Niño (Feldman <u>et</u> <u>al</u>., 1984; Kogelschatz <u>et</u> <u>al</u>., 1985), the productivity of the offshore waters was significantly reduced (Barber and Chavez, 1983).

Patterns of Production and the Seasonal Cycle

The satellite ocean color images presented in this paper demonstrate the degree of spatial and temporal variability of phytoplankton biomass around the Galapagos Islands. One of the first things evident from the satellite images is that some areas around the Galapagos appear richer than others. To quantify this observation, I divided the archipelago into nine sampling regions (Figure 6). Since,

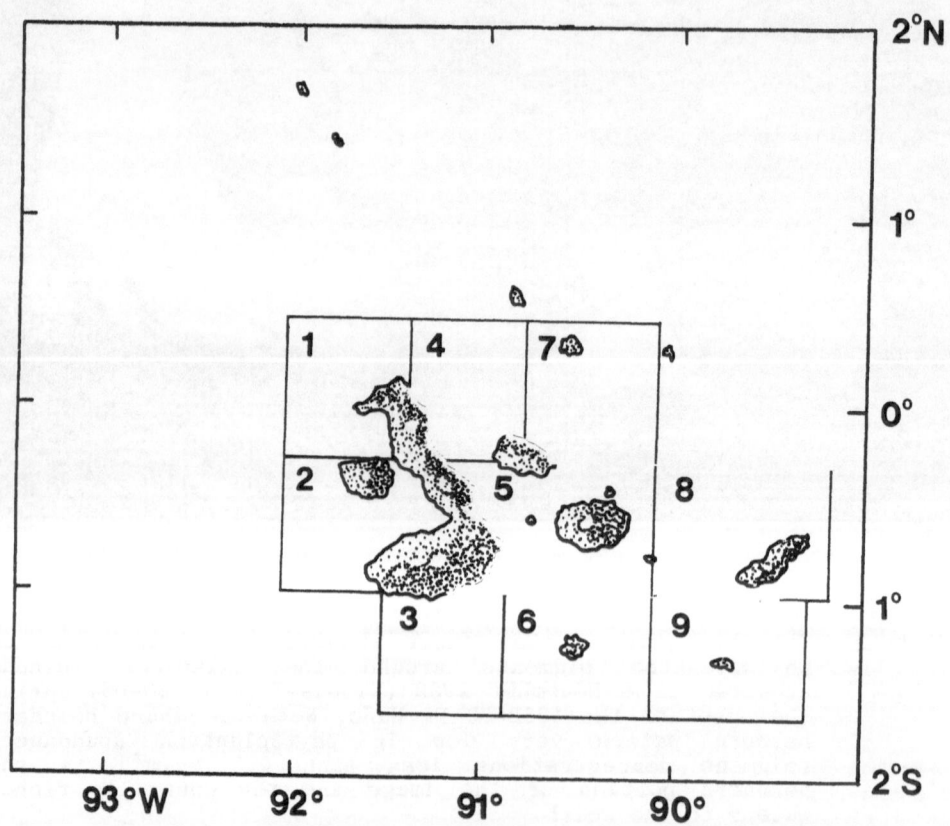

Figure 6. Chart of the Galapagos Islands showing the location of the nine mesoscale sampling regions used to assess the spatial and temporal variability of phytoplankton biomass around the archipelago.

as has been previously noted, all the images are remapped to the same
spatial coordinants, the geographic area covered by each sampling
region is the same in each image. The mean pigment concentration for
each sampling region could then be calculated.

Several additional kinds of information can be obtained through
this approach. First, by comparing the individual regions within a
single scene, the spatial variability in phytoplankton biomass around
the islands can be determined. Second, a given sampling region or
group of regions can be compared from one image to the next, giving
an indication of the temporal variability. Third, images can be
composited to develop an overall as well as seasonal mean. Twelve
CZCS images of the Galapagos Islands spanning the period from December 1978 through November 1981, including the images presented in
this paper, were composited to produce the data presented in Table 1,
in which the overall mean pigment concentration for each sampling
region, along with the minimum and maximum values are given.

The range of mean pigment concentrations for each of the
sampling regions indicates that the greatest variability was found in
the maximum values. The minimum values for all nine regions were
surprisingly similar (0.12-0.21 mg/m^3) whereas the maximum values
ranged from 0.25-2.05 mg/m^3. Individually, sampling regions 6, 8,
and 9 had the lowest overall mean concentrations and the smallest
ranges. The sampling regions in the eastern portion of the archipelago appear more oceanic in character, with low, and relatively
uniform pigment concentrations. The description of the western
regions of the archipelago as being the most productive (e.g.
Maxwell, 1974; Jiminez, 1981) is confirmed by the satellite data,
particularly when any seasonal influences are not considered as is
the case in the composited data just described. The seasonal
influence on the distributions and abundances of phytoplankton around
the Galapagos, however, appear to be significant.

Seasonal variability in the oceanic and atmospheric parameters
affecting the Galapagos Islands has been described by many authors;
however, the transition periods from one season to the next are often
poorly defined and to a large extent based upon the data set being
described. Palmer and Pyle (1966) noted a distinct annual variation
between a wet season from January to April and a dry season during
the rest of the year. The wet season is characterized by increased

Table 1. Overall mean phytoplankton pigment concentrations (mg/m^3) for each of the nine mesoscale sampling regions derived by compositing twelve CZCS images of the Galapagos Islands for the period December 1978 through November 1981. The maximum and minimum mean pigment concentration for each region is also given.

	Sampling Region								
	1	2	3	4	5	6	7	8	9
MEAN	0.41	0.83	0.29	0.26	0.38	0.21	0.25	0.23	0.18
MAX	0.95	2.05	0.50	0.43	0.74	0.31	0.48	0.35	0.25
MIN	0.16	0.16	0.15	0.15	0.21	0.13	0.14	0.17	0.12

air and sea temperatures and a weakening of the southeast Trade Winds which, during this season, are often replaced by calms or periods of westerly winds. The wet season, also referred to as the warm season by some authors, is the time when heavy rains fall on the typically arid Galapagos. The dry season is characterized by cooler air and sea temperatures, strong southeast Trade Winds and less frequent periods of rainfall. Distinct annual variations in the intensity and duration of both seasons have also been noted. Houvenaghel (1978) places the warm season from February until April; Hayes (1984) from February to May; while others (Maxwell, 1974; Kogelschatz et al., 1984) extend it from January through May.

The oceanographic conditions around the Galapagos, in particular the two main current systems which influence this region, also exhibited pronounced seasonality. The strength of the Equatorial Undercurrent varies annually (Wyrtki, 1974; Lukas, 1981; Leetma and Molinari, 1984) and is generally stronger during the early part of the year. Halpern (1983) found that the near-surface flows of the South Equatorial Current have a strong annual cycle with predominant westward flows from July-December, eastward flows during March-May (Fig. 5), with February and June being transition periods between the two.

Although there are reports of the interannual variability in phytoplankton biomass and productivity around the Galapagos Islands (Barber and Chavez, 1983; Kogelschatz et al., 1985) the identification of a distinct seasonal cycle has been hampered by the difficulty in making synoptic measurements of this highly dynamic region using traditional ship sampling techniques. Maxwell (1974) reported that seasonal differences in productivity were evident in Galapagos waters with the cold season (June-December) being more productive than the warm season (February-May). Houvenaghel (1978), however, reported that no distinct seasonal cycle in either phytoplankton biomass or productivity could be deduced.

The satellite data presented here provide both the synoptic perspective and the temporal coverage required to address more effectively the question of whether or not a seasonal cycle in phytoplankton abundance does exist, and if so, to what factors may it be related. The mean pigment concentrations for each of the nine sampling regions for the twelve CZCS images (December 1978-November

Seasonal Distribution

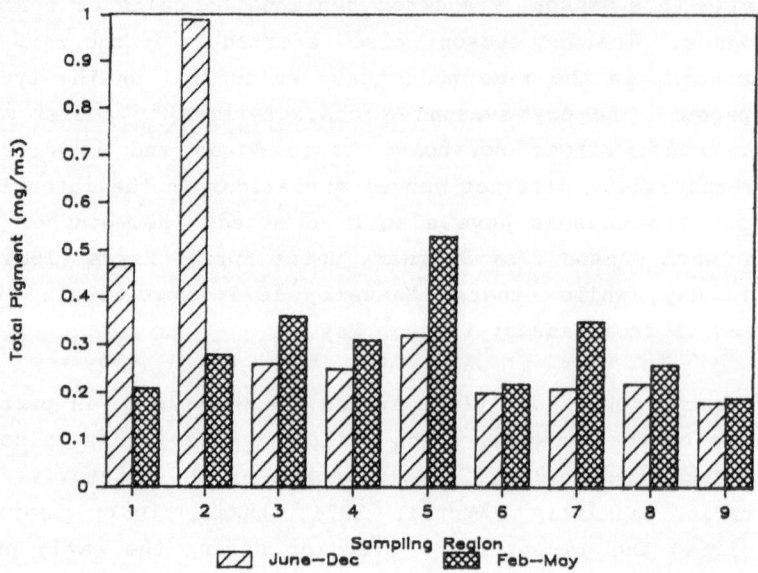

Figure 7. Seasonal mean phytoplankton pigment concentrations (mg/m³)
for each of the nine mesoscale sampling regions based upon
a seasonal grouping of twelve CZCS images of the Galapagos
Islands for the period December 1978 through November 1981.

1981) were grouped according to season and the seasonal mean
concentration for each region was calculated. These values are
presented in Figure 7.

The first conclusion that can be drawn from the data is that
there appears <u>not</u> to be a statistically significant difference in
pigment concentrations between seasons for the archipelago as a
whole. The mean concentration for the June through December period
(0.34 mg/m³) was just slightly greater than that for the February
through May period (0.31 mg/m³). The major difference between the
two seasons, however, is the dramatic and definitely significant
decrease in pigment concentrations in the western half of the
archipelago (regions 1 and 2) from the June through December to the
February through May period. The mean concentration in region 2
decreased nearly 4-fold while the mean concentration in region 1
during February through May was less than half of that recorded
during June through December. Also seen in Figure 7 along with the
decrease in pigment concentrations to the west of the archipelago

during February through May, is a corresponding increase in concentrations for all regions to the east.

It is this seasonal redistribution or, more specifically, a shift in the areas of increased production, that the satellite images have allowed us to identify. From the data presented, it appears that the patterns of phytoplankton distributions revealed in the satellite ocean color observations reflect features of, and changes in, the large-scale oceanic and atmospheric circulation systems. Although to draw conclusions based upon twelve CZCS images spanning 2.5 years may be tenuous, it appears as if interannual variability is seen predominantly by an increase or decrease in pigment concentrations over the region as a whole, while seasonal variability more specifically affects the distribution of phytoplankton around the islands. Obviously, increasing the number of images from which these kinds of analyses are based will greatly improve our understanding of these relationships. Also, the increasing amount of meteorological and oceanographic data that will be gathered as part of the ten year observational and modeling program designed to study the interrelationships between the tropical oceans and the atmosphere (TOGA) will vastly expand our knowledge of the physical environment of this region.

The Productive Habitat of the Galapagos Archipelago

To what extent do the Galapagos Islands enhance the production of the surrounding waters? Making use of both the qualitative and quantitative information contained in the CZCS data, it is possible to determine the spatial extent of the area of enhanced biological production, referred to here as the productive habitat (Feldman, 1985), associated with the islands. The relationships among biomass, growth, and production are complex. Yentsch (1984) has argued that the pigment distributions observed by the CZCS reflects the net growth processes of phytoplankton and not merely the redistribution of abundance. He hypothesizes that it is vertical mixing, resulting in the periodic injection of nutrients into the surface waters, that is responsible for the growth and, therefore, the abundance of phytoplankton evident in the satellite images. The colder, subsurface waters around the Galapagos are generally rich in nutrients; thus the appearance of cold water at the surface is, in this case, an indicator of nutrient rich water. The relationship between temperature and the distribution and abundance of phytoplankton presented in this paper (i.e. the newly upwelled water seen in Plate 1b and Figure 3) support Yentsch's hypothesis. In general, it is fair to say that the areas of increased phytoplankton biomass seen in the satellite ocean color images reflect a period of increased phytoplankton production.

If the open ocean phytoplankton concentrations observed in areas not influenced by probable island-induced enhancement represent nutrient-limited conditions (Simpson et al., 1982), then the increased nutrient supply generated by mixing processes around the Galapagos should produce regions of enhanced phytoplankton production. Such regions are evident in all the CZCS images, but in particular, the twin plumes extending nearly 600 km downstream from the islands in the 20 August 1980 scene (Plate 3) are particularly impressive.

To calculate the area of enhancement associated with the Galapagos, it was assumed that the regions in the CZCS images where pigment concentrations were less than 0.4 mg/m^3 (blue) represented open ocean, unenhanced conditions. The next step was to select a pigment concentration value which delineated most clearly the boundary separating the enhanced conditions (productive habitat) from the open ocean region. The actual plumes appeared to be very well defined by the yellow (>0.7 mg/m^3) contour. However, aside from the plumes themselves, there appeared to be somewhat less distinct regions of enhanced phytoplankton abundances obviously associated with the islands, perhaps representing the diffusing outer edges of the plumes. These regions were best delineated by the 0.4 mg/m^3 contour and are colored green in the images. The areas of the productive habitat calculated from five large-scale CZCS images of the Galapagos Islands are given in Table 2. What is particularly striking about these findings are the size and variability of the region influenced by the Galapagos and the ability of the satellite ocean color data to assess it.

It is perhaps more meaningful to compare the actual areal extent of the productive habitats, rather than their percentage of the total because the cloud-free surface area varied considerably between images. A visual inspection of the images reveals for instance that the areal extent of the plume (>0.7 mg/m^3) in the August 1980 image may have been underestimated by at least 12,000 km^2 because of cloud

Table 2. Total cloud-free surface area (10^3 km^2), the areas of ocean surface containing phytoplankton pigment concentrations greater than 0.4 mg/m^3 (wake) and 0.7 mg/m^3 (plume), and the overall mean pigment concentration (mg/m^3) and Standard Deviation (SD) computed from five large-scale Coastal Zone Color Scanner images of the Galapagos Islands. The numbers in parentheses are the percentage of total surface area represented by each value.

Date	Area (10^3km^2) >0.4	Area (10^3km^2) >0.7	Total cloud-free area (10^3km^2)	Overall mean pigment concentration
2-19-79	55 (21%)	9 (3%)	267	0.30 (0.29)
11-24-79	110 (57%)	18 (10%)	192	0.45 (0.36)
4-26-80	46 (14%)	9 (3%)	318	0.29 (0.43)
8-20-80	80 (26%)	25 (8%)	313	0.37 (0.69)
6-11-81	31 (10%)	6 (2%)	317	0.28 (0.44)

cover. The largest plumes and consequently, the largest productive habitats (November 1979, August 1980) appear to be associated with times of maximum trade wind strength and strong westward flows of the South Equatorial Current. Buoy displacements (Pazos and Paul, 1984) indicate that surface flows of approximately 50 cm/sec were found in the vicinity of the northern plume during the August 1980 period. This would mean that a water parcel brought to the surface at the Galapagos, would take between 10-15 days to reach the westernmost tip of the plume.

It is interesting to speculate as to whether or not a steady state system exists, with nutrients being supplied by mixing processes at the Galapagos supporting enhanced phytoplankton production along the axis of the plume. The horizontal (downstream) extent of the plume may be limited by several factors including (1) rate of nutrient supply, (2) advective and diffusive losses, (3) nutrient depletion resulting from phytoplankton uptake, (4) grazing by zooplankton, or (5) sinking losses. There is also the possibility that the residence time in the plume may be longer than estimated because of the complex circulation patterns and numerous trapping mechanisms, including eddies that form on the downstream sides of islands (Hamner and Hauri, 1981; Wolanski et al., 1982). The topic of island plumes and wakes is of considerable interest both from a physical and biological perspective and deserves further study. An island plume model (Okubo, pers. comm.) is being developed which will make use of the satellite data to study the processes associated with plume formation and decay and to more fully assess the impact of islands and island groups on the productivity of the surrounding waters.

REFERENCES

Arcos, F. 1981. A dense patch of Arcartia levequei (Copepoda calanoida) in upwelled equatorial undercurrent water around the Galapagos Islands. In: Coastal Upwelling, pp. 427-432. F.A. Richards (ed.), American Geophysical Union, Washington, D.C.

Barber, R.T. and F.P. Chavez. 1983. Biological consequences of El Nino. Science 222: 1203-1210.

Barber, R.T. and J.H. Ryther. 1969. Organic chelators: factors primary production in the Cromwell Current upwelling. J. Exp. Mar. Biol. Ecol. 3: 191-199.

Beebe, C.W. 1926. The Arcturus Adventure. Putnam, NY.

Bubnov, V.A. and V.C. Egorikhin. 1981. Central Pacific equatorial currents. Tropical Oceans-Atmosphere Newsletter No. 7: 1-3.

Doty, M.S. and M. Oguri. 1956. The island mass effect. J. Cons. Perm. Intl. Expl. Mer. 222: 33-37.

Eppley, R.W. 1984. Relations between primary production and ocean chlorophyll determined by satellites. In: Global Ocean Flux Study, Proceedings of a Workshop, pp. 85-102. Board on Ocean Science and Policy, National Research Council, National Academy Press, Washington, D.C.

Feldman, G., D. Clark and D. Halpern. 1984. Satellite color observations of the phytoplankton distribution in the eastern equatorial Pacific during the 1982-1983 El Nino. Science 226: 1069-1071.

Feldman, G.C. 1985. Variability of the productive habitat in the eastern equatorial Pacific. In: Proc. Symp. Vertical Motion in the Equatorial Upper Ocean and Its Effects Upon Living Resources and the Atmosphere. In press.

Fitzroy, R. 1839. Narrative of the Surveying Voyages of His Majesty's Ships Adventure and Beagle Between the years 1826 and 1836, London.

Gilmartin, M. and N. Revelante. 1974. The 'island mass' effect on the phytoplankton and primary production of the Hawaiian Islands. J. Exp. Mar. Biol. Ecol. 16: 181-204.

Gordon, H.R., C.K. Clark, J.W. Brown, O.B. Brown, R.H. Evans and W.F. Broenkow. 1983. Phytoplankton pigment concentrations in the Middle Atlantic Bight: comparison of ship determinations and CZCS estimates. Appl. Optics 22: 20-36.

Halpern, D. 1983. Annual and interannual current and temperature fluctuations in the Eastern Pacific Upper Ocean, 1980-1982. Tropical Oceans-Atmosphere Newsletter, No. 16.

Hamner, W.M. and I.R. Hauri. 1981. Effects of island mass: water flow and plankton pattern around a reef in the Great Barrier Reef lagoon, Australia. Limnol. Oceanogr. 26: 1084-1102.

Hansen, D.V. and C.A. Paul. 1984. Genesis and effects of long waves in the Equatorial Pacific. J. Geophys. Res. 89: 10431-10440.

Hayes, S.P. 1985. Sea level and near surface temperature variability at the Galapagos Islands, 1979-1983. In: Galapagos 1982-1983: A Chronicle of the Effects of El Nino. G. Robinson and E. del Pino (eds.). In press.

Hayes, S.P. and D. Halpern. 1984. Correlation of current and sea level in the Eastern Equatorial Pacific. J. Phys. Oceanogr. 14: 811-824.

Houvenaghel, O.T. 1978. Oceanographic conditions in the Galapagos Archipelago and their relationships with life on the islands. In: Upwelling Ecosystems, pp. 181-202. R. Boje and M. Tomczak (eds.). Springer-Verlag, New York.

Hovis, W. 1981. The Nimbus-7 Coastal Zone Color Scanner Program. In: Oceanography from Space, pp. 213-225. J.F.R. Gower (ed.). Plenum Press, New York.

Jimenez, R. 1981. Composition and distribution of phytoplankton in the upwelling system of the Galapagos Islands. In: Coastal Upwelling, pp. 327-338. F.A. Richards (ed.). American Geophysical Union, Washington, D.C.

Kogelschatz, J., L. Solorzano, R. Barber and P. Mendoza. 1985. Oceanographic conditions in the Galapagos Islands during the 1982/1983 El Nino. G. Robinson and E. del Pino (eds.). In press.

LaFond, E.C. and K.G. LaFond. 1971. Oceanography and its relation to marine organic production. In: Fertility of the Sea, pp. 241-265. J.D. Costlow (ed.). Gordon and Breach.

Leetmaa, A. and R.L. Molinari. 1984. Two cross-equatorial sections at 110°W. J. Phys. Oceanogr. 14: 255-263.

Lukas, R. 1981. The termination of the equatorial undercurrent in the eastern Pacific. Ph.D. dissertation, University of Hawaii.

Maxwell, D.C. 1974. Marine primary productivity of the Galapagos Archipelago. Ph.D. dissertation, Ohio State University.

Palmer, C.E. and R.L. Pyle. 1966. The climatological setting of the Galapagos. In: The Galapagos, pp. 108-122. R. Bowman (ed.). University of California Press, Berkeley.

Pazos, M.C. and C.A. Paul. 1984. Drifting buoy data from the equatorial Pacific Ocean for the period of August 31, 1980 through April 30, 1982. NOAA Tech. Memorandum ERL AOML-60.

Philander, S.G.F., D. Halpern, D. Hansen, R. Legeckis, L. Miller, C. Paul, R. Watts, R. Weisberg and M. Wimbush. 1985. Long waves in the Equatorial Pacific Ocean. EOS 66: 154.

Platt, R. and A. Herman. 1983. Remote sensing of phytoplankton in the sea: surface-layer chlorophyll and primary production. Internatl. J. Remote Sensing 4: 343-351.

Simpson, J.H., P.B. Tett, M.L. Argote-Espinoza, A. Edwards, K.J. Jones and G. Savidge. 1982. Mixing and phytoplankton growth around an island in a stratified sea. Cont. Shelf Res. 1: 15-31.

Smith, R.C., R. Eppley and K. Baker. 1982. Correlation of primary production as measured aboard ship in southern California coastal waters and as estimated from satellite chlorophyll images. Mar. Biol. 66: 281-288.

Steele, J.H. 1978. Some comments on plankton patches. In: Spatial Patterns in Phytoplankton Communities. J.H. Steele (ed.). Plenum Press, pp. 1-20.

Stevenson, M.R. and B.A. Taft. 1971. New evidence of the equatorial undercurrent east of the Galapagos Islands. J. Mar. Res. 29: 103-115.

White, W.B.   1973.   An oceanic wake in the equatorial undercurrent downstream from the Galapagos Archipelago.   J. Phys. Oceanogr. 3: 156-161.

Wolanski, E., J. Imberger and M.L. Heron.   1984.   Island wakes in shallow coastal waters.   J. Geophys. Res. 89: 10,553-10,569.

Wyrtki, K.   1974.   Sea level and seasonal fluctuations of equatorial currents in the western Pacific Ocean.   J. Phys. Oceanogr. 4: 91-103.

Yentsch, C.S.   1983.   Remote sensing of biological substances.   In: Remote Sensing Applications in Marine Science and Technology, pp. 263-297.   A.P. Cracknell (ed.).   Reidel Publ. Co., Boston.

Yentsch, C.S.   1984.   Satellite representation of features of ocean circulation indicated by CZCS colorimetry.   In: Remote Sensing of Shelf Sea Hydrodynamics.   J. Nihoul (ed.).   Elsevier, Amsterdam.

THE STIMULATION OF PHYTOPLANKTON PRODUCTION
ON THE CONTINENTAL SHELF WITHIN VON KARMAN VORTEX STREETS

M. J. Bowman

Marine Sciences Research Center
State University of New York

Figure 5. Visible photograph of Von Karman vortices in the lee of Guadalupe Island off the coast of Baja California (from Pitts et al., 1977).

In recent years many physical mechanisms have been identified and studied for generating variability in biological production and concentration in coastal seas. Included in this list would be coastal upwelling, circulation dynamics near and within fronts, mixing around islands (the so-called island mass effect), and over shoals, entrainment and mixing in coastal jet currents, tidal mixing variations in estuaries and on the continental shelf, storm driven turbulent mixing, and mixing within plumes near river mouths.

In each case, localized growth enhancement can be attributed to increased nutrient fluxes into the euphotic zone in a region of significant upwelling or mixing, often followed by a period or region of restabilization where the residence time of phytoplankton in a favorable growth environment is sufficient to allow significant cell accumulations to occur.

In this chapter results are presented of a study of mixing and cyclogenesis downstream of a mid-shelf shoal on the northwestern continental shelf of the South Island, New Zealand. Upwelling and mixing in the turbulent wake of the shoal is sufficient to promote and sustain a tightly coupled food chain which supports a major Japanese squid fishing industry. The apparent regularity of eddy shedding from these Kahurangi Shoals prompted an investigation of candidate mechanisms responsible for this enhanced productivity. Several mechanisms appear to be important in inducing mixing and upwelling: i) tidal and/or coastal current driven turbulent mixing in nearshore water and over the shoals; ii) coastal upwelling driven by wind and/or bottom Ekman pumping; iii) periodic shedding of vortices from the Kahurangi Shoals. As in other unusually productive marine ecosystems, it is often a combination of mechanisms that

Lecture Notes on Coastal and Estuarine Studies, Vol. 17
Tidal Mixing and Plankton Dynamics. Edited by J. Bowman, M. Yentsch and W. T. Peterson
© Springer-Verlag Berlin Heidelberg 1986

(16°45'N, 169°31'W). The wake system moved downstream at 45 cm s^{-1} with a period of 4 days, a wavelength of 160 km and a width between vortex rows of 55 km. Radar was used for detecting ship drift within the wake, as well as the drift and set of long line fishing gear. From these observations he was able to calculate a number of important parameters associated with the classical theory of Von Karman vortex streets.

Observations of vortex streets in the atmosphere have been made, for example, by Lyons and Fujita (1968) downstream of the Aleutian Islands, by Chopra and Hubert (1964, 1965) in the Grand Canary, Tenerife and Madeira Islands, and by Johnston (1977) in the lee of Guadalupe Island. In the atmosphere, cyclonic vortices usually grow preferentially over anticyclonic, and Warren (1967) has suggested that planetary vorticity may influence the trajectory of the wake's eddies and perhaps suppresses formation of vortices of one sign until the relative vorticity exceeds f (Pohle et al., 1965), thus making the critical value of Reynolds number for the onset of wake instability a function of latitude.

A leading question which arises in all of these geophysical studies of vortex wakes, and not satisfactorily explained, is why should a stratified, rotating geophysical flow with kinematic Reynold's numbers $\sim 10^{10}$ so strikingly resemble those laboratory experiments with homogeneous, non rotating flows at Re $\sim 10^{2}$?

OBSERVATIONS IN THE WAKE PATTERN OF KAHURANGI SHOALS

The patterns of surface temperature, salinity, nitrite-nitrate and chlorophyll a mapped during one period of an austral summer survey of the Cape Farewell upwelling zone (Bowman et al., 1983) consisted of a train of cold, saltier, nutrient enriched eddies and meanders, apparently streaming off from the locality of the Kahurangi Shoals (Figs. 1-4). The deviations from background values of each of these variables was a measure of the intensity of mixing and upwelling induced in the wake of the shoals. It was these vertical motions that produced the surface signatures that were conveniently

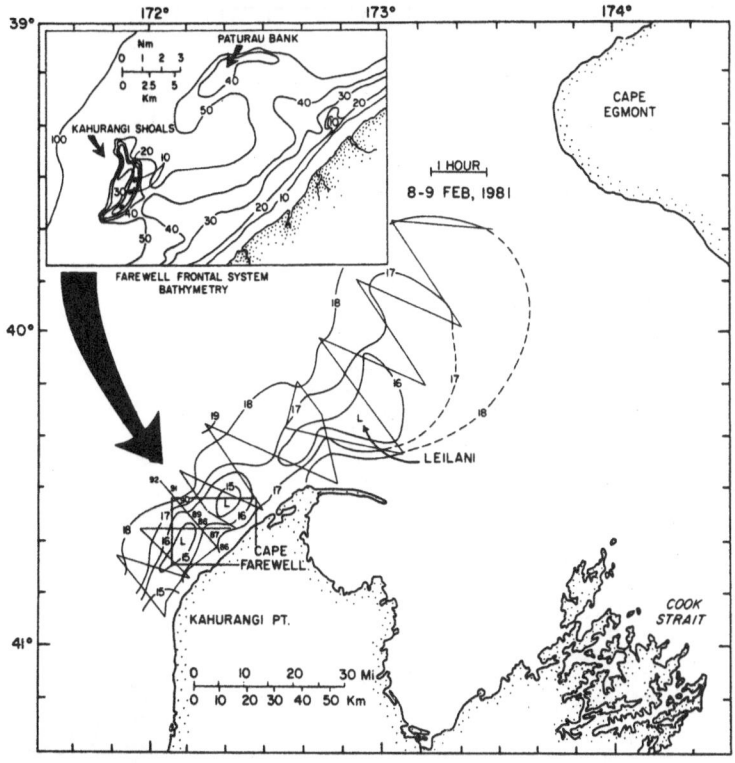

Figure 1. Locator map, local bathymetry (inset) and surface
temperature (°C) near and downstream of Kahurangi Shoals,
8-9 February, 1981.

mapped using water continuously drawn from the survey ship's cooling
intake.

Cold core (cyclonic) eddies are associated with clockwise
rotations in the southern hemisphere. Turning in the bottom Ekman
frictional layer leads to a spiraling inwards of bottom water and
associated upwelling in the center of the eddy, the so called "Ekman
pumping" effect. The opposite occurs in anticyclonic rotations and
the associated downwelling leaves no surface signature. This is why
only cyclonic motions are apparent in Figs. 1-4. An analogous
situation exists in the atmosphere (Fig. 5) where cyclonic vortices
dominate vortex streets (counterclockwise in the northern
hemisphere).

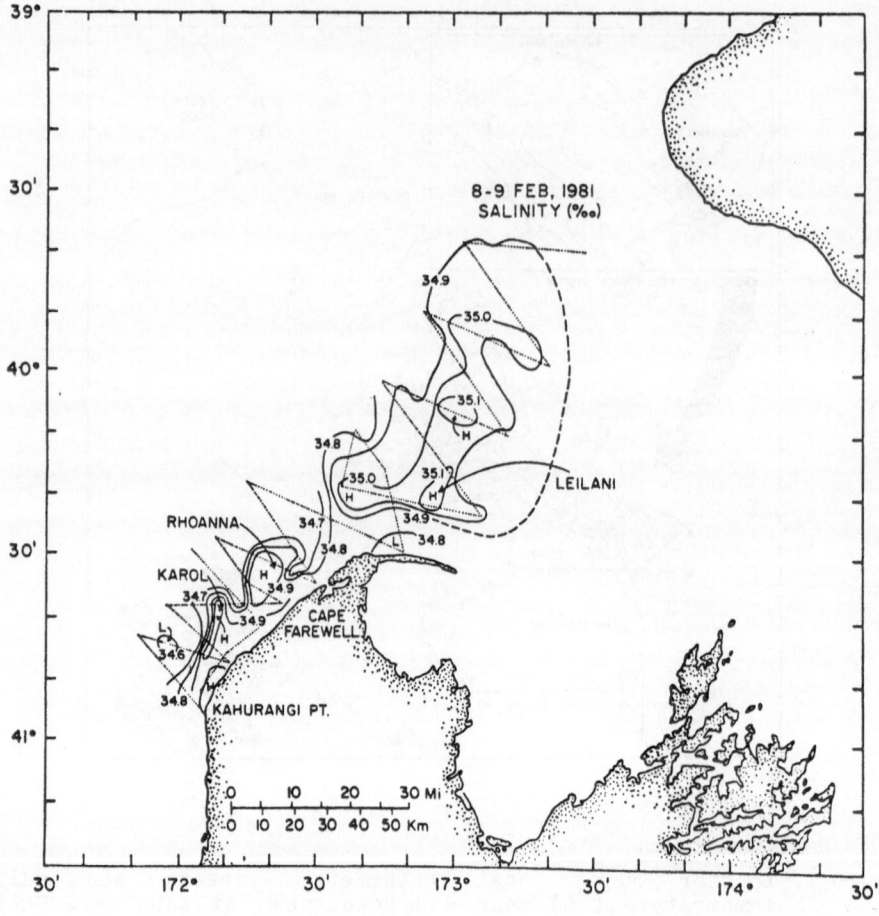

Figure 2. Surface salinity near and downstream of Kahurangi Shoals,
8-9 February, 1981.

The laboratory experiments of Boyer and Davies (1982) on a β
plane also showed that cyclonic vortices dominate even in unstrati-
fied flows (Simpson and Tett; this volume) so it is not simply an
ability to detect upwelling and downwelling signatures. Experiments
on highly stratified but non rotating flows past submerged obstacles
by Brighton (1978) showed that in the lowest layers, flow around the
obstacle moved almost as in potential flows, but further up in the
water column the flow separated from the sides of the obstacle and
under appropriate conditions, eddies were shed, where the axes of the
vortices leant downstream (Fig. 6). Further downstream the influence
of the vortices propagated upward from region IV. Whether this model

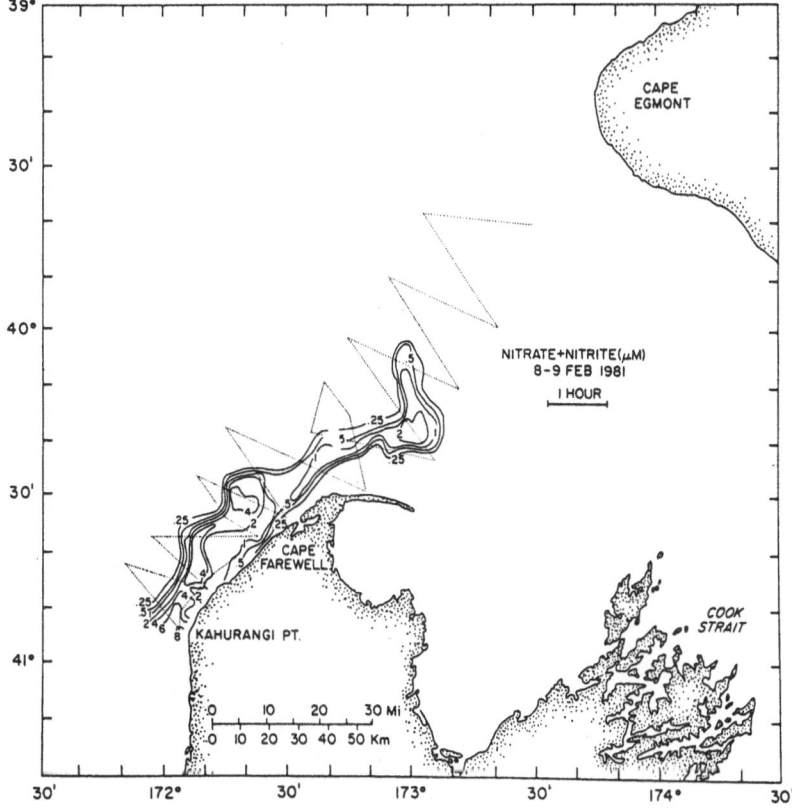

Figure 3. Surface nitrate-nitrite near and downstream of Kahurangi
Shoals, 8-9 February, 1981.

is representative of flow past Kahurangi Shoals remains to be investigated, but is useful at least as a conceptual model of subsurface
motions around such obstructions.

COASTAL CURRENT MODELING

As a first step in understanding the nature of the flow past
Kahurangi Shoals, a steady coastal current was simulated with a two
dimensional (vertically integrated) model (Chiswell, 1983). Nested
models with grid lengths of 4, 1 and 1/3 nautical miles, respectively, were run in serial fashion. The solutions of the lower
resolution models were used sequentially to obtain boundary
conditions for the higher resolution models. Further details are

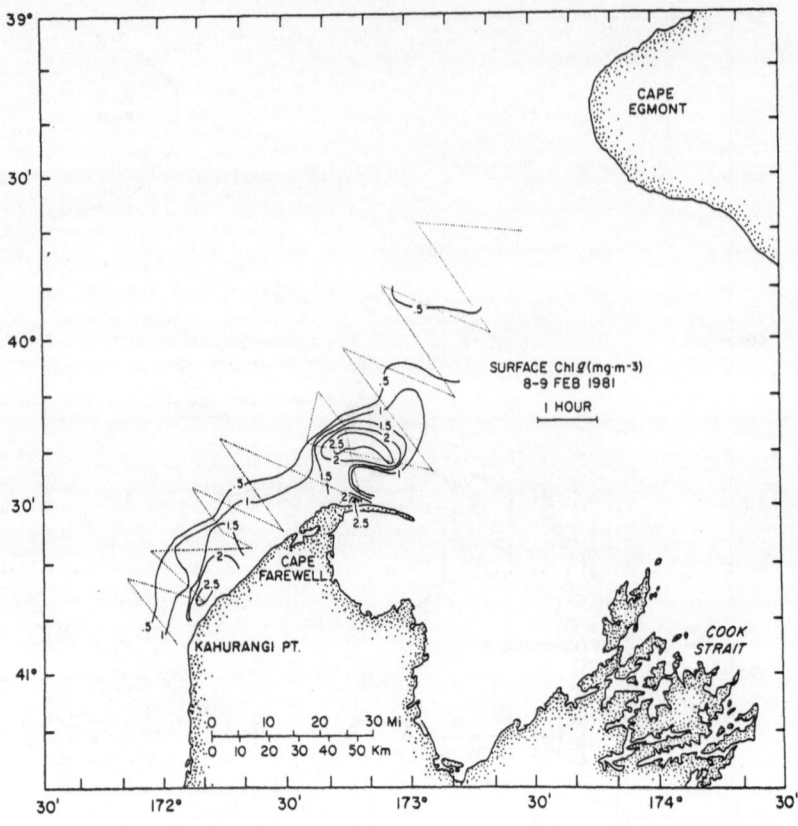

Figure 4. Surface chlorophyll a near and downstream of Kahurangi
Shoals, 8-9 February, 1981.

discussed in Chiswell (1983). The bathymetry of the highest resolution model (1/3 mile), encompassing the region of the Shoals with
superimposed selected streamlines is shown in Fig. 7.

An important attribute of the flow is its relative vorticity
(Fig. 8) because of its relation to Ekman pumping. The inner and
outer flanks of the shoals (which are aligned at about 45° to the
undisturbed flow) are characterized by lobes of high negative and
positive vorticity (for example streamline (d) in Fig. 7 passes
through the core of the seaward zone of negative vorticity shown in
Fig. 8). Numerical simulations of low Reynolds number, steady flows
past oblique plates (elliptic cylinders) produce similar results
(Fig. 9; Lugt, 1983). The streamline that intersects the tips of the

Figure 5. Visible photograph of Von Karman vortices in the lee of Guadalupe Island off the coast of Baja California (from Pitts et al., 1977).

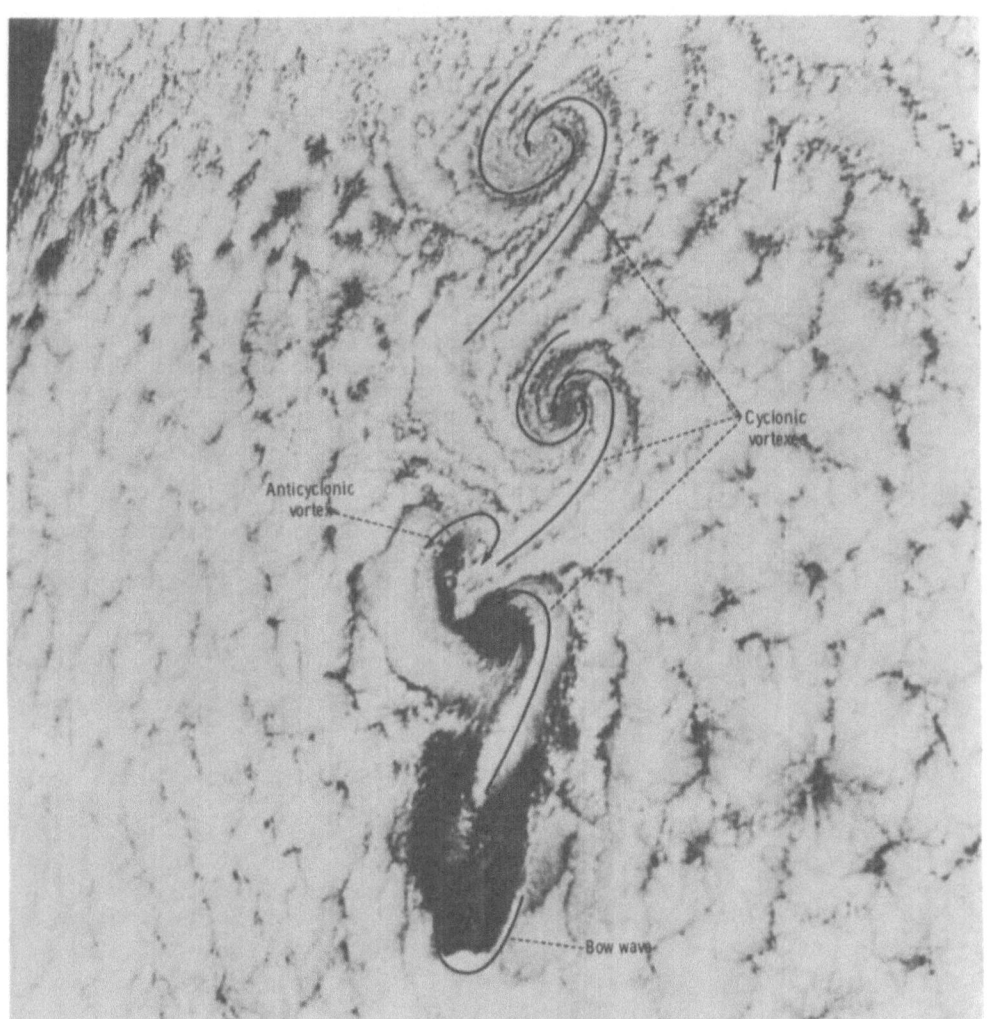

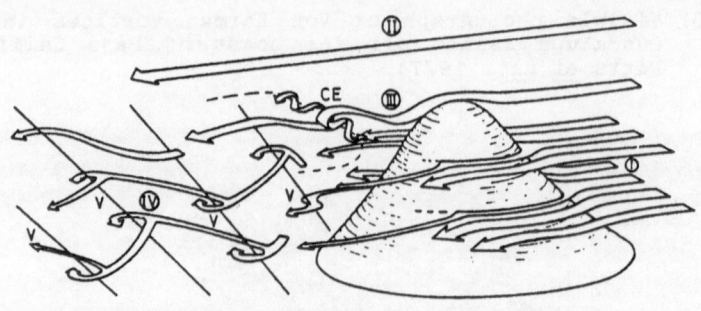

Figure 6. Sketch of dominant features of strongly stratified, non
rotating flow past a bluff body in a laboratory flow.
Region I is the upstream undisturbed flow; Region II is the
little affected flow well above the top of the obstacle;
Region III is very near the summit and is characterized by
lee waves; Region IV is where vortices (V) are formed. The
axes of these vortices lead forwards downstream. CE stands
for cowhorn eddy (from Brighton, 1978).

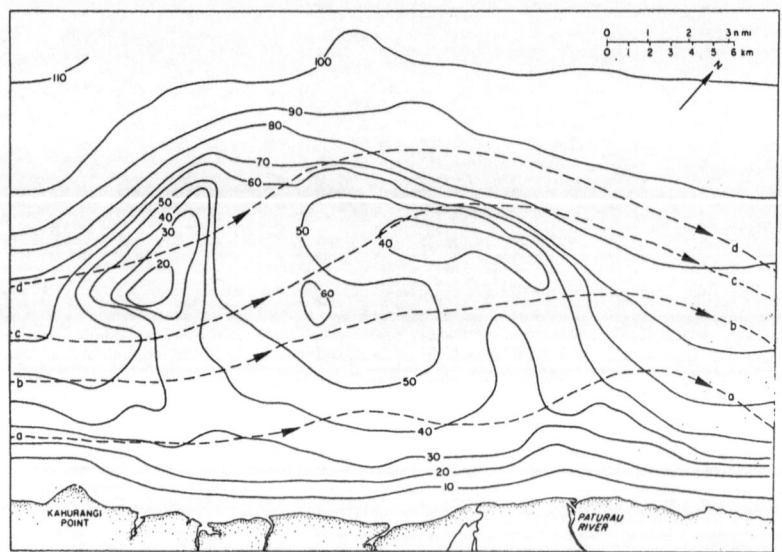

Figure 7. Bathymetry (m) and selected streamlines from the numerical
simulation of steady flow over and near Kahurangi Shoals
(from Bowman et al., 1983).

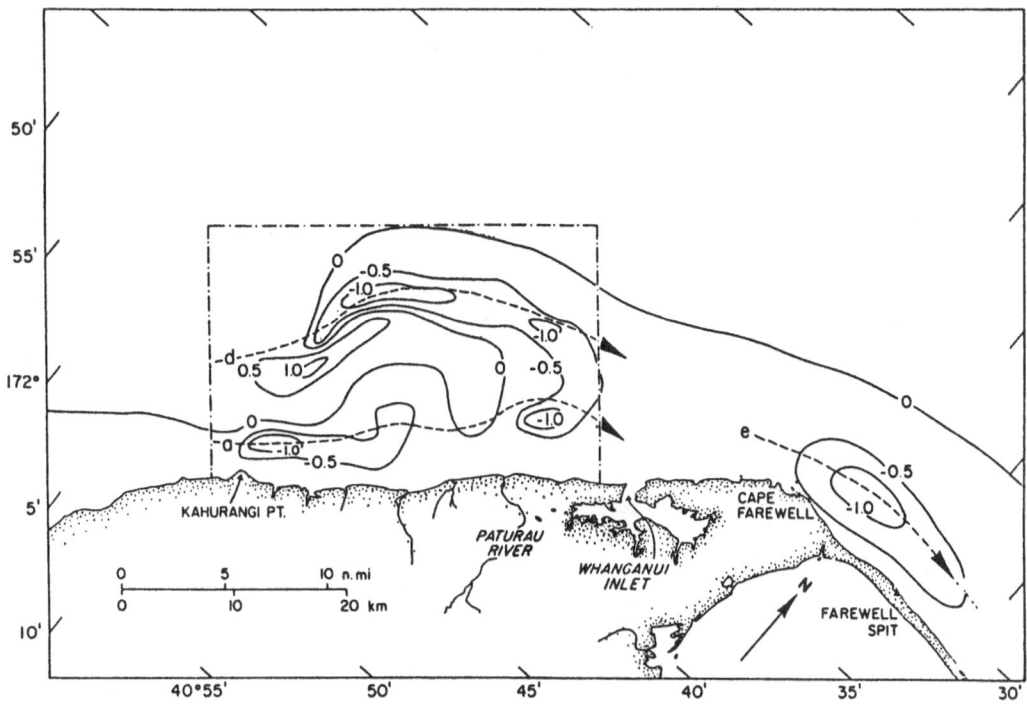

Figure 8. Relative vorticity in the simulated steady flow over and
near Kahurangi Shoals (from Bowman et al., 1983).

obstacle, aligned at 45° to the undisturbed flow (Fig. 9a), divides
the flow into two semi-infinite planes of positive and negative
vorticity (Fig. 9b). However, if the flow rate is increased then
Lugt (1983) showed that a vortex street develops at Reynolds numbers
in the range Re = 45 to Re = 200. Three vortices can be recognized
in the "snapshot" of Fig. 10, two of which have separated and are
swimming away with the flow. Further images of streamlines and
vorticity contours equispaced in time throughout a shedding cycle are
shown in Fig. 11 for Re = 200 (Lugt, 1983). These show how vortices
develop, grow and shed from such obstacles.

INTERPRETATION OF THE SURFACE HYDROGRAPHY AS A VORTEX STREET

Using the surface map of salinity as a template (salinity is
considered the most conservative property measured), a Von Karman
vortex street consisting of two rows of counter rotating eddies has

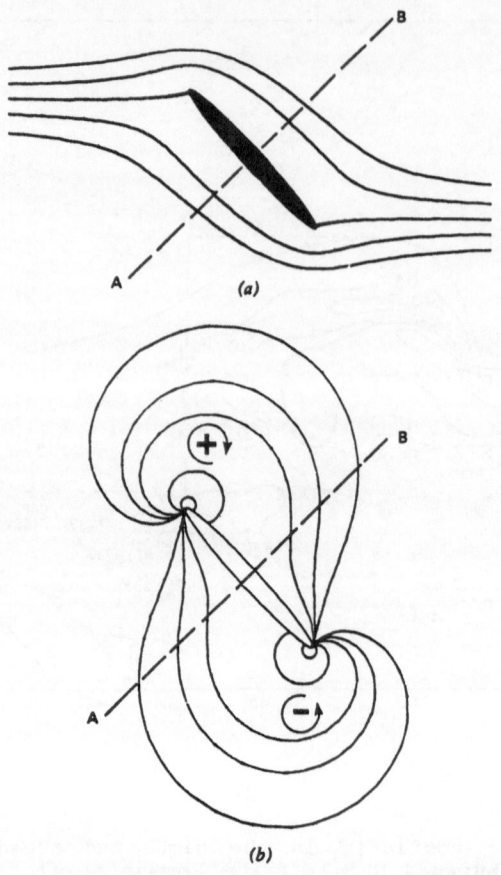

Figure 9. a) Streamlines and, b) contours of vorticity around a plate
         imbedded obliquely in a steady parallel flow for Re << 1
         (from Lugt, 1983).

been superimposed (Fig. 12). The evidence for the cyclonic eddies is
more apparent for reasons discussed above; three vortices are suggested, a number similar to that found in observations of atmospheric
vortex streets. Note that the width of the street slowly increases
downstream as the vortices diffuse into the ambient fluid, and
finally, each other.

Eddy Leilani was subsequentially followed for about 4 days as it
propagated into the western approaches to Cook Strait. It was there
that fishing for squid was particularly intense and favorable (Bowman
et al., 1983).

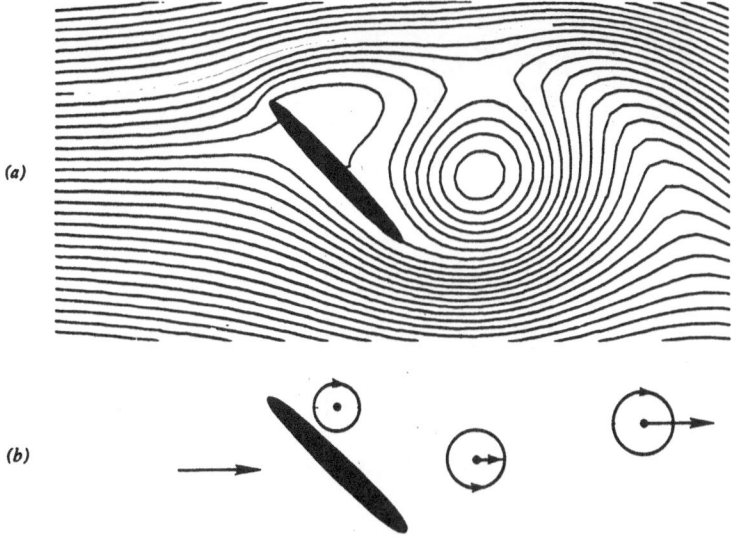

Figure 10. a) Streamlines around an oblique plate at one instant of time behind which a vortex street develops. Three vortices can be recognized, whose locations and scale are indicated in (b) (from Lugt, 1983).

Analysis

In the absence of current measurements, the interpretation of the observed upwelling on February 8 and 9, 1981, rests on an analysis of the regularity and spacing of the meandering surface isohalines redrawn in Fig. 12. The analysis to follow, then, is based on the geometry of the street and a knowledge of the upstream current velocity. The latter was estimated by tracking a number of surface drifters in the vicinity of the shoals (Chiswell, 1983).

Viscosity plays two roles in Von Karman vortex streets:
(i) Viscosity leads to the formation of boundary layers in which vorticity is generated and the vortex pairs are formed, the vortices being shed alternately near each edge of the obstacle; and
(ii) The eddies are subsequentially dissipated by viscosity as they spin down and swim away with the flow.
In other respects the fluid may be considered inviscid.

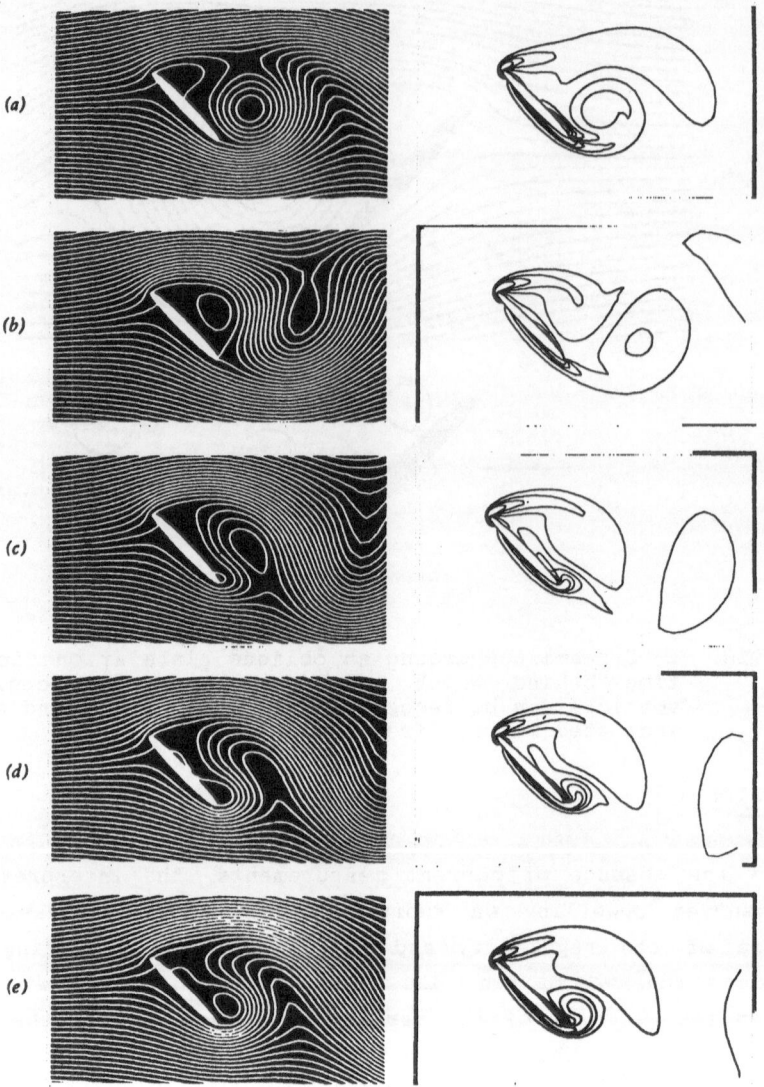

Figure 11.   Numerical simulations of flow past an oblique plate for
Re = 200.   The left hand panels are streamlines (the top
left hand panel corresponds to Fig. 10) and the right
hand panels are contours of vorticity (from Lugt, 1983).

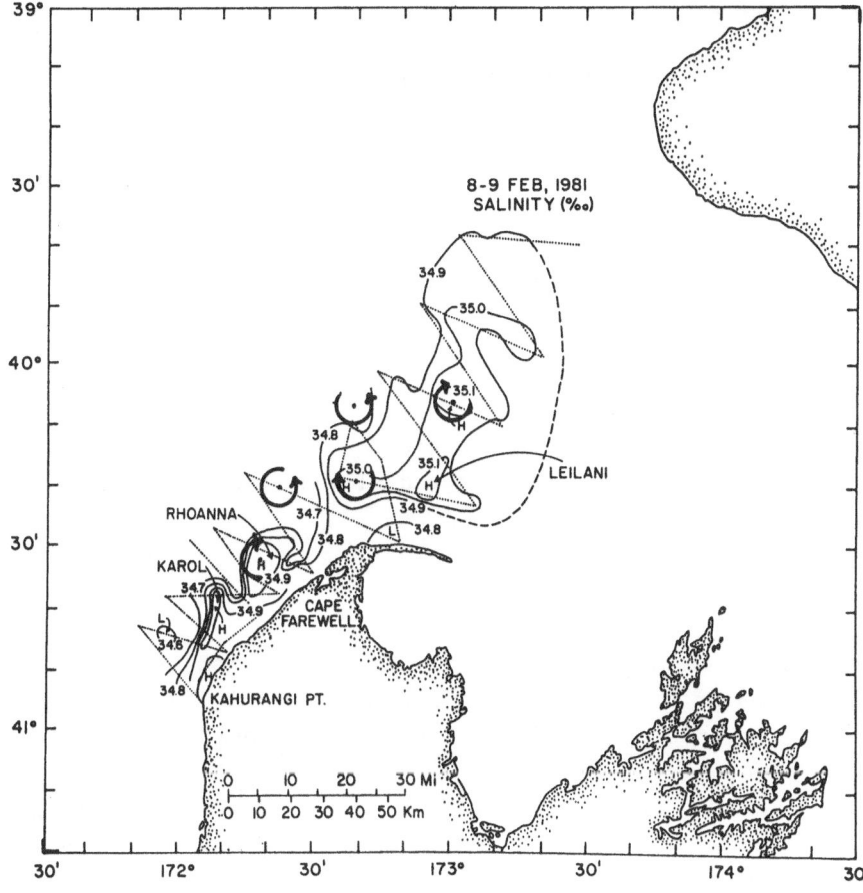

Figure 12. Surface salinity, 8-9 February 1981, with apparent positions of cyclonic (inshore) and anticyclonic (offshore) vortices superimposed. The location of Kahurangi Shoals is shaded black.

For an initial point source of vorticity of strength k, the diffusive decay of vorticity ζ follows from the diffusion equation

$$\frac{\partial \zeta}{\partial t} = \nu \left(\frac{\partial^2 \zeta}{\partial r^2} + \frac{\partial \zeta}{\partial r} \right) \qquad (1)$$

which has the solution

$$\zeta(r,t) = \frac{k}{4\pi\nu t} \exp - (r^2/4\nu t) \tag{2}$$

where ν is the kinematic viscosity,
   r is the eddy radius,
   t is the elapsed time, and
   k is the circulation, defined as the integration of the rela-
      tive vorticity over the surface area of the eddy; viz.,

$$k = \int_0^\infty \zeta\, 2\pi r\, dr.$$

Thus vorticity at the eddy center $\zeta(0,t) = k/4\pi\nu t$ decreases inversely with time, and vorticity diffuses radially during an interval of time t over an e-folding radius

$$r = 2(\nu t)^{\frac{1}{2}} \tag{3}$$

As mentioned earlier, in geophysical studies of atmospheric and oceanic studies of Von Karman vortex streets, ν is considered equivalent to an effective horizontal eddy viscosity A_h.

Lamb (1932, p. 225) showed that for a system consisting of 2 parallel rows of equidistant vortices, arranged so that each vortex in one row is opposite the center of the interval between two conse- cutive vortices in the other row, the general velocity of advance of the configuration in the negative x direction is given by

$$u = \frac{k}{2a} \tanh \frac{\pi h}{a} \tag{4}$$

where k is the strength of circulation within each vortex, a is the wavelength along each row, and h is the spacing between the rows.

Thus if the vortices are imbedded in a flow of ambient velocity u_o. The general velocity of advance is

$$u_e = u_o - (\frac{k}{2a}) \tanh (\pi h/a) \tag{5}$$

Hence, the faster the eddies spin, the slower they progress.

The drag, D, per unit height on the obstacle is given by two formulae:

$$D = 2\rho h \, u_e(u_o - u_e) \coth (\pi h/a) \tag{6}$$

based on momentum flux from the mean stream into the turbulent wake, and

$$D = (k\rho/2\pi a) \{k + 2\pi h \, (2u_e - u_o)\} \tag{7}$$

based on the drop in pressure across the obstacle.

These are equivalent when

$$\coth (\pi h/a) = \pi h/a = 1.2$$

$$\text{i.e. } h/a = 0.39 \tag{8}$$

This gives a theoretical spacing of the eddy rows normalized by their longitudinal spacing (i.e. the aspect ratio).

The dimensionless drag coefficient C_D is defined by

$$C_D = D/\tfrac{1}{2}\rho d u_o^2 \tag{9}$$

where ρ is the density of the fluid. The value of C_D depends on the shape of the obstacle and is known only for certain well defined shapes. For example, within the range of Reynold's numbers encountered in laboratory studies of Von Karman eddies formed behind circular obstacles, $C_D \sim 1.25$ for Re = 100 (Batchelor, 1967, p. 261). It can also be shown (Birkhoff and Zarantonello, 1957) that

$$C_D \stackrel{\sim}{\sim} h/d \tag{10}$$

the normalized width of the wake behind the obstacle.

Thus, based on a knowledge of the geometry of the eddies h/d can be measured and equations (10) and (9) substituted in equation (6) to obtain an estimate of u_e/u_o

$$\text{viz., } \quad u_e/u_o = (1 + (1 - a/\pi h)^{\frac{1}{2}})/2 \tag{11}$$

From equation (5), the shedding rate, N, of vortex pairs is given by

$$N = \frac{u_e}{a}$$

$$= \frac{1}{a} \left\{ u_o - \frac{k}{2a} \tanh (\pi h/a) \right\}$$

Now u_o/N represents the distance a particle of water moves during one shedding cycle. Thus $u_o/dN = 1/S$ is the distance this particle moves in one shedding cycle in distance units of an obstacle diameter. Thus the Strouhal number S can be thought of as the time for the particle to traverse one obstacle diameter in units of time of the shedding period $1/N$.

Another important dimensionless parameter in the study of flows past obstacles is the Lin parameter

$$\beta = S/Re \qquad\qquad (12)$$

$$= \frac{Nd_o/u_o}{u_o d/\nu}$$

$$= \frac{\nu N}{u_o^2} \qquad\qquad (13)$$

Note that β is independent of the dimensions of the obstacle and depends only on the parameters ν, N and u_o, i.e. it is an inherent property of the vortex street.

Lin (1954) found that recognizable Von Karman vortex streets in laboratory flows are produced only for a very limited range of β satisfying $10^{-3} < \beta < 2.5 \times 10^{-3}$.

The lifetime τ of an eddy is given by

$$\tau = L/u_e \qquad\qquad (14)$$

where L is the observed length of the street and u_e is calculated from equation (11).

The eddy radius (equation 3) can be rewritten in terms of the Lin parameter by eliminating ν from equations (3) and (13)

viz., $r = 2u_o (\beta Nt)^{\frac{1}{2}}/N.$ (15)

When the radius of the furthest downstream eddy equals half the width of the street, viz.,

$r = h/2,$

then eddies start to overlap and become unstable and annihilation of the street begins.

This gives another estimate for the lifetime τ of the eddies $t = \tau$ as from equation (15) as

$$\tau = \frac{Nh^2}{16u_o^2\beta}$$ (16)

The velocity of eddy drift u_e is the product of the shedding rate N and their wavelength a:

$$u_e = Na$$ (17)

Thus from (16) and (17)

$$\beta = \frac{1}{16\tau N} \left(\frac{h}{a}\right)^2 \left(\frac{u_e}{u_o}\right)^2$$ (18)

The equation (18) can be used to estimate the Lin parameter β from a knowledge of street geometry and using equations (11), (14) and (17), assuming u_o and L are known.

The above analysis considers four dimensionless parameters C_D, h/a, u_e/u_o and β. If one of these is known, the others can be calculated.

The coefficient of drag depends on the shape of the obstacle and is known only for a few regular shapes. The velocity ratio and the Lin parameter require *a priori* assumptions, so h/a is the known, measured parameter.

For comparison, typical values for the above dimensionless parameters determined from laboratory experiments are:

$$0.28 < h/a < 0.52 \qquad\qquad 0.7 < u_e/u_o < 0.9$$
$$35 \;\; < Re \;\; < 2500 \qquad\qquad 10^{-3} < \;\; \beta \;\; < 2.5 \times 10^{-3}$$

APPLICATION TO KAHURANGI SHOALS

The eddy pattern shown in Fig. 12 gives the following results:

$$d \sim 8 \text{ km} \qquad h \sim 15 \text{ km} \qquad a \sim 35 \text{ km}.$$

Thus $h/a = 0.43$.

From equation (10),

$$h/d = 1.9 = C_D$$

From equation (11),

$$u_e/u_o = 0.75.$$

From equation (17),

$$N = 1.1 \times 10^{-5} \text{ s}^{-1} \text{ for } \mu_o = 50 \text{ cm}^{-1};$$

i.e. the shedding period $1/N = 26$ hours.
(note: the inertial period $12/\sin\theta \sim 18.4$ hour where θ is the latitude).

The eddy lifetime τ, from equation (14) is
$$\tau = 78 \text{ hours}$$
and from equation (18)
$$\beta = 2.1 \times 10^{-3}$$
The Strouhal number
$$S = Nd/u_o = 0.17$$
and from (12)
$$Re = \frac{S}{\beta} = 81.$$

Finally, from (13)

$$\nu = \frac{\beta u_o^2}{N} = 0.49 \times 10^6 \ \text{cm}^2 \ \text{s}^{-1}.$$

This value of eddy viscosity is consistent with Okubo's (1976) empirical formula for scale dependent viscosity

$$\nu = 0.068 \ r^{1.15} \qquad \text{(cgs units)} \tag{19}$$

Substituting the value of ν above into equation (19) yields a value for r of

$$r = 9.2 \ \text{km},$$

which is consistent with the observed scale of the wake pattern (Fig. 12), suggesting that the calculated value of ν is a reasonable one.

A representative value of relative vorticity at the eddy center after formation can be derived from equations (2), (5), and (8);

$$\zeta(o,t) = \frac{k}{4\pi\nu t} \quad \frac{2a(u_o - u_e)\coth(\pi h/a)}{4\pi\nu t} \tag{20}$$

$$= 6.6 \times 10^{-4} \ \text{s}^{-1} \ \text{when} \ t = 26/2 \ \text{hours, i.e. after}$$

half of a shedding cycle T/2.

Thus $\frac{\zeta}{f}$ (0,T/2) = 6.8 where f is the Coriolis parameter. (21)

The above results are summarized in Table I along with corresponding values obtained by Chopra and Hubert (1965) for atmospheric wakes behind the Madiera Islands and by Barkley (1972) for oceanic wakes behind Johnston Atoll.

From equation (2) and the vorticity equation

$$\zeta = \frac{v}{r} + \frac{\partial v}{\partial r}$$

where v is the tangential swirl velocity, it can be shown that (see Fig. 13)

$$v = \frac{k}{2\pi r} \{1 - \exp - r^2/(4\nu t)\}. \tag{22}$$

At the end of a half eddy-pair shedding period $t = T/2 \sim 13$ hours, the values of k and ν quoted in Table I when substituted into equation (22) lead to an estimate of maximum tangential velocity v = 0.35

TABLE I

Comparison of properties of vortex street wakes in the ocean and atmosphere
(from Chopra and Hubert, 1965; Barkley, 1972) with those of Kahurangi Shoals.

Variable	Symbol	Johnston Atoll	Madeira Island	Kahurangi Shoals
Effective obstacle diameter	d	26 km	40 km	8 km
Wavelength or pitch	a	160 km	190 km	35 km
Width between eddy rows	h	55 km	83 km	15 km
Ratio h/a	h/a	0.34	0.43	0.43
Speed of incident flow	u_o	0.6 m s^{-1}	10 m s^{-1}	0.5 m s^{-1}
Speed of wake translation	u_e	0.45 m s^{-1}	7.5 m s^{-1}	0.38 m s^{-1}
Period of eddy pair formation	$T = 1/N$	96 h	7.2 h	26 h
Period of half pendulum day	-	42 h	22 h	18 h
Frequency of eddy pair formation	N	2.9x10^{-6} s^{-1}	3.9 x 10^{-5} s^{-1}	7.1x10^{-5} s^{-1}
Strouhal number	S	0.12	0.16	0.17
Reynolds number (based on eddy viscosity)	R_e	70	90	81
Eddy viscosity	ν	2.2x10^2 m^2 s^{-1}	44x10^2 m^2 s^{-1}	.49x10^2 m^2 s^{-1}
Drag coefficient	C_D	2.1	2.1	1.9
Circulation strength	k	6.1x10^4 m^2 s^{-1}	1.1x10^6 m^2 s^{-1}	1.0x10^4 m^2 s^{-1}
Drag per unit height	D	9.8x10^6 kg s^{-2}	5.0x10^6 kg s^{-2}	1.9x10^6 kg s^{-2}

TABLE I CONTINUED

Comparison of properties of vortex street wakes in the ocean and atmosphere (from Chopra and Hubert, 1965; Barkley, 1972) with those of Kahurangi Shoals.

Variable	Symbol	Johnston Atoll	Madeira Island	Kahurangi Shoals
Vorticity at eddy center (t=T/2)	ζ_o	1.3×10^{-4} s^{-1}	16.6×10^{-4} s^{-1}	6.6×10^{-4} s^{-1}
Ratio ζ_o/f	—	3.1	21	6.8
Decay e-folding time	$t_e = r^2/4\nu$	125h (r=20km)	6.3 h (r=20km)	26 h(r=3km)
Age, oldest eddy observed	$nt_e = \tau$	375 h	20 h	78 h
Distance downstream, oldest eddy	L	~600 km	540 km	105 km
Number of eddies visible	n	3	3	3
Lin parameter	β	1.75×10^{-3}	1.7×10^{-3}	2.1×10^{-3}

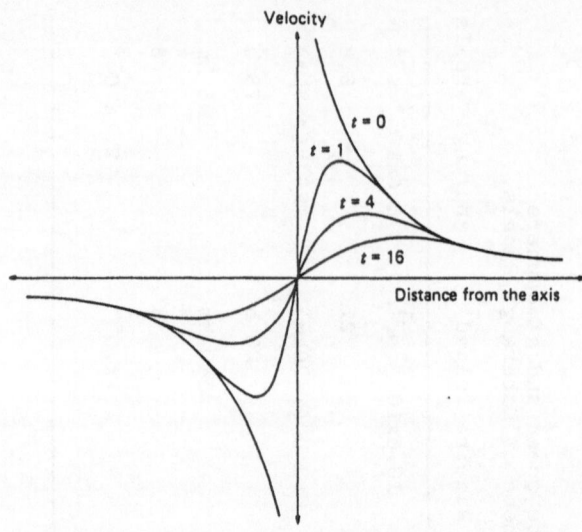

Figure 13. Radial distribution of tangential velocity around a potential vortex (see equation 22). Note how near the vortex center the flow behaves like a rotating solid body, and the radius of maximum velocity increases with time (from Lugt, 1983).

m s^{-1} located at a radius r = 3.5 km. Thus the period of rotation at this radius is $\frac{2\pi r}{v}$ = 17 hr which is very close to the period of an inertial oscillation but in the reverse sense of rotation for the cyclonic eddies.

SOME COMMENTS ON THE COMPARISONS WITH OTHER VORTEX STREET OBSERVATIONS

The remarkable similarity between the numerical values of the parameters (Re, β, h/a, S, and u_e/u_o) calculated for the Kahurangi Shoals and those appropriate to the much larger scale Johnston Atoll and Madiera Island wakes (and much smaller laboratory studies) lends strong support to the interpretation of the observed surface fields of temperature, salinity and nutrients downstream of Kahurangi Shoals as being signatures of a classical Von Karman vortex street. The appropriate value of effective eddy viscosity consistent with the above parameter values is close to that expected for horizontal scale dependent diffusion in the ocean.

The calculated periods of eddy rotation and velocity are close to what was observed (Bowman et al., 1983). The tangential velocity of 0.35 m s^{-1} calculated from equation (22) is very close to the eddy wake translation speed u_e = 0.37 m s^{-1}. Thus to an observer near the coast looking down on the cyclonic eddy train, the eddies would almost appear to be rolling along the coast with little slippage, while the velocities at their outer edge would be $\sim$0.7 m s^{-1} relative to the coast.

The fact that the Rossby number $R_o = \zeta_o/f$ is considerably greater than one in each of the eddy streets examined (see Table I) suggests that Coriolis effects are relatively unimportant to the generation of the wake, and may help to explain why the classical theory for a non rotating fluid works as well as it does.

UPWELLING AND MIXING ASSOCIATED WITH THE EDDY TRAIN

If it were not for bottom Ekman pumping, one might not even detect from surface measurements the presence of the vortex street in the encounter by the lower layer of a strongly stratified coastal current with a submerged obstacle. However, strong upwelling velocities w are expected in the cores of the cyclonic eddies of the order

$$w = (A_v/2f)^{\frac{1}{2}}\zeta$$

(Pedlosky, 1979, p. 198) where A_v is a vertical eddy viscosity and ζ is the vorticity associated with the interior water column (assumed geostrophic) current. For example, a modest value of vertical eddy viscosity A_v = 10 cm^2 s^{-1} gives an upwelling velocity w $\sim$ 5 m h^{-1} at t = 13 hours, which helps explain the strong uplift isopleths illustrated in Fig. 14.

In high Rossby number eddies where centrifugal effects become important, even stronger bottom convergence and upwelling can exist (the atmospheric analog of the tornado clearly illustrates this). This effect has been investigated by Pingree (1978) in the Portland Bill headland eddies off the southern coast of Devon, England. Pingree claims that bottom convergence in the cyclonic gyres is

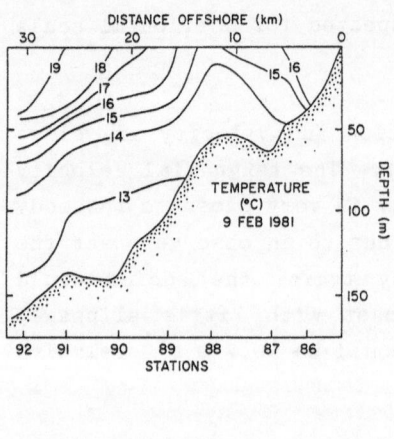
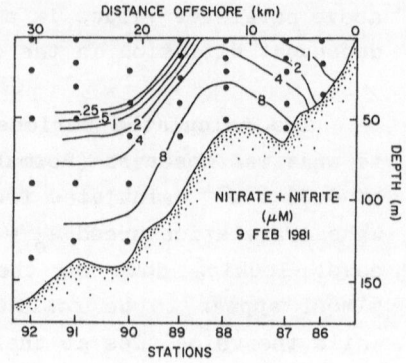
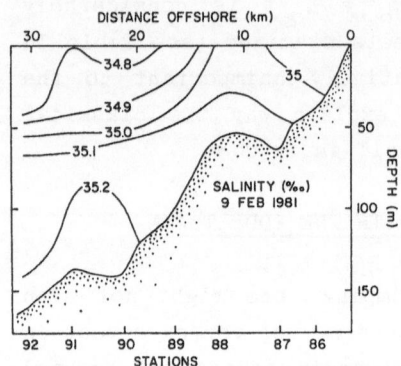
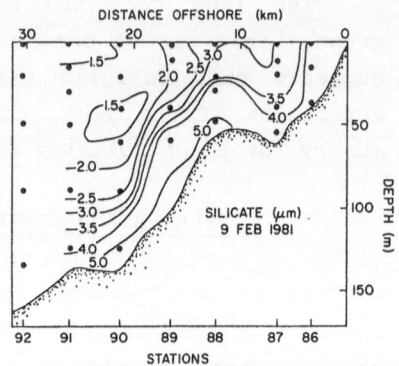
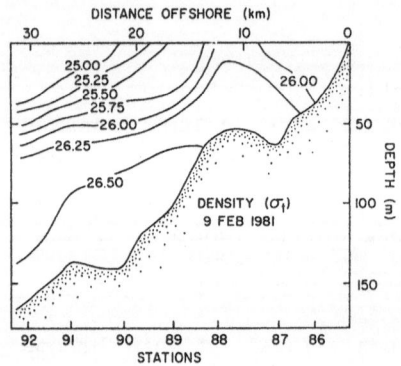
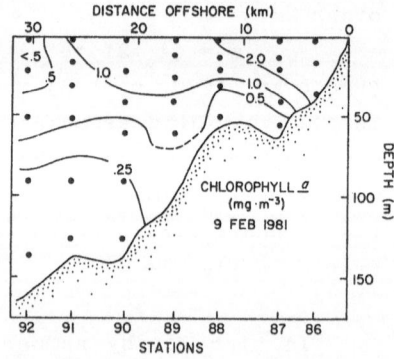

Figure 14. Vertical hydrographic sections across the wake, 9 February 1981. The locations of the stations are given in Fig. 1.

strong enough to produce preferential sediment accumulation in the eddy center over that found in the companion anticyclonic eddy.

Little has been said up to now about the role of vertical mixing near the Kahurangi Shoals in stimulating primary production. The Simpson-Hunter (1974) stratification index $s = \log_{10} H/u^3$, where H is the local water depth and u the amplitude of the current is shown for the Kahurangi Shoals region in Figure 15. This was derived from the numerical model of the *steady* coastal current - the addition of tides (typically ~10-20 cm s^{-1}) to the model would reduce the s values somewhat but would not change the basic patterns. Two areas of low s (<1.5), shaded in Fig. 15 lie over the Kahurangi Shoal and off Cape Farewell. We can expect, then, that vertical mixing over the Shoals and off Cape Farewell to also be important in contributing to the appearance of cool surface water in these regions.

NOTES ON THE DISTRIBUTION OF CHLOROPHYLL A

Surface chl_a concentrations in the upwelling zone were increased some 4 to 6 times background levels farther offshore (Fig. 4). These patches were usually located near to but not necessarily coincidental with the cores of the upwelling cells. Possibly production was greatest in the fronts around the eddies where marginal stratification coupled with an adequate light regime and plentiful nutrients led to favorable growth conditions. Copepod abundance was at least four times higher within and around the shoreward perimeters of the eddies than offshore. Larval stages of euphausiids were present in increased concentrations at the southern edge of eddy Leilani (Fig. 1) during subsequent measurements on February 12. Increased concentrations of fecal pellets from grazing copepods and sinking phytoplankton in frontal boundaries of senescent eddies are thought to attract the krill, and as the krill move towards the surface at night, the squid follow (B. Foster, pers. comm.).

UPWELLING NEAR POINTS OF FLOW SEPARATION

In addition to the mechanisms of vortex shedding and tidal mixing near flow obstacles on the shelf, negative vorticity and associated upwelling is also linked with flow separation near convex coastal bends (Fig. 16). In practice, it is difficult to determine the exact locations of separation points, but Figs. 1 to 3 suggest

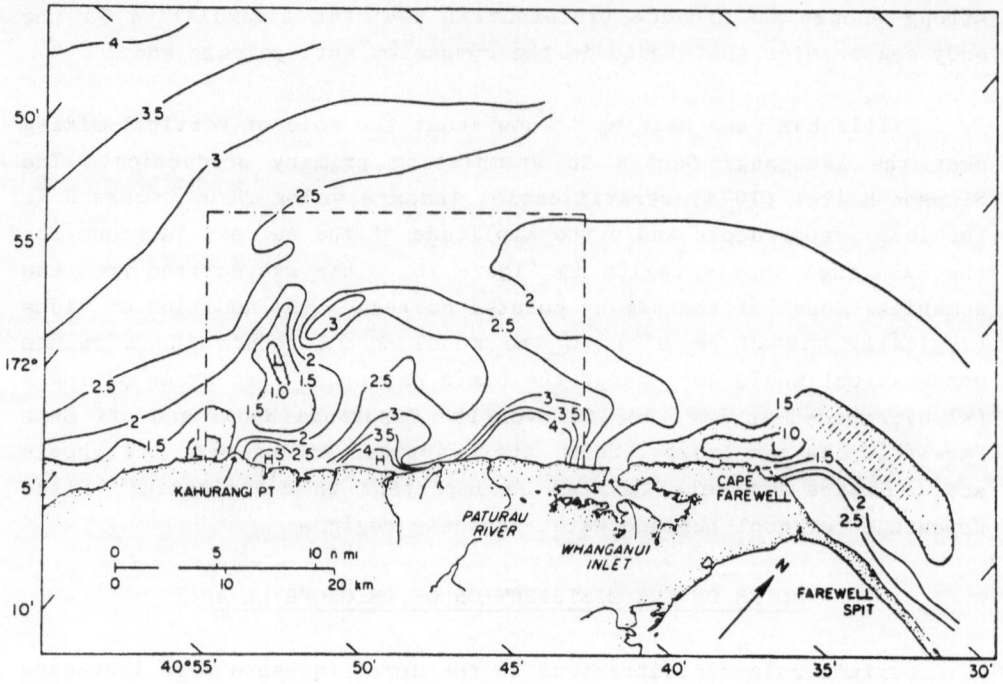

Figure 15.   Contours of the stratification index, s, on a $\log_{10}$ scale
in cgs units based on a numerical simulation of the <u>mean</u>
coastal flow (from Bowman <u>et al</u>., 1983).

that separation is occurring at Kahurangi Point (also see Bowman <u>et
al</u>., 1983; Fig. 2). At other times, the separation point moves north
to the base of Farewell Spit, in response to changing wind patterns.
In each case, surface signatures of upwelling seem to be associated
with the (assumed) negative vorticity zone along the coast north of
the separation point (Fig. 16). Note that the upwelling driven by
Ekman pumping from the bottom boundary layer is additive to any wind
induced coastal upwelling that may be occurring.

ENHANCEMENT OF REGIONAL PRODUCTIVITY BY THE CYCLONIC EDDY TRAIN

Apart from local stimulation of production within and around the
frontal boundaries of the cyclonic eddies, it is of interest to
investigate what the regional influence on shelf production of the
vortex street might be. Depending on local winds and the set of the
D'Urville Current which sweeps into the western approaches to Cook
Strait, the cyclonic eddies often migrate across the relatively flat

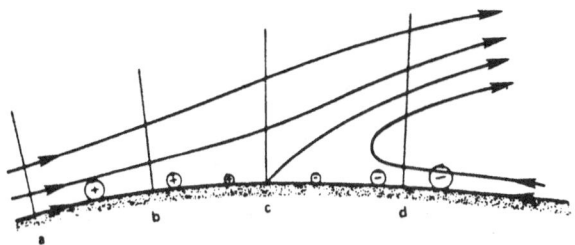

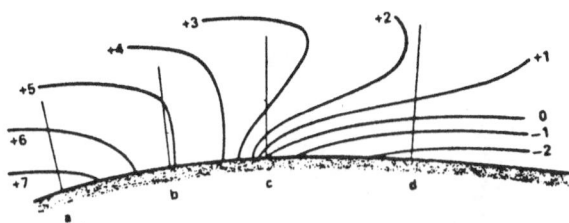

Figure 16.   Upper: streamlines of a coastal flow near a separation
point c.   Lower: equivorticity lines for the same flow
(from Lugt, 1983).

Egmont Terrace (∿100 m depth) before they finally dissipate in the
South Taranaki Bight or are mixed away in the turbulent narrows of
Cook Strait.   The water overlying the western approaches is oligo-
trophic Tasman Sea surface water of subtropical origin, and is
sufficiently stratified to be nutrient depleted in summer.

The semi-enclosed summer stratified sea bounded by a straight
line stretching from Kahurangi Point and Cape Egmont to the west, the
north coast of the South Island, a segment of the North Island, and
the stratification index contour s = 2 (Fig. 17) was chosen for a
first order nutrient budget analysis.   The waters outside the s < 2
contours represent tidally mixed waters which ensure an adequate
supply of nutrients to the euphotic zone.   The s values were derived

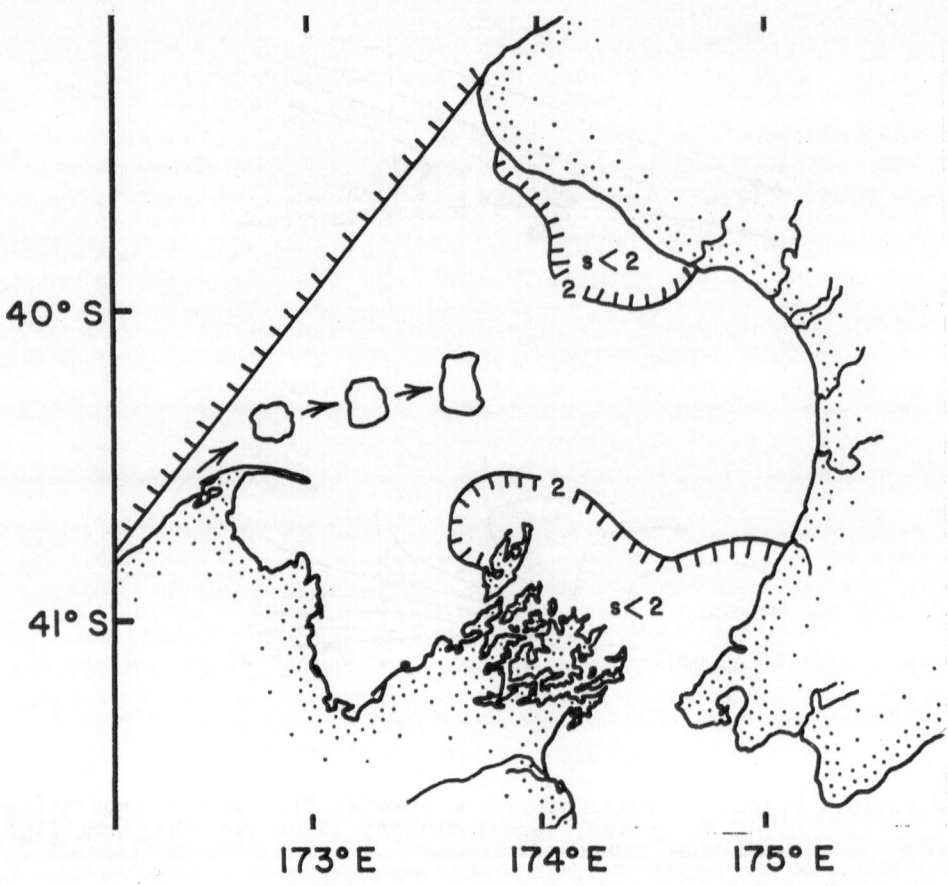

Figure 17.   Sketch for modeling of contribution of eddy transport to
the nitrate budget of the stratified waters of the South
Taranaki Bight.   The s = 2 contours are derived from a
numerical model of Greater Cook Strait (Bowman et al.,
1980) and separate the stratified waters to the west
(s>2) from the mixed waters to the east and south (s<2).
Cold core eddies are injected from the southwest corner
of the region at a rate of about one per day and migrate
into the western approaches withi the D'Urville Current
before dissipating over a time scale of a few days.

from a numerical model of the M_2 tides in greater Cook Strait (Bowman et al., 1980).

The basis of the nutrient (nitrate budget) calculation is as follows: First the nitrate demand of the shelf sea is estimated on an (assumed) uptake rate of nitrate by the ambient primary production, diffusing vertically through the thermocline. Second a steady stream of cyclonic eddies emanating from Kahurangi Shoals migrates into the area. These eddies transport nutrients not only as the slug of elevated concentrations contained within each eddy at their time of crossing the border, but also by active upwelling within the eddies as they spin down and dissipate.

The nitrate demand F_1 in the stratified region (s > 2) is:

$$F_1 = kc_u \, A \, h_u$$
$$= (1/3 \text{ days})(0.25 \text{ μM})(26,400 \text{ km}^2)(50 \text{ m})$$
$$= 79 \text{ kg s}^{-1}$$

where k is the first order uptake rate (i.e. a slug of nutrient would be reduced to 1/e of its initial concentration in 3 days, c_u is the ambient surface concentration in the stratified shelf sea of surface area A, and h_u is the depth of the euphotic zone.

The lateral flux of nitrate F_2 is given by the product of the volume of the eddy (upper 50 m) V_e, the nutrient concentration c_e and the entry rate across the boundary N, i.e.,

$$F_2 = V_e \, c_e \, N$$

$$= (\pi/4)(10^4 \text{m})^2 (50\text{m})(2\text{μM})(1\text{day}^{-1}) = 6 \text{ kg s}^{-1}$$

where the eddy diameter is taken as 10 km (Fig. 12), c_e = 2 μM (Fig. 3) and the entry rate N as one per day (T = 1/N ∿ 26 hours; Table I).

The upwelling flux F_3 of nitrate within the eddies is given by the product of the upwelling velocity w, the eddy surface area A_e, the concentration of nitrate within the bottom waters of the upwelling core c_b, and the number of eddies n present at any one time:

$$F_3 = w \, A_e \, c_b \, n$$

A conservative value of w is taken as the value appropriate to the oldest eddy observed ($\tau \sim 78$ hours; Table I) after 3 e-folding vorticity spin down times from an initial value of 5 m h^{-1}, c_b is taken as 6 μM (Fig. 14), and n as 3 (Table I).

Thus $F_3 = (7 \times 10^{-5}$ m $s^{-1})(\pi/4)(10^4 m)^2(6 \mu M)$ (3)

$$= 6 \text{ kg s}^{-1}$$

Thus the sum of F_2 and F_3 represents a supply rate equal to about 15% of the total demand over the shelf sea. Although the estimates are necessarily tentative, it appears that vortex shedding from the Shoals must seriously be considered as an important contributor to the regional productivity and thus to the squid fishery.

DISCUSSION

The circulation over the shelf seas of central New Zealand are quite variable, and it would be misleading to suggest that the shedding of Von Karman vortices is a continuous phenomenon. In fact, following the experiment, the coastal current diminished in intensity, and eddies Karol and Rhoanna (Fig. 2) stalled and coalesced to form Karolanna (Bowman et al., 1983; Fig. 11), while eddy Leilani migrated into the western approaches and was subsequentially tracked for more than 3 days. However, there is sufficient satellite thermal IR imagery of the area to confirm that coastal and midshelf upwelling near Kahurangi Point and Shoals is a regular feature of the coastal dynamics. Clearly more research is needed to arrive at more accurate estimates of the sustained high shelf production by the various mechanisms of tidal mixing, coastal upwelling and vortex shedding in this interesting and commercially important shelf sea.

ACKNOWLEDGEMENTS

This work was supported by NSF grant OCE 81-8283 and the Maui Development Environmental Study (MDES). The MDES was commissioned by Shell BP Todd Oil Services acting on behalf of Maui Development Ltd., and was coordinated by Prof. A.C. Kibblewhite at the University of Auckland.

Mr. B.H. Olsson, Director of the Defence Scientific Establishment is thanked for the use of HMNZS "Tui" and thanks are also extended to Lt. Comm. G.C. Wright, RNZN, for assistance at sea.

The research program was facilitated by a memorandum of understanding in Marine Sciences between the State University of New York at Stony Brook and the University of Auckland. Contribution 498 of the Marine Sciences Research Center (MSRC).

REFERENCES

Barkley, R.A. 1972. Johnston Atoll's wake. J. Mar. Res. 30: 201-216.

Batchelor, G.K. 1967. An Introduction to Fluid Dynamics. Cambridge Univ. Press. 615 pp.

Birkhoff, G. and E.H. Zarantonello. 1957. Jets, Wakes, and Cavities. Academic Press, NY.

Bowman, M.J., S.M. Chiswell, P.L. Lapennas, R.A. Murtagh, B.A. Foster, V. Wilkinson and W. Battaerd. 1983. Coastal upwelling, cyclogenesis and squid fishing near Cape Farewell, New Zealand. In: Coastal Oceanography. H.G. Gade, A. Edwards and H. Svendsen (eds.). Plenum, NY. 582 pp.

Bowman, M.J., A.C. Kibblewhite and D.E. Ash. 1980. M_2 tidal effects in greater Cook Strait, New Zealand. J. Geophys. Res. 85: 2728-2742.

Boyer, D.L. and P.A. Davies. 1982. Flow past a circular cylinder on a β-plane. Phil. Trans. R. Soc. Lond. A306: 533-556.

Boyer, D.L. and M.L. Kmetz. 1983. Vortex shedding in rotating flows. Geophys. Astrophys. Fluid Dynamics 26: 51-83.

Brighton, P.W.M. 1978. Strongly stratified flow past three-dimensional obstacles. Quart. J.R. Met. Soc. 104: 289-307.

Chiswell, S.M. 1983. Vorticity generation and upwelling near an isolated topographic feature on the continental shelf. Ph.D. Thesis, Marine Sciences Research Center, SUNY, Stony Brook.

Chopra, K.P. and L.F. Hubert. 1964. Karman Vortex Streets in earth's atmosphere. Nature 203: 1341-1343.

Chopra, K.P. and L.F. Hubert. 1965. Karman Vortex Streets in wakes of islands. J. AIAA 3: 1941-1943.

Lamb, H. 1932. Hydrodynamics. Dover Publ., NY. 738 pp.

Lin, C.C. 1954. On periodically oscillating wakes in the Oseen approximation. In: Studies in Mathematics and Mechanics Presented to Richard von Mises, pp. 170-176. Academic Press, NY. 353 pp.

Lugt, H.J. 1983. Vortex Flow in Nature and Technology. John Wiley and Sons, NY. 297 pp.

Lugt, H.J. and H.J. Haussling. 1974. Laminar flow past an abruptly accelerated elliptic cylinder at 45° incidence. J. Fluid Mech. 65: 711-734.

Lyons, W.A. and T. Fujita. 1968. Mesoscale motions in oceanic stratus as revealed by satellite data. Mon. Weath. Rev. 96: 304-314.

Okubo, A. 1976. Remarks on the use of 'diffusion diagrams' in modeling scale-dependent diffusion. Deep-Sea Res. 23: 1213-1214.

Pedlosky, J. 1979. Geophysical Fluid Dynamics. Springer Verlag, NY. pp. 624.

Pingree, R.D. 1978. The formation of the shambles and other banks by tidal stirring of the seas. J. Mar. Biol. Ass. U.K. 58: 211-226.

Pitts, D.E., J.T. Lee, J. Sein, Y. Sasaki, K. Wagner and Johnson, R. 1977. Mesoscale cloud features observed from Skylab. In: Skylab Explores the Earth. NASA Special Publ. 380: 479-501. Washington, D.C.

Pohle, J.F., A.K. Blackadar and H.A. Panofsky. 1965. Characteristics of quasi-horizontal mesoscale eddies. J. Atmos. Sci. 22: 219-221.

Simpson, J.H. and J.R. Hunter. 1974. Fronts in the Irish Sea. Nature 250: 404-406.

Warren, B.A. 1967. Notes on translatory movement of rings of current with applications to Gulf Stream eddies. Deep-Sea Res. 14: 505-524.

DYNAMICS OF PHYTOPLANKTON PATCHES ON NANTUCKET SHOALS: AN EXPERIMENT INVOLVING AIRCRAFT, SHIPS AND BUOYS

J.W. Campbell and C.S. Yentsch
Bigelow Laboratory for Ocean Sciences
McKown Point
W. Boothbay Harbor, ME 04575

W.E. Esaias
Code 671
Goddard Space Flight Center
Greenbelt, MD 20771

INTRODUCTION

The concept of using aircraft as research vessels in biological oceanography is relatively new. With recent advances in airborne remote sensor technology, it is clear that aircraft have a role distinct from, yet complementary to, that of ships, buoys, and satellites. One distinction lies in the technology itself. Measurement capabilities of airborne remote sensors generally fall between those of ships and satellites.

Roles of the various platforms are also distinct with respect to their time-space sampling capabilities. Esaias (1980) compared the sampling domains of various platforms with the space and time scales of biological and physical processes (Fig. 1). In Figure 1-a, adapted from Steele (1978), the various trophic levels are depicted as lying along a diagonal with phytoplankton having the shortest space and time scales, and zooplankton and fish progressively longer scales.

The sampling capabilities of satellites, aircraft, balloons, ships and buoys are represented in Figure 1-b. Each platform is associated with a distinct region of the time-space domain. Ship coverage, for example, lies to the right and below a line constrained by the maximum speed of ships, such that increased areal coverage is attained only at the expense of less frequent measurements. When the two parts of Figure 1 are superimposed, it is clear that several platforms are required to study interdependencies with an ecosystem.

Lecture Notes on Coastal and Estuarine Studies, Vol. 17
Tidal Mixing and Plankton Dynamics. Edited by J. Bowman, M. Yentsch and W. T. Peterson
© Springer-Verlag Berlin Heidelberg 1986

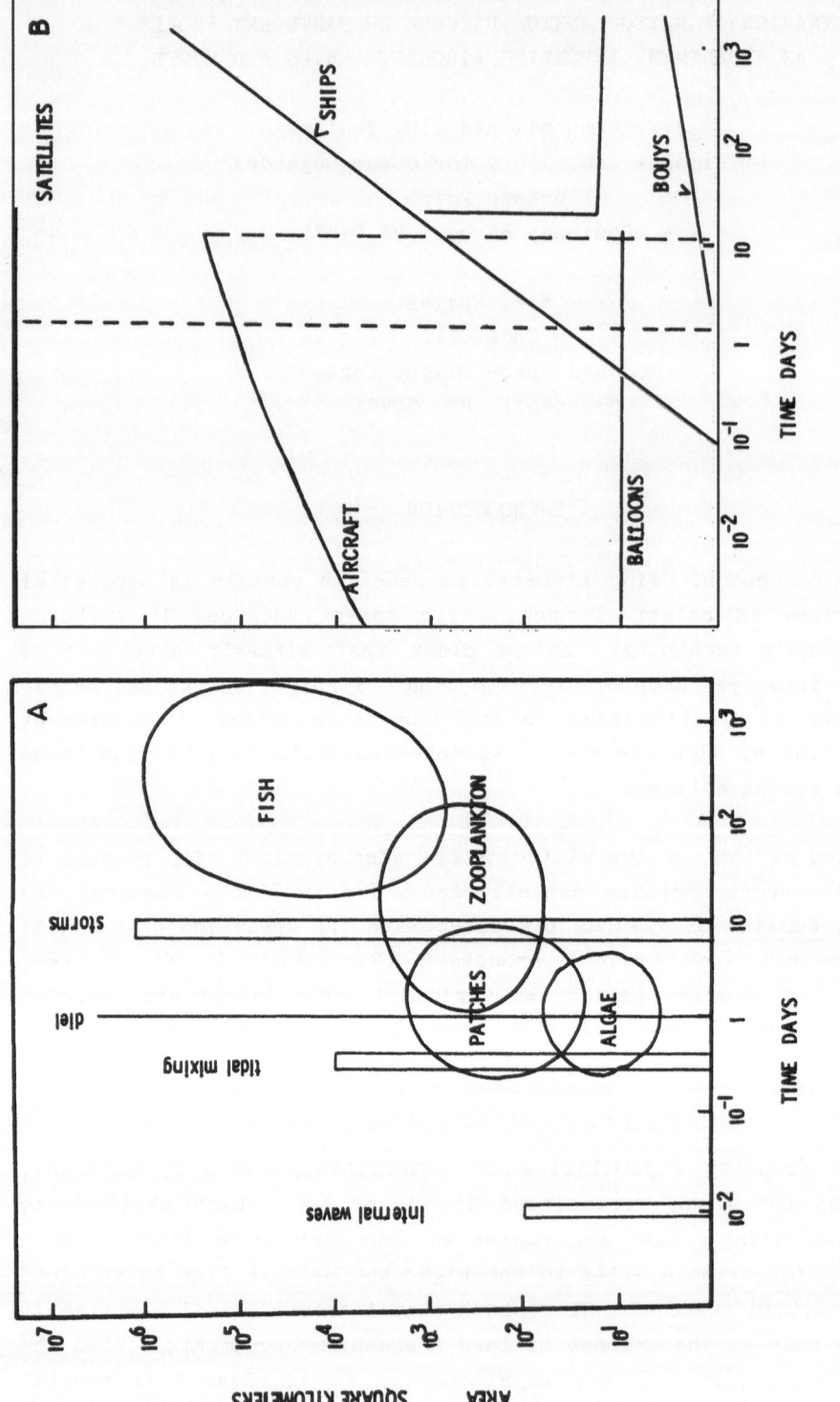

Figure 1. Time-space domains for oceanic phenomena. A. Processes, indicating periods for physical forcings and excursion-generation times of some biological components. B. Sampling capabilities of various platforms. From Esaias (1980).

The space-time sampling capability of aircraft is especially suited to the study of phytoplankton dynamics in a tidally mixed ecosystem. Aircraft can survey an area of 10^2-10^4 km^2 every few hours. This capability is required to characterize temporal variability in spatial distributions controlled by tides and winds, with operative time scales <24 h.

In an effort to test these concepts and thus demonstrate the role of airborne remote sensors in such studies, the National Aeronautics and Space Administration (NASA) sponsored a multidisciplinary field experiment in May 1981. The Nantucket Shoals ecosystem (Fig. 2) was the site of this experiment, which involved surface vessels, moorings and drifters in addition to a suite of remote sensors flown on NASA aircraft. The scientific objective was to investigate the dynamic coupling between phytoplankton growth and tidally driven nutrient input and transport over the Shoals.

The general patterns of circulation and phytoplankton distributions on Nantucket Shoals were investigated on a series of cruises in 1978-79 (Limeburner et al., 1980; Limeburner and Beardsley, 1982). During those cruises, waters were well mixed in regions of depth <40-60 m, and thermal fronts or local temperature minima existed along the northeastern boundary of the Shoals. Local chlorophyll concentration maxima were also present, but displaced somewhat from these fronts. The existence of a non-tidal residual flow, as measured with current meters, suggested that the phytoplankton "patches" lay downstream of the frontal areas.

Although correlation between thermal fronts and phytoplankton growth has been well documented, transport and subsequent growth has not. Further understanding of these phenomena requires synoptic measurements of the study area taken at subtidal frequencies to avoid tidal aliasing of the data. Because of the spatial ($\sim 10^4$ km^2) and temporal (~ 12 h) scales involved, the sampling requirements clearly make the use of aircraft remote sensing an attractive possibility.

In this chapter we present results of the NASA-sponsored experiment conducted in May 1981. The experiment plan was based on a conceptual model of phytoplankton dynamics that was formulated from results of the 1978-79 cruises. We first describe this model, and

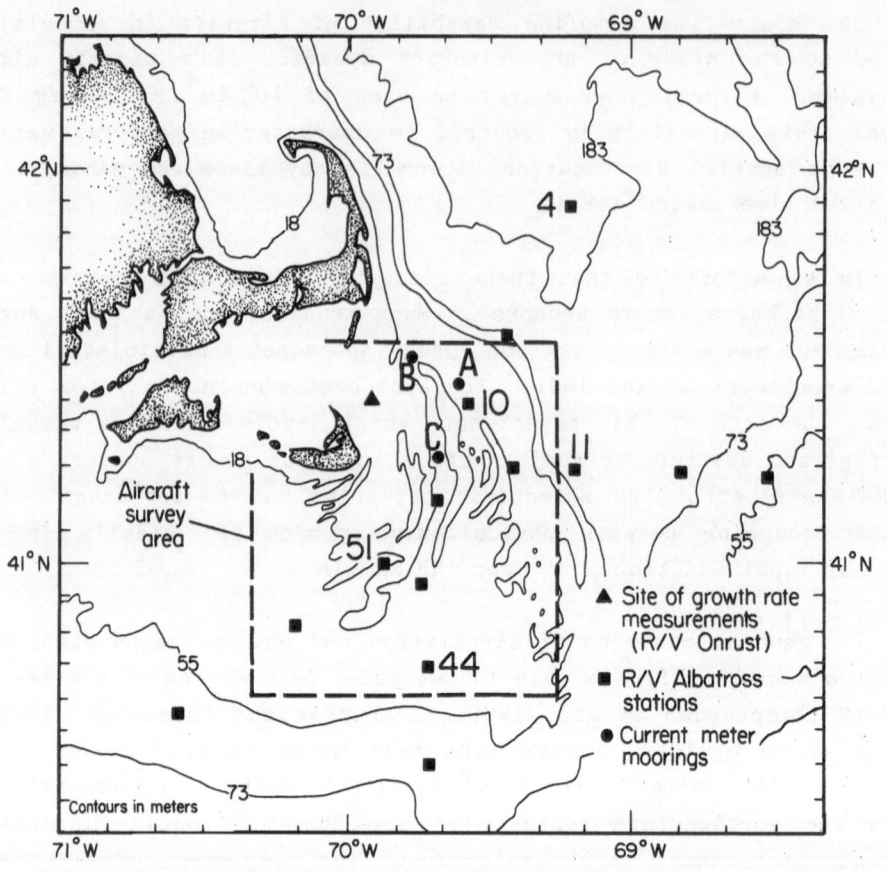

Figure 2. Site of the Nantucket Shoals Experiment, May 6-15, 1981,
showing aircraft survey area, buoy and ship sampling
stations. Growth rate was measured at position ▲ and
squares ■ indicate principal Albatross IV stations.

later address the validity of the model in light of the 1981 results.
We conclude with a discussion of the role of airborne remote sensors
in general, and lessons learned through this experience concerning
strategies for combining aircraft remote sensing data with more
conventional oceanographic measurements for the more effective study
of tidal mixing and plankton dynamics in shallow seas.

CONCEPTUAL FRAMEWORK

Background

Beginning in May 1978, six bimonthly cruises were conducted to study spatial and temporal patterns in temperature and salinity over the Nantucket Shoals (Limeburner and Beardsley, 1982). In addition, current meter moorings were deployed to study currents in the vicinity of the Shoals, and chlorophyll and nutrient data were collected on 5 of the 6 cruises (Limeburner et al., 1980).

Limeburner and Beardsley (1982) described the Nantucket Shoals as a leaky boundary between the western Gulf of Maine and eastern New England Shelf. They estimated that water flows over the Shoals from northeast to southwest with an average velocity of 10 cm s^{-1} (8.6 km d^{-1}), and they found the water column well mixed to bottom depths of 40-60 m.

Figure 3 shows surface temperature and chlorophyll patterns on May 28-June 2, 1978, and May 19-23, 1979, redrawn from Limeburner et al. (1980). The local temperature minima or "cold spots" to the north and east of the Shoals are believed to reflect tidal mixing of stratified waters. Based on the measured non-tidal flow, it was believed that these waters were being advected toward the southwest where they were warmed by solar heating and by mixing with Nantucket Sound waters.

The areas of relatively high chlorophyll immediately east of Nantucket Island are believed to have resulted from nutrients supplied in the Gulf of Maine source waters (i.e. cold spots). Nutrient distributions were consistent in that the highest nutrient concentrations corresponded with the coldest waters. The spatial displacement between the phytoplankton patch and the cold spot was attributed to the time lag required for growth.

Nutrient-Growth Displacement Model

It was hypothesized that phytoplankton patches on Nantucket Shoals develop as a result of nutrients found in waters transported onto the Shoals by tidal mixing along the northern and eastern boundaries; that these waters are subsequently advected toward the

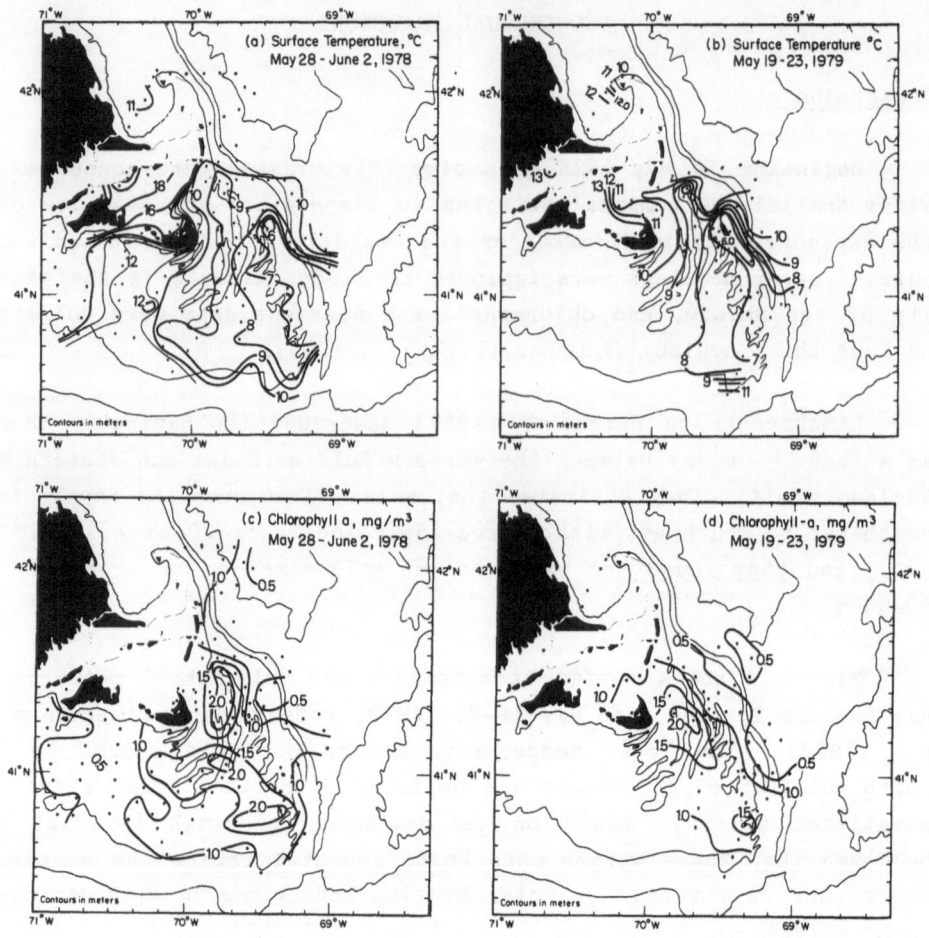

Figure 3. Sea surface temperature and chlorophyll maps, 1978-79 (from Limeburner et al., 1980). (a) Sea surface temperature (°C), May 28-June 2, 1978. (b) Sea surface temperature (°C), May 19-23, 1979. (c) Chlorophyll a (mg m^{-3}), May 28-June 2, 1978. (d) Chlorophyll a (mg m^{-3}), May 19-23, 1979.

southwest where, in a shallow well-mixed water column, there is sufficient light for phytoplankton growth; and, therefore, phytoplankton patches develop downstream of cold spots. Further downstream, nutrients become depleted, losses exceed growth, and phytoplankton concentrations decrease.

If this concept is valid, then spatial relationships, readily measurable by remote sensors, should contain information on growth rates and rates of non-tidal advection. Specifically, the distance between the temperature low and chlorophyll high would be

$$D = \frac{U}{\mu} \log_2 (C_1/C_0) \tag{1}$$

where U is the rate of nontidal advection, μ is the growth rate expressed as chlorophyll doublings per day, and C_0 and C_1 are chlorophyll concentrations in the source water and patch, respectively.

Growth rates were not measured on the cruises in 1978 and 1979. However, a rate of $\mu = .9 \ d^{-1}$ would be consistent with the measured mean current $U = 8.6 \ km \ d^{-1}$ and the spatial patterns shown in Figure 3 ($D \cong 20 \ km$, $C_1/C_0 \cong 4$). Rates on the order of 1 doubling per day are reasonable in natural populations given adequate light and nutrients for growth.

OBSERVATIONAL TECHNIQUES AND METHODS

The Nantucket Shoals Experiment of May 1981 called for (1) airborne remote sensors to provide synoptic measurements of surface temperatures and chlorophyll patterns at subtidal frequencies; (2) ships to collect data on growth rates, productivity, and nutrient concentrations, and to observe vertical structure and verify remote sensing data; (3) current meters moored at locations along the northeastern boundary of the Shoals to record tidal and nontidal currents; and (4) drifters deployed in the vicinity of the moorings for Lagrangian measurements of water transport over the Shoals.

Aircraft Data

The remote sensing data were obtained by a suite of sensors flown on a NASA P-3 aircraft based at Nantucket Island for the period of the experiment. Previous experience with this suite of sensors had been obtained in 1980 during the Chesapeake Bay Plume Experiment (Campbell and Thomas, 1981), and the documentation of that experiment contains detailed descriptions of the sensors.

Dates and times of the aircraft flights are given in Table 1 and the area surveyed ($\sim 7 \times 10^3 \ km^2$) is shown in Figure 2. The standard

mapping pattern consisted of a series of east-west flight legs spaced
~9 km apart and ranging in length from 50 to 80 km. These flight
legs were flown at a nominal altitude of 150 m. With a ground speed
of 100 m s^{-1}, it took less than 3 hours to cover the standard grid.
An onboard Loran C system was used to record the aircraft position
every 9 seconds.

The aircraft data were correlated with shipboard measurements of
surface temperature and chlorophyll. Details of these correlations
are given in Campbell and Esaias (1985). Sea surface temperature was
obtained using Barnes PRT-5 infrared radiometers. The absolute
accuracy of the data is estimated to be $\pm$ 0.5°C and its relative
precision $\pm$ 0.2°C.

Chlorophyll concentrations were derived by 2 methods. One was
based on passive spectral radiances measured by the Multichannel
Ocean Color Sensor or MOCS (Campbell and Esaias, 1983); the other was
based on induced fluorescence produced by the Airborne Oceanographic
Lidar or AOL (Hoge and Swift, 1981). The remote chlorophyll
measurements are estimated to be accurate to within 30%.

Ship Data

Three research vessels -- the Edgerton (WHOI), Onrust (SUNY,
Stony Brook) and Gloria Michelle (NMFS) -- made daily trips from
Nantucket Island. Their sampling patterns were concentrated in the
area north and east of Nantucket Island, and southwest of moorings A
and B. A larger vessel -- the R/V Albatross IV (NMFS) -- covered the
Nantucket Shoals and surrounding waters during an extended 10-day
cruise (Thomas, 1981). Principal Albatross IV stations are indicated
on Figure 2, with station numbers given only if referred to subsequently.

Standard oceanographic methods (Strickland and Parsons, 1972)
were used in the shipboard measurements which included both discrete
and continuous sampling. Details of these methods may be found in
separate data reports (Malone et al., 1981; Phinney and Kilpatrick,
1981; Thomas, 1981). The measurements included temperature, salinity,
chlorophyll and nutrients (nitrate, nitrite, silicate, phosphate and
ammonium).

The rate of carbon doubling by phytoplankton was estimated using radioactive ^{14}C techniques and 24 hour light-dark incubations of natural surface populations (Malone et al., 1981). Samples from a station located near the entrance to Nantucket Shoals (see Fig. 2) were incubated at surface water temperatures using sunlight and neutral density filters. A growth rate was estimated by the formula

$$\mu = \log_2 \frac{C:CHL + PP:CHL}{C:CHL} \qquad (2)$$

where μ = doublings of phytoplankton carbon per day; C:CHL = carbon to chlorophyll a ratio taken as 50; and PP:CHL = ratio of integrated euphotic zone productivity (mg C m^{-2} d^{-1}) to integrated chlorophyll a (mg Chl m^{-2}).

One of the weakest aspects of the experiment, as a test of the conceptual model, concerns the difficulty in estimating phytoplankton doubling rates. Direct growth estimates are complicated by a host of problems which cannot be discussed here. Aside from inaccuracy in the method (which may be as much as 100%), the doubling rate derived at one station may not have been representative of growth rates elsewhere on the Shoals.

Current Meters and Drifters

Three taut-wire moorings were located at positions A, B, and C shown in Figure 1 for the period May 6-15, 1981 (Wilson, 1981). Each mooring held 2 Endeco Type 174 recording current meters, one positioned 6 m below the surface and the other 15-20 m below the surface. The meters recorded current speed and direction, temperature, and conductivity at 2 minute intervals.

A separate study (Limeburner, 1981) involved Lagrangian current measurements. At 0800 EDT on May, 1981, 4 passive drogues were deployed in the vicinity of mooring A (41°39.3'N, 69°36.7'W) and tracked both visually and via radar for a period of 102 hours.

Winds

Data on wind speed and direction were obtained from hourly records made at the Nantucket Island airport. Comparisons with wind information in the R/V Albatross IV bridge log for the same period indicate sufficient spatial coherence to consider the Nantucket Island winds representative of the Shoals area.

RESULTS

Physical Forcings

Tidal phase and wind conditions at the time of the aerial surveys are listed in Table 1. Tides were both semidiurnal (period = 12.42 h) and diurnal (period = 24 h), with the semidiurnal tide dominant. Tidal excursions, based on the current meter records, ranged from 14 to 22 km.

Wind stress patterns (i.e. vector magnitudes proportional to square of windspeed) are shown in Figure 4 for the period April 26-May 15, 1981. The experiment followed a storm that had persisted for 3 days with winds up to 50 knots from the northeast. Weather began to clear on May 6, the day the moorings were set out, but northeast winds continued until the morning of May 8. Thereafter, a steady pattern of moderate ($\sim$20 knots) southeast winds persisted for several days.

Spatial Patterns in Temperature and Chlorophyll

Sea surface temperature patterns from the first 4 aerial surveys are shown in Figure 5. In general, SST patterns were similar to those seen during the 1978-79 cruises (Fig. 3). Cold spots appeared along the northern and eastern boundaries. There was a gradual warming from northeast to southwest and evidence of mixing with warm Nantucket Sound waters flowing around Nantucket Island.

These maps reveal a high level of temporal variability on a time scale of hours. Forcings of tides and winds affected the position of the cold spots as well as the intensity of vertical mixing as indicated by the minimum temperatures. The two afternoon flights (5-a and 5-d) coincided approximately with high tide at Nantucket Island, but different meteorological conditions existed (Table 1). The 2 morning flights (5-b and 5-c) were at low tide and under similar meteorological conditions.

Table 1. Dates and times of aircraft surveys of Nantucket Shoals,
          May 1981.

Flight	Date (May 1981)	Times (EDT)	Tides	Winds
1	7	1200–1511	high at 1500	northeast
2	8	0830–1042	low at 0900	variable
3	9	0851–1112	low at 1000	southeast
4	9	1414–1658	high at 1700	southeast
5	11	1130–1425	low at 1200	southeast
6	13	1320–1713	low at 1400	southwest

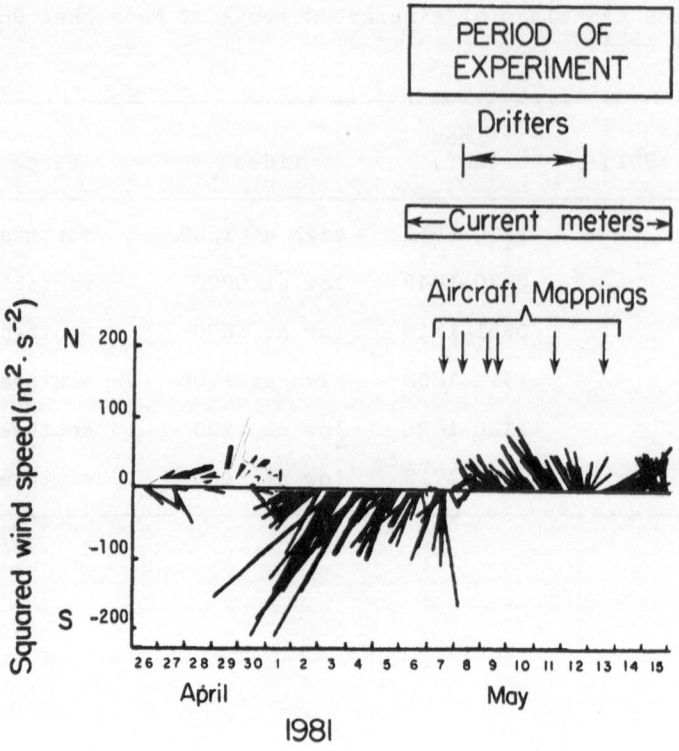

Figure 4. Winds at Nantucket Airport, April-May, 1981. Vector magnitudes are square of wind speed which is proportion to wind
stress. Period of current meter and drifter deployments,
and aerial surveys are indicated.

Spatial patterns in chlorophyll for the same 4 flights are shown
in Figure 6. Chlorophyll concentrations remained stable throughout
the period of the experiment, with the highest levels staying at
about 3 mg m^{-3}. There were no patches directly east of Nantucket
Island as there had been in the earlier cruises. The only area of
high chlorophyll (>2.5 mg m^{-3}) over the Shoals was located along the
southeastern flank where water depths ranged from 30 to 50 m.
Phytoplankton patches were observed in the same vicinity in the May
cruises of 1978 and 1979 (Fig. 3) and also in September, 1978 (Limeburner et al., 1980).

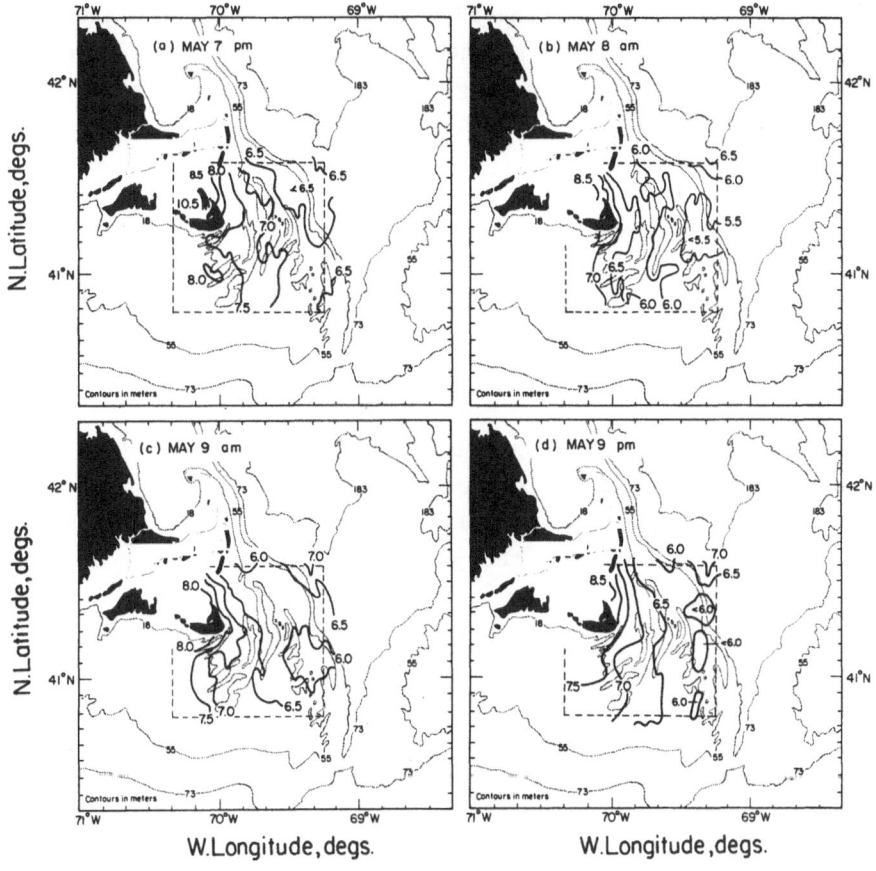

Figure 5. Aerial maps of sea surface temperature (°C) based on first
      four mappings, May 7-9, 1981.

On the first 3 aircraft mappings, a second area of high surface
chlorophyll was observed north of the Shoals in deeper waters of
Wilkinson Basin. Flight lines were later shifted to the south to
concentrate on the southern region, and the northern region was not
sampled on later flights (e.g. 6-d).

Like the temperature patterns, chlorophyll exhibited temporal
variability on a time scale of hours but its variation was more

systematic. Phytoplankton behaved as a passive substance advected by rotary tidal currents. The southern patch moved a distance of 15-18 km during a tidal cycle, returning to the same location at the end of the cycle. This agrees with the tidal excursions (14-22 km) estimated from current meter records.

Ship profiles revealed the water column to be well-mixed over the Shoals. Albatross stations 10 and 11 (see Fig. 2) north and east of the Shoals were located in regions of deep vertical mixing, and the transition from well mixed to stratified began just south of station 44.

The ship sampling was too sparse to reveal the localized chlorophyll high seen in aerial surveys along the southeastern flank (Fig. 6). A map of surface chlorophyll derived from Albatross stations (Thomas, 1981) did not show the patch, although there were measurements of 2.34 and 2.81 mg chl m^{-3} in that vicinity, and lower values (0.78 and 1.16 mg chl m^{-3}) immediately north and south of there.

Advective Transport

Mean currents from the 6 current meters are compared in Figure 7 with mean current vectors from the 1978-79 cruises (Limeburner and Beardsley, 1982). In general, flow patterns in May 1981 were consistent with the earlier data, although rates of non-tidal advection in 1981 were somewhat less. A closer examination of the current meter data reveals 2 distinct periods associated with the different wind patterns.

This may be seen in the progressive vector diagrams for the upper and lower current meter records at mooring A (Fig. 8). The residual flow was greater for the first 2 days when winds were from the northeast, and it slowed considerably during the period of southeast winds. This pattern appeared in all three mooring records. Estimates of mean flow for these two periods are listed in Table 2 (cols. 3 and 4), along with mean flow for the whole period (col. 5).

The drifters deployed at mooring A traveled a distance of 17 km over the 102 h deployment and thus had a mean velocity of 4 km d^{-1}. Only 2 drifters were recovered at the end of this period, and these

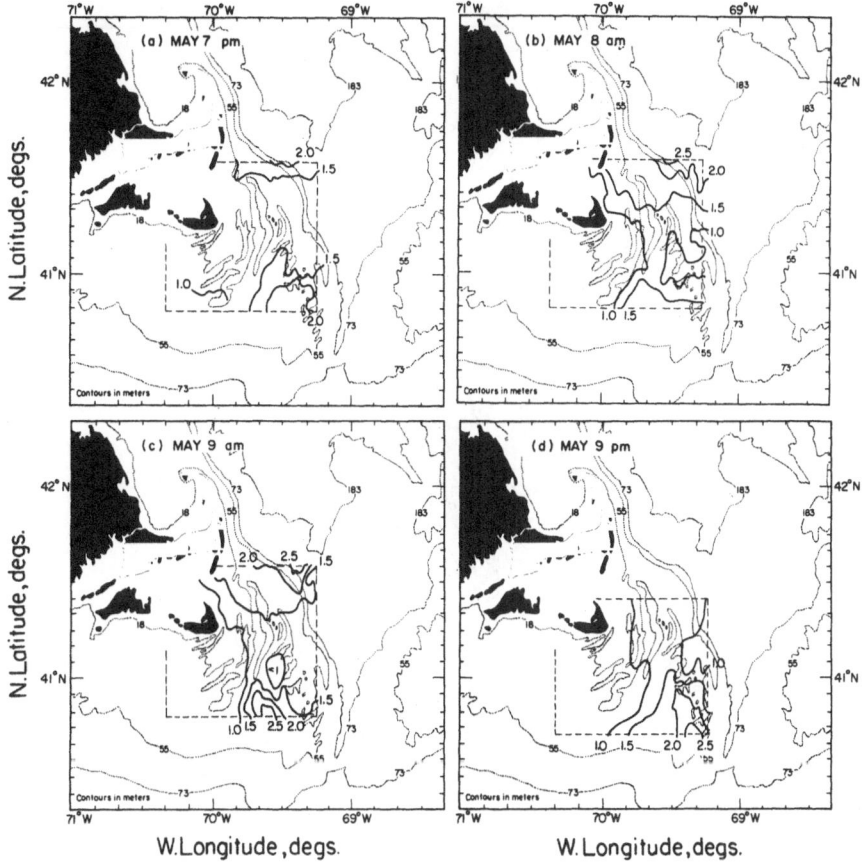

Figure 6. Aerial maps of chlorophyll concentration (mg m^{-3}) based on first four mappings, May 7-9, 1981. (Note: The area mapped was shifted southward on the last flight.)

were nondispersed. They followed a westward course, apparently influenced by the steady southeast winds. Assuming the rate of advection due to winds was 3% of windspeed, the entire displacement could be attributed to windage (Limeburner, 1981).

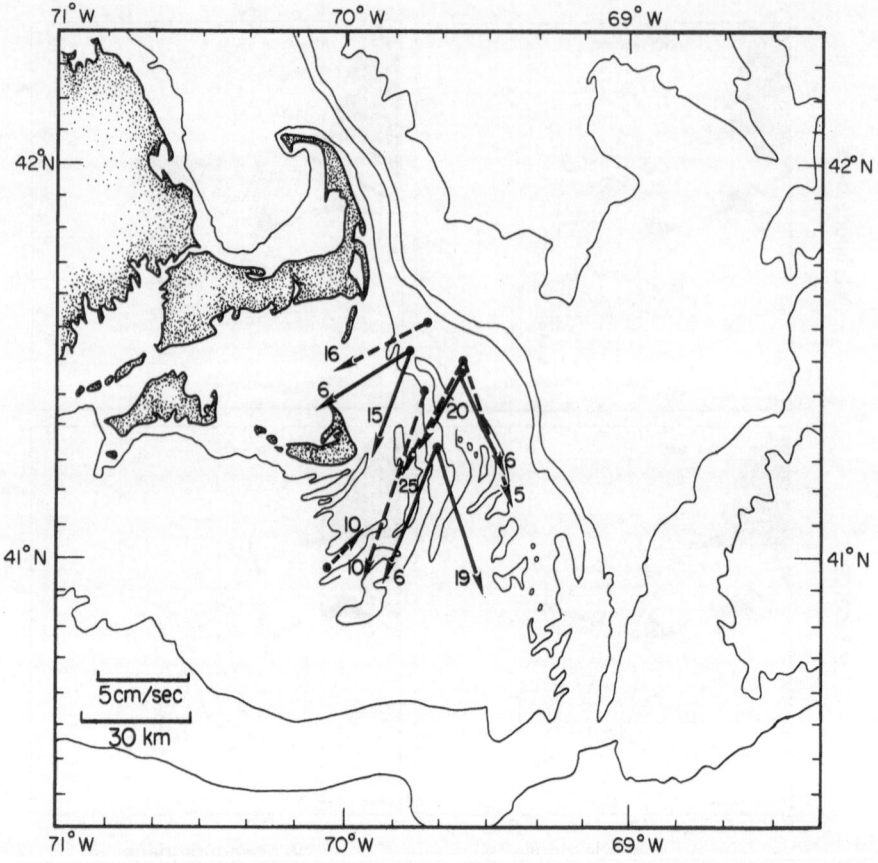

Figure 7. Mean currents from moored current meters. Solid vectors
          are mean currents from May 6-15, 1981 (Limeburner, 1981)
          and dashed vectors are from moorings in January-March and
          July-August, 1979 (Limeburner and Beardsley, 1982).
          Numbers at end of vectors are depth (m) of current meter.

Nutrients

        Nutrients were lower over the Shoals as compared to surrounding
waters. Column averaged nitrate + nitrite concentrations, for
example, ranged from 3.41 mg-at m^{-3} at station 11 to .09 mg-at m^{-3} at
stations 44 and 51. Concentrations decreased monotonically in the
direction of the mean flow (i.e. northeast to southwest).

156

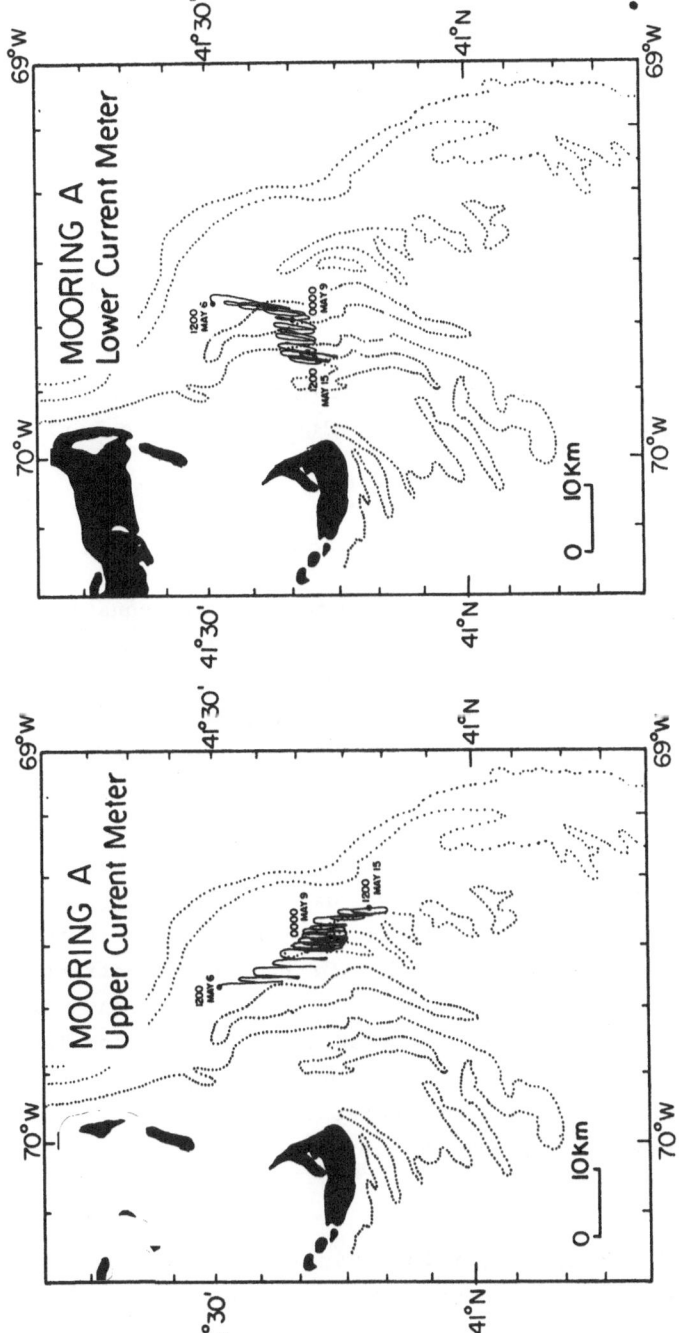

Figure 8. Progressive vector diagrams of current meter records at mooring A. (a) Upper meter and (b) lower meter. Before 0000 h on May 9, there were strong winds from the northeast. Afterwards, winds shifted to be from the southeast at 20 kts (see Figure 4).

Growth Rates

The 24-h incubation experiments yielded PP:CHL values of 8.6 mg C $(mg \: chl \: d)^{-1}$ on May 9 and 12.5 mg C $(mg \: chl \: d)^{-1}$ on May 12. These converted to carbon-specific doubling rates (C:CHL = 50) of 0.23 d^{-1} and 0.32 d^{-1}, respectively, for an average of 0.28 d^{-1}.

To gain perspective as to whether these rates applied to other locations on the Shoals, we examined the variability in light satur-ated photosynthesis rates (P_{max}) measured at each of the stations in Figure 2. In samples collected at approximately the same time of day, there existed as much as a factor of 4 difference in P_{max} at different stations. Thus, spatial variability may have been signi-ficant. Accordingly, we will use the value μ = 0.28 d^{-1} with some caution.

Displacements: A Test of the Conceptual Model

To test the nutrient-growth displacement model (equation 1), we took station 10 (1.2 mg chl m^{-3}) and station 11 (1.3 mg chl m^{-3}) to be representative of source waters, so that C_0 = 1.25 mg m^{-3}, and we let C_1 = 3.0 mg m^{-3} be the maximum chlorophyll concentration in the patch. If the chlorophyll doubling rate was μ = 0.28 d^{-1}, then 4.5 days were required for chlorophyll levels to increase from C_0 to C_1.

Figure 9 shows the position of the patch on May 8-9 relative to the location of cold spots on those days. The displacement is approximately 45 km in the direction of the mean current at mooring C (upper meter). Meteorological conditions had been constant for a period of at least 5 days prior to May 8. There were strong northeast winds and non-tidal currents on the order of 10 km d^{-1} (see Table 2). If the position of the cold spot shown in Figure 9 is representative of the "daily average" position of source waters during this period, then the displacement of 45 km is consistent with our hypothesis (equation 1).

Over the next 5 days (May 9-13), winds were from the southeast and residual currents slowed considerably. If the mean current was 4 km d^{-1} during this interval, then by May 13, a patch of the same concentration should have been located 18 km downstream of the source. This is not what we observed in the May 13 mapping. At that

Table 2. Mean currents, May 6-15, 1981.

| Mooring | Depth (m) | Mean currents (km d^{-1}) | | |
		May 6-8	May 9-15	May 6-15 (average)
A	6	8.8	3.1	4.9
A	20	7.4	2.0	3.4
B	6	6.6	3.9	4.5
B	15	6.4	3.7	4.4
C	6	14.1	3.7	6.7
C	18	11.2	5.4	6.9

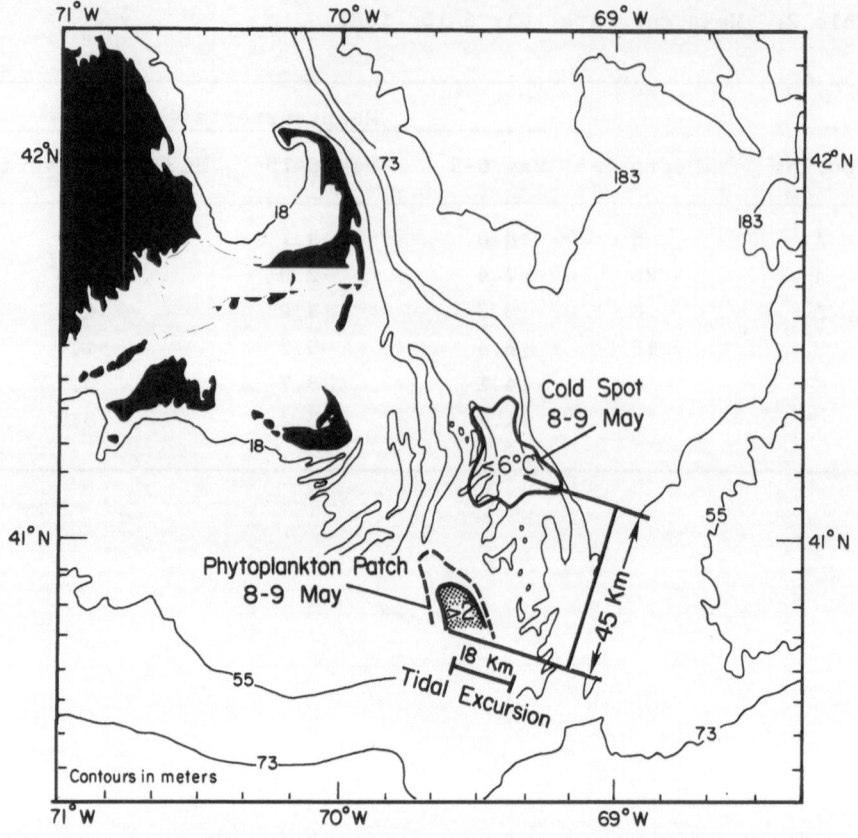

Figure 9. Positions of phytoplankton patch and cold spots on May 8-9, 1981. Displacement of 45 km is consistent with 10 km d^{-1} advection rate and .26 chlorophyll doublings per day (see equation 1).

time, a chlorophyll maximum of approximately the same concentration existed midway between the position of the patch at low and high tide on May 9. Thus, we found no evidence of any significant change in position or abundance of phytoplankton between May 8 and May 13.

CONCLUSIONS

We began with the hypothesis that phytoplankton patches develop downstream of localized surface temperature minima ("cold spots") that mark the site of deep vertical mixing along the boundary of the Shoals. What we observed, however, was not fully consistent with this concept. In general, there was a lack of consistency between the variability observed in certain physical processes and that observed in chlorophyll distributions. Specifically:

(1) The only phytoplankton patch observed over the Shoals behaved as a passive substance advected by rotary tidal currents. Its mean position and chlorophyll level remained nearly constant over the 8 days of measurements.

(2) Local minima in surface temperatures appeared along the northern and eastern boundaries of the Shoals as predicted, but their spatial patterns changed on a time scale of hours. According to our hypothesis, the variability in position and intensity of tidal mixing would imply a similar variability in the supply of nutrients, and, hence, in the distribution of phytoplankton.

(3) The temporal variability observed in residual currents was, likewise, not consistent with the relative stability of phytoplankton distributions. At the current meter moorings, the direction of non-tidal advection remained southwesterly throughout the period of deployment (May 6-15, 1981), but the rate of flow changed signifi-cantly with a change in meteorological conditions on May 8. If currents in the vicinity of the phytoplankton patch were similar, then our hypothesis would predict a corresponding translation in the patch position between May 8 and May 13.

The relative stability of the phytoplankton distribution along the southeastern flank of the Shoals suggests a local source of nutrients in that vicinity rather than the hypothesized displacement. Since we did not measure currents or growth rates in the patch, we cannot be certain that the hypothesis was wrong. The hypothesis could still be valid if non-tidal currents were slower and/or growth faster than those rates measured elsewhere on the Shoals.

ROLE OF AIRCRAFT REMOTE SENSORS

Airborne remote sensors can address the dynamics of phytoplank-ton on a time scale comparable with the biological and physical processes affecting growth (Esaias, 1980). Spatial patterns in temperature and chlorophyll derived from the ship surveys in 1978-79, though similar in appearance to the synoptic maps, contained tidal aliasing. For example, what appeared to be 2 phytoplankton patches along the southern portion of the Shoals in the May-June 1978 data may actually have been the same patch sampled at different phases of the tide.

Aircraft are the only remote sensing platforms capable of providing subtidal coverage with adequate spatial resolution to study phytoplankton dynamics on scales of 10-100 km. Subtidal coverage was required to measure the tidal excursions of patches. This motion had to be estimated and thus accounted for before net non-tidal change could be assessed. Similarly, subtidal coverage revealed the dynamic nature of tidal mixing processes along the perimeter of Nantucket Shoals.

A major advantage of aircraft is their flexibility. They are best used in an interactive program with other platforms to help direct ship sampling and to position moorings. Ironically, it was lack of flexibility in the Nantucket Shoals Experiment that resulted in inconclusive results. Because of the number of investigators involved (over 50 scientists, engineers and technicians), the expense of operating a large turboprop aircraft, and other factors that complicated the logistics, modifications to sampling plans made in advance of the experiment were difficult to effect.

We expected to find a phytoplankton patch east of Nantucket Island as there had been in 1978 and 1979. The mooring positions were selected to be upstream of this patch, and the small vessels based at Nantucket Island could have thoroughly sampled the patch and surrounding waters. After the aerial surveys determined the only patch to be located along the southeastern flank, the moorings had already been deployed and the patch was out of range of the smaller vessels.

A better strategy would have been to have made several aerial surveys at subtidal frequencies before establishing the mooring positions and ship sampling plans. Based on these surveys, a much wiser in situ sampling plan could have been devised. Fewer surface vessels, fewer investigators and a smaller aircraft (thus more hours of air time for same cost) would make this strategy more practicable.

For airborne remote sensors to become an integral component of field experiments in biological oceanography, operational systems have to be developed that are less complex and less expensive to use than those systems in current use today. It is well within present state-of-the-art technology to develop such a system. The sensors used to provide temperature and chlorophyll in the Nantucket Shoals Experiment could be integrated into a small system, together with a Loran C and data acquisition computer. The system would require only one operator and could fly on a twin-engine aircraft. Operational costs could be reduced by an order of magnitude or more.

In summary, there is great potential for using airborne remote sensors to study plankton dynamics in tidally mixed systems or in other systems where time scales are on the order of hours. These sensors have the unique capability of measuring temporal variability in spatial patterns. They are best used in conjunction with ships and other platforms, to help establish sampling plans, and their operation should be simplified to allow for flexibility.

REFERENCES

Campbell, J.W. and J.P. Thomas. (Eds.) 1981. Chesapeake Bay Plume Study: Superflux 1980. NASA CP-2188, 522 p.

Campbell, J.W. and W.E. Esaias. 1983. Basis for spectral curvature algorithms in remote sensing of chlorophyll. Appl. Optics. 22(7): 1084-1093.

Campbell, J.W. and W.E. Esaias. 1985. Spatial patterns in temperature and chlorophyll over Nantucket Shoals, May 7-9, 1981. J. Mar. Res. 43: 139-161.

Esaias, W.E. 1980. Remote sensing of oceanic phytoplankton: Present capabilities and future goals. In: Primary Productivity in the Sea, pp. 321-337. P.F. Falkowski (ed.). Plenum Press.

Hoge, F.E. and R.N. Swift. 1981. Airborne simultaneous spectroscopic detection of laser-induced raman backscatter and fluorescence from chlorophyll and other naturally occurring pigments. Appl. Optics 20(18):

Limeburner, R. 1981. Nantucket Shoals current drifter study. Final Rept. on Contract between NASA and WHOI.

Limeburner, R. and R.C. Beardsley. 1982. The seasonal hydrography and circulation over Nantucket Shoals. J. Mar. Res. 40(suppl.): 371-406.

Limeburner, R., R.C. Beardsley and W.E. Esaias. 1980. Biological and hydrographic station data obtained in the vicinity of Nantucket Shoals, May 1978–May 1979. WHOI Tech. Rept. WHOI-80-7, 87 pp.

Malone, T.C., P.G. Falkowski and T.E. Whitledge. 1981. Support study for dynamics of phytoplankton patches on Nantucket Shoals. Final Rept. Brookhaven Natl. Lab., Upton, NY, 111 p.

Phinney, D. and K. Kilpatrick. (Eds.) Nantucket Shoals Experiment, R/V Albatross IV, May 5-6, 1981 (Data report). Bigelow Lab. for Ocean Sciences Tech. Rept. #21. ISSN 0273-2149. 109 pp.

Steele, J.H. 1978. Spatial Patterns in Plankton Communities. Plenum Press, NY, 1-20.

Strickland, J.D.H. and T.R. Parsons. 1972. A Practical Handbook of Seawater Analysis. Fish. Res. Bd. Can., Bull. 167, Ottawa, Canada.

Thomas, J.P. 1981. Cruise Results, NOAA R/V Albatross IV, Cruise No. AL IV 81-04 (Data report). NOAA/NMFS Sandy Hook Laboratory, Highlands, NJ, 49 pp.

Wilson, R.E. 1981. Current observations on Nantucket Shoals, May 6, 1981 to May 15, 1981. Final Rept. on Contract between NASA and Marine Science Research Center, SUNY, Stony Brook, NY.

THE INFLUENCE OF TIDAL MIXING ON THE TIMING OF THE SPRING PHYTOPLANKTON DEVELOPMENT IN THE SOUTHERN BIGHT OF THE NORTH SEA, THE ENGLISH CHANNEL AND ON THE NORTHERN ARMORICAN SHELF

R.D. Pingree and G.T. Mardell
Marine Biological Association of the United Kingdom
Citadel Hill, Plymouth PL1 2PB, U.K.

P.C. Reid and A.W.G. John
Institute for Marine Environmental Research
Prospect Place, The Hoe, Plymouth PL1 3DH, U.K.

INTRODUCTION

Hydrographic surveys of water properties have been conducted in March for two consecutive years (1983 and 1984) to examine the timing of the spring phytoplankton development in relationship to environmental parameters. A wide range of physical conditions was encountered in the two March surveys which stretched from the clear open ocean conditions of the Bay of Biscay to the shallow turbid waters of the Southern Bight of the North Sea. Over the shelf region surveyed, the water depth (Fig. 1) changed by an order of magnitude (20 m - 200 m), the beam attenuation coefficient had a 20 fold change (0.4 m^{-1} - 9 m^{-1}) and the magnitude of the M_2 tidal currents (Fig. 2) (Pingree, 1983) varied by a factor of 25 (10 cm s^{-1} to 250 cm s^{-1}). Vertical differences of salinity, ΔS, varied from mixed to $\Delta S \sim 3$ $^{o}/oo$. The observed chlorpohyll 'a' distributions are discussed in terms of regional changes in these variables which effectively control the availability of light with a view to providing some sea truth for the future development of realistic models. Numerical models simulating the evolution of the temperature structure in the English Channel (see for example, Agoumi et al., 1983) are well advanced. However, corresponding models illustrating the regional development of phytoplankton still require representative distributions of chlorophyll 'a' (and their evolution in time) as well as the range of physical conditions likely to be encountered in the region.

Lecture Notes on Coastal and Estuarine Studies, Vol. 17
Tidal Mixing and Plankton Dynamics. Edited by J. Bowman, M. Yentsch and W.T. Peterson
© Springer-Verlag Berlin Heidelberg 1986

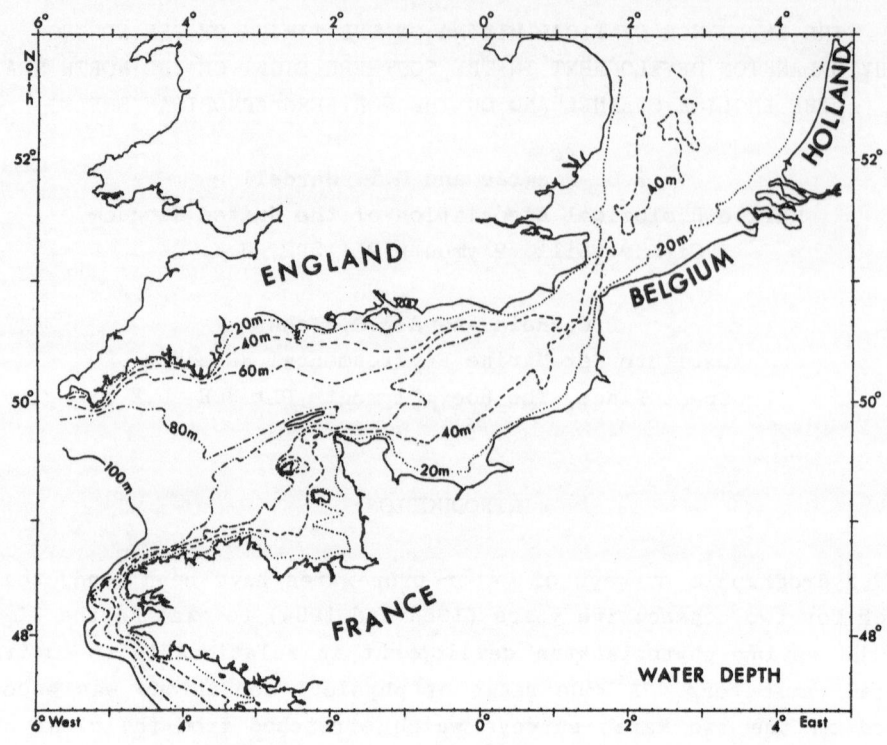

Figure 1. Chart showing water depths (m) in the English Channel and
Southern Bight of the North Sea.

METHODS

Surveys aimed at determining the timing of the spring phytoplankton outburst in the Southern Bight of the North Sea and the
English Channel were made from the R/V Frederick Russell between 7-18
March 1983. Further surveys were made from the R/V Frederick Russell
between 5-13 March 1984, and covered a region from the Bay of Seine
in the east to the open oceanic regions of the Bay of Biscay in the
west.

Continuous underway measurements of sea-surface temperature and
salinity were made obtained by pumping water (from a depth of 3.5 m)
into a deck tank containing an STD probe. A further deck tank
contained a Sea Tech 25 cm beam transmissometer which used a light
emitting diode to provide a light source with a wavelength of 660 nm.

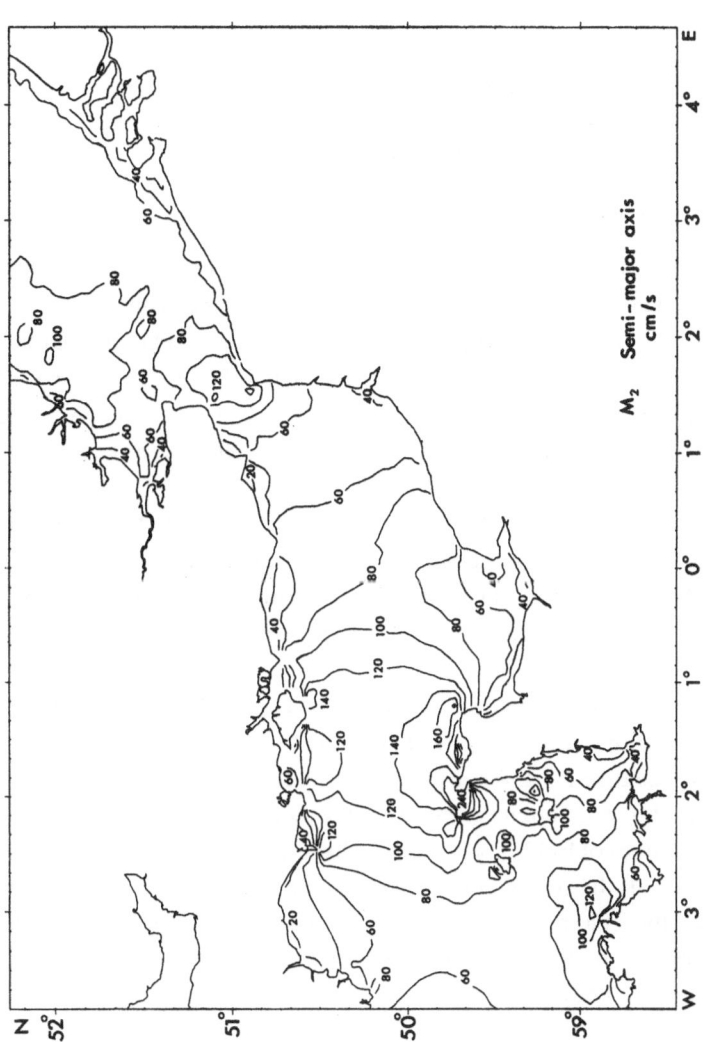

Figure 2. Semi-major axis (cm s^{-1}) of the M$_2$ tidal currents (vertically integrated numerical model).

Some of the pumped water was also passed through a Turner Model III fluorometer, fitted with filters appropriate for detecting chlorophyll 'a' fluorescence, and through an analyzer for continuous measurements of nitrite and reactive silicate.

RESULTS

Contour maps of the measured properties and the cruise track of the R/V Frederick Russell over the period of the measurements from 7-18 March 1983 are presented in Figs. 3 and 4.

In the Western English Channel the sea surface temperature is about average for this time of year, though in the Southern Bight of the North Sea the values are about $\sim 0.5^{\circ}C$ cooler than average (Dietrich, 1962). The salinity distribution shows the characteristic salty tongue stretching along the Channel though the salinity maximum in the eastern English Channel is somewhat below average since the $35^{\circ}/oo$ contour usually extends as far east as the Strait of Dover.

Regions with increased levels of chlorophyll 'a' occurred along the eastern side of the Bay of Seine, apparently associated with the run-off from the Seine, in the shallow waters near the Bay of Somme (and the nearby crescent shaped bank "Bassure de Bass") and over an extended area in the offshore regions of the Southern Bight of the North Sea (Gieskes and Kraay, 1975; Holligan et al., 1976). In general, the surface values of chlorophyll 'a' are representative of the whole water column. However, at station 15 (on 12 March 1983, see Fig. 3) in the northern part of the Southern Bight of the North Sea, levels of chlorophyll 'a' near the sea floor exceeded surface values, reaching as high as 17 mg m^{-3}. The bloom here was probably well advanced by this time, the subsurface structure possibly indicating previous higher concentrations of phytoplankton which had now sunk to the sea bed.

The reactive silicate distribution showed an inverse correlation with salinity with highest levels indicating the outflow of the river Seine in the English Channel and the combined outflows of the Rhine, Meuse and Schelde in the Southern Bight of the North Sea. The area with the lowest values of reactive silicate approximately coincided with the zone where maximum levels of chlorophyll 'a' were found in

168

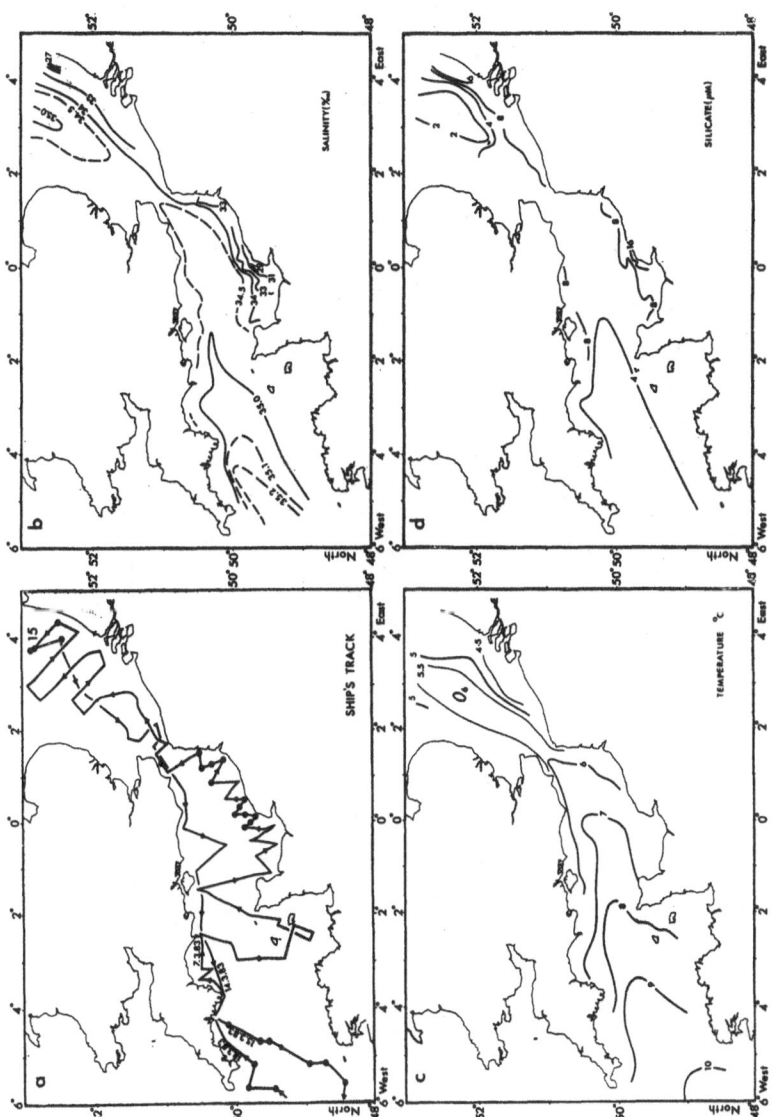

Figure 3. (a) Track of the R.V. Frederick Russell (March 1983). Dots show station positions. (b) Surface salinity (^{o}/oo). (c) Surface temperature (^oC). (d) Surface reactive silicate (μM).

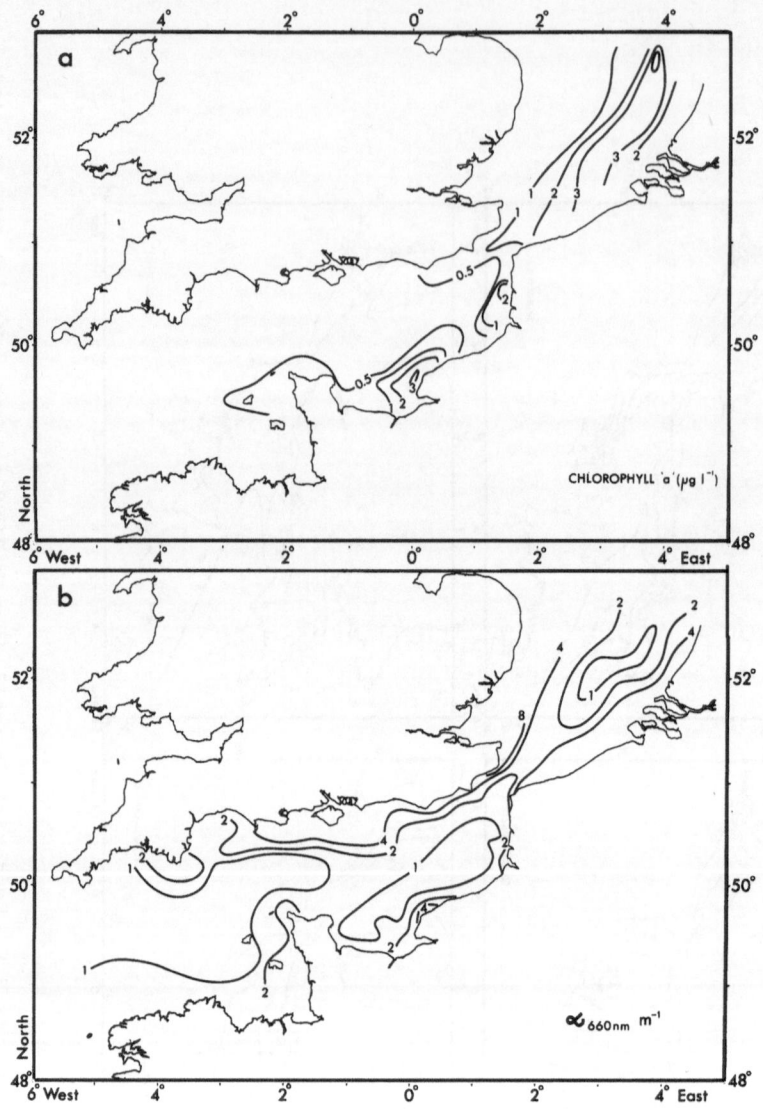

Figure 4. (a) Surface chlorophyll 'a' (μg ℓ$^{-1}$). (b) Surface beam
attenuation coefficient, α, for red light.

the Southern Bight, supporting the idea that the bloom was already
well advanced in this region.

When considering the potential for growth by photosynthesis in
mixed waters, the relevant light measurement is the scalar irradiance

and its associated vertical attenuation coefficient. The data on the optical properties of sea water, however, were limited to beam transmission from which the beam attenuation coefficient α (m^{-1}) was derived. The beam transmissometer effectively measures the loss of photons in a narrow parallel beam of light (at 660 nm) due to absorption and scattering. The light transmission measurements are continuous so the data are compatible with the other continuously monitored sea-surface properties. The beam attenuation levels show the expected increases in shallower water where more particles in suspension in the water column increase the attenuation by scattering. However, the values have not been corrected for forward scattering so may underestimate the true attenuation by about 25% for this instrument in very turbid conditions (see Bartz et al., 1978). The overall distribution is thought to represent conditions at the start of the productive season though it is clear there will be temporal variations due to tidal state and local weather conditions. In the absence of phytoplankton blooms, the coastal values of attenuation coefficient would tend to be highest in the winter period when increased swell and run off are likely to make a significant contribution to the material in suspension. The minimum in the values of beam attenuation coefficient in the eastern English Channel is displaced towards the coast of France and the English coastal regions are observed to be much more turbid (less transparent) than the French coastal regions. The turbidity minimum in the Southern Bight (Lee and Folkard, 1969) is also displaced towards the coast of Holland and is some 30 km to the east of the salinity maximum. Clearly the levels of phytoplankton in the water column will affect the beam attenuation coefficient. The values of the beam attenuation coefficient in the regions of increased chlorophyll 'a', between the shallow turbid waters and the deeper clear waters, would tend to be lower at the start of the bloom when there is less absorption and scattering by the phytoplankton themselves.

March 1984

In the English Channel, the March 1984 sea surface contours of temperature and salinity (Fig. 5) showed similar patterns to the March 1983 distributions, though the isohalines indicated reduced run off compared with the previous year with salinity values about $\sim 1^{o}/oo$ higher coastally in the Bay of Seine. The mean discharge from the Seine for the winter months, December, January and February, was

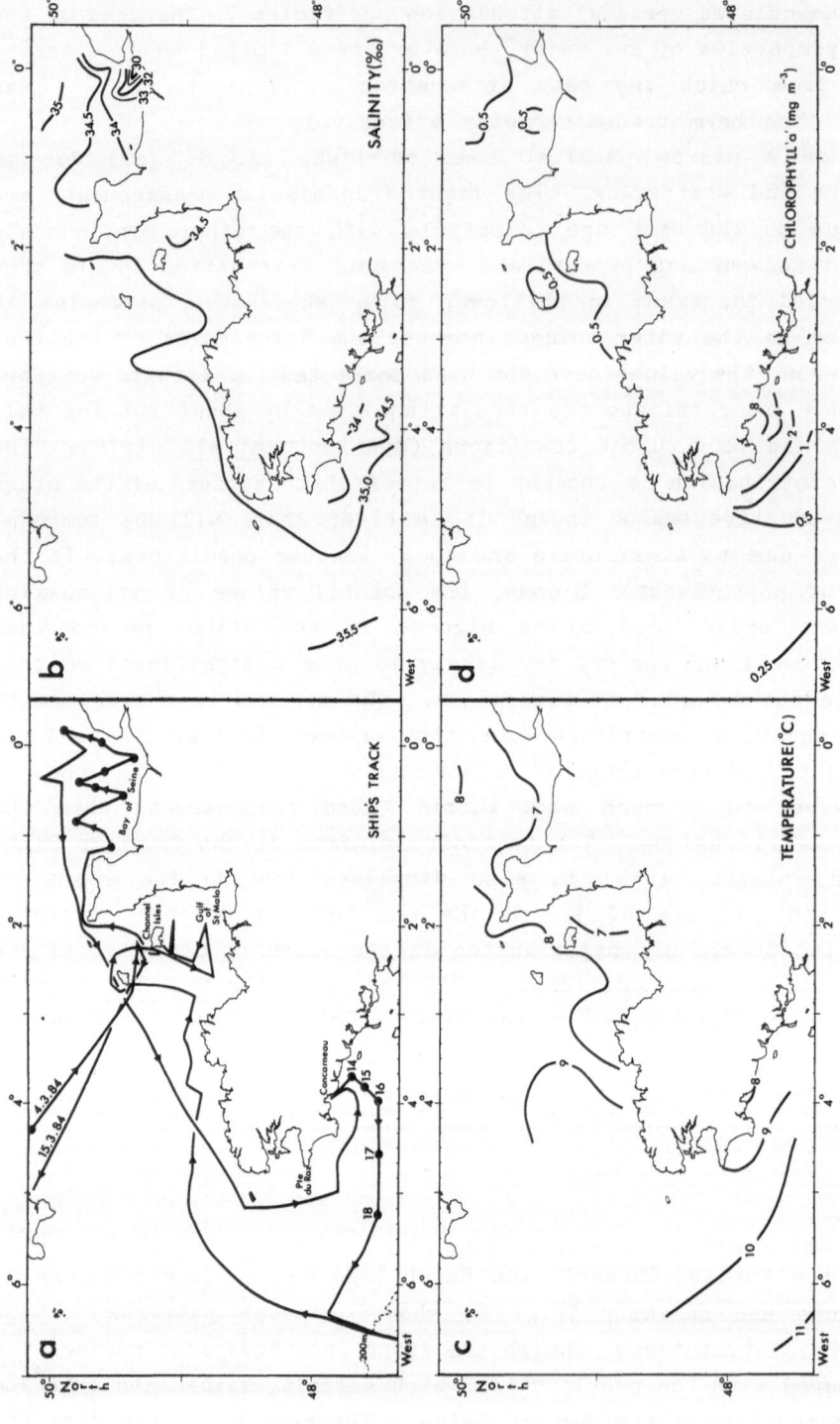

Figure 5. (a) Track of the R/V Frederick Russell (March 1984). Dots show station positions. (b) Surface salinity (⁰/oo). (c) Surface temperature (°C). (d) Surface chlorophyll 'a' (mg m⁻³).

about 750 m^3 s^{-1} in 1983 compared with about 600 m^3 s^{-1} in 1984. The influence of run off was marked coastally along the Armorican shelf as far north as Pte du Raz where the cold fresh surface waters indicated the northward penetration of run off from the river Loire.

The chlorophyll 'a' sea surface distributions (Fig. 5) showed some marginal increases in the shallowest regions of the Channel Isles. In the Bay of Seine the chlorophyll 'a' levels only reached 0.6 mg m^{-3} so there were clearly some differences in timing between the years 1983 and 1984 since both these surveys were conducted on the same days (8 and 9 March). The most noticeable increases of chlorophyll 'a' however, occurred coastally along the northern part of the Armorican shelf near Concarneau where surface levels reached 8 mg m^{-3}. The subsurface structure in this region showed maximum values of chlorophyll 'a' on the halocline at a depth of about 15 m where values of 20 mg m^{-3} chl 'a' were recorded at St. 14 near the coast (Fig. 6). At this station the surface to bottom salinity differences were 1°/oo and the temperature increased with depth by 1.3^{o}C. The influence of this layering due to run off from the river Loire extended to St. 16 on the 100 m contour where the isohalines outcropped at the sea surface about 40 km from the coast. Nitrate and silicate values were relatively high in the surface waters so the increased levels of chlorophyll 'a' at the halocline could not be a response to the availability of nutrients and might reflect sinking of a surface bloom.

The light transmission data (Fig. 7) showed, in agreement with the 1983 results, the clear ($\alpha \sim$ 1 m^{-1}) central waters of the English Channel in marked contrast to the coastal regions of the Gulf of St Malo and Bay of Seine where the beam attenuation coefficient, α, exceeded 4 m^{-1}. These results were also supported with further measurements of beam attenuation coefficient at 430 nm. In 1984 coastal conditions in the Bay of Seine were rather more turbid than in 1983 despite lower levels of chlorophyll 'a' and reduced run off, and this is attributed to the fact that the 1984 observations were made a few days after spring tides whereas the 1983 results were obtained at neaps. The clearest coastal waters were found on the Armorican shelf near the bloom region showing that values of chlorophyll 'a' as high as 8 mg m^{-3} had a much smaller effect on beam attenuation than the levels of material in suspension in the other more tidally mixed areas. At the shelf break the beam attenuation

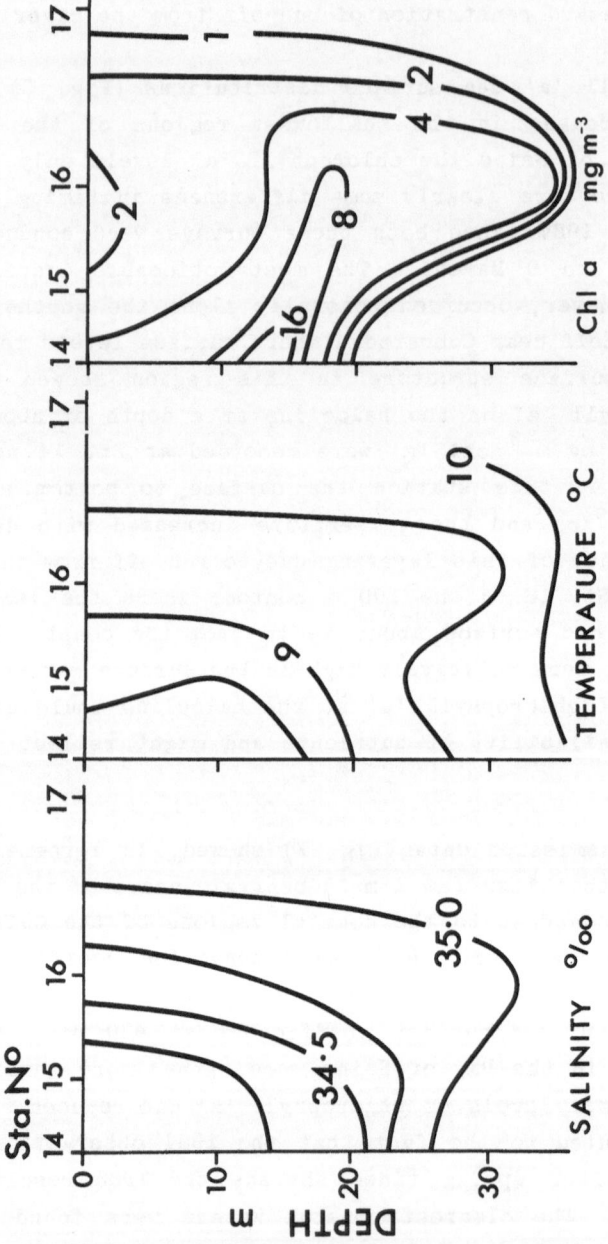

Figure 6. Vertical sections of salinity, temperature and chlorophyll 'a'. See Figure 5(a) for positions of stations 14,15,16, and 17.

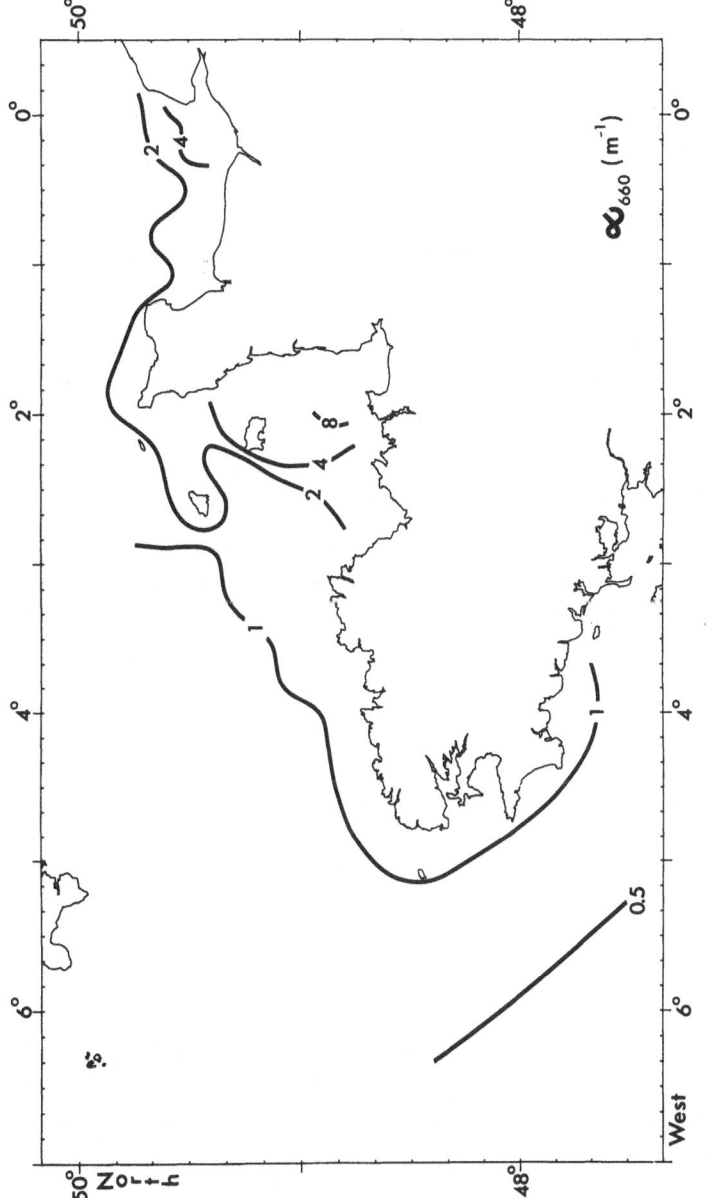

Figure 7. Surface beam attenuation coefficient, α (for red light), for the March 1984 measurements made along the track shown in Figure 5(a).

fell to values as low as 0.4 m^{-1} tending towards the absorption coefficient ($\sim$ 0.35 m^{-1}) for pure water at this wavelength.

DISCUSSION

Light

For a given surface irradiance the mean light levels experienced by the phytoplankton depend on the depth of the surface mixed layer, h_m, and the vertical extinction coefficient, k. These two parameters are in part related to tidal mixing since the strength of the tides determines both whether the whole water column is mixed and how much material there is in suspension and therefore the mean ambient light levels.

(a) Optical measurements

From the range of values of the beam attenuation coefficient encountered it is possible to make an estimate of the changes in the vertical extinction coefficient that might have occurred over the regions surveyed. Although this requires some approximations it is pursued here for illustrative purposes which can be refined when actual in situ measurements in different parts of the visible spectrum become available.

A correlation existed between the depth, D, at which a Secchi Disc just disappears from sight and the beam attenuation coefficient, α (Højerslev, 1977). Figure 8 shows that the relationship between beam attenuation coefficient and Secchi Disc can be approximately represented by the curve

$$\alpha D = 11 \tag{1}$$

There was also a close linear correlation between the beam attenuation coefficient at 660 nm (α_r) and the beam attenuation coefficient measured at 430 nm (α_b) so a similar relationship could be derived for blue (430 nm) light.

Basically the illumination of a submerged Secchi Disc depends on the extinction coefficient, k_D, for the downwelling light and on the beam attenuation, α, for the light coming from the disc to the

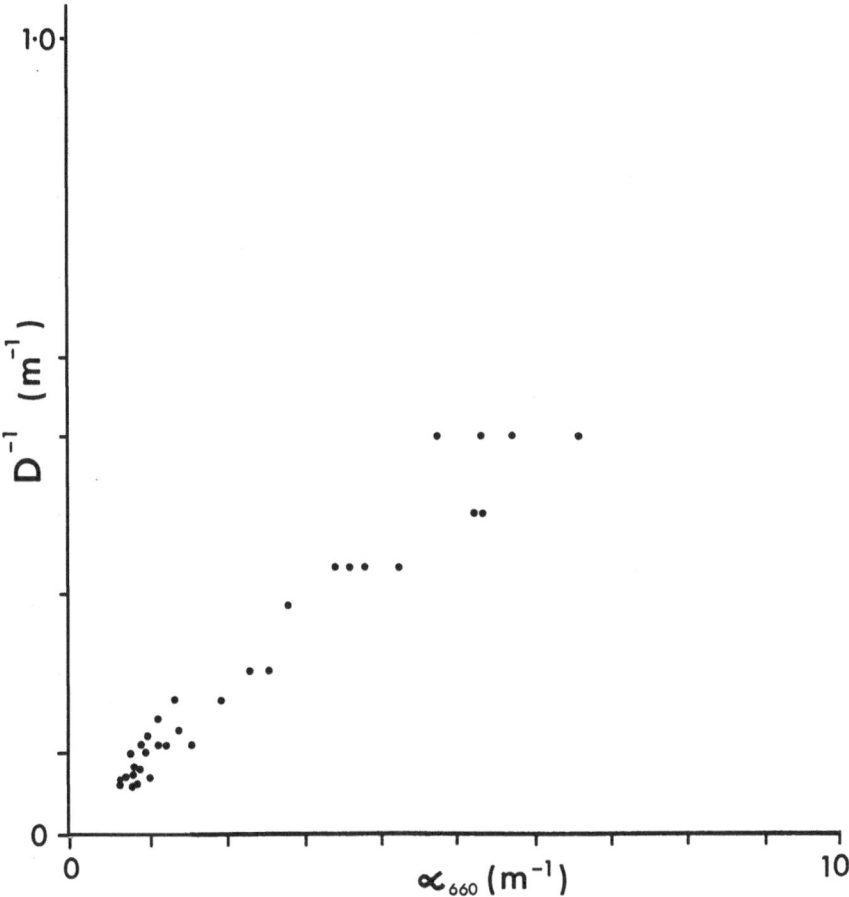

Figure 8. Plot of I/D against α.

observer so the visibility of the disc depends on the sum $(\alpha + k_D)$. Unless α and k_D are proportional there cannot be a universal relationship between k_D and D. Le Grand (1939) showed on theoretical grounds that

$$k_D D = \gamma \qquad (2)$$

where γ was in the range of $1.5 < \gamma < 4$.

This result is largely supported by the work of Holmes (1971) ($\gamma \sim 1.3$ in very turbid water with $D \sim 2$ m), Poole and Atkins (1929) ($\gamma \sim 1.7$ English Channel with $D \sim 14$ m) and Tyler (1968) ($\gamma \sim 2$, Manning Cruise 36 data with $D \sim 30$ m). Thus empirically, very approximately (2) becomes

$$k_D D \sim 0.2 \, D^{\frac{1}{2}} + 1 \tag{3}$$

where D is the numerical value of the Secchi Disc in meters and k_D is the numerical value of the vertical extinction coefficient in m^{-1}.

Substituting (1) in (3) gives a relationship between extinction coefficient and beam attenuation coefficient

$$k_D \sim 0.6\alpha^{\frac{1}{2}} + 0.09 \, \alpha \tag{4}$$

suggesting that a 4-fold change of α from about 1 to 4 m^{-1} in Fig. 4b might correspond with similar, but slightly reduced, change in k_D ($\sim$ 3 fold).

Since the changes in α are as large as changes in water depth, h, the productive season would not necessarily start first nearest the coast. In the Southern Bight of the North Sea, for example, the mean light level in a mixed water column actually increases with increasing water depth reaching a maximum about 50 km from the Dutch coast where the highest chlorophyll 'a' values were recorded in the March 1983 cruise.

(b) Suspended material

The material in suspension (seston) in the water column, which largely determines the light levels at a fixed depth, was measured by filtering fixed quantities of sea water drawn from samples taken at two hourly intervals. Subsamples were also extracted for phytoplankton and Coulter Counter analysis.

Over a limited region a broad correlation exists between the material in suspension and the beam attenuation coefficient, α (Fig. 9). In the March 1984 cruise there were, very approximately, about 6 gm of seston per m^3 in the more turbid regions near the Channel Isles where $\alpha \sim 6 \, m^{-1}$ (Fig. 7). Much higher levels of seston were encountered in some regions on the March 1983 cruise. A value of 18 gm m^{-3} was found near the Isle of Wight and a value of 20 gm m^{-3} was recorded off North Foreland near the entrance to the Thames estuary. About 70% of this material was inorganic.

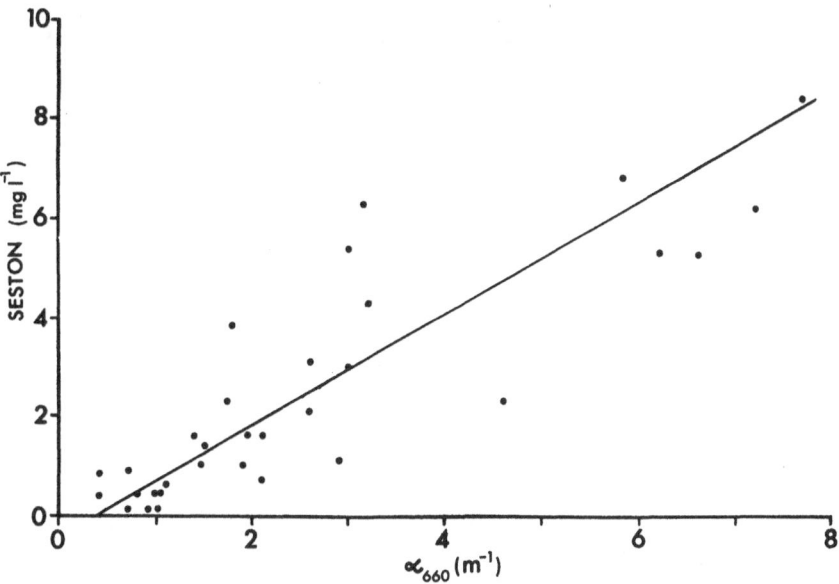

Figure 9. Weight of suspended material (seston) per litre against beam attenuation coefficient, α, for the March 1984 results.

A striking visual correlation between the beam attenuation coefficient (Fig. 4b) and the inorganic content of the suspended material can be observed simply by placing the ashed filters on a map at the positions where the samples were collected (Fig. 10). All the light shaded filters occur where the beam attenuation coefficient is low and the contrast between the English coast (where the filters are dark) and the French coast is marked.

Coulter counter analysis shows that the darker filters had higher particle counts and that the concentration of the suspended material in a given size range increased as the particle size decreased. Some representative particle size distributions for the English coastal waters, the central Channel and French coastal regions (see samples W, B, H, Fig. 10) are shown in Fig. 11. In the Southern Bight of the North Sea similar distributions occurred (S, P, L) only now the continental coastal waters were more turbid than the French coastal Channel waters (compare H and L, Figs. 10, 11).

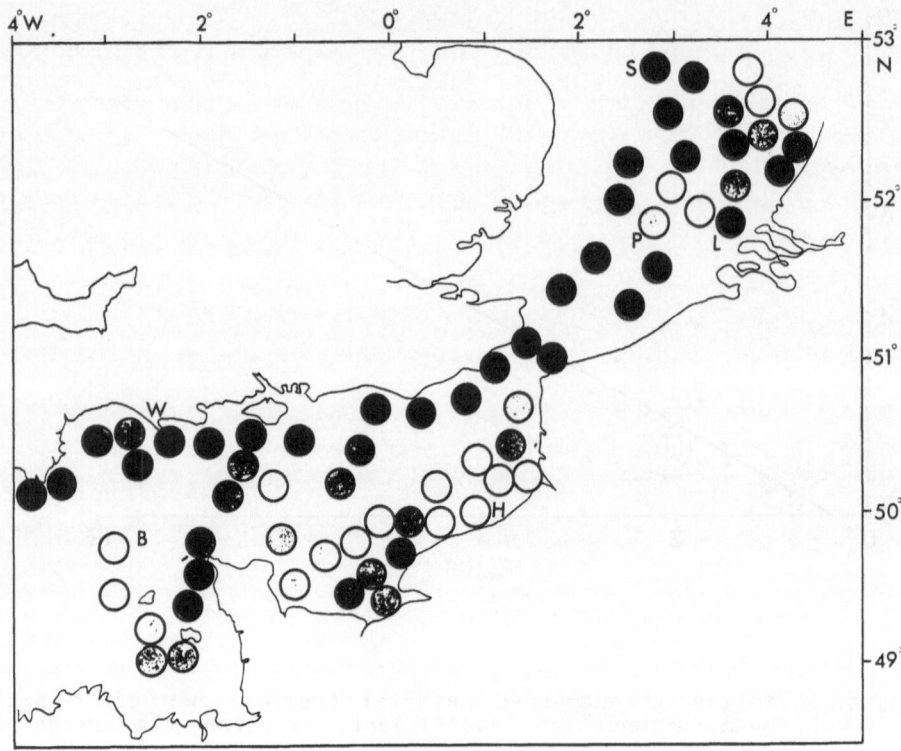

Figure 10. Ashed seston filters placed on a map at the positions where the samples were drawn (see Fig. 3(a) for ship's track). The dark filters were reddish brown near the English Coast and light brown near the continental coast. Note the visual correlation with Fig. 4(b). See Figure 11 for particle size analysis of samples W,B,H,S,P,L.

It is noteworthy that the turbidity in the English Channel does not reflect the size distributions of the bottom sediments which are known to relate to the strength of the tidal currents, with gravel in regions of high bed stress (central English Channel), ranging through sand to mud in low tidal current environments like Lyme Bay (Bureau de Recherches Géologiques et Minières, 1979). However, in the Southern Bight, the zone of minimal turbidity might partly coincide with distribution of medium sands as suggested by Lee and Folkard (1969).

(c) Vertical Salinity Structure

Whilst in March the net heat flux through the sea surface is not sufficient to produce any significant permanent stratification of the

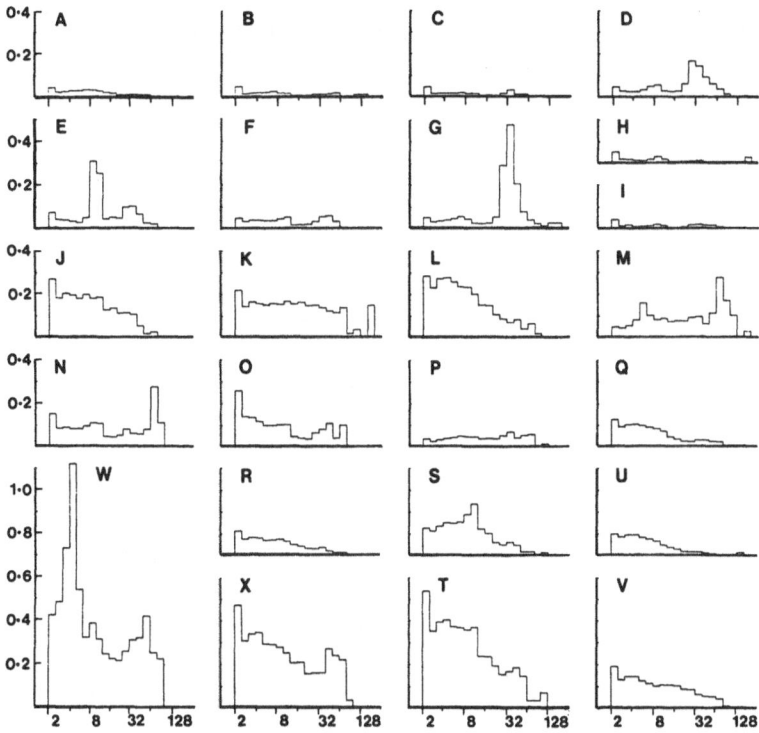

Figure 11. Size spectra of suspended particles taken at the positions
marked in Fig. 12. Horizontal axis: equivalent spherical
diameter (μm). Vertical axis: volume of particles in ppm
by volume.

water column, the stabilizing effect due to run off reaches its maximum amplitude at this time of year (Maddock and Pingree, 1982). Thus
in weaker tidal regions, in the vicinity of rivers with large run
off, an input of buoyancy at the surface can result in the formation
of a shallow surface mixed layer with a corresponding increase in
mean light level in the wind mixed layer.

The surface to bottom salinity differences, $\Delta S(^{o}/oo)$ for the
combined 1983 and 1984 cruises are shown in Fig. 13. In the English
Channel these salinity differences can result in a shallow halocline
(∿<10 m depth) coastally and would favor the development of phytoplankton blooms in the Bay of Seine, the Bay of Somme and towards the
Dutch coast in the Southern Bight of the North Sea. In accordance
with these considerations all these regions showed increases of
surface chlorophyll 'a' in March 1983 (Fig. 4).

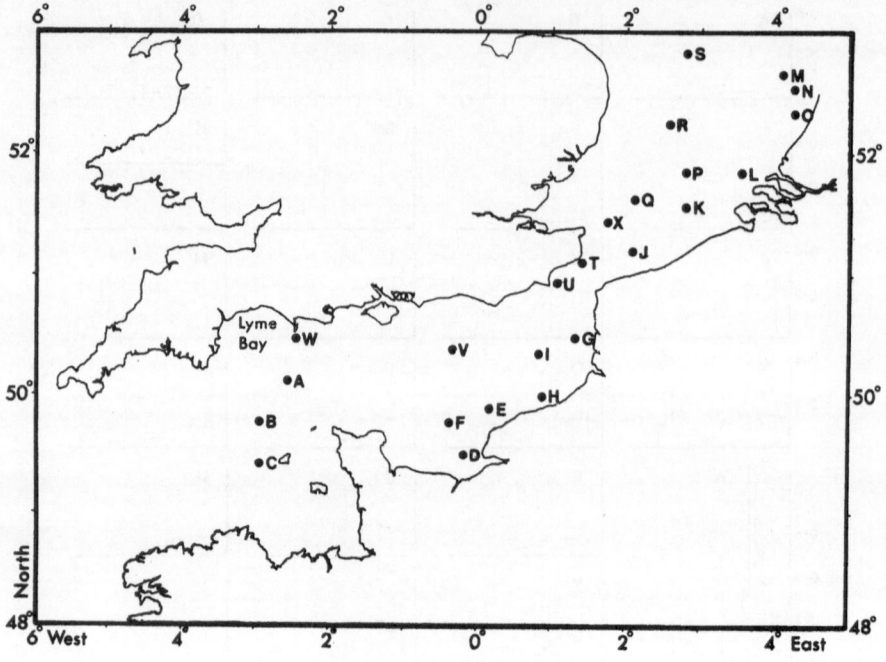

Figure 12. Map showing the positions from which the particle counts in Fig. 11 were taken.

A marked salinity layering ($\Delta S \sim 1^\circ/oo$) also occurred on the northern part of the Armorican shelf near the coast where the coastal surface waters had their lowest values of beam attenuation coefficient. In this region the tidal currents ($M_2 \sim 10$ cm s^{-1}) are the weakest in the whole area surveyed and an early start of the bloom here (Fig. 5) would appear to be indirectly related to the absence of mixing due to tides. What is not yet clear, however, is why the levels of chlorophyll 'a' show even higher values on the halocline at a depth of about 14 m, though this might be attributed to the sinking of a surface bloom.

The possible extent of the region of the surface layering on the Armorican shelf shows up clearly in some CZCS satellite images at this time of year (Plate I).

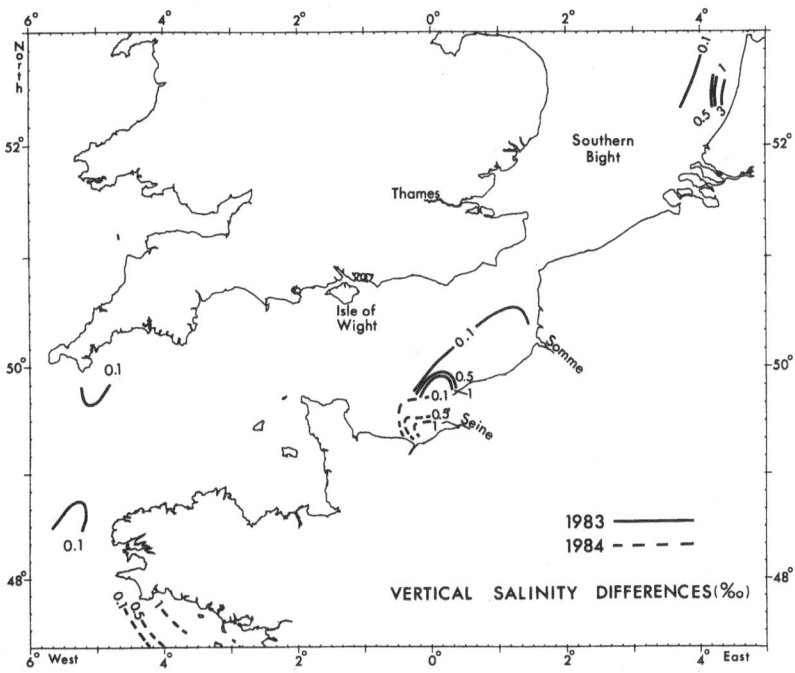

Figure 13. Vertical difference in salinity (^{o}/oo) between the surface
(2 m) and the bottom.

(d) The Phytoplankton March 1983

Samples for phytoplankton were preserved in Lugols iodine and
analyzed on an inverted microscope using the technique of Utermöhl
(Hasle, 1978). The results are discussed in relation to particle
counts which were made on the same samples soon after they were
collected and while the ship was on passage. A model TAII Coulter
Counter with 100 μm and 400 μm tubes was used in the analysis of this
material (Sheldon and Parsons, 1967). Representative graphs of the
particle counts are given in Fig. 11 with the positions of the
samples marked on Fig. 12. The total volume of suspended matter in
each sample, summed flagellates and diatom counts and levels of
chlorophyll are given in Table I.

Samples from the central, western Channel and as far south as
the west of Guernsey (Fig. 11A,B,C) had low particulate counts with a
phytoplankton dominated by Cryptophyceae and other unidentified

Table 1. Chlorophyll, phytoplankton and Coulter counts from selected stations in March 1983 (see Figures 10-12).

	Suspended Matter[†] $\mu m^3 \cdot 10^6\ ml^{-1}$	Total Flagellate Count $10^3 l^{-1}$	Total Diatom Count $10^3 l^{-1}$	Chlorophyll 'a' $mg \cdot m^{-3}$
A	0.40	242	22	0.08
B	0.34	NA	NA	0.32
C	0.29	170	10	0.65
D	0.95	108	38	1.77
E	1.30	2192	52	6.29
F	0.58	36	6	0.28
G	1.64	30	100	0.64
H	0.29	146	14	0.44
I	0.26	78	4	0.36
J	2.31	34	188	2.50
K	2.66	8	168	3.60
L	2.77	363	637	2.02
M	1.84	NA	114	NA
N	1.62	200	456	2.56
O	1.58	395	670	1.04
P	0.74	NA	NA	1.32

Table 1. Continued.

	Suspended Matter[+] $\mu m^3 \cdot 10^6\ ml^{-1}$	Total Flagellate Count $10^3\ l^{-1}$	Total Diatom Count $10^3\ l^{-1}$	Chlorophyll 'a' $mg \cdot m^{-3}$
Q	1.00	218	105	1.28
R	0.87	93	48	0.56
S	1.77	331	419	0.65
T	4.57	NA	NA	0.40
U	0.80	32	72	0.73
V	1.70	109	44	0.40
W	6.76	NA	NA	0.32
X	4.26	NA	NA	1.28

NA = not analyzed

[+]See text: (d) The phytoplankton March 1983

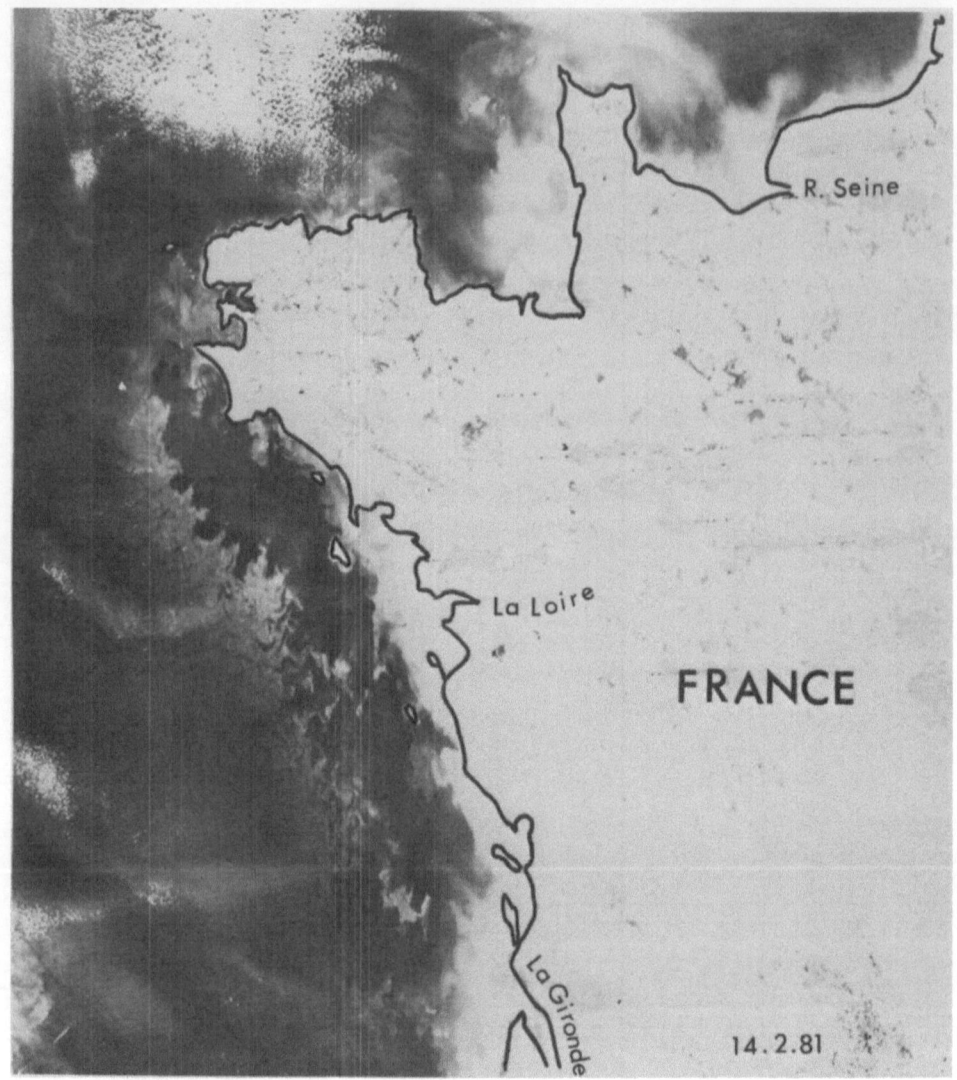

Plate I. Coastal Zone Color Scanner (CZCS) image (14.2.81, Channel
          3) probably illustrating the westward limit of reduced
          surface salinities on the Armorican shelf. The regions is
          defined by a zone of high surface reflectance, perhaps the
          result of a frontal bloom. In the north, vertical mixing
          due to strong tidal streaming, limits the northward pene-
          tration of this surface layer. In the English Channel,
          high reflectance is also associated with the fresh water
          run-off from the River Seine.

flagellates (<5 μm in size) with few diatoms. On the return leg of
the cruise phytoplankton counts south of Lyme Bay had increased to
236,000 flagellates and 36,000 diatoms per liter but were still
dominated by flagellates <5 μm in size.

On the western side of the Bay of Seine a mixed assemblage of flagellates was dominated by <u>Katodinium</u> sp. and Cryptophyceae with an increasing preponderance of <u>Thalassiosira</u> (30-50 µm diameter) as the Seine estuary was approached (Fig. 11D). In a narrow band immediately to the east of the mouth of the river Seine and coinciding with a chlorophyll maxima and salinities of 29°/oo the flagellate counts reached $2.2.10^6$ cells 1^{-1} and were largely composed of <u>Katodinium</u> sp. 8 µm long (Fig. 11E). On the outer edge of the Bay of Seine the phytoplankton counts were reduced to levels similar to the central Channel, but with an increase in the number of unidentifiable organic detrital particles (Fig. 11F).

Coinciding with the chlorophyll maxima in the Bay of Somme was a phytoplankton assemblage which was dominated by large <u>Thalassiosira</u> spp. (Figure 11G). Flagellates and especially <u>Katodinium</u> sp. were more important towards the coast. Between the Bays of Seine and Somme is a region with low chlorophyll values which reaches close inshore. This is reflected in the Coulter results (Figure 11H,I) and the moderate to low counts of diatoms and flagellates.

To the west of the Rhine and south of the central axis of the Southern Bight (Figure 11,J,K,L) but approximately coinciding with the chlorophyll maxima in Figure 4A, the phytoplankton is predominantly composed of small diatoms including <u>Fragilaria</u> sp., <u>Asterionella</u> <u>glacialis</u>, <u>Skeletonema</u> <u>costatum</u> and small 'centric' species. Sample L (Figure 11) from off the Rhine had the largest volume of suspended matter of the samples which were not from the highly turbid locations (which are conveniently categorized with α >4 m^{-1}); it also contained a rich phytoplankton assemblage with more than twenty species of diatoms and an abundance of small flagellates <5 µm. In the area of higher chlorophyll in the Southern Bight (Figure 11M,N, O), larger diatoms predominated comprising large "centrics" 30-100 µm in diameter, <u>Ditylum</u> <u>brightwellii</u>, <u>Rhizosolenia</u> <u>delicatula</u>, and <u>R.</u> <u>hebetata</u> <u>semispina</u> with <u>Cylindrotheca</u> <u>closterium</u> and chains of <u>Skeletonema</u> <u>costatum</u>, <u>Hyalochaetes</u> and <u>Phaeoceros</u>. At the highest levels of chlorophyll recorded (St. 15, see Figure 3), <u>R.</u> <u>delicatula</u>, <u>Hyalochaetes</u> and <u>Nitzschia</u> <u>delicatissmia</u> were predominant in both surface and subsurface samples. Flagellates and especially <u>Katodinium</u> sp. were again more important towards the coast (Figure 11 N,O) with specimens of the freshwater phytoplankton <u>Asterionella</u> <u>formosa</u> in sample O.

Just to the north of the chlorophyll maximum off the Rhine and coinciding with clear waters defined by beam attenuation, the particle content in the water was low (Figure 11P). Phytoplankton samples from this area were clear with little detritus and were characterized by high numbers of small flagellates <5 μm with some small diatoms such as C. closterium, A. glacialis, 'centrics' (10 μm) and Fragilaria sp.

In the turbid water to the north of this clear region (Figure 11Q,R,S) clastic detrital particles were a dominant feature of the particulate content of the Lugols samples with a progressive decline in the contribution of the different size fractions from small to larger grains. There were few diatoms in the samples and the majority of these were empty frustules, but small flagellates were occasionally abundant.

Samples from the northern coast of the Channel (Figure 11U,T,V) also contained much clastic material and little phytoplankton. In very mixed and turbid locations (α >4 m^{-1}), as for example in sample W just off Portland Bill and sample X to the east of Kent, a double peak was evident in the size frequency graphs with the suspension of larger size fractions of clastic detrital particles.

March 1984

Counts of flagellates in the Bay of Seine were higher in 1984 than for the same period in 1983. However, they were comprised mainly of small species <5 μm and Cryptophyceae which together contributed little to the chlorophyll. Katodinium sp. was relatively unimportant, though there was an increase in the numbers of this species and of flagellates in general towards the mouth of the Seine. In the shallow water area to the east of the Seine which exhibited high levels of chlorophyll in 1983, counts of phytoplankton were especially low. Reduced numbers of phytoplankton were also evidence in the shallow waters of the Gulf of St. Malo and in the central English Channel off the Bay of Seine. Small flagellates <5 μm were moderately abundant to the west of Jersey and occurred in high numbers 1-2 10^6 cells 1^{-1} in association with a rich diatom flora of Thalassiosira spp., Cylindrotheca closterium, Chaetoceros spp. and Navicula spp. in coastal samples south of the Pointe du Rax. A concentration of small Thalassiosira spp. (T. conferta ~8 μm, T. angulata ~14 μm) occurred in the halocline at 12 m off Concarneau with numbers >2.5 10^6 cells 1^{-1} (see Figure 6, St. 14, 15, and 16).

The most consistent pattern to emerge from phytoplankton analyses was the abundance of the small dinoflagellate Katodinium sp. in samples from areas with reduced salinities. This genus is characteristic of cold brackish conditions and has been reported as occurring in bloom concentrations at temperatures of $9^{\circ}C$ in Norwegian waters (Tangen, 1979), and also in dense concentrations in the spring in Scottish sea lochs (Tett and Droop, 1980). The reduced numbers of this genus and the low chlorophyll levels in the Bay of Seine in 1984 may therefore be largely attributable to the higher salinities in this area in 1984.

Variability in the levels of total suspended matter and hence turbidity of the water seems likely, and this will affect the timing of the spring phytoplankton development. Eisma and Gieskes (1977) recorded higher levels of total suspended matter (>2 based on Coulter counts) in both clear and the more turbid regions of the Southern Bight. Dickson and Reid (1983) presented evidence for the great variability in the transparency of the waters for this area and Reid et al. (1983) attributed an exceptional bloom of diatoms in the Southern Bight in early 1977 to the enhanced clarity and stability of the water under the unusual hydrographic conditions which prevailed at that time.

SUMMARY

The results of the 1983 and 1984 March cruises, aimed at determining the relationship between environmental conditions and the start of the productive season, can be conveniently summarized on an S-H diagram (Pingree, 1978), where

$$S = \log_e \left(\frac{h}{D}\right)_{M_2} \tag{1}$$

(h is the depth of the water column and D_{M_2} is mean rate of dissipation of M_2 tidal energy due to bottom friction, h/D_{M_2} is in cgs units) and $H = k_D h_m$. S effectively determines the potential of a water column to stabilize (Simpson and Hunter, 1974) under the influence of a surface buoyancy flux and H is the optical depth of the mixed layer. Equivalently, $e^{-k_D h_m}$ is the fraction of the incident light found at the base of the mixed layer, or at the sea floor if the water column is well mixed. More significantly, the reciprocal $(k_D h_m)^{-1}$ is a measure of the mean light level in a mixed water column. With values of S > 1.5 the water column will stabilize during the seasonal heating cycle; then the mixed layer depth becomes less than the water depth, h_m < h, resulting in a corresponding increase in the mean light level of the mixed layer.

The points S B (shelf break), W E (western entrance to the English Channel), C E (central English Channel), and E E (eastern English Channel) are depicted schematically on the S-H diagram in Figure 14. These regions have the lowest mean light levels and therefore have the lowest potential for a phytoplankton population increase. Although the central and eastern English Channel waters (C E and E E) have lower values of H, the productive season may well start first on the shelf (W E to S B) where S has values greater than

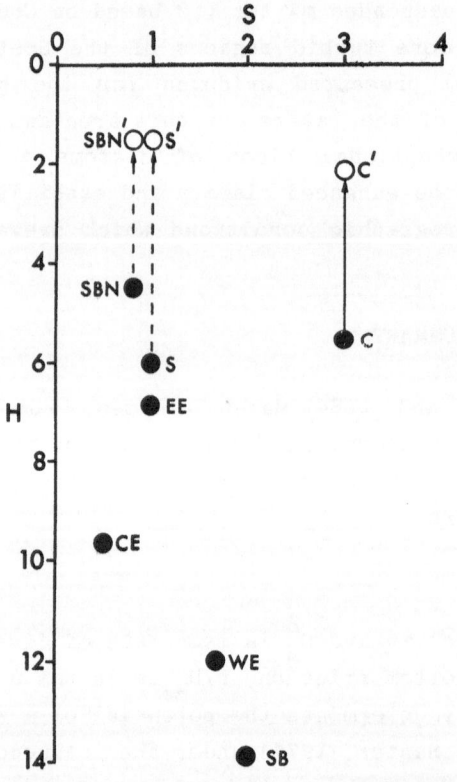

Figure 14.   S-H diagram (for explanation see text).

1.5 indicating that the water column will stabilize when the net heat flux becomes positive ($\sim$ i.e. March-April), whereas at C E and E E the whole water column will remain mixed throughout the year.

The bloom regions are depicted as S B N (Southern Bight of the North Sea, S (eastern side of the Bay of Seine) and C (coastal region of the Armorican shelf in the neighborhood of Concarneau). On the basis of the mixed layer representing the whole water column (i.e. h_m = h), the Southern Bight (S B N) with H = 4.5 would appear to have the most favorable environment for an early start to the spring bloom. However, as we have seen, all these regions are subject to stratification due to run-off. This results in a reduced mixed layer depth and so the points SBN, S, and C may move to S B N ', S' and C' as depicted by the arrows.

The points S B N and S are in regions with smaller values of $\log_e (\frac{h}{D_{M_2}})$ and have intermittent stratification (depicted by broken arrows) depending on whether it is neaps or springs, levels of run-off and wind stress and so the precise timing of the productive season might depend on all these factors. The point C however with $\log_e (\frac{h}{D_{M_2}})$ as high as 3 would still tend to stratify even under much reduced levels of run-off and so would be predicted to have the earliest start to the spring bloom in the whole region surveyed, as was indeed observed.

ACKNOWLEDGEMENTS

It is a pleasure to thank Dr. H.H. Bottrell (Institute for Marine Environmental Research, Plymouth) for developing the software for the Coulter Counter analyses.

REFERENCES

Agoumi, A., M.J. Enderle and R.A. Gras. 1983. Modélisation du Régime thermique de la Manch. Oceanol. Acta 6: 393-406.

Bartz, R., J.R.V. Zaneveld and H. Pak. 1978. A transmissometer for profiling and moored observations in water. S.P.I.E. 160, Ocean Optics V: 102-108.

Bureau de Recherches Géologiques et Minières. 1979. Map of surficial sediments of the English Channel, 1/500 000. With booklet of explanatory notes in French and English (28 p). Orléans, B.R.G.M.

Dickson, R.R. and P.C. Reid. 1983. Local effects of wind speed and direction on the phytoplankton of the Southern Bight. J. Plank. Res. 5: 441-455.

Dietrich, G. (ed.). 1962. Mean Monthly Temperature and Salinity of the Surface Layer of the North Sea and Adjacent Waters from 1905-1954. Conseil Permanent International pour l'exploration de la Mer, Service Hydrographique, Charlottenlund Slot, Denmark.

Eisma, D. and W.W.C. Gieskes. 1977. Particle size spectra of non-living suspended matter in the southern North Sea. Interne Verslagen, Nederlands Instituut voor Onderzoek der Zee, Texel: 1-22.

Gieskes, W.W.C. and G.W. Kraay. 1975. The phytoplankton spring bloom in Dutch coastal waters of the North Sea. Neth. J. Sea Res. 9: 166-196.

Hasle, G. 1978. The inverted-microscope method. In: Phytoplankton Manual, A. Sournia (ed.), pp. 88-96. UNESCO, Paris.

Højerslev, N.K. 1977. Spectral daylight irradiance and light transmittance in natural waters measured by means of a Secchi Disc only. International Council for Exploration of the Sea C.M./C: 42 Plankton Committee.

Holligan, P.M., R.D. Pingree, P.R. Pugh and G.T. Mardell. 1976. The hydrography and plankton of the eastern English Channel in March 1976. Annal. Biolog. 33: 69-71.

Holmes, R.W. 1971. The Secchi Disc in turbid coastal waters. Limnol. Oceanogr. 15: 688-694.

Le Grand, Y. 1939. La Pénétration de aa Lumière dans la Mer. Annal. Inst. Oceanogr. 19: 393-436.

Lee, A.J. and A.R. Folkard. 1969. Factors affecting turbidity in the Southern North Sea. J. Cons. int. Explor. Mer 32: 291-302.

Maddock, L. and R.D. Pingree. 1982. Mean heat and salt budgets for the eastern English Channel and the Southern Bight of the North Sea. J. mar. Biol. Ass. U.K. 62: 559-575.

Pingree, R.D. 1978. Mixing and stabilization of phytoplankton distributions on the Northwest European continental shelf. In: Spatial Pattern in Plankton Communities, J.H. Steele (ed.), pp. 181-200. Plenum Press, New York and London.

Pingree, R.D. 1983. Spring tides and quadratic friction. Deep-Sea Res. 30: 929-944.

Poole, H.H. and W.R.G. Atkins. 1929. Photo-electric measurements of submarine illumination throughout the year. J. mar. biol. Ass. U.K. 16: 297-324.

Reid, P.C., H.G. Hunt and T.D. Jonas. 1983. Exceptional blooms of diatoms associated with anomalous hydrographic conditions in the Southern Bight in early 1977. J. Plank. Res. 5: 755-765.

Sheldon, R.W. and T.R. Parsons. 1967. A practical manual on the use of the Coulter Counter in marine science. Coulter Electronic Sales Company, Toronto, 66 pp.

Simpson, J.H. and J.R. Hunter. 1974. Fronts in the Irish Sea. Nature (London) 1250: 404-406.

Tangen, K. 1979. Dinoflagellate blooms in Norwegian waters. In: Toxic Dinoflagellate Blooms, D.L. Taylor and H.H. Seliger (eds.), pp. 179-182. Elsevier/North Holland, New York.

Tett, P. and M.R. Droop. 1980. Phytoplankton Ecology in Loch Striven, 21 June 1979. In: Phytoplankton and Fish Kills in Loch Striven, pp. 99-107.

Tyler, J. 1968. The Secchi Disc. Limnol. Oceanogr. 13: 1-6.

ARE RED TIDES CORRELATED TO SPRING-NEAP TIDAL MIXING?:
USE OF A HISTORICAL RECORD TO TEST MECHANISMS
RESPONSIBLE FOR DINOFLAGELLATE BLOOMS

W.M. Balch
Institute of Marine Resources, A-018
Scripps Institution of Oceanography
University of California, San Diego
La Jolla, CA 92093

INTRODUCTION

Vertical mixing is generally considered to be one of the most influential factors affecting phytoplankton abundance in the ocean (Sverdrup, 1953; Gran and Braarud, 1935). Changes in both phytoplankton abundance and species composition are thought to occur as a function of turbulence (Margalef, 1978). Research has therefore been directed at the sources of water column turbulence in nature and how these sources affect phytoplankton community dynamics.

Tidal mixing is one source of turbulence which has received considerable attention. Simpson and Hunter (1974) formulated a stratification parameter, S, which is a log function of water column depth, a frictional parameter (C_S) and the mean cubed semi-diurnal tidal current (U). Since the tidal current appears to the third power, the stratification parameter is quite sensitive to changes in tidal current velocities. Most of the applications of this model have concerned semi-diurnal (M_2) tidal currents but there is growing evidence that the fortnightly (M_f) and lunar (M_m) tidal components are important in creating turbulence as well. For example, Simpson and Bowers (1981) have shown with models that thermal fronts in shelf seas would be expected to move on a fortnightly cycle. Simpson (1981) used satellite imagery to demonstrate that frontal advancement was greatest two days following spring tides in European shelf waters. Webb and D'Elia (1980) showed that nutrient and oxygen distributions were vertically homogeneous during spring tides in a tributary of Chesapeake Bay (see also Haas, 1977). Resuspension of estuarine sediments has also been observed to occur during spring tides (Postma, 1967; Balch et al., 1983). Holligan and Harbour (1977) suggested that such tidal mixing events might affect phytoplankton abundance.

Lecture Notes on Coastal and Estuarine Studies, Vol. 17
Tidal Mixing and Plankton Dynamics. Edited by J. Bowman, M. Yentsch and W. T. Peterson
© Springer-Verlag Berlin Heidelberg 1986

The prediction of Holligan and Harbour (1977) has proven correct. Although not discussed in great detail, the data of Winter et al. (1975) (their Figure 5) show an effect of the spring-neap tidal cycle on the water column stability, chlorophyll concentration and primary production. Sinclair (1978) also found a statistically significant correlation between primary production and density stratification using the data of Winter et al. (1975). Moreover, he also observed strong fluctuations in biomass and production in the St. Lawrence Estuary that appeared to be related to the spring-neap tidal variability. Haas et al. (1980) made similar observations in the Chesapeake while Balch (1981) saw lunar period fluctuations in the surface abundance and community structure of phytoplankton at a shelf break station in the Gulf of Maine. All of these observations suggest that phytoplankton bloom dynamics can be affected by the M_f and M_m tidal components.

One question which has not been answered in great detail is "Do phylogenetically different algae predominate during specific phases of the lunar cycle"? Margalef's mandala (Margalef, 1978, 1979) predicts that in shallow seas and estuaries, diatoms will be favored in well-mixed waters during spring tides, then, as waters stratify during neap tides and nutrients are depleted, dinoflagellates will increase in relative abundance. If the condition arose in which the water column was stratified and turbulence was low but there was a high concentration of nutrients in the upper layer (i.e., just after a spring tide), then dinoflagellates might be favored. Haas et al. (1980) observed a red water bloom of Cochlodinium heterolobatum associated with a spring tide destratification event in a tributary of the Chesapeake Bay. (The term "bloom" is used to describe the dominance of one species of phytoplankton over all other species. It frequently, but not exclusively, involves water discoloration.) The bloom intensified as stratification resumed during the subsequent neap cycle. Diatom blooms coincident with the spring tides and dinoflagellate blooms following the spring tides were observed in the Gulf of Maine (Balch, 1981).

Questions concerning the effect of lunar periodicity on the abundance of algal taxa can be answered, in part, by using the historical record of red water blooms. Red tides (usually due to dinoflagellates) have been well documented because 1) they are often easily observed, 2) they are often associated with fish mortality and

human affliction (Taylor and Seliger, 1979) and 3) they are often
followed by massive algal die-off and putrification (Brongersma-
Sanders, 1957). Dinoflagellate blooms have been documented sporadi-
cally up to the mid-19th century and quite well documented thereafter. The data are global in coverage and span 150 years. In this
paper I will test the hypotheses: (1) dinoflagellate red tides occur
evenly through the spring-neap tidal cycle, and (2) (relating to
Margalef's mandala) dinoflagellate red tides occur in stratified
waters during neap tides. I will discuss segregating the data
according to specific coastal regions and particular species responsible for the blooms.

METHODS

Data on red tides were collected from published accounts as well
as from a polling of dinoflagellate ecologists around the world. The
date of first sighting, the species responsible for the red water,
the location, and the date of the previous new moon (taken from the
Nautical Almanac, Washington, D.C.) were tabulated. The lunar phase
was calculated by setting the date of the new moon as day 29 and the
day <u>after</u> a new moon as day 1. The lunar cycle is approximately 29.5
days in length and it varies several hours about this mean value in
any given month (Nautical Almanac). If the lunar cycle begins after
mid-day, then the lunar cycle can encompass 30 integer days rather
than 29. Due to the fact that bloom observations were only specified
to a given day (unlike the lunar cycles which were specified to the
minute) units of "days" were used when comparing the day a bloom
occurred to the phase of the lunar cycle. This calculation is therefore not exact and all data concerning the phase of the lunar cycle
during which a bloom occurred have a confidence interval of $\pm$ 1 day.

The influence of the lunar cycle was examined by plotting the
data as 29 day circular histograms. Two out of the 226 blooms
tabulated occurred on day 30 of the lunar cycle. These data were
eliminated from the data set (this did not affect the acceptance or
rejection of the null hypothesis). Distributions were tested against
even distributions using the Chi square test, the Kolmogorov-Smirnov
test and statistical tables (according the Zar, 1974). Regional bias
was examined by comparing the total data set to the data from the
Southern California Bight, the Gulf of Mexico, the W. European
continental shelf, New Guinea, Chile (20-30° latitude), and Gulf of

Maine/New England waters. The data base for the specific locations was, however, too small to use the Chi square test. The Kolmogorov-Smirnov ("KS") test remains robust at small sample size, however (Zar, 1974), so the data were pooled into six 4 day intervals and one 5 day interval and tested against an even distribution with the KS test (the single 5 day interval was weighted appropriately to make it comparable to the 4 day interval.) The Rayleigh test (Zar, 1974) was used to check for a significant mean direction in the circular histograms. The median angle was also calculated (Zar, 1974) which, when compared with the mean angle, allows detection of symmetrical distributions versus asymmetrical distributions.

A 62 year record of daily sea surface temperature anomalies at the Scripps pier was examined for lunar periodicity using time series averaging (Bendat and Piersol, 1971). All manipulations were performed with an HP9845B computer. The anomaly was calculated by averaging the daily temperatures over 62 years and subtracting each daily temperature from the 62 year average for that day (e.g. mean temp for Nov. 5 over 62 years - temp for Nov. 5, 1938 = anomaly for Nov. 5, 1938). Time series averaging involved fitting the temperature anomaly for each day of the last 62 years (January 1920 through December 1981) to a day of the lunar cycle, where day 1 was the day after a new moon. Then all of the anomalies for each day of the lunar cycle were averaged and the 29 day record was examined for periodicity. It was also possible to scan the 62 year record for anomalies greater or less than some designated temperature, and note their timing during the lunar cycle.

RESULTS

The data set used in this study is given in Appendix 1. There were 226 observations of red tides and 62 observations of fish mortality or human affliction related to red tides (more data were received after this writing and are included in the appendix but not in the calculations). Figure 1 shows a world map with locations of sightings noted by their numerical designations (Appendix 1). It is apparent that the sampling distribution is not uniform. 59.4% of the observations were made from North American shores, 12.1% from Central and South America, 8.6% from Western Europe and Iceland, 1.4% from the Black Sea, 2.4% from Japan, 4.5% from India, 5.2% from Africa and 6.2% from the Phillipines and New Guinea. Figure 2 shows the

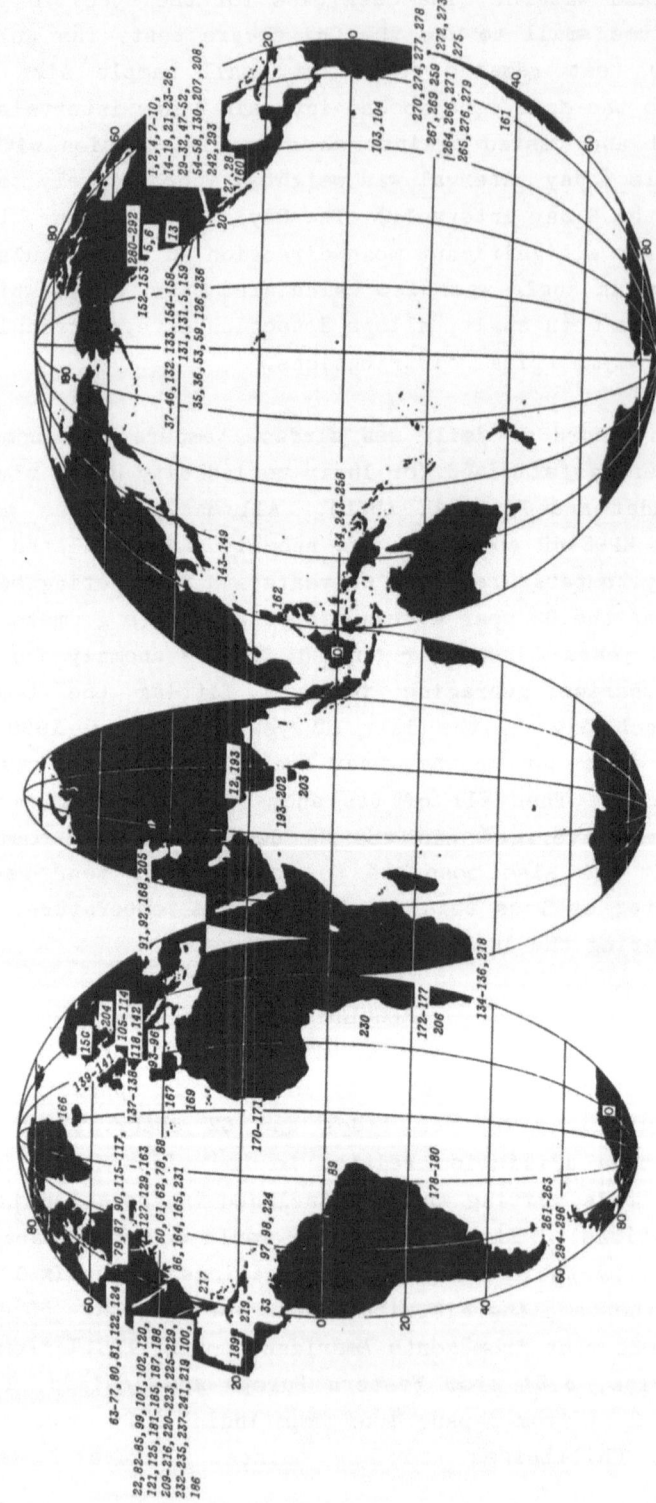

Figure 1. Global map showing locations of red tide sightings. Numbers refer to the list of sightings in Appendix 1.

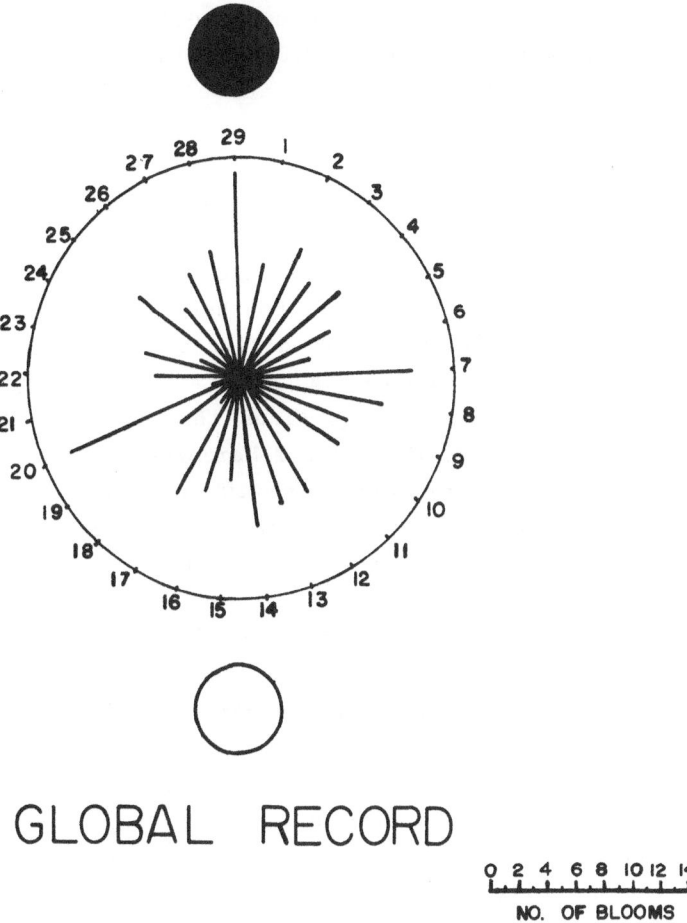

GLOBAL RECORD

NO. OF BLOOMS

Figure 2. 29 day circular histogram showing the distribution of red
tide sightings through the lunar cycle. The data set is
global in coverage. Phase of the moon is noted on the
perimeter of the histogram. The scale appears in the lower
left.

circular histogram for the complete data set. There was an average
of 7.83 red tide observations per day of the lunar cycle (standard
deviation = 2.87, n = 29 days). The data were normally distributed
(Kolmogorov-Smirnov test against a normal curve, P <0.05). Chi square
and Kolmogorov-Smirnov tests showed this distribution of blooms over

the lunar cycle not to be significantly different from an even distribution at an alpha effort of 0.05. Using the Rayleigh test, there was no significant direction to the circular distribution in Fig. 2 (P<0.05).

These data were treated separately to examine any regionality in bloom timing. Figure 3 shows circular histograms for red tides from 6 geographic regions. The small sample size necessitated pooling the data into 4 day increments except in Southern California. None of these distributions are significantly different from an even distribution (KS test, P<0.05) although a qualitative examination shows some evidence of lunar bias in red tides that was not the same at each location. For example, the Gulf of Maine/New England data (Fig. 3B) showed a bimodal distribution with 50% of the blooms observed from days 2-5 or days 18-21 of the lunar cycle (28% of the cycle). On the W. European Shelf, 32% of the observations were made between days 14 and 17 of the lunar cycle (10% of the cycle). Forty percent of the Gulf of Mexico blooms were documented between days 27 and 1 or between days 6 and 9 of the lunar cycle (28% of the cycle). In Southern California, 17% of the blooms occurred on days 28 and 29 (7% of the lunar cycle).

Figure 4 shows the data broken down by region but the bloom observations were plotted against tidal range rather than day of lunar cycle. The tidal ranges varied considerably amongst these locations. Red tides clearly occur in regions of variable tidal range (0.2 up to 4 m).

Proper interpretation of Figure 4 is complicated by the fact that a very high or low tidal range is seen less frequently than an average tidal range (due to the sinusoidal nature of the tidal range over a fortnightly tidal cycle). Therefore, if red tides occur randomly with respect to tidal range, then more blooms would occur when there was an average tidal range than during those few occasions when the tidal range was very high or low and we might mistakenly conclude that there was a relationship between tidal range and red tide formation. The data in Figure 4 were replotted (not shown) after the "number of blooms" had been weighted by the frequency of occurrence of a given tidal range, but no relation between tidal range and red tides was observed.

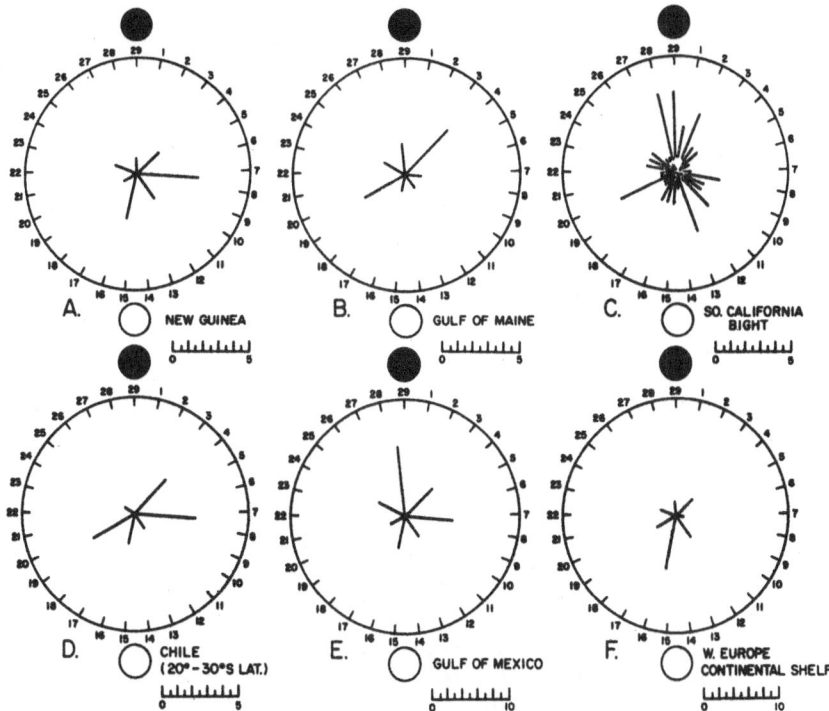

Figure 3. Circular histograms showing the distribution of red tide
sightings through the lunar cycle at six general locations
around the globe: a) New Guinea (n = 15); b) Gulf of Maine
and New England waters (n = 14); c) S. California Bight (n
= 60); d) Chile (20-30°S latitude, n = 14); e) Gulf of
Mexico (n = 35); and f) the W. European Shelf (n = 22).
All histograms but that for S. California Bight involved
small data sets which necessitated pooling the data into
six-4 day increments and one-5 day increment. The number
of observations in the 5 day increment was weighted by 4/5
to make it comparable to the other 4 day increments.

Southern California Bight Observations

As already illustrated in Figures 3C and 4C, red tides show
little correlation to the lunar phase or tidal range in the Southern
California Bight. Nevertheless, the 62 year temperature record from
the same region shows evidence of temperature perturbations which
appear to be correlated to the lunar cycle. Figure 5 shows the
results of the time series averaging of Scripps Pier temperature
anomalies from the 62 year record. It is apparent that there was a

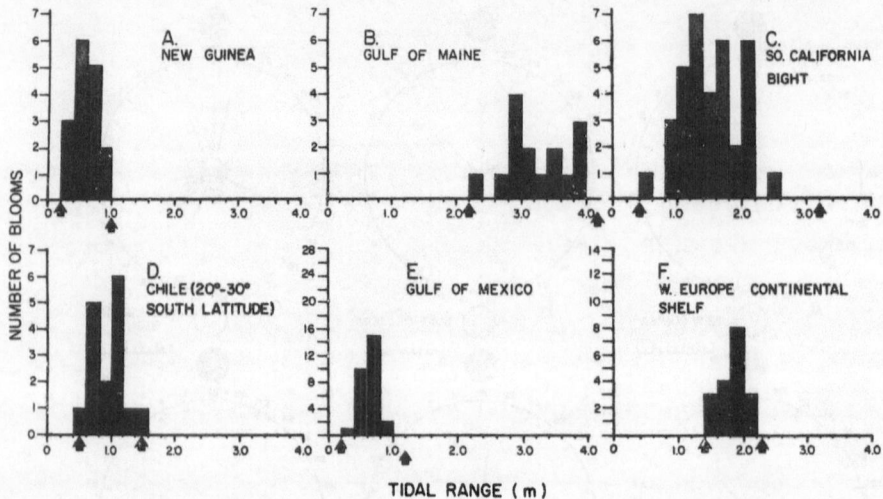

Figure 4. Histograms of number of bloom observations vs. tidal range
for the 6 regions given in Fig. 3. "Tidal range" refers to
the maximum tidal range observed on the day of a reported
bloom. Arrows on the abscissa designate the minimum and
maximum tidal range for each particular locality (given in
the U.S. Department of Commerce tidal tables).

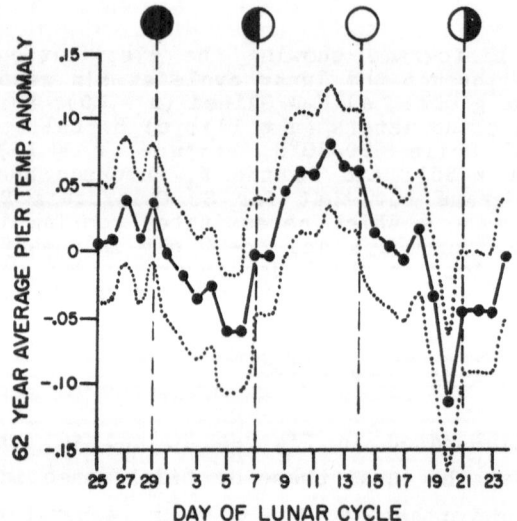

Figure 5. Time series averages of SIO pier temperature anomalies
through the lunar cycle. Data are from a 62 year daily
record of sea surface temperatures. Lines on either side
of the central average line represent 1 standard error.
The lunar phase is designated below the abscissa.

cyclic nature to the average temperature anomaly through the lunar cycles. The general trend of the anomaly was that the maximum temperature anomalies led the lunar cycle by 0 to 2 days. The maximum peak to trough difference in the anomalies was 0.196°C. Highest temperatures were observed at day 27 and day 12 and coldest temperatures were seen at days 5-6 and 20. There is no obvious explanation for the two peaks at days 27 and 29. However, they are not significantly different from the anomalies of the surrounding days. If the cycle of temperature anomalies is analyzed using the period between successive "troughs," it shows a frequency of 2 cycles per 29 days (or 0.069 cycles/day).

In any sampling of a time series, aliasing by other important frequencies must be considered. The SIO pier temperatures were sampled every 24 hours. The data were checked to see whether a semidiurnal tidal cycle of temperature (1.935 cpd; Pond and Pickard, 1978) sampled every 24 hours could appear as a fortnightly cycle with a frequency of 0.069 cpd. The equation of Bendat and Piersol (1971) was used to calculate the alias frequencies and none were found with a frequency of 0.069 cpd. The closest harmonic to 0.069 cpd is at 0.0645 cpd (15.5 days per cycle).

Figure 6 shows evidence that large negative temperature anomalies in the Southern California Bight have occurred non-randomly relative to the lunar tidal cycle. The general trend in the histograms is cyclic and the maximum cold (>-4°C) anomalies lags the lunar cycle by approximately 1 to 4 days (Fig. 6B). Anomalies exceeding -4°C were most frequent on days 1-5 and days 16-20 of the lunar cycle and least frequent between days 7-14 and 25-27 of the lunar cycle. The anomalies exceeding -3°C were most commonly seen on days 20 and 28 of the lunar cycle. The -3°C anomalies were least frequently seen on days 7-10 and day 24 of the lunar cycle. The distribution in Figure 6A is significantly different from an even distribution using the Chi square test (P<0.05) while the distribution in Figure 6B is not significantly different from an even distribution (P≤0.05). The two histograms are similar in shape, however.

Several major blooms on the Southern California continental shelf can be defined as cell concentrations sufficient to yield 50 µg chlorophyll a ℓ^{-1} (the minimum chlorophyll a concentration for visible red water discoloration is about 5 µg ℓ^{-1}; Holmes et al., 1967).

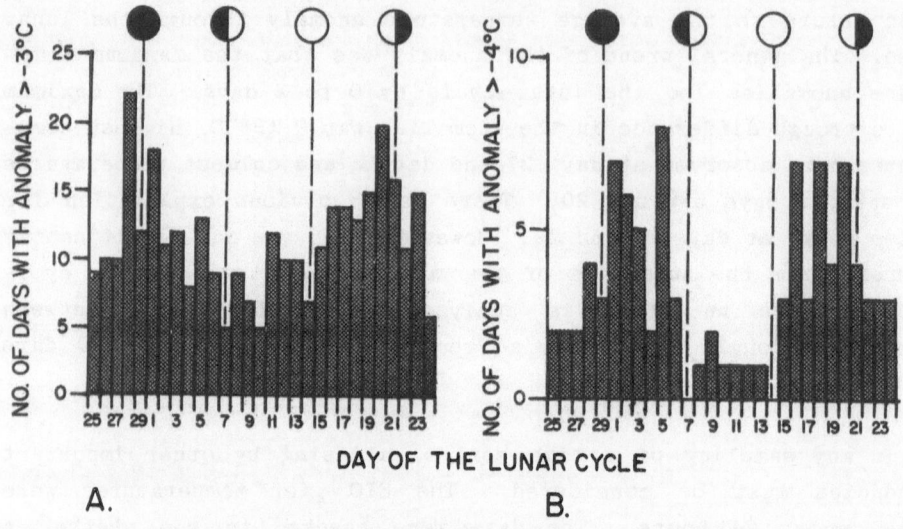

DAY OF THE LUNAR CYCLE

A.   B.

Figure 6. Histograms showing the number of observations of a designated temperature anomaly in the S. California Bight vs.
the phase of the lunar cycle. Panels A and B refer to
negative temperature anomalies of $>-3^{\circ}C$ and $>-4^{\circ}C$, respectively. Phase of the moon is also shown below the X axis.
Data are from the 62 year daily record of sea surface
temperature at the SIO pier.

A total of 9 of these major blooms have been recorded since 1920
(Table 1; refs. in Appendix 1). Each of these, save one, has taken
place immediately after an abrupt fall in sea surface temperature
measured at the SIO pier (Fig. 7). These sharp temperature declines
imply nutrient-rich surface waters since several nutrient concentrations vs. temperature relations have shown waters colder than $14-15^{\circ}C$
to contain measurable nitrate while warmer waters do not (Jackson,
1983) except in special locations with impeded circulation as within
kelp beds (Jackson, 1977). (If we were sure the same water parcel
was observed over time we could say the blooms followed mixing events
when the waters were once again becoming stratified. As already
illustrated in Fig. 6A and B, major negative temperature anomalies in
the S. California Bight tend to lag the new and full moon by 1-4
days.)

The entire data set was analyzed with respect to species to
answer the question, "Are there single species which bloom at a

Table 1. Dinoflagellate blooms off Southern California with chlorophyll a levels greater than 50 µg ℓ⁻¹. (Chlorophyll per cell taken as 30 pg for G. polyedra, 8 pg for P. micans.)

Date first observed	Day of lunar cycle	Dominant species	Approximate chlorophyll a at maximum density µg ℓ⁻¹	Appendix reference number
17 May 1933	23	Prorocentrum micans	>50	9
30 May 1938	1	Gonyaulax polyedra	90	7
13 Sept 1945	7	Gonyaulax polyedra	180	8
9 Mar 1954	4	P. micans	95	24
15 Aug 1958	29	G. polyedra	>500	30
30 June 1960	6	G. polyedra	63	297
13 Apr 1961	13	P. micans	110	31
10 May 1964	28	mixed	170 measured	298
1-2 June 1965	2	G. polyedra	420 measured	299

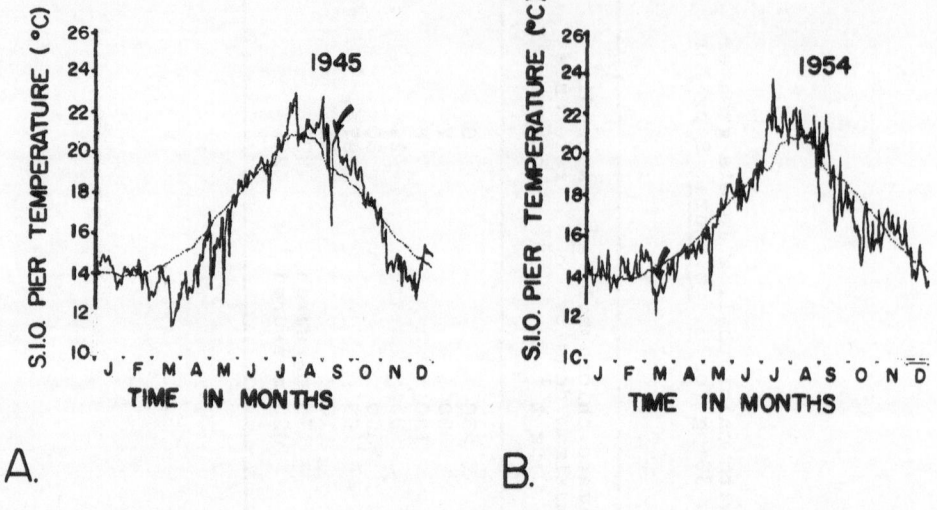

Figure 7. SIO pier temperature record for two years in which major
red tides were observed (in which chlorophyll levels were
>50 μg ℓ⁻¹). The solid black line represents the recorded
temperature and the dotted line represents the 62 year
daily average temperature. Arrows denote the dates that
the major red tides were first observed.

particular phase of the lunar cycle?" Only Gonyaulax polyedra,
Prorocentrum micans, and Gymnodinium breve showed any bias in the
times that red tides were observed. Forty-five percent of the red
tides of Gymnodinium breve (n = 24) were observed from day 25 to 29
of the lunar cycle (17% of the total cycle), 24% of the Prorocentrum
red tides (n = 17) were observed on day 28 of the lunar cycle (3% of
the total cycle) and 40% of the Gonyaulax polyedra red tides (n = 25)
were observed from day 25 through day 1 of the lunar cycle (21% of
the total cycle). There were not enough data of the other 25 species
to reliably examine for a lunar bias.

The annual cycle of dinoflagellate red tides is shown in Fig. 8;
northern hemisphere blooms are shown in the upper panel, southern
hemisphere blooms in the lower panel. It is apparent that red tides
are usually observed during summer months in the northern hemisphere
with 56% of the observations between June and September. In the
southern hemisphere, the red tides showed a bimodal distribution
through the annual cycle with 68% of them occurring between December
and May.

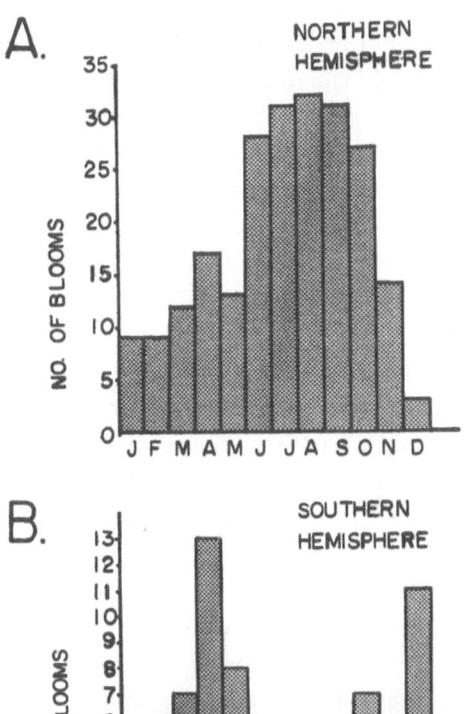

Figure 8. Histograms showing the annual cycle of dinoflagellate blooming in a) the northern hemisphere, and b) the southern hemisphere. Letters under the abscissa refer to months of the year.

DISCUSSION

Two types of mechanisms can be invoked to explain dinoflagellate red tides; they are not mutually exclusive. A biological mechanism involves a growth response to some exogenous or endogenous factor (e.g. dinoflagellate excystment in response to a temperature cue, vertical migration, etc.) while a physical mechanism involves some sort of physical concentration of the cells (e.g. Langmuir cells). Combinations of these mechanisms are particularly appealing, such as physical turbulence transporting nutrients into the mixed layer and a subsequent growth response of the algae, or stratification with diel vertical migration across the thermocline leading to horizontal concentration (Kamykowski, 1979). However, it is known the physical concentrating mechanisms must operate in some blooms because more nutrient can be found in the cells than was present in the water column before the bloom (Ketchum and Keen, 1948; Holmes et al., 1967).

Most workers who have observed a relation between the lunar cycle and phytoplankton blooming have noted that spring tidal currents increase the thickness of the bottom mixed layer (or surface mixed layer; see Hayward et al., this volume), break down the seasonal thermocline and inject nutrients into the surface mixed layer (Pingree et al., 1976, for example). Subsequent stratification during neap tides could give the red tide dinoflagellates a competitive advantage if a) diatoms fail to remain suspended following stratification of the water column, or b) if the dinoflagellates can migrate into a deepening nutrient-rich layer at night and return to a nutrient-poor surface layer during the day (Eppley and Harrison, 1975; Harrison, 1976; Haas et al., 1980). Paasche et al. (1984) showed, however, that such vertical migration was not required for European blooms of Prorocentrum minimum and Gyrodinium aureolum. Nevertheless, for migrating dinoflagellate species, the original conceptual model formulated to describe the expected results for this study on a global basis is shown in Figure 9A. It predicts 1) peak tidal amplitude at new and full moon, 2) negative temperature anomalies (on average) associated with the increased spring tidal mixing, 3) positive anomalies (on average) during the more stratified neap tides and 4) most of the dinoflagellate blooms associated with the more stratified neap tides. The latter is not supported by the global view of Figure 2 so this conceptual model must be revised.

Pooling the data into a global data set may be presumptuous, particularly if one is invoking a tidal mechanism of bloom formation. Tides vary globally as a function of bottom topography, basin resonance and general geographic location, not only with the phase of the moon (Pond and Pickard, 1978). Garret and Munk (1971) pointed out that the peak tidal range is not always synchronous with the new or full moon but lags it by one or two days. They coined the term "age of the tide" which refers to the lag between full or new moon and

A.   B.

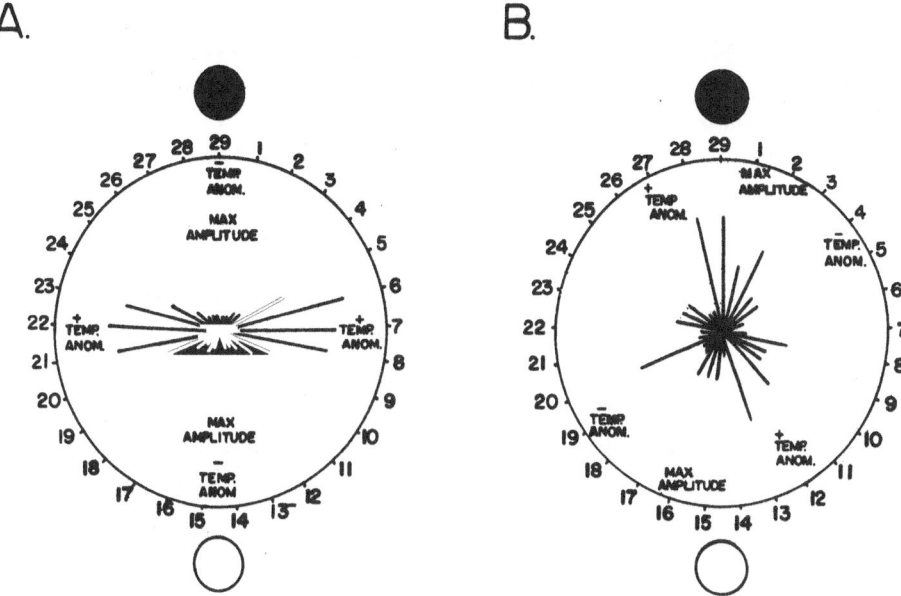

Figure 9. A. Idealized conceptual model depicting the appearance of
            dinoflagellate blooms through the lunar cycle (based on the
            Margalef mandala). The circular histogram is a representa-
            tion of what would be expected if dinoflagellates bloomed
            during stable neap tidal periods. This model incorporates
            1) peak tidal amplitude at new and full moon, 2) negative
            temperature anomalies associated with the increased spring
            tidal mixing and 3) positive temperature anomalies asso-
            ciated with more stratified neap tides. Lunar phase is
            shown at the perimeter of the circular histogram.

            B. Data from the Southern California Bight to compare to
            idealized conceptual model. Data from Figures 3 and 5 were
            used to make this figure. Tidal amplitude is shifted in
            this figure according to Garret and Munk (1971).

peak tidal amplitude. Thus, the rationale for examining the dinoflagellate blooms from 6 specific areas (Figs. 3,4) was to see whether
the variance due to geographical factors could be reduced. If so,
then this would have argued for a physical component of bloom
formation.

    Figure 9B depicts the observed red tides, temperature anomalies
and tidal amplitudes from a specific area, the Southern California
Bight, which can be directly compared with the original conceptual
model in Figure 9A. It is apparent that the days of the lunar cycle

with the significantly high number of red tides (days 28 and 29) occur 1-2 days after the maximum of the time-averaged temperature anomalies. However, there were other blooms throughout the lunar tidal cycle. Perhaps more surprising is the closer association of the average positive temperature anomalies with the spring tides and the average negative anomalies with the neaps. This is opposite to the idealized model in Figure 9A but is nevertheless consistent, in part, with Margalef's mandala. When the temperature anomaly is, on average, positive (i.e. the water column is more stratified), more red tides are observed. The mechanism of bloom formation must be more complicated, however, since 1) blooms are observed throughout the lunar tidal cycle, and 2) plotting the data relative to tidal range rather than lunar phase fails to lower the variance of the data set (the blooms are observed when the tidal range is low and high) (see Fig. 4).

Kamykowski (1981a) argues that dinoflagellate red tides result from horizontal concentration of dinoflagellates migrating diurnally through internal wave fields. Depending on the differences in current velocities, basin dimensions and migrational characteristics, his simulated populations form concentrated bands of cells parallel to the shore, 2-10 km apart. Essentially, the red tide formation relies on superposition of internal wave periods of 21.5 h, 12.5 h and 0.42 h with a 24 h diurnal pattern of dinoflagellate migration through a baroclinic water column. Conceivably, the fortnightly and lunar internal tides are important in this context as well. Kamykowski's model (1979; 1981a) is very appealing since several of the most frequently observed red tide organisms in California are strong vertical migrators (Gonyaulax polyedra, Prorocentrum micans, Gymnodinium splendens) (Kamykowski, 1981b). It is not known whether Gymnodinium breve, a very frequent cause of red tides off the west coast of Florida, is a vertical migrator. It is curious that G. polyedra, P. micans, and G. breve show a strong lunar bias in bloom formation (see Results section).

Given the preceding results, it is clear that the hypothesis that "dinoflagellate red tides are observed evenly through the spring-neap tidal cycle" cannot be negated on a global basis (Fig. 2). The hypothesis cannot be rejected even if the data are analyzed by geographic region, albeit the data in Figure 3 suggest some evidence for a lunar influence. These results for bloom-forming dinoflagellates are in contrast to the data of Winter et al. (1975), Sinclair (1978), Haas et al. (1980), Balch (1981) and other workers that observed changes in abundance of phytoplankton, not necessarily large-celled dinoflagellates, related to the lunar tidal cycle. The reasons for the discrepancy are not obvious but may be due to a suite of factors which increase the variance of the data set.

Perhaps the largest amount of variance in the data set results from the fact that the data are based on human observation. The "date of first sighting," although the best way to compare between observations, assumes that the bloom was not observable on the previous day, yet there might not have been an observer present. Moreover, it is known that red tides usually represent a later successional stage in the growth of dinoflagellate populations. The initiation of a bloom may occur weeks previous to the initial observation of a red tide. Alternatively, red tides could be advected into a region which might create an "observational lag."

One solution to these problems might be to intensively sample at one sight. When this is done (e.g. La Jolla, CA, USA), the pattern of bloom observation still does not correlate well to the tidal cycle. Nonetheless, in this one region 1) red tides are most frequently observed during that part of the lunar cycle when the sea surface temperature anomaly is, on average, positive, and 2) major red tides (with chlorophyll a levels >50 µg/ℓ) are often observed 1-2 days after a large negative temperature spike when the water is warming. Perhaps the combination of a sharp mixing event and slight stratification provides the necessary, but not always sufficient, conditions for bloom formation. Such events take place several times each year without bloom formation.

Variance due to the wide variety of species involved can be discerned. Not all species of dinoflagellates appear to be vertical migrators yet they can form surface slicks (Paasche et al., 1984). Thus the mechanism responsible for dinoflagellate aggregations proposed by Kamykowski (1979) is probably only one of several explanations for red tides. Other purely physical concentration mechanisms were reviewed by Ryther (1955). The fact that a non-motile diatom can form surface slicks (see Appendix, reference 6) is tribute to the fact that such physical mechanisms are important in bloom dynamics. Species specificity in bloom timing is further implied when the data are segregated taxonomically (see Results section). Indeed, Allen (1940) observed species specific seasonality in the appearance of dinoflagellates in the Southern California Bight.

One conclusion is particularly obvious from this work. It is that phytoplankton "red tides" cannot be treated as a single entity. Phytoplankton bloom formation is affected by a host of biological factors, namely, species specific migrational capabilities, growth rates, and grazing potential as well as physical factors such as lunar tidal cycles, semidiurnal internal tides, ephemeral mixing events, seasonality, etc.

Historical data are quite useful in the context of red tide prediction, particularly toxic red tide prediction. For example, merely knowing that the locality of concern is in the northern hemisphere allows one to predict that 13-14% of the red tides occur during each of the months of June, July, August and September (see Fig. 8). Recognition of the seasonality in toxic dinoflagellates many years ago resulted in the "mussel quarantine" each summer imposed by the California Department of Health in S. California. Given a larger data base, predictive abilities might be sufficiently enhanced to allow species specific bloom forecasting on shorter time scales than seasons. Factors such as the previous temperature record are clearly important for predicting large blooms. Moreover, particular species show a greater likelihood of blooming during certain phases of the lunar cycle (e.g. one quarter of the Prorocentrum blooms occurred on day 28 of the lunar cycle).

One goal of this symposium series is to outline the types of techniques and observational strategies needed to validate the relationship between physical structure, circulation and mixing to biomass and production. Future studies on red tide formation should be more temporally comprehensive in order to catch the initial stages of bloom formation. This is neither easy nor cheap since red tides are very sporadic in time and space (which is why I used the historical data base). Efforts should be made to remove the observational bias since this undoubtedly increases the variance of the data set. Remote sensing provides a partial solution to this problem but adequate water sampling is still required. Since red tides are difficult to sample, the most reasonable approach towards understanding them remains to study the specific red tide organisms and their distributions and natural history. Their abundance can then be examined in relation to the average physical environment to allow broader insights into the physical forcing functions important to bloom dynamics. The paper of Yentsch et al. (this volume) is an example of this approach.

SUMMARY

An historical, global record of dinoflagellate red tides is used to examine whether dinoflagellate blooms are correlated with the lunar (M_m) or fortnightly (M_f) tidal components. Blooms are tabulated with respect to the phase of the moon as well as tidal range. The data are analyzed with respect to taxa and geographic location.

Red tides are not well correlated, on a global or regional basis, to lunar phase or tidal range, in contrast to the findings of some other workers. These results are made more surprising by the fact that at one location, the Southern California Bight, a 62 year temperature record shows evidence for a lunar cycle of stratification and destratification. Large negative temperature anomalies are also associated with particular phases of the lunar cycle. Major blooms in this region (>50 µg chl a ℓ^{-1}) usually follow a large negative temperature anomaly suggesting the importance of upwelled water to bloom formation. These historical data suggest that several different mechanisms are probably responsible for red tides. Future work should focus on the natural history of specific organisms and their typical physical environments.

ACKNOWLEDGEMENTS

Many people have been instrumental in the completion of this work. Thanks go to researchers at SIO and elsewhere for reporting red tides to me. Betsy Stewart kindly provided the SIO pier temperature data and much computer assistance. Members of the Scripps Aquarium and Institute of Marine Resources took the daily pier temperatures. Drs. J.R. Enright, R.W. Eppley, L.W. Haas, G.A. Jackson and D. Lang provided valuable discussion on data interpretation. Dr. R.W. Eppley provided data and ideas on the major blooms in the S. California Bight region and kindly reviewed the manuscript along with Dr. C.M. Yentsch and two reviewers. M.A. Ogle typed the manuscript.   W.M.B.   is   supported   by   NSF   grants   OCE81-20773   and OCE80-08308.

REFERENCES

Allen, W.E.   1940.   Twenty years of statistical studies of marine plankton dinoflagellates of southern California.   Amer. Midl. Nat. 26: 603-635.

Balch, W.M.   1981.   An apparent lunar tidal cycle of phytoplankton blooming and community succession in the Gulf of Maine.   J. exp. mar. Biol. Ecol. 55: 65-77.

Balch, W.M., P.C. Reed and S.C. Surrey-Gent.   1983.   Spatial and temporal variability of dinoflagellate cyst abundance in a tidal estuary.   Can. J. Fish. Aquat. Sci. (Suppl.) 40: 244-261.

Bendat, J.S. and A.G. Piersol.   1971.   Random data: analysis and measurement procedures.   Wiley-Interscience, New York.   407 p.

Brongersma-Sanders, M.   1957.   Mass mortality in the sea.   Mem. Geol. Soc. Am. 67: 941-1010.

Eppley, R.W. and W.G. Harrison.   1975.   Physiological ecology by _Gonyaulax polyedra_, a red water dinoflagellate off Southern California, p. 11-22.   _In_ V.R. LoCicero (ed.).   The First International Conference on Toxic Dinoflagellate Blooms.   Mass. Sci. Techn. Found., Wakefield, MA, USA.

Garrett, C.J.R. and W. Munk.   1971.   The age of the tide and the "Q" of the ocean.   Deep-Sea Res. 18: 493-503.

Gran, H.H. and T. Braarud.   1935.   A quantitative study of the phytoplankton in the Bay of Fundy and the Gulf of Maine including observations on hydrography, chemistry, and turbidity.   J. Biol. Bd. Can. 1: 219-467.

Haas, L.W.   1977.   The effect of the spring-neap tidal cycle on the vertical salinity structure of the James, York and Rappahannock Rivers, Virginia, U.S.A.   Est. Coast. Mar. Sci. 5: 485-496.

Haas, L.W., S.J. Hastings and K.L. Webb.   1980.   Phytoplankton response to a stratification-mixing cycle in the York River Estuary during late summer.   _In_ B.J. Neilson and L.E. Crown (eds.).   Estuaries and Nutrients.   The Humana Press Inc., Clifton, NJ.

Harrison, W.G.   1976.   Nitrate metabolism of the red tide dinoflagellate _Gonyaulax polyedra_.   J. exp. mar. Biol. Ecol. 21: 199-209.

Hayward, D., L.W. Haas, J.D. Boon, III, K.L. Webb and K.K. Friedland. A regression model of neap-spring tidally associated stratification variation in the York River estuary, Maryland.   _In_ M. Bowman, C.M. Yentsch and W.T. Peterson (eds.), Tidal Mixing and Plankton Dynamics.   Springer-Verlag.   This volume.

Holligan, P.M. and D.S. Harbour. 1977. Vertical distribution and succession of phytoplankton in the western English Channel. J. mar. biol. Assn. U.K. 5: 1075-1093.

Holmes, R.W., P.M. Williams and R.W. Eppley. 1967. Red water in La Jolla Bay, 1964-1966. Limnol. Oceanogr. 12: 503-512.

Jackson, G.A. 1977. Nutrients and production of giant kelp, Macrocystis pyrifera off southern California. Limnol. Oceanogr. 22: 979-995.

Jackson, G.A. 1983. The physical and chemical environment of a kelp community, pp. 11-37. In W. Bascom (ed.). The Effects of Waste Disposal on Kelp Community. Southern California Coastal Water Research Project, Long Beach, CA. 328 p.

Kamykowski, D. 1979. The growth response of a model Gymnodinium splendens in stationary and wavy water columns. Mar. Biol. 50: 289-303.

Kamykowski, D. 1981a. The simulation of a southern California red tide using characteristics of a simultaneously-measured internal wave field. Ecological Modeling 12: 253-265.

Kamykowski, D. 1981b. Laboratory experiments on the diurnal vertical migration of marine dinoflagellates through temperature gradients. Mar. Biol. 62: 57-64.

Ketchum, B. and D.J. Keen. 1948. Unusual phosphorous concentrations in the Florida "Red Tide" seawater. J. Mar. Res. 7(1): 17-21.

Margalef, R. 1978. Life-forms of phytoplankton as survival alternatives in an unstable environment. Oceanol. Acta 1: 493-509.

Margalef, R. 1979. Functional morphology of organisms involved in red tides as adapted to decaying turbulence, pp. 89-94. In D.L. Taylor and H.H. Seliger (eds.). Toxic Dinoflagellate Blooms. Elsevier/North Holland.

Nautical Almanac, 1900-1983. United States Nautical Almanac Office, U.S. Naval Observatory, Washington, D.C.

Paasche, E., I. Bryceson and K. Tangen. 1984. Interspecific variation in dark nitrogen uptake by dinoflagellates. J. Phycol. 20: 394-401.

Pingree, R.D., P.M. Holligan, G.T. Mardell and R.N. Head. 1976. The influence of physical stability on spring, summer and autumn phytoplankton blooms in the Celtic Sea. J. mar. biol. Assn. U.K. 56: 845-873.

Pond, S. and G.L. Pickard. 1978. Introductory dynamic oceanography. Pergamon Press, New York. 241 p.

Postma, H. 1967. Sediment transport and sedimentation in the estuarine environment, pp. 158-179. In G.H. Lauff (ed.). Amer. Assoc. Adv. Sci., Publication 83.

Ryther, J.H. 1955. Ecology of autotrophic marine dinoflagellates with reference to red water conditions, pp. 387-414. In F.H. Johnson (ed.). The Luminescence of Biological Systems. Amer. Assoc. Adv. Sci., Washington, D.C.

Simpson, J.H. 1981. The shelf-sea fronts: implications of their existence and behavior. Phil. Trans. R. Soc. Lond. A302: 531-546.

Simpson, J.H. and D. Bowers. 1981. Models of stratification and frontal movement in shelf seas. Deep-Sea Res. 28: 727-738.

Simpson, J.H. and J.R. Hunter. 1974. Fronts in the Irish Sea. Nature 250: 404-406.

Sinclair, M. 1978. Summer phytoplankton variability in the lower St. Lawrence Estuary. J. Fish. Res. Bd. Can. 35: 1171-1185.

Sverdrup, H.U. 1953. On conditions for the vernal blooming of phytoplankton. J. Cons. Perm. Internatl. Explor. Mer. 18: 287-295.

Taylor, D.L. and H.H. Seliger. (eds.) 1979. Toxic Dinoflagellate Blooms. Elsevier/North Holland, New York. 505 p.

U.S. Dept. Commerce, NOAA National Ocean Survey. Tide tables high and low water prediction. U.S. Govt. Printing Ofc., Washington, D.C.

Webb, K.L. and C.F. D'Elia. 1980. Nutrient and oxygen redistribution during a spring neap tidal cycle in a temperate estuary. Science 207: 983-985.

Winter, D.F., K. Banse and G.C. Anderson. 1975. The dynamics of phytoplankton blooms in Puget Sound, a fjord in the Northwestern United States. Mar. Biol. 29: 139-176.

Yentsch, C.M., P.M. Holligan, W.M. Balch and A. Tvirbutas. Tidal stirring vs. stratification: Phytoplankton dynamics with special reference to toxic dinoflagellates. In M. Bowman, C.M. Yentsch and W.T. Peterson (eds.). Tidal Mixing and Plankton Dynamics. Springer-Verlag. This volume.

Zar, J.H. 1974. Biostatistical analysis. Prentice-Hall Inc., N.J. 620 p.

REF #	RED TIDE OR POISONING	REFERENCE	LOCATION	YEAR	DATE FIRST SITED	DATE OF PREVIOUS NEW MOON	DAY LUNAR CYCLE	SPECIES OR DESCRIPTION
1	R	Allen, W.E., Amer. Midland Nat. 26:603-635	La Jolla, CA U.S.A.	1917	4-VI	20-V	15	P. micans
2	R	Allen, W.E., S.I.O. Bull. Tech Series No. 1: 347-356	"	1924	I-VI	3-V	30	"
3	R	W.F. Allen, Naut. Gazette, 26 Aug 1939	"	1939				
4	R	W.E. Allen, Trans of Amer. Microp. Soc. 62: 262-264	"	1942	11-IX	10-IX	1	G. polyedra
5	R	W.E. Allen, Science, 88:55-56	Copalis Beach, WA U.S.A.	1938	14-V	30-IV	14	Aulacodiscus kittoni
6	R	"	Seaside, WA, N. Columbia River	1938	15-V	30-IV	15	Asterionella Kariana
7	R	"	La Jolla, CA U.S.A.	1938	"Later May"	29-V	1?	G. polyedra
8	R	W.E. Allen, Trans. of Amer. Microp. Soc. 65:149-153	"	1945	⁻13-IX	6-IX	7	G. polyedra
9	R	W.E. Allen, Science, 78:12-13 (1933)	"	1933	17-V	24-IV	23	C. tripos P. micans
10	R	W.E. Allen, Science, 82:325-326 (1935)	"	1935	30-VII	30-VII	29	Yellow flagellate
11	R	Mead, Science, 8:707-709 (1898)	Providence, R.I. U.S.A.	1898	8-IX			Peridinium ap.
12	R	"	Khattywar, India	1949	27-X			"
13	R	Torrey, Am. Nat. 36: 187-192 (1902)	San Pedro Harbor U.S.A.	1902	7-VII	5-VII	2	Gonyaulax ap.
14	R	Cullen et al., J. Exp. Mar. Biol. Ecol. 63:67-80 (1982)	La Jolla, CA U.S.A.	1980	21-VII	12-VII	9	Yellow tide
15	R	W. Balch - P.O.?	"	1981	30-X	27-X	3	?
16	R	W. Balch and L. Vakasian, P.O.	"	1981	10-XI	27-X	14	G. polyedra?
17	R	W. Balch - P.O.	Del Mar, CA U.S.A.	1981	26-XI	26-XI	29	?
18	R	D. Goodman - P.O.	La Jolla, CA U.S.A.	1982	11-I	17-I	16	?
19	R	W. Balch - P.O.	"	1982	19-I	17-I	2	?
20	R	W. Balch, R. Cowen - P.O.	80 km S. of Tijuana, Mexico	1982	25-I	17-I	8	?
21	R	J. Nelson, W. Balch	La Jolla, CA U.S.A.	1982	30-I	17-I	13	?
22	R	Finucane et al., Fla. Board of Conservation Rep.	Tampa Bay, FL U.S.A.	1963	11-IV	25-III	17	G. breve
23	R	B. Sweeney, 1st Int. Conf. on Toxic Dino. Blooms, Mass. Sci. Tech. Foundation, 1975	La Jolla, CA U.S.A.	1952	11-VII	22-VI	19	P. micans
24	R	"	"	1954	9-III	5-III	4	"
25	R	"	"	1954	26-IV	3-IV	23	"
26	R	"	"	1955	21-IV	24-III	28	"
27	R	"	Ensenada, Mexico	1958	17-VI	17-VI	29	G. polyedra
28	R	"	"	1958	4-VII	17-VI	17	"
29	R	"	Imperial Beach, CA, U.S.A.	1958	21-VII	30-VII	5	"
30	R		La Jolla, CA U.S.A.	1958	15-VIII	15-VIII	29	"
31	R	"	"	1961	13-IV	16-III	28	P. micans
32	R	"	"	1961	28-IV	15-IV	13	G. polyedra
33	R	"	Oyster Bay, Jamaica	1966	13-I	22-XII	22	P. bahamense
34	R	"	Bostrain Bay, New Guinea	1969	13-X	11-X	2	G. polygramma

REF #	RED TIDE OR POISONING	REFERENCE	LOCATION	YEAR	DATE FIRST SITED	DATE OF PREVIOUS NEW MOON	DAY LUNAR CYCLE	SPECIES OR DESCRIPTION
35	R	"	Santa Monica CA U.S.A.	1970	4-X	30-IX	4	G. polyedra
36	R	"	"	1971	30-VII	22-VII	8	"
37	R	"	Rincon, CA U.S.A.	1971	4-VIII	22-VII	13	"
38	R	"	"	1971	7-VIII	22-VII	16	"
39	R	"	Goleta Bay, CA U.S.A.	1971	16-VIII	22-VII	25	"
40	R	"	"	1971	18-VIII	22-VII	27	"
41	R	"	"	1972	8-IX	7-IX	1	"
42	R	"	"	1972	15-IX	7-IX	8	"
43	R	"	"	1973	22-X	26-IX	26	"
44	R	"	Long Beach, CA U.S.A.	1974	2-VIII	19-VII	13	"
45	R	"	Ventura, CA U.S.A.	1974	10-VIII	19-VII	21	"
46	R	"	Rincon, CA U.S.A.	1974	19-VIII	17-VIII	2	"
47	R	Eppley & Harrison, 1st Int. Conf. on Toxic Dino. Blooms, Mass. Sci. Tech. Found. 1975	Seal Beach, CA U.S.A.	1967	3-IX	6-VIII	28	Ceratium sp.
48	R	"	Sorrento Slough, CA, U.S.A.	1968	15-II	29-I	17	?
49	R	"	La Jolla, CA U.S.A.	1968	27-III	28-II	28	G. polyedra P. micans
50	R	"	"	1968	17-V	27-IV	20	Ceratium sp. P. micans
51	R	"	"	1969	26-II	16-II	10	G. polyedra
52	R	"	"	1969	27-V	16-V	11	P. micans
53	R	"	Newport Beach, CA U.S.A.	1970	3-IX	31-VIII	3	?
54	R	"	La Jolla, CA U.S.A	1970	22-X	30-IX	22	G. polyedra
55	R	"	"	1970	9-XII	28-XI	11	"
56	R	"	"	1971	22-VIII	20-VIII	2	Mesodinium C. furca
57	R	"	La Jolla, CA U.S.A.	1972	24-IV	13-IV	11	Dino. mixture
58	R	"	"	1974	8-VIII	19-VII	20	G. polyedra
59	R	"	El Segundo, Los Angeles, CA, U.S.A.	1974	16-IX	16-IX	29	?
60	R	Hartwell, 1st Int. Conf. on Toxic Dino Blooms. Mass. Sci. Techn Found. 1975	Essex, MA U.S.A	1974	20-V	22-IV	28	G. tamarensis
62	R	"	Offshore Ipswich Bay, MA, U.S.A.	1972	4-IX	9-VIII	26	"
63	R	Zubkoff & Warinner Ibid	Chesapeake Bay U.S.A.	1973	11-V	2-V	9	?
64	R	"	"	1973	16-V	2-V	14	?
65	R	"	"	1973	11-IX	28-VIII	14	?
66	R	"	"	1973	28-IX	26-IX	2	?
67	R	"	"	1973	31-X	26-X	5	?
68	R	"	"	1974	21-II	23-I	29	?
69	R	"	"	1974	4-III	22-II	10	?
70	R	"	"	1974	27-III	23-III	4	?
71	R	"	"	1974	11-IV	23-III	19	?
72	R	"	"	1974	7-V	22-IV	15	?
73	R	"	"	1974	24-V	21-V	3	?
74	R	"	"	1974	10-VI	21-V	20	?
75	R	"	"	1974	2-VII	20-VI	12	?
76	R	"	"	1974	12-VIII	19-VII	24	?
77	R	"	"	1974	16-IX	16-IX	29	?
78	R	Yentsch & Salvagio, Ibid	Ipswich Bay, MA U.S.A.	1972	14-IX	7-IX	7	G. tamerensis

REF #	RED TIDE OR POISONING	REFERENCE	LOCATION	YEAR	DATE FIRST SITED	DATE OF PREVIOUS NEW MOON	DAY LUNAR CYCLE	SPECIES OR DESCRIPTION
79	R	"	Boothbay Harbor, MA, U.S.A.	1974	8-IX	17-VIII	22	"
80	R	Seliger, Ibid	Chesapeake Bay U.S.A.	1970	19-VI	4-VI	15	P. minimum
81	R	"	"	1971	18-VI	24-V	15	"
82	R	Wardle et al., Ibid	Galveston, TX U.S.A.	1971	25-VIII	20-VIII	5	G. monilata
83	R	"	"	1972	12-VIII	9-VIII	3	"
84	R	Habas & Gilbert, Ibid	St. Petersburg, FL, U.S.A.	1973	22-X	26-IX	26	G breve
85	R	"	Redington Beach, FL, U.S.A.	1973	3-IV	3-IV	29	"
86	R	Jensen, Ibid	Long Is. Sound, NY, U.S.A.	1972	4-X	7-IX	27	Gonyaulax Peridinium Ceratium
87	P	Hurst, Ibid	Cape Elizabeth, ME, U.S.A.	1974	30-V	21-V	9	G. tamarensis
88	P	"	Hampton, N.H. U.S.A.	1974	21-VIII	17-VIII	4	"
89	R	Machado, in "Toxic Dino. Blooms" Proc. of 2nd Int. Conf of Toxic Dino Blooms, Elsevier-N. Holland, NY 1979	Hermenegilde, S. Brazil	1978	1-IV	9-III	23	Gymnodinium sp.
90	R	Yentsch & Mague, Ibid	Monhegan Is., ME U.S.A.	1977	1-X	13-IX	18	G. tamarensis
91	R	Bodeanu & Usurelu, Ibid	Romanian Black Sea	1976	28-VI	27-VI	1	Exuviella
92	R	"	"	1976	4-VIII	27-VII	8	"
93	R	Fraga & Sanchez, Ibid	NW Spain	1977	30-IX	13-IXZ	17	Amphidinium
94	R	"	Rio de Vigo, Spain	1952	States that Outbreaks Coincided		14	"
95	R	"	"	1953	With Full		14	"
96	R	"	"	1954	Moon		14	"
97	R	Reyes, Vasquez et al., Ibid	Margarita Is. NE Venezuela	1977	30-IV	18-IV	12	G. tamerensis
98	P	"	Sucre Stae Venezuela	1977	31-VII	16-VII	15	Cochlodinium
99	R	Zotter, Ibid	Galveston, TX U.S.A.	1976	18-XII	21-XI	27	Exuviella
100	R	Roberts, Ibid	Venice, FL U.S.A.	1976	22-IX	25-VIII	28	G. breve
101	R	"	"	1976	7-X	23-IX	14	G. breve
102	R	"	Sarasota, FL U.S.A.	1977	3-X	13-IX	20	G. breve
103	R	Blasco, Ibid	Cabo Nazoo, Peru	1978	4-IV	9-III	26	?
104	R	"	"	1978	10-IV	7-IV	3	?
105	R	Kat, Ibid	Dutch Coast, N. Sea	1971	5-VII	22-VI	13	P. micans
106	R	"	"	1971	20-VII	23-VI	28	P. redfieldii
107	R	"	"	1971	3-VIII	22-VII	12	C. fusus
108	R	"	"	1972	5-VII	11-VI	24	Noctiluca miliaris
109	R	"	"	1972	21-VI	11-VI	10	"
110	R	"	"	1973	2-VII	30-VI	2	"
111	R	"	"	1973	18-IX	26-IX	20	P. redfieldii P. micans
112	R	"	"	1974	7-VIII	19-VII	19	C. fusus
113	R	"	"	1974	20-VI	20-VI	29	N. miliaris
114	R	"	"	1975	8-IX	5-IX	3	P. redfieldii P. micans
115	P	Hurst, Ibid	SW Maine, U.S.A.	1976	30-IV	29-IV	1	G. tamerensis
116	P	"	"	1976	30-VI	27-VI	3	"
117	P	"	Moose Cove, ME U.S.A.	1976	14-VII	27-VI	17	"
118	R	Holligan, Ibid	Ushant, W. Eng. Channel	1976	31-VII	27-VII	4	G. aureolum

REF #	RED TIDE OR POISONING	REFERENCE	LOCATION	YEAR	DATE FIRST SITED	DATE OF PREVIOUS NEW MOON	DAY LUNAR CYCLE	SPECIES OR DESCRIPTION
119	R	Wyatt, Ibid	Endeavor Cruise	1768	7-XI			
120	R	Haddad & Carder, Ibid	Cape Romano, FL U.S.A.	1977	28-X	12-X	16	G. breve
121	R	"	"	1977	23-XI	11-XI	12	"
122	R	Zubkoff et al., Ibid.	York River Chesapeake, U.S.A.	1976	28-VII	27-VII	1	G. splendens
123	R	"	"	1976	2-VIII	27-VII	6	"
124	R	Clark, Ibid	Mill Creek, Chesapeake, U.S.A.	1977	15-VIII	14-VIII	1	G. nelsonii
125	R	Mueller, Ibid	Dry Tortugas, Carribean	1977	13-X	12-X	1	G. breve
126	R	Morey Gaines, Ibid	Los Angeles Hbr, CA, U.S.A.	1976	16-VIII	27-VII	20	Ceratium sp.
127	R	White, Ibid	Bay of Fundy, Canada	1977	15-IX	13-IX	2	G. tamarensis
128	R	Balch, J. Exp. Mar. Biol Ecol.	Gulf of Maine U.S.A.	1980	30-VIII	10-VIII	20	G. tamarensis
129	R	"	"	1978	25-VI	5-VI	20	"
130	R	A. Krikos, P.O.	La Jolla, CA U.S.A.	1982	15-II			Intense Bioluminesence
131	R	Baker & Dustan,	San Diego to San Fransciso, U.S.A.	1980	21-VI	12-IV	9	?
131	R	D. Redalje, P.O.	La Jolla, CA U.S.A.	1982	6-III			?
132	R	G. Kleppel, P.O.	Pt Arena To Pt. Reyes	1980	25-VI	12-VI	13	?
133	R	"	"	1980	11-VII	28-VI	13	?
134	R	Grindley and Nel, S. Afr. Divi. Fish. Bull. 6:36-55 (1970)	Elands Bay, S. Africa	1966	18-XII	12-XII	6	G. grindleyii
135	R	"	Adjacent to Elands Bay, S. Africa	1966	16-XII	12-XII	4	"
136	R	"	Ysterfontein S. Africa	1967	24-I	10-I	14	?
137	R	Ballantine & Smith, Br. Phycol. J. 8:233-238	N. Wales, Conwy, U.K.	1971	5-X	19-IX	16	?
138	R	"	Llandudno, U.K. (?)	1971	5-X	19-IX	16	?
139	R	Hickel & Drebes, Helgo. Wiss. Meeresunteres 22:401-416	N. Sea Helgoland Bight	1968	14-VIII	25-VII	20	G. breve
140	R	"	"	1968	30-VIII	23-VIII	7	"
141	R	"	Esbjerg (N. Sea)	1968	14-X	22-IX	22	"
142	R	Pingree. Nature 258:672-677 (1975)	Ushant, W. English Channel	1975	26-VII	9-VII	17	?
143	R	Iizuka & Irie, Bull. Plank. Soc. Japan 16:99-115 (1969)	Omura Bay, Japan	1967	31-VIII	6-VIII	25	G. breve
144	R	"	"	1968	20-IX	23-VIII	27	"
145	R	Ujeno, Bull. Plank. Soc. 16:89-98 (1969)	Japan Estuary	1967	10-VII	7-VII	3	S. costatum
146	R	"	"	1968	7-VII	25-VI	12	"
147	R	"	"	1968	21-IX	23-VIII	29	G. breve
148	R	Iizuka, Bull. Plank. Soc. Japan 19:22 (1972)	Omura Bay, Japan	1967	1-VIII	7-VII	25	Gymnodinium sp.
149	R	"	"	1968(?)	7-VIII	25-VII	13	"
150	R	Tangen, Sarsia 63:128-133 (1977)	Stavanger, Norway (?)	1976	15-XI	23-X	23	G. aureolum
152	R	Brongersma-Sanders, Geol. Soc. of America 67:941-1010 (1957)	Brit. Columbia	1933	28-IV	24-IV	4	Meso. rub
154	P	"	Cal. U.S.A.	1896	End Feb			?
155	R	"	"	1917	4-VI	20-V	15	?
156	R	"	"	1907	Early Aug.			?
157	P	"	San Francisco, CA U.S.A.	1938	5-IX	25-VIII	11	?
158	P	"	Fort Bragg, U.S.A.	1938	12-IX	25-VIII	18	?
159	P	"	San Luis Obispo, CA. U.S.A.	1946	19-VI	30-V	20	?

REF #	RED TIDE OR POISONING	REFERENCE	LOCATION	YEAR	DATE FIRST SITED	DATE OF PREVIOUS NEW MOON	DAY LUNAR CYCLE	SPECIES OR DESCRIPTION
160	R	?	Angel de la Gardia Is., Gulf of Cal.	1937	20-III	12-III	8	N. scintillans
161	R	"	Chile	1895	7-II			?
162	P	"	Manila, Phillipines	1767	22-IX			?
163	R	"	Gulf of Maine, U.S.A.	1882	1-VIII			?
164	R	"	Naragansett Bay, R.I., U.S.A.	1886	18-VI			?
165	R	"	"	1886	1-VI			?
166	R	"	Iceland	1904	18-VII	12-VII	6	Mesodin. rubrum
167	R	"	Portugal	1845	3-VI			"Protococcus"
168	R	"	Zaton, Black Sea	1913	21-VII	3-VII	18	Yellow-red water
169	R	"	Cape Blanco, Morocco, Afr.	1951	17-VIII	2-VIII	15	?
170	P	"	Dakar	1944	17-II	25-I	23	?
171	P	"	"	1946	25-III	3-III	22	?
172	R	"	S.W. Africa	1950	25-XII	9-XII	16	Glenodinium
173	R	"	"	1851	5-XII			?
174	R	"	"	1880	21-XII			?
175	R	"	"	1920	10-XII	9-XII	1	?
176	R	"	"	1925	23-XII	15-XII	8	?
177	R	"	"	1950	25-XII	9-XII	16	?
178	R	"	Rio de Janeiro, Brazil	1945	24-I	14-I	10	Glenodinium
179	R	"	"	1946	26-VIII	26-VIII	29	?
180	R	"	"	1948	16-IV	9-IV	7	?
181	R	?	Florida, U.S.A.	1882	20-VII			?
182	R	"	"	1946	20-XI	24-X	27	?
183	R	"	"	1947	20-VI	18-VI	2	?
184	R	"	"	1947	18-VII	18-VII	29	?
185	R	"	"	1952	25-X	18-X	7	?
186	R	"	"	1952	11-XI	18-X	24	?
187	R	"	Dry Tortugas, Caribbean	1953	3-I	31-XII-52	3	?
188	R	"	"	1953	22-I	15-I	7	?
189	P	"	Vera Cruz, Mexico	1797	10-XI			?
190	R	"	Yemen, Red Sea	1947	23-VIII	16-VIII	7	?
191	R	"	Cape Madraba, S.E. Arabia	1936	27-IX	15-IX	12	?
192	R	"	20°N x 59°E	1950	24-X	11-X	12	?
193	R	"	Kathiawar, N.W. coast India	1849	27-X			?
194	R	"	Malabar, W. India	1507	25-VIII			?
195	R	"	"	1861	1-X			?
196	R	"	"	1916	"Last wk August"	28-VIII		?
197	R	"	"	1916	25-IX	28-VIII	28	?
198	R	"	"	1916	9-X	26-X	13	?
199	R	"	"	1922	22-IX	20-IX	2	?
200	R	"	"	1944	3-XI	17-X	17	?
201	R	"	Malabar, W. India	1946	20-IX	26-VIII	25	?
202	R	"	"	1946	31-X	24-X	7	?
203	R	"	Gulf of Manor, India	1942	17-V	15-V	2	?
204	P	"	Baltic	1931	5-XI	11-X	25	?
205	P	"	Bulgarian Black Sea	1940	24-VII	5-VII	19	?

REF #	RED TIDE OR POISONING	REFERENCE	LOCATION	YEAR	DATE FIRST SITED	DATE OF PREVIOUS NEW MOON	DAY LUNAR CYCLE	SPECIES OR DESCRIPTION
206	P	Wyatt. J. Cons. Int. Exp. Mer 39:1-6	S.W. Africa	1880	21-XII	*	15 or 29?	?
207	R	B. Balch & D. Redalje, P.O.	La Jolla, CA U.S.A.	1982	4-IV			G. polyedra
208	R	"	"	1982	22-IV	23-IV	28	G. polyedra
209	P	Rounsefell & Nelson, U.S. Dept. of Interior, Fish and Wildlife Svc. Bureau Comm. Fish No. 535, 1966	Texas, U.S.A.	1935	27-VI	1-VI	26	?
210	P	Chew, Ibid. 1955	Dry Tortugas, Caribbean	1953	22-I	15-I	7	?
211	R	Funicane, Ibid, 1960	St. Petersberg, FL, U.S.A.	1959	29-IX	3-IX	26	G. breve
212	P	"	Egmont Key, FL, U.S.A.	1959	22-X	2-X	20	?
213	P	"	"	1963	3-IV	25-III	9	?
214	P	Glennan, U.S. Fish Comm. Bull. 1887	"	1885	28-X			?
215	R	Gunter, Sci 105:256	Dry Tortugas, Caribbean	1947	19-I	23-XII	27	Gymnodinium
216	P	Gunter, Zool. Monogr. 18:309	Naples, FL	1946	20-XI	9-XI	11	?
217	P	"	Cape Fla., FL, U.S.A.	1947	2-IV	22-III	11	?
218	R	Hart, J. Nature Nature 134:439	Cape Penninsula, S. Afr.	1934	Mid July	11-VII		?
219	R	Hela. Bull. Mar. Sci. Gulf Carib. 5:269	Dragovich, Caribbean?	1954	24-XI	26-X	29	G. breve
220	R	Hutton, R.F., Quart. Fla. Acad. Sci. 23:163	Egmont Key, FL, U.S.A.	1966	23-III	26-II	25	"
221	R	"	Tampa Bay, FL, U.S.A.	1966	27-III	27-III	29	G. splendens P. micans
222	P	Ingersoll, Proc. U.S. Nat. Mus. 3:74-80	Egmont	1880	17-X			?
223	R	Jefferson, J.P., Ibid 1:363	Dry Tortugas, Caribbean	1878	20-XI			?
224	R	Lackey, J.B., Quart. J. Fla. Acad. Sci. 19:71	Trinidad	1955	30-III	24-III	6	G. breve
225	P	"	Longboat Key, FL, U.S.A.	1953	18-IX	8-IX	9	?
226	P	"	Midnight Pass, FL, U.S.A.	1953	18-XII	6-XII	12	?
227	P	"	Sanibel Light, FL, U.S.A.	1954	18-III	5-III	13	?
228	R	"	Big Sarasota Bay, FL, U.S.A.	1953	3-IX	9-VIII	25	G. breve
229	R	"	Lemon Bay, FL U.S.A.	1954	18-VI	1-VI	17	?
230	R	Numann, Arch. Fischereiwiss 8:204-209	Angola	1951	27-VII	4-VII	23	?
231	R	Pomeroy et al., Limnol. Oceanogr. 1:54-60 (1956)	Delaware Bay, U.S.A.	1955	21-IV	24-III	28	Gymnodinium sp.
232	R	Slobodkin, J. Mar. Res. 12:148	26°N x 84°W	1952	2-VI	23-V	10	Mesodinium
233	R	"	Fort Myers, FL U.S.A.	1947	20-VI	18-VI	21	?
234	R	"	Boca Grand, FL, U.S.A.	1916	3-X	26-IX	8	?
235	R	"	"	1952	25-X	18-X	7	?
236	R	Sommer & Clark, Cal. Fish and Game 32:100 (1946)	Santa Monica, CA, U.S.A.	1946	19-VI	30-V	20	Ceratium
237	P	Walker, Proc. U.S. Nat. Mus. 6:105-109 (1884)	Bird Key, FL, U.S.A.	1880	20-XI	30-V	20	Ceratium
238	R	Obum et al., Bull. Mar. Sci. Carib. 1955	Sanibel Is., FL, U.S.A.	1954	30-IV	3-IV	27	G. breve
239	R	"	Sarasota Pt., FL, U.S.A.	1954	12-VIII	29-VII	14	"

REF #	RED TIDE OR POISONING	REFERENCE	LOCATION	YEAR	DATE FIRST SITED	DATE OF PREVIOUS NEW MOON	DAY LUNAR CYCLE	SPECIES OR DESCRIPTION
240	R	"	Englewood, FL, U.S.A.	1954	23-VII	30-VI	23	"
241	R	"	Sanibel Is., FL, U.S.A.	1953	18-IX	8-IX	10	G. breve
242	R	B. Balch & S. Horrigan, P.O.	La Jolla, CA U.S.A.	1982	15-V	22-V	21	?
243	P	McLean, Papau New Guinea Agri. Jour. 24:131-138	Talasea, Papua, New Guinea	1961	20-IV	15-IV	5	?
244	P	"	Manus Is. New Guinea	1962	23-V	4-V	19	?
245	R	"	Talasea, Papau New Guinea	1963	29-VI	21-VI	8	?
246	R	"	Port Moresby, New Guinea	1970	16-XI	30-X	17	?
247	R	"	Lae, New Guinea	1971	7-IV	26-III	12	?
248	R	"	Marshall Lagoon, New Guinea	1971	18-IV	26-III	23	?
249	R	"	Salamana, New Guinea	1971	31-V	24-V	7	?
250	R	"	Port Moresby, New Guinea	1972	13-III	15-II	27	?
251	R	"	"	1972	27-IV	13-IV	14	P. bahamense
252	P	"	"	1972	12-V	13-IV	30	?
253	R	"	"	1973	15-II	3-II	12	?
254	R	"	Amazon Bay, New Guinea	1973	20-II	3-II	17	?
255	R	"	Kapa Kapa to Hood Pt. New Guinea	1973	9-III	5-III	4	?
256	R	"	Talasea, New Guinea	1973	10-V	2-V	8	?
257	R	"	Ruliger Pt., Milne Bay, W. New Britain	1973	11-V	2-V	9	?
258	R	"	Milne Bay, New Guinea	1973	26-VI	1-VI	25	P. bahamense
259	R	S. Placier, P.O.	Arica-Iquique, Chile	1956	18-IV	11-IV	7	P. micans
260	R	"	Valparaiso, Chile	1968	9-III	28-II	10	Mesodinium rubrum
261	R	"	Punta Arenas, Chile	1972	22-X	7-X	15	G. catenella
262	R	"	"	1973	23-I	4-I	19	Amphidinium sp.
263	R	"	"	1975	10-III	11-II	27	Mesodinium rubrum
264	R	"	Valparaiso, Chile	1975	9-III	11-II	26	"
265	R	"	Chañaral, Chile	1975	12-XII	3-XII	9	"
266	R	"	Mejillon, Chile	1976	6-I	1-I	5	C. tripos
267	R	"	Antofagasta, Chile	1976	25-II	31-I	25	P. micans
268	R	"	Mejillon, Chile	1976	22-IV	30-III	23	C. furca
269	R	"	Antofagasta, Chile	1976	7-IX	30-III	8	P. micans
270	R	"	Arica, Chile	1976	15-V	29-IV	16	Gymnodinium sp.
271	R	"	Antofagasta, Chile	1976	28-X	23-X	5	?
272	R	"	"	1976	30-XI	21-XI	9	"
273	R	"	"	1977	10-I	21-XII-1976	20	"
274	R	"	Arica, Chile	1977	10-I	21-XII-1976	20	Glenodinium sp.
275	R	"	Ayman, Chile	1978	16-III	9-III	7	Mesodinium rubrum
276	R	"	Valparaiso, Chile	1979	7-V	26-IV	11	P. micans
277	R	"	Arica to Antofagasta, Chile	1980	27-XI	7-XI	20	M. rubrum
278	R	"	Iquique, Chile	1980	23-XII	7-XII	16	P. graale
279	R	"	Valparaiso, Chile	1981	8-IV	4-IV	4	Scrippsiella trochoides

REF #	RED TIDE OR POISONING	REFERENCE	LOCATION	YEAR	DATE FIRST SITED	DATE OF PREVIOUS NEW MOON	DAY LUNAR CYCLE	SPECIES OR DESCRIPTION
280	P	Quayle, D.B. Fish. Res. Bd. Can. 168:1-67	British Columbia Canada	1793	15-VI			?
281	P	"	British Columbia, Canada	1942	2-V	15-IV	17	?
282	P	"	"	1943	20-VII	2-VII	18	?
283	P	"	"	1948	1-IX	5-VIII	27	?
284	P	"	"	1949	13-IX	24-VIII	20	?
285	P	"	"	1957	23-X	8-X	15	?
286	P	"	"	1963	12-VII	21-VI	21	?
287	P	"	"	1963	20-VI	23-V	28	?
288	P	"	"	1965	1-VI	30-V	2	?
289	P	"	"	1958	4-II	19-I	16	?
290	P	"	"	1943	12-III	6-III	6	?
291	P	"	"	1963	18-XI	16-XI	2	?
292	P	"	"	1967	19-VI	8-VI	11	?
293	P	B. Balch & F.T. Haxo, P.O.	La Jolla, CA U.S.A.	1983	9-IV	12-IV	25	?
294	R	Guzman & Campodonica, Interciencia 3:144-151 (1972)	Tierra del Fuego S. America	1972	22-X	7-X	15	G. catenella
295	R	Guzman et al., Apartado Analis del Instituto de la Patagonia 6:173-183	"	1972	11-XI	6-XI	5	"
296	R	"	"	1972	28-XI	6-XI	22	"
297	R	B. Sweeney & A. Dodson, P.O.	La Jolla, CA U.S.A.	1960	30-VI	24-VI	6	G. polyedra
298	R	"	"	1964	10-V	12-IV	28	Mixed species
299	R	"	"	1965	1-VI	30-V	2	G. polyedra
300	R	Robinson & Brown, Can. J. Fish. Aquat. Sci. 40:2135-2143	Esquimalt Lagoon Juan de Fuca Strait	1974	14-X	16-IX	28	?
301	R	"	"	1976	5-X	23-IX	12	G. sanguineum
302	R	"	"	1976	16-XI	23-X	24	"
303	R	"	"	1976	21-XII	21-XII	29	"
304	R	"	"	1977	27-IX	13-IX	14	Gymnodinium sp.
305	R	"	"	1978	14-IX	2-IX	12	G. sanguineum
306	R	"	"	1978	2-X	2-X	29	"
307	R	"	"	1979	15-X	21-IX	24	Gymnodinium sp.
308	R	"	"	1980	24-VI	12-VI	12	"
309	R	"	"	1980	6-VIII	12-VII	25	"
310	R	"	"	1980	20-VIII	10-VIII	10	"
311	R	"	"	1980	22-X	9-X	13	G. sanguineum

†"P.O." means "Personal Observation"

* Reads, "...it being spring tide"

TIDAL STIRRING VS. STRATIFICATION:
MICROALGAL DYNAMICS WITH SPECIAL REFERENCE TO
CYST-FORMING, TOXIN PRODUCING DINOFLAGELLATES

C.M. Yentsch
Bigelow Laboratory for Ocean Sciences
McKown Point
W. Boothbay Harbor, ME   04575 USA

P.M. Holligan
Marine Biological Association
Citadel Hill
Plymouth PL1 2PB
United Kingdom

W.M. Balch
Institute of Marine Resources A018
Scripps Institute of Oceanography
La Jolla, CA   92093 USA

A. Tvirbutas
Charles S. Draper Laboratory
555 Technology Square
Cambridge, MA   02138 USA

INTRODUCTION

The early literature on the distribution of toxic dinoflagellates indicates that saxitoxin-producing microalgae of the Gonyaulax type are confined to temperate waters with greater than 24° North or South latitude (e.g. Schantz, 1970). However, since that publication toxin-producing Gonyaulacoid dinoflagellates have been recorded in Matzalan, Mexico (B.C. Abbott, per. comm.), Venezuela (Ferraz-Reyes et al., 1979), and India (Karunasagar et al., 1984). These observations fail to support the claim that distribution patterns are related to latitude. We ask, if conditions of temperate waters are not the governing forces, then what are? Do toxic Gonyaulacoid dinoflagellates occur more generally in regions of high productivity associated, for example, with upwelling and frontal boundaries?

Lecture Notes on Coastal and Estuarine Studies, Vol. 17
Tidal Mixing and Plankton Dynamics. Edited by J. Bowman, M. Yentsch and W. T. Peterson
© Springer-Verlag Berlin Heidelberg 1986

Phytoplankton species are frequently grouped functionally into those which dominate in a) a well-mixed, high energy environment, b) more stable stratified waters, and c) waters characterized by fluctuating turbulence (e.g. temperate regions at the time of vernal warming, or due to spring-neap tidal cycles). Margalef and co-workers summarized theory and observation into a phytoplankton mandala, with red tide dinoflagellates adapted to decaying turbulence (Margalef et al., 1979; see also Bowman et al., 1981; Pingree et al., 1979), therefore potentially nutrient-rich but stable water masses. Dinoflagellates become an important component of phytoplankton once there is a subsidence of vertical mixing.

Our hypothesis is that cyst-forming dinoflagellates, in particular the subset of easily traced, toxin-producing Gonyaulacoid-type which cause shellfish toxicity, occur in regions of seasonal tidal fronts where bottom tidal mixing and associated advective flow cause the resuspension or introduction of benthic cysts, and surface stratification in the summer favors the growth of the planktonic motile stage. Hydrographic observations in frontal regions provide some insight into factors that may control the occurrence of toxic blooms, and the accessory framework for further studies of toxin or species distribution. Accordingly, a series of observations are brought together on the global, mesoscale, and local scale. Relevant information on the life history and toxins of these dinoflagellates, and on the physical effects of tidal mixing is first summarized.

Life History and Toxin Production

Gonyaulax-type dinoflagellates are characterized both by a 30x45 μm-sized ovoid cyst stage, which permits year-to-year survival of the population in any one place, and by a 20-50 μm-sized motile planktonic stage which, through vertical migrations, can exploit situations where a shallow steep pycnocline separates nutrient-depleted surface water from nutrient-rich deep water (Harrison, 1976; Eppley and Harrison, 1978).

Dale (1983) has claimed that even the basic elements of dinoflagellate so-called swimming strategy in contrast to the so-called sinking strategy of diatoms (or the possible combination of both as shown by the coccolithophorids) remains largely undefined in terms of possible range of behavioral patterns. Cyst-forming dinoflagellates have evolved discrete life stages, one planktonic and one benthic, the combination of which offers a more viable "life strategy." Studies on just the planktonic stages have given an incomplete picture of their ecology. Early work on cysts, carried out primarily by paleontologists, reveals that a benthic view of dinoflagellate ecology offers new insights into the factors concerning species distribution. These have been summarized by Dale (1983) who concludes that

Cyst production is a much more widespread and important process than the phycological literature suggests, and

The subsequent transport of cysts once formed depends very much on the physical environment.

In addition, relevant facts about dinoflagellate cysts are:

Under unfavorable conditions for excystment (e.g. low temperature below 'temperature window' 5-7°C), cysts may remain viable for several years.

Anoxic conditions seem to pose no threat to the cysts during normal resting period.

A period of dormancy is required.

Cysts behave as fine silt particles in the sedimentary regime.

In shallower neritic waters, cysts often sink quickly in the water column to concentrate in the unconsolidated flocculative upper layers of bottom sediments.

The toxins are contained in cysts as well as motile cells (Dale et al., 1978). They are accumulated (thus integrated over time) in filter-feeding shellfish, and can be considered as semi-conservative tracers reflecting the past distribution of the source organisms. However, the interpretation of shellfish toxicity data in this way presents a particular problem.

Toxin is generally assayed in mussel tissue. While mussels are ubiquitous and a near ideal test organism (Incze and Yentsch, 1979), great care must be taken over sampling strategies with respect to hydrographic factors that largely determine the distributions of dinoflagellate cysts and motile cells. By examining toxin data, it is clear that there are several patterns which may be important in the preparation of any predictive index for the occurrence of so-called red tides. For example, there are 200 sampling stations along the 3,000 mile Maine coastline. Highest toxin levels are generally witnessed on offshore islands, with intermediate levels on the tips of headlands and lowest levels in the bays. There are some regions of no toxicity sandwiched between regions of high toxicity (Hurst and Yentsch, 1981). Along the entire coast, the toxic events are seasonal, occurring for the most part between mid-May and mid-September. Despite variability from one year to the next, there are broad geographic regions where the phenomenon occurs regularly over spatial scales of tens of kilometers. Toxicity persists in cysts over several months (Selvin et al., 1984). In another study, cyst abundance correlated rather well with the toxin levels which had occurred the previous autumn, but did not correlate well with the toxin levels the subsequent spring (Thayer et al., 1983). Thus, information on cyst distribution/abundance is unlikely to have much predictive value.

When toxin data from several years are compared for one station, the time of the onset of the toxin levels in excess of the quarantine level of 80 µg saxitoxin-equivalents/100 g tissue, reveals that the problems generally arise at one of three times: 1) the onset of stratification, 2) any disruption of the stratification due to storm events, or 3) the time of destratification (Yentsch, 1984). An increase in mussel toxicity in the spring is common at inshore stations, which is interpreted as the timing of excystment in local embayments (J.W. Hurst, pers. comm.). This early spring rise has not been demonstrated at offshore stations perhaps due to the depth of the water column.

Far greater numbers of samples are assayed in regions where there is both an ample shellfish resource suitable for harvesting and a socio-economic concern for public safety upon the consumption of shellfish. These do not necessarily correspond to areas where hydrographic conditions favor dinoflagellate growth, and there are no means by which to evaluate the extent of possible bias with available data.

There are many genera of toxic dinoflagellates. The discussions here refer strictly to the Gonyaulacoid-type cyst-forming dinoflagellates. These represent a subset of both toxin-producing and cyst-forming types, but are the main source of PSP (as opposed to DSP).

Tidal Energy Dissipation

An account of tides in general can be found in most physical oceanographic textbooks. Much of the tidal energy from the deep oceans is dissipated in shallow continental shelf water through bottom friction. Tidal flow erodes water column buoyancy resulting from heating or introduction of fresh water, and promotes the upward mixing of nutrient rich bottom water across the pycnocline. The ratios of the major to minor spring tide velocities or height and width of tidal excursion path are important in defining tidal frontal stability over the M_f cycle and thus lead to variations in the availability of made nutrients over this time period. We believe that this dissipation of tidal energy upon broad continental shelves, which is an extremely predictable feature in the physical environment, is significant within the toxic dinoflagellate context.

On a global basis, tidal energy dissipation has been quantitatively analyzed by Miller (1966). A major percentage of the energy of the tides is dissipated in a very minor percentage of the surface area of the ocean. Each continental shelf region shows considerable spatial heterogeneity in the pattern of tidal flows largely due to local topography. Such flow patterns have been extensively investigated both by direct observation and by theoretical modelling. More recently, they have been analyzed in relation to vertical mixing (Simpson and Hunter, 1974), generally in terms of the log H/U^3 parameter where H is the water depth and U the vertically integrated M_2 tidal current. The magnitude of this parameter largely determines the degree of thermal stratification during the summer months, with the formation of tidal fronts between relatively cool, well mixed water and relatively warm, stratified water. Models of H/U^3 have now been developed for several shelf regions (e.g. Pingree and Griffiths, 1978; Garrett et al., 1978), and the tidal fronts that they define are recognized as biologically important boundary zones due to favorable light and nutrient conditions for phytoplankton growth (Pingree et al., 1978; Simpson et al., 1982). Table I shows regions of high tidal dissipation and associated toxic dinoflagellate occurrence.

Table I. Regions of high tidal dissipation (adapted from Simpson and Bowers, 1981) and associated toxic dinoflagellate occurrence.

Region	% of total	Latitude	Toxic dinoflagellate(s)	Cyst?	Reference
1. Sea of Okhotsk	13.8	55°N			
2. United Kingdom shelf seas	12.5	55°N	Protogonyaulax Gyrodinium aureolum	yes ?	Holligan et al., 1983
3. Australia north coast shelf	11.8	10°S			
4. Patagonian shelf of S. America	8.5	45°S	Protogonyaulax	yes	Argentina Rept., P. Davison, Uruguay
5. Hudson Strait	7.9	60°N			
6. Malacca Strait	4.6	5°N			
7. Yellow Sea	3.9	30°N			
8. Bay of Bengal	3.9	20°N			
9. Northeast coast of S. America	3.2	equator	Protogonyaulax	yes	Karunasagar et al., 1984; Ferraz-Reyes, 1979
10. Gulf of California	2.3	30°N	Protogonyaulax	yes?	Abbott Morey-Gaines
11. Bering Sea	2.0	55°N	Gonyaulax catanella	yes	Hall, 1984
12. Cook Straits	1.6	40°S			
13. Bay of Fundy Gulf of Maine	1.3	45°N	Gonyaulax tamarensis var. excavata	yes	Yentsch, 1984; Anderson, 1984; White, 1984

In relation to studies on dinoflagellates, the most important features of the tides are:

(a) the main tidal forces show semi-diurnal (M_2) and fortnightly (M_f) cycles;

(b) within a fixed topographic framework, the timing, height and range of the tides are highly consistent and predictable;

(c) accordingly, this makes models which predict critical values of H/U^3 for defining tidal effects and frontal positions very reliable;

(d) due to drag, the tides act as a bottom mixing force (especially relevant to cyst resuspension) which interacts with surface stabilization due to heating or fresh water; and

(e) horizontal gradients in H/U^3 (closeness of contours) and ratios of the major to minor spring tide velocities or height may also be important in defining frontal stability over the M_f cycle, and therefore how much nutrient can be made available by mixing processes over this time scale.

'Deep-water' (greater than approximately 60 meters) tidal fronts are a special type of front in several respects. Namely: (a) their spatial/temporal occurrence is extremely regular, even at small scales (e.g. time of year, around islands, headlands, etc.); (b) their general hydrographic properties (salinity, nutrient levels, density gradients) are consistent year-to-year and are suitable (especially high nutrients in deeper water) for plankton growth; and (c) to the stratified side, bottom conditions are usually low energy and therefore are often favorable for the accumulation of fine muds and survival of cysts -- this is not always the case as advection and the relatively strong tidal mixing required to maintain frontal boundaries in deep water may prevent any accumulation of fine muds.

Other types of fronts rarely possess all these features - e.g. 'shallow water' tidal fronts, due to the tendencies for the mixed water to become nutrient depleted and for the pycnocline to extend to the bottom rather than the surface, provide less suitable nutrient conditions; estuarine fronts are hydrographically variable; shelf break fronts do not show sharp shallow pycnoclines. One exception may be the 'negative estuarine condition' (Margalef et al., 1979) where red tides appear to develop in response to bottom inflow of cold, nutrient-rich offshore water (Tyler and Seliger, 1980). Additionally of interest are the enclosed bay systems and fjords which may be affected by eutrophication (Tangen, 1979).

The Global Patterns

In an attempt to address some of the most general questions concerning the occurrences of Gonyaulax-type dinoflagellates, their global distribution is compared with global primary production and tidal dissipation as shown in Figures 1 to 3. Gonyaulax distribution does not strictly reflect the pattern of global primary production, but is more in concert with the pattern of tidal dissipation. This correspondence, we believe, is due to factors which operate to provide a favorable environment for the initiation, growth and maintenance of Gonyaulax populations. The most important aspect is that the tidal energy dissipation phenomenon leads to conditions of marginal stratification, or of alternating stratification and destratification. In turn, this provides a favorable nutrient regime plus water column stability for autotrophic dinoflagellates. Tidal stirring at the bottom is also active in resuspending cysts which have previously settled out of the water column. However, it is impossible to provide conclusive evidence for a cause and effect relationship for the association of toxic dinoflagellates with the regions of high tidal dissipation.

Let us examine regions with near zero $erg \cdot sec^{-1}$ contours. Namely, these are: eastern Africa, southern peninsula of India, southwest Malaysia, southern Australia, northern New Guinea, western Central America, the Peru coast, southern Brazil, Cape Florida to the New York Bight on the eastern coast of N. America, Labrador, the east coast of Greenland, northern Norway and the Mediterranean and western coast of Africa. To the knowledge of the authors, records of toxic dinoflagellates in these regions reveal very low abundance of only few, if any, toxic cyst-forming species. When found, they are confined to semi-enclosed regions where tidal energy dissipation is locally increased.

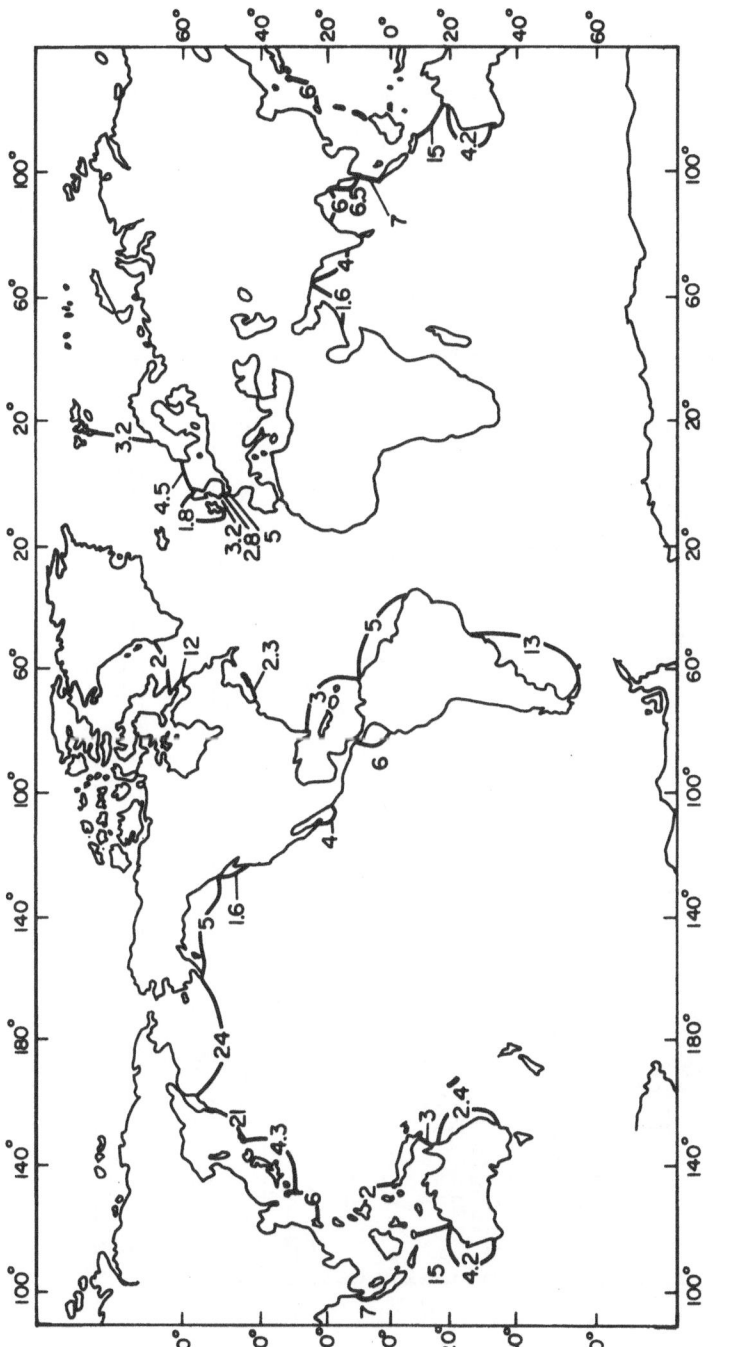

Figure 1. Major global areas of high tidal dissipation (flux of tidal energy) out of the deep oceans in units of 10^{17} ergs·sec^{-1} based on estimates of frictional dissipation. Contours less than 1.4 ergs·sec^{-1} are deleted. These are based on the semi-diurnal lunar (M_2) tidal energy flux derived from harmonic analysis of tidal height and tidal currents, as opposed to earlier methods of frictional dissipation. Global representation adapted from Miller, 1966.

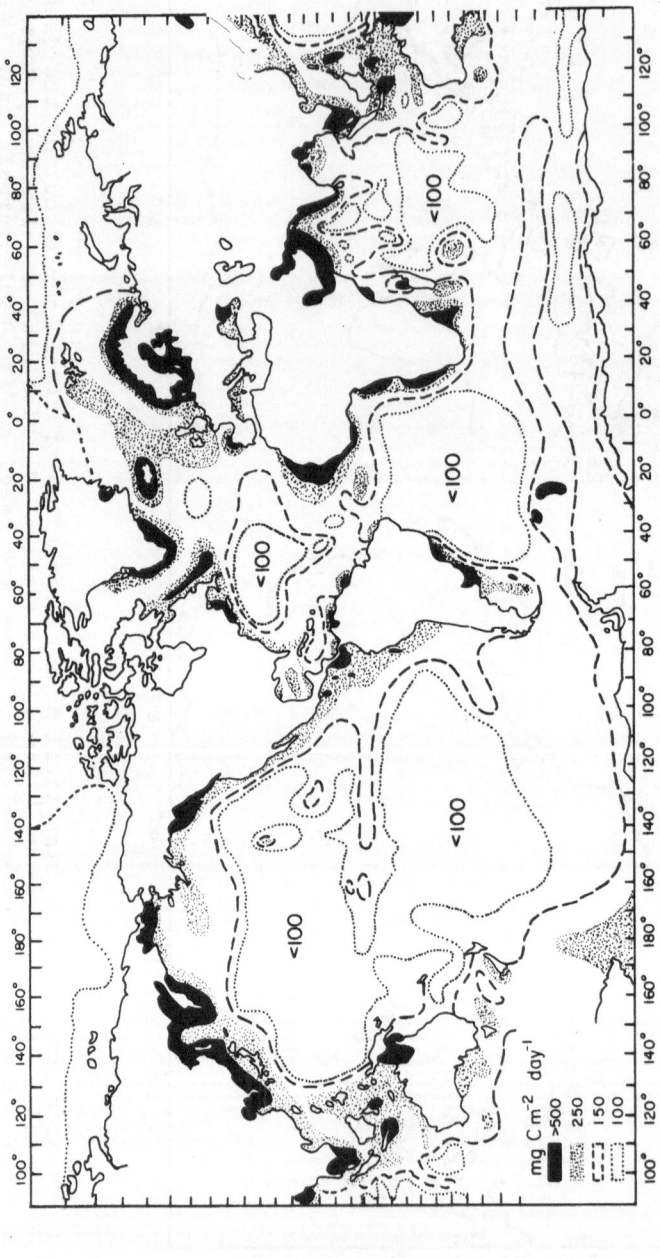

Figure 2. Major global areas of high primary productivity expressed as mg C·m⁻²·day⁻¹. Toxic dino-flagellates are most abundant in areas >250 mg C·m⁻²·day⁻¹. Figure courtesy of J. Campbell as redrawn from Koblentz-Mishke et al. (1970).

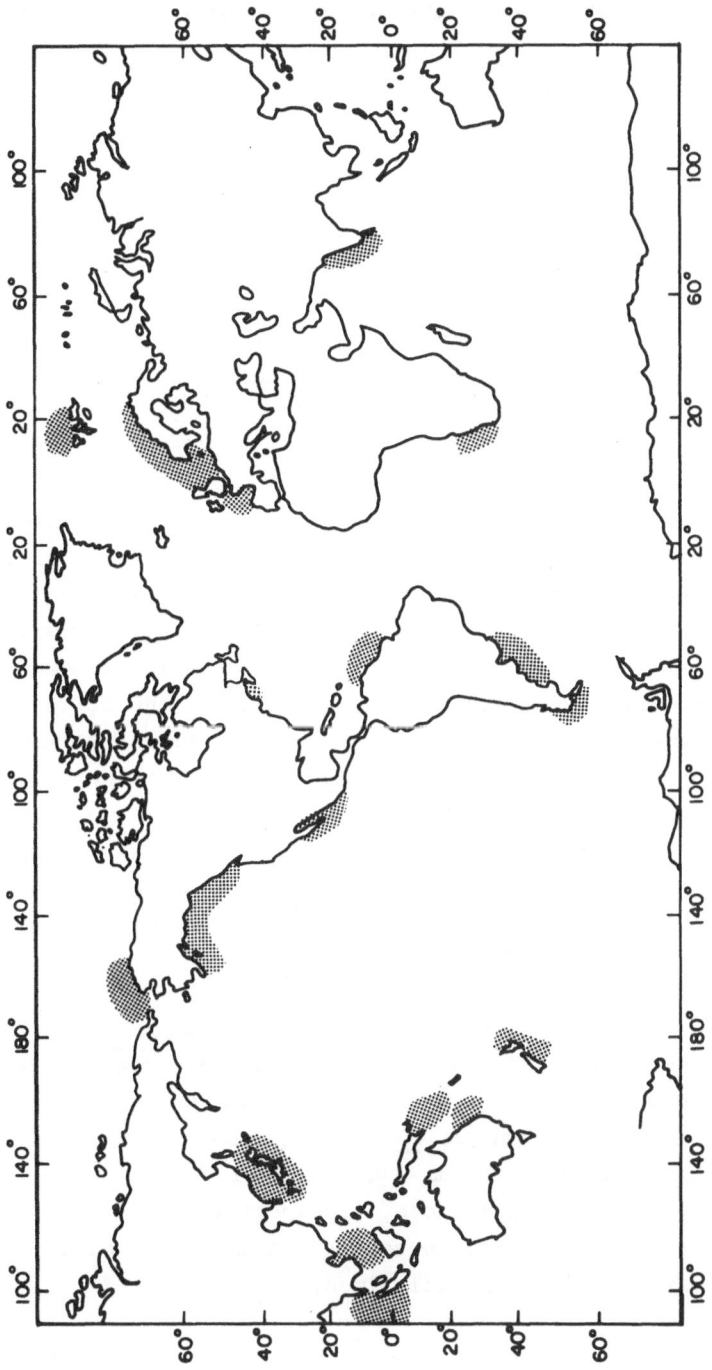

Figure 3. Major reported global areas of shellfish toxin/toxic cyst-producing Gonyaulacoid dinoflagellates and closed related species. Adapted from F.J.R. Taylor (1984).

Upwelling ecosystems, especially Peru and N.W. Africa, have been the site of major studies of phytoplankton dynamics, especially in reference to the physical and chemical environment. While dinofla- gellates are common at times of relaxation of the winds and upwelling (Margalef et al., 1979), the dominant forms are species which are generally neither toxic nor thought to be cyst-forming.

By way of contrast, regions such as the Patagonian Shelf and Sea of Okhotsk are regions of extreme tidal dissipation and do have toxic and cyst-producing forms. These regions (which have been relatively little studied for dinoflagellate distributions), would make good test sites for the general hypothesis proposed here.

Evidence from the Mesoscale

While there is some information for the northwest European Shelf and the Bering Sea the distribution of toxic dinoflagellates is best known for the Gulf of Maine, and therefore information presented here is from this region. The extent of the Gulf of Maine, approximately 100,000 km^2, and relative isolation from oceanic influences by both Georges Bank to the southeast, and Browns Bank to the northeast, are sufficient to allow a consistent succession of plankton communities each year; tidal mixing is weak enough in the central part to permit formation of an intermediate water layer (Hopkins and Garfield, 1979) which clearly segregates the nutrient regeneration in the upper part of the water column from that in the surface sediments. During the summer fronts are common along the shelf break roughly located where the water depth reaches 100 m.

Tools for an assessment of mesoscale phenomena rest primarily with a) numerical models of tidal mixing predicting frontal location, b) remote sensing of surface chlorophyll and temperature, and c) toxin as a tracer. Garrett et al. (1978) have constructed a tidal mixing model on a scale of 1 km. Regions which show H/U^3 <2.0 are considered to be mixed at all times, irrespective of seasonal strati- fication criteria. Basically, these regions are more shoal, so that tides drive vertical mixing at all times.

Three main points emerge from studies on the distribution of Gonyaulax. First, not all potentially favorable areas appear to be affected equally. Secondly, for any particular area there is consi-

derable year-to-year variability. Reasons may include the presence/
absence of seed populations such as the possible introduction of
Gonyaulax tamarensis var. excavata cysts into the southern Gulf of
Maine in September 1972. Thirdly, hydrographic conditions affecting
cell growth and encystment and excystment may differ. For example,
frontal systems along eastern coasts (Gulf of Maine and Northeast
England) may be significantly affected by upwelling due to prevailing
offshore winds. Factors affecting cyst occurrence and survival are
not well known, but are certain to be related to gradients in hydrographic properties. An additional consideration must be the competition from ecologically similar species (e.g. other dinoflagellates
such as Ceratium spp. and Gyrodinium aureolum).

Field observations (Yentsch and Garfield, 1981) have shown that
satellite imagery can be deciphered in terms of sea surface temperature variations and the positions of tidal fronts. Satellite imagery
has become generally accessible only recently, yet images already
exist that could be useful for interpreting surface temperature and
phytoplankton distributions in many regions of the world's coast. It
is important that these are processed and made available for areas of
tidal dissipation and toxic dinoflagellate blooms.

The Gulf of Maine, a semi-enclosed sea in the temperate latitudes, becomes stratified during maximum solar insolation, character-
istically from mid-May to mid-September. The effects of stratification can be easily traced by satellite imagery, either with the
thermal infrared bands (AVHRR) or with wavelengths which depict ocean
color (CZCS) from which pigment concentrations can be calculated.
The sequence in Figures 4a-d shows changes in sea surface temperature
and the abundance of all pigment-containing organisms (of which the
cyst-producing toxic dinoflagellates are only a fraction). The
images define the critical window (both with regard to time and
space) during which hydrographic conditions are favorable for phytoplankton bloom and demonstrate the important implications of
satellite information.

The spring bloom (diatoms) develops in April-May at the time of
the onset of seasonal stratification. By the summer a lens of warm
(maximum, 20°C), nutrient impoverished water forms in the central

Figure 4. CZCS and AVHRR representation of chlorophyll and sea
surface temperature, respectively, for the Gulf of Maine
for 1979. A. January, B. May, C. June, and D. August.
Colors are used to indicate the following intervals:

chlorophyll	sea surface temperature
(top panels)	(bottom panels)
blue = <1 $\mu g \cdot \ell^{-1}$	blue = <8°C
green = 1-4 $\mu g \cdot \ell^{-1}$	green-yellow = 8-16°C
yellow = >4 $\mu g \cdot \ell^{-1}$	orange-red = >16°C

Satellite derived phytoplankton pigment and temperature annual
series. For the time period November 1978 to December 1979, a series
of monthly NIMBUS-7 CZCS scenes were obtained from NASA Goddard Space
Flight Center. The scenes were obtained in form of Level 1 computer
compatible tapes (CCT's) except the September and November scenes
which were in the form of Level 2 CCT's (Hovis et al., 1980). The
Level 1 data was processed to Level 2 by the Charles S. Draper
Laboratory (CSDL) using the atmospheric corrections and phytoplankton
pigment algorithms proposed by Gordon (1978) and Gordon et al.
(1983). Compensation for the decrease of sensor degradation was
accomplished using method IV proposed by SASC. This produced a
series of 12 phytoplankton pigment scenes (including September and
November) which were processed with a consistent atmospheric
correction algorithm and phytoplankton pigment algorithm, and a
consistent sensor degradation model. Table 1a indicates the dates,
times and orbits of pigment series. The series of pigment scenes
shown in Figure 1 were obtained by coordinate transforming the
satellite data to a latitude-longitude coordinate frame. The images
were color coded, and land and clouds were delineated by use of
homogeneous color to aid interpretation.

The annual temperature series shown in Figure 2 consists of scenes
obtained by processing of AVHRR HRPT data from the NOAA Polar
Orbiter. In all cases, only Channel 4 data were used. Table 1b
indicates the satellite, time and orbit of each scene. The calibration of the scenes was accomplished in two steps: 1) converting
the thermal Channel 4 data to brightness temperature; 2) adjusting
the temperature from 1 for each scene by a bias correction derived
from buoy data.

The time of the buoy data was selected to be approximately $\pm$ 2
hours to the time of the satellite overpass. The bias correction was
assumed to be equal to the average of the differences between
corresponding satellite and buoy data points for each scene. The
corrections ranged from 0.3^{o}C to 3.1^{o}C and were generally lower in
the winter months.

The temperature scene was coordinate transformed to a coordinate
frame common with pigment series by the Massachusetts Institute of
Technology under contract to the Northeast Fisheries Center.

The scenes for the temperature and pigment annual series were
produced by the Charles S. Draper Laboratory (CSDL) under contract to
the Northeast Fisheries Center, National Marine Fisheries Service,
NOAA.

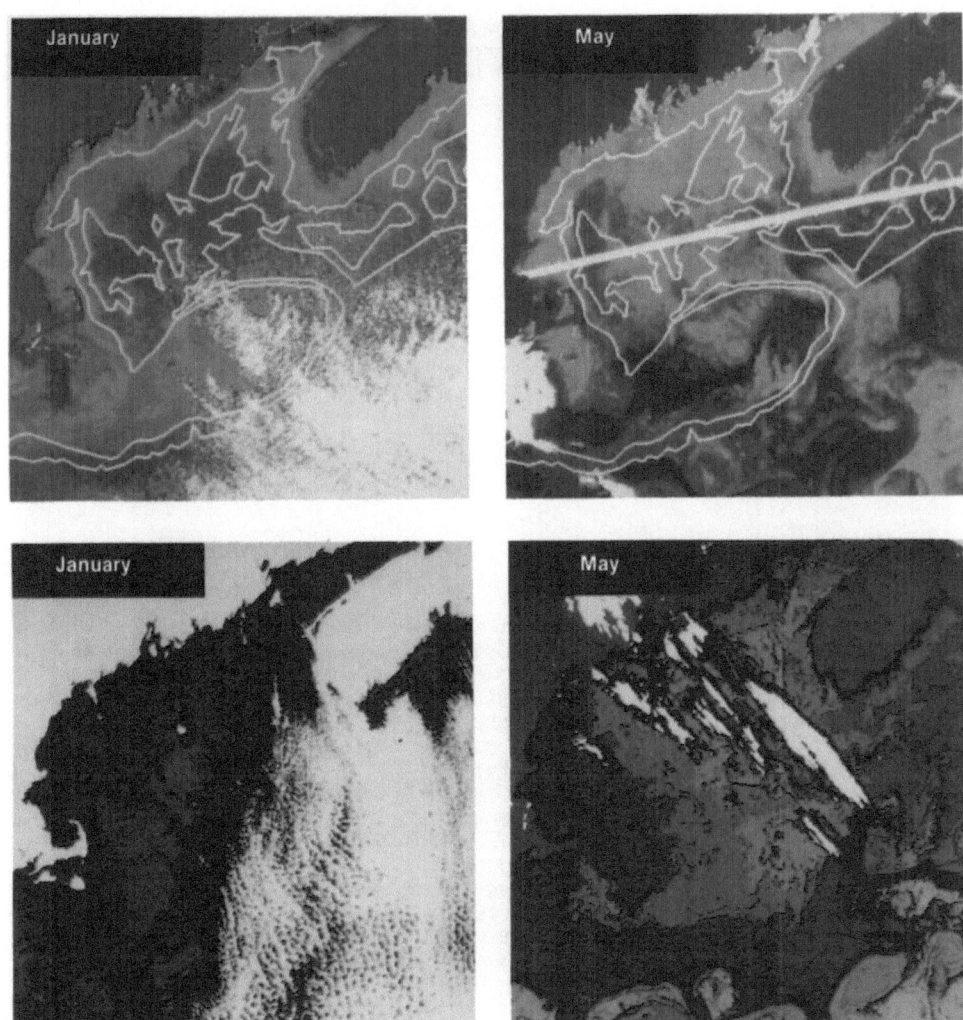

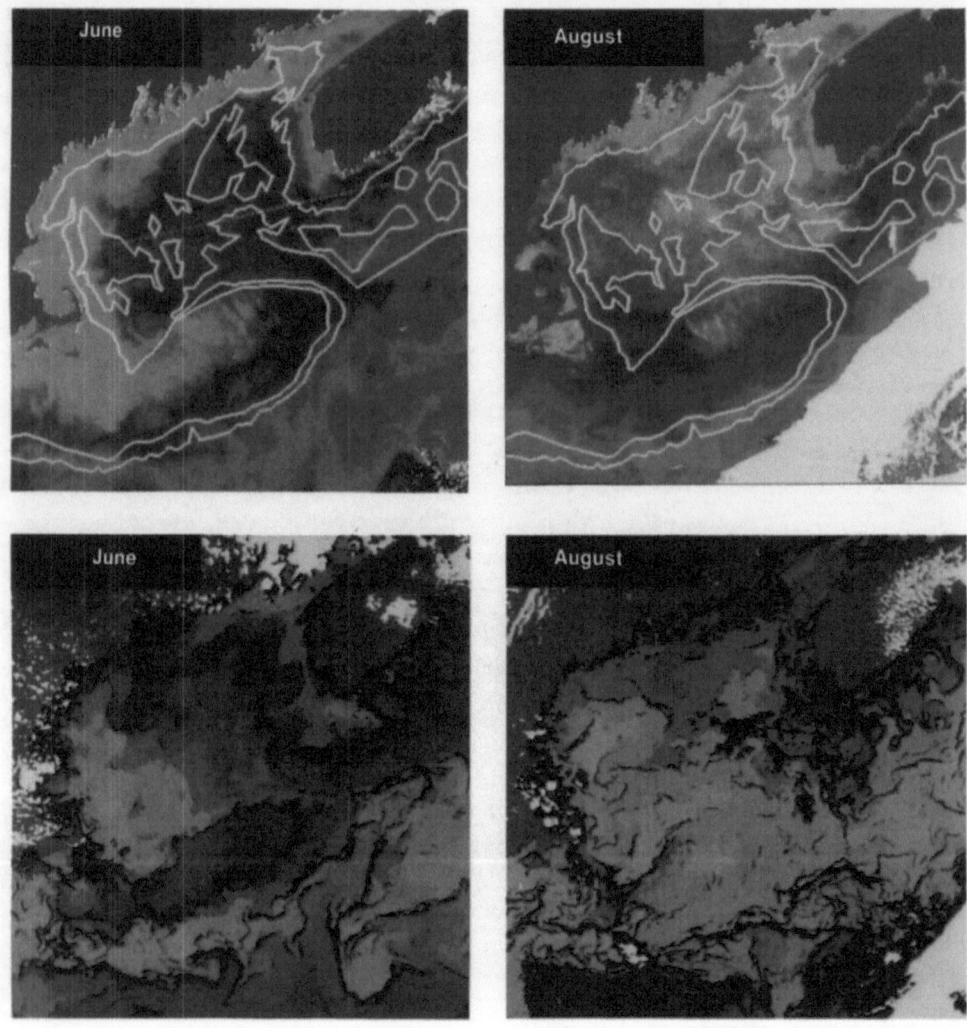

Gulf of Maine, the edge of which is subjected to tidal erosion. The extensive area of high phytoplankton biomass seen in May, which corresponds to the main development of the spring diatom outburst, becomes confined through the summer months to specific coastal regions within the 100 m contour. (The two depth contour intervals delineated on the chlorophyll images are 100 m and 200 m). These areas include those where Gonyaulax populations are found.

Satellite imagery can be used to estimate chlorophyll (an index of productivity), yet there is no indication where one can expect the toxic vs. non-toxic cyst-forming dinoflagellates. Many of these species co-occur (Yentsch et al., 1978) and regional differences have been clearly documented within the Gulf of Maine (Hurst and Yentsch, 1981). While we have used toxin reports integrated over the 1900's in our global approach, the relevant time scale for averaging the mesoscale is a year/season. Toxin data collected over several years appears to be instructive here. Figure 5 a,b,c are charts of the Gulf of Maine with the size and shade of the circle reflecting maximum toxin levels reached over a season in a general sampling region (as numbered) per year. The years of 1982, 1983, and 1984 are presented.

It is clear that there are "non-toxic" areas sandwiched between highly toxic regions. For clarification of information, local studies are essential on spatial scales of meters to kilometers and time scales of hours to days.

Evidence from the Local Scale

Much information about the distributions of dinoflagellates has come from studies of local features. Methods include daily sampling for toxin as well as species enumeration, pump profiling and aircraft remote sensing.

In the Gulf of Maine, temporal and spatial variations in species composition are largely related to hydrographic conditions which are dictated in large part by the complex bathymetric configuration. The detailed results presented here are from a region to the east of Boothbay Harbor extending to some 25 miles offshore. This is an area with extremely variable bottom topography (Vermersch et al., 1979),

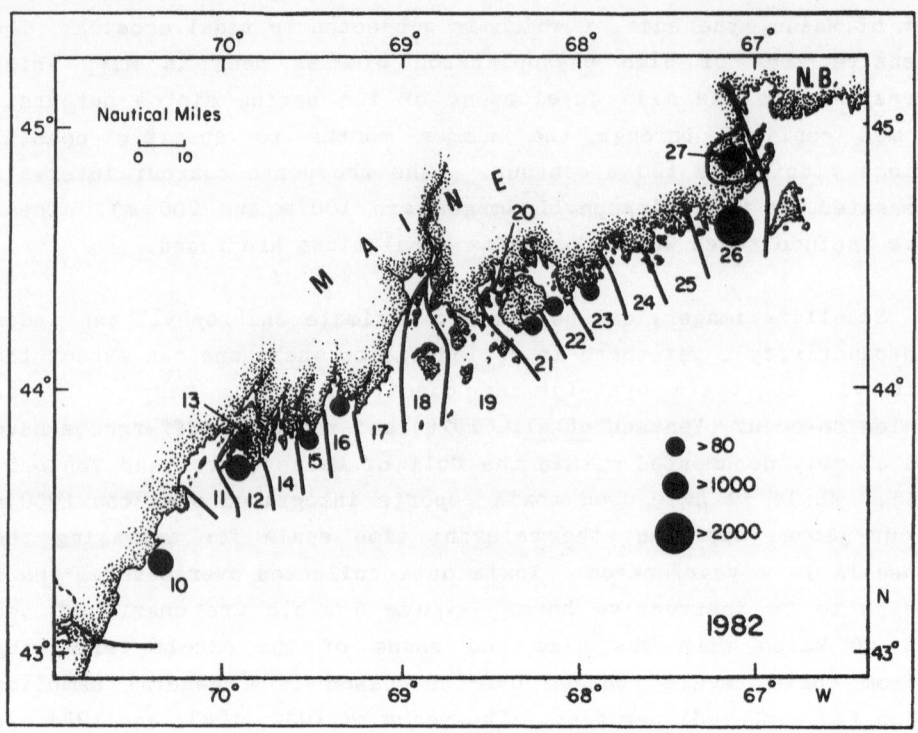

Figure 5. A.   For 1982 season.   Maximum levels of shellfish toxin
detected in shellfish by the Maine Shellfish Monitoring
Program separated by region.   ● >80 µg saxitoxin-equiva-
lents·100 g^{-1}; ● >1000 µg saxitoxin-equivalents·100 g^{-1}; ●
>2000 µg saxitoxin-equivalents·100 g^{-1}.

with inshore waters being tidally mixed, offshore waters well stratified during the summer season, and the fronts occurring at the
boundary of the two water types.

There is a series of islands located at the edge of the shelf
beyond which the bottom drops rapidly to >100 m. Observations around
Monhegan Island (43°45'N; 69°20'W) appear to support the models of
Pingree and Maddock (1979) and Simpson et al. (1982) to produce
significantly higher biomass of phytoplankton in an already productive region (Townsend et al., 1983, Figure 6). This appears to be a
consistent feature of island systems in shallow continental shelf
seas where tides dominate the water motion. In general, a condition
of tidal mixing may augment plankton growth by a factor of 5 times

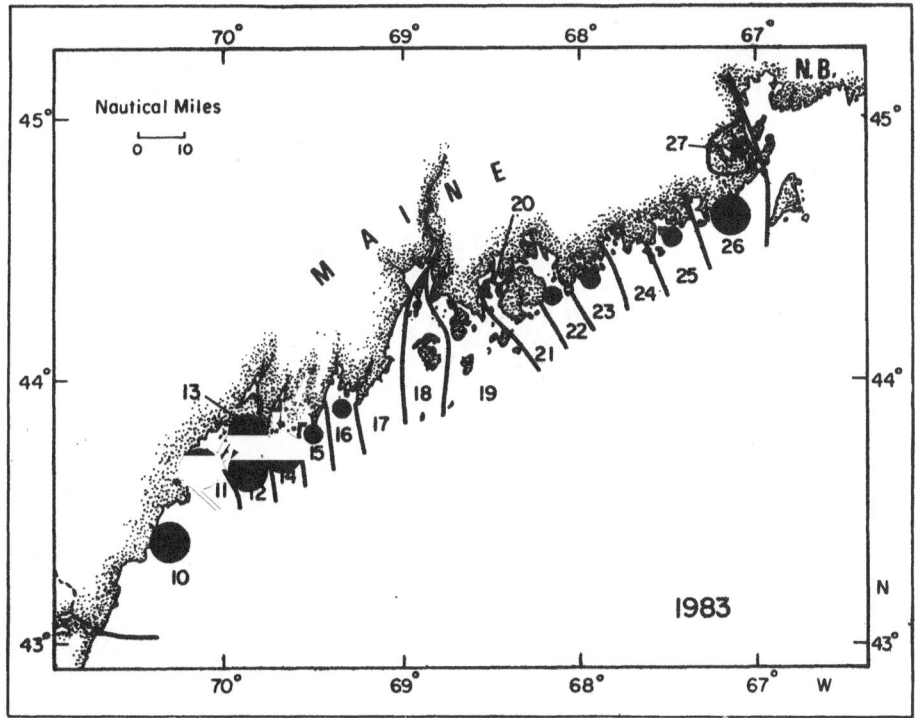

Figure 5. B. For the 1983 season.

over regions of 20 times the island area (Simpson et al., 1982).
Whatever the genesis, it certainly plays a role in developing a
variety of niches, many of which are suitable for toxic dinoflagellate proliferation (Fig. 7).

The tides at Monhegan Island have characteristic M_2 (two high
tides daily) as well as a semi-monthly cycle with two spring tides
(M_f) per month. The average major spring, minor spring, and neap
tidal rages are 4.0, 2.9 and 2.4 m, respectively. Tidal current data
specifically around Monhegan Island showed current velocities of ~20
$cm \cdot s^{-1}$ at depth for peak ebb and flood tides (Balch, 1981; Townsend
et al., 1983). An apparent lunar tidal cycle of phytoplankton
blooming and community succession has been described (Balch, 1981).

In addition to phytoplankton populations associated with island
mixing, a strong subsurface chlorophyll maximum is generally present
offshore throughout the summer. This can be dominated by Gonyaulax

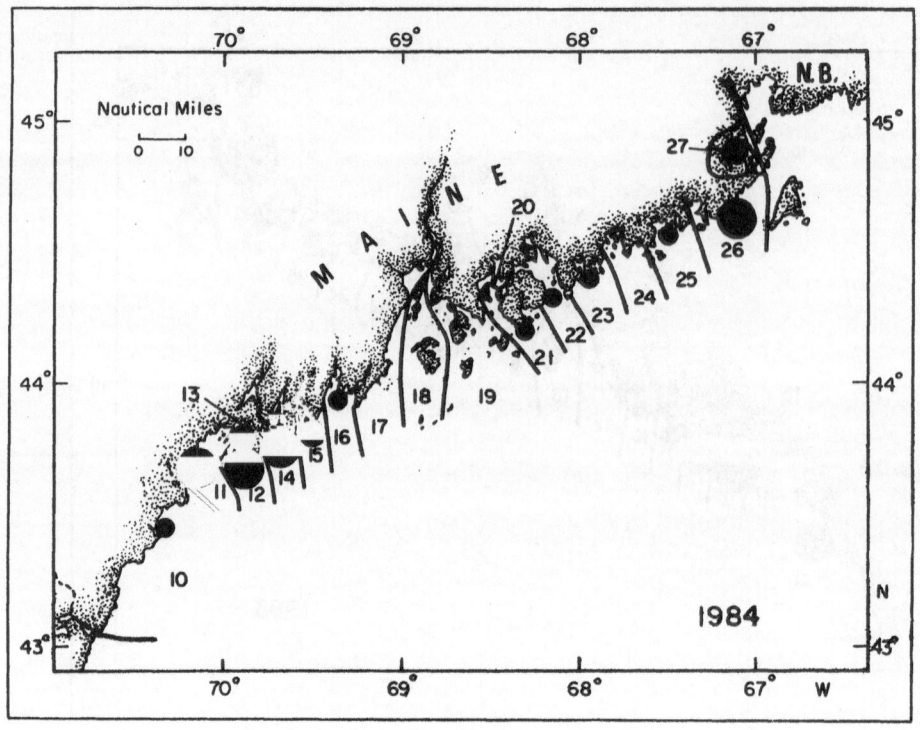

Figure 5. C.   For 1984 season.   Monhegan Island is in Region 15.
Note two apparent dynamic systems, one in the southwest and
one in the northeast (continuous with Canadian toxifica-
tions), separated by non-toxic areas.   These are hypothesis
based on consistent hydrographic differences which should
be tested.   Data courtesy of J.W. Hurst.

as shown in Figs. 8 and 9.   Especially note Station #3 at which
Gonyaulax formed almost 50% of the total phytoplankton biomass.
Insert C in Fig. 9 describes the pattern of chlorophyll and Gonyaulax
with depth.   Further details can be found in Holligan et al., 1984.

   The horizontal and vertical variations in taxonomic composition
suggest that active growth and accumulation of particular motile
species are generally important in the formation and persistence of
the subsurface chlorophyll maximum as opposed to passive mechanisms
controlled by physical mixing processes and sinking properties of the
cells.   It seems likely that these offshore dinoflagellate populations
give rise to sporadic inshore "blooms" when hydrographic conditions
favor their shoreward transport.

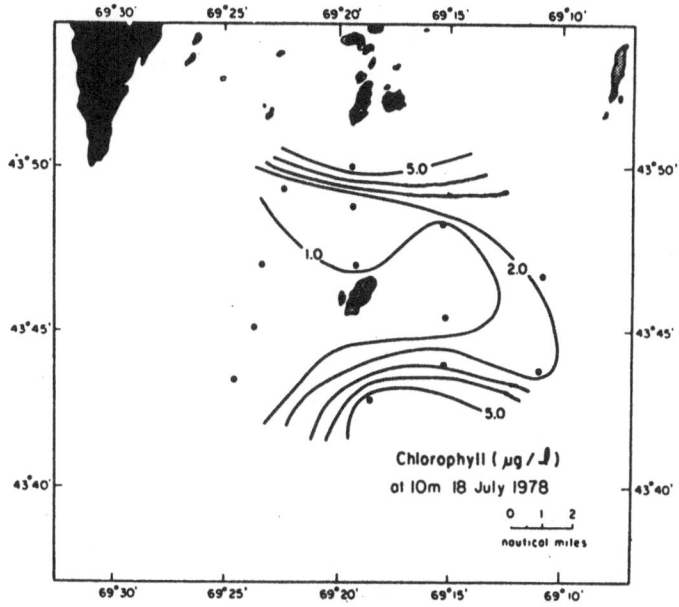

Figure 6. Chlorophyll ($\mu g \cdot \ell^{-1}$) at 10 m depth from 18 July 1978 around Monhegan Island depicting increase in biomass resulting from island mixing. Details in Townsend et al. (1983).

CONCLUDING REMARKS

The association of toxic dinoflagellates with tidal fronts is based on the premise that regions of high tidal dissipation provide (1) local stirring necessary for re-entry of a cyst or dormant resting stage into the pelagic environment, (2) a near-continuous (semi-diurnal) flux of nutrients necessary for division of the motile stage into the near-surface photic zone as well as bimonthly spring-neap pulse extremes, and (3) sufficient residence time in a semi-stable environment to permit incubation times compatible with cell division rates to result in massive accumulations as observed (Yentsch, 1984). Furthermore, endogenous rhythms may explain an active role both in annual/seasonal and spring-neap tidal occurrence (Auclair et al., 1982).

This hypothesis is difficult to test experimentally, but there is a growing suite of observations which support the importance of dinoflagellates in tidal and estuarine frontal systems. Admittedly, there are many uncertainties. Yet there exists both reason for, and evidence of, the association of toxic cyst-forming dinoflagellates in

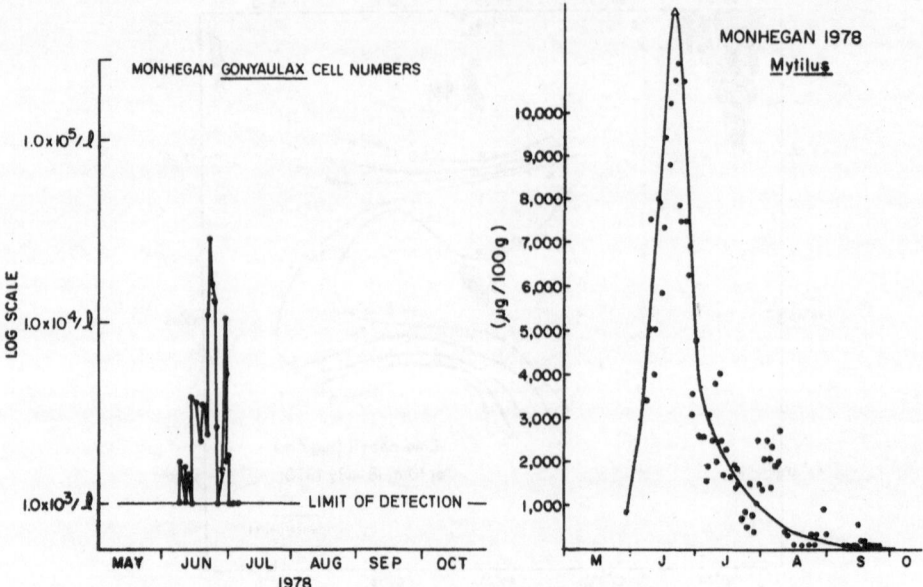

Figure 7. Day-to-day fluctuations in cell numbers of Gonyaulax tamarensis var. excavata expressed per liter on a log scale vs. time and shellfish toxin present in blue mussels Mytilus edulis expressed as µg saxitoxin-equivalents per 100 g shellfish tissue on a linear scale vs. time. Quarantine level (the level at which shellfish beds must legally be closed to the harvest of shellfish) is 80 µg saxitoxin-equivalents per 100 g shellfish tissue. Note that toxin levels exceeded 12,000 µg per 100 g with the maximum cell number detection being about 2 x 10^4 cells per liter -- orders of magnitude lower levels of cell numbers than results in any visible red discoloration of the seawater.

regions of high tidal dissipation. The organisms of interest exploit the limits of nitrogen and light as ingredients for new production which occurs at the lower reaches of the euphotic zone, frequently, but not always in association with the pycnocline. These large dinoflagellates appear to be one of the final/terminal species in the summer succession at the subsurface chlorophyll maximum, which out-persist (vs. out-compete with high uptake rates or high division rates) other microalgal species (Fig. 10).

Satellite imagery is useful in defining the temporal and spatial window in which one might expect to find the toxic motile form to

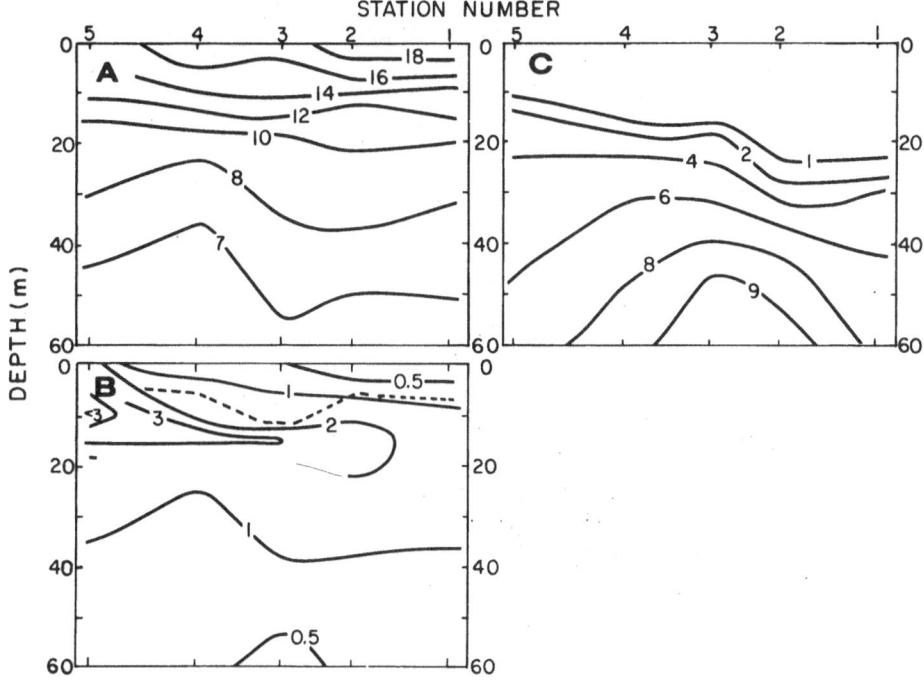

STATION NUMBER

Figure 8. A. temperature (°C), B. chlorophyll (µg·ℓ⁻¹) and C. nitrate
concentrations (µmoles) with depth for a section progres-
sing from nearshore to offshore (as charted in 9B), located
a few miles southeast of Monhegan. Details in Holligan et
al. (1984).

flourish. Historical shellfish toxin data are useful tracers and can
help in predicting on a regional/local basis. Careful monitoring of
the organisms in the water column and toxin in the shellfish are
exceedingly useful in understanding bloom dynamics, species distribution and interspecific replacement of similar ecotypes. Regions
identified as ideal for hypothesis testing should be examined in
detail.

ACKNOWLEDGEMENTS

    The work presented here was supported in part by FDA 223-77-
2314; HEW/NIH #5RO1ESO1329-02; NOAA/NEFC NA80AAD00119, and the State
of Maine. P. Boisvert and P. Colby prepared the manuscript, and J.
Rollins and K. Knowlton prepared the illustrations.

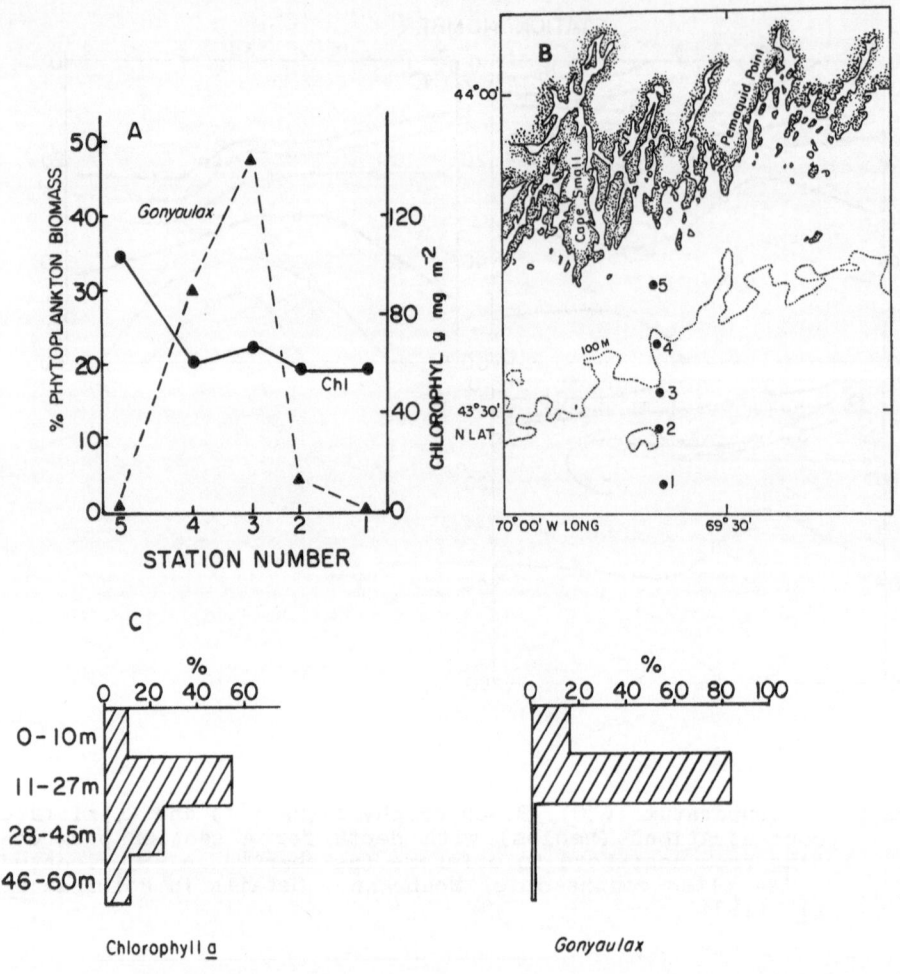

Figure 9. Chlorophyll ($\mu g \cdot \ell^{-1}$) and cell numbers (from the previous section) of <u>Gonyaulax</u> <u>tamarensis</u> var. <u>excavata</u> integrated over the upper 60 m of the water column. Notice peak in <u>Gonyaulax</u> numbers at station #3. Adapted from Holligan <u>et al</u>. (1984).

The authors thank the funding agencies and the above named individuals. Additionally, fruitful discussions with J. Campbell, J.W. Hurst, K. Richardson, J. Simpson and C.S. Yentsch are acknowledged.

This is Bigelow contribution number 84033.

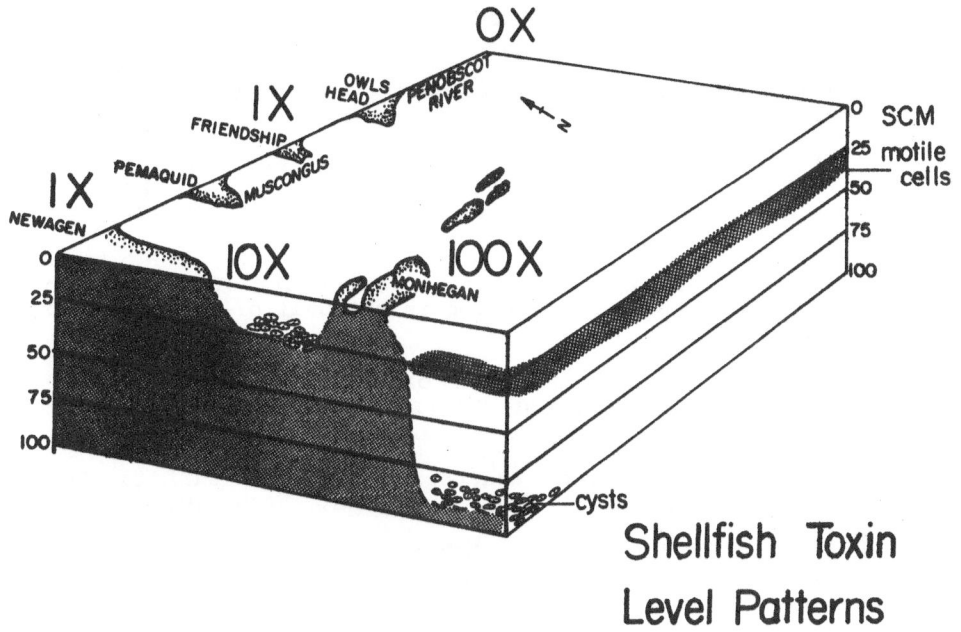

Shellfish Toxin
Level Patterns

Figure 10.   Summary schematic diagram depicting shellfish toxin
level patterns, namely 100 times "base level" concentration of toxin at the Monhegan Island system, 10 times
base level concentrations at headlands and base level
concentrations in bays. The Penobscot River and other
estuaries frequently show zero toxicity. Cyst accumulations are noted, again with the greatest abundances
offshore, accumulating with other sedimentary particles
of equal density. The subsurface chlorophyll maximum
(SCM), a general feature at the frontal boundaries
between summer stratified waters offshore and tidally
mixed waters inshore, is depicted at approximately 10 m
depth.

REFERENCES

Anderson, D.M. 1984. In: Seafood Toxins. E. Ragelis (ed.). ACS Symposium Volume.

Auclair, J.C., S. Demers, M. Frechette, L. Legendre and C.L. Trump. 1982. High frequency endogenous periodicities of chlorophyll synthesis in estuarine phytoplankton. Limnol. Oceanogr. 27: 348-352.

Balch, W.M. 1981. An apparent lunar tidal cycle of phytoplankton blooming and community succession in the Gulf of Maine. J. exp. mar. Biol. Ecol. 55: 65-77.

Bigelow, H.B. 1926. Plankton of the offshore waters of the Gulf of Maine. U.S. Dept. Commerce, Bull. U.S. Bur. Fish., Wash. 40: 1-509.

Bigelow, H.B. 1927. Physical oceanography of the Gulf of Maine. U.S. Dept. Commerce, Bull. U.S. Bur. Fish., Wash. 40: 511-1027.

Bowman, M.J., W.E. Esaias and M.B. Schnitzer. 1981. Tidal stirring and the distribution of phytoplankton in Long Island and Block Island Sounds. J. Mar. Res. 39: 587-603.

Brosche, P. and J. Sundermann. 1978. Eds. Tidal Friction and the Earth's Rotation. Springer-Verlag, New York.

Brown, W.S. and R.C. Beardsley. 1978. Winter cooling in the western Gulf of Maine: Part 1. Cooling and water mass formation. J. Phys. Oceanogr. 8: 265-277.

Dale, B. 1983. Dinoflagellate resting cysts: "benthic plankton". In: Survival Strategies of the Algae. G.A. Fryxell (ed.). Cambridge Univ. Press, NY.

Dale, B., C.M. Yentsch and J.W. Hurst. 1978. Toxicity in resting cysts of the red-tide dinoflagellate Gonyaulax excavata from deeper water coastal sediments. Science 201: 1223-1225.

Eppley, R., and Harrison. 1978. In: Proc. 1st Int. Conf. on Toxic Dinoflagellate Blooms.

Ferraz-Reyes, E., G. Reyes-Vasquez and I.B. Bruzual. 1979. Dinoflagellate blooms in the Gulf of Cariaco, Venezuela. In: Toxic Dinoflagellate Blooms, 2nd Intnl. Conf. on Toxic Dinoflagellate Blooms, pp. 155-160. D.L. Taylor and H.H. Seliger (eds.). Elsevier North Holland, NY.

Garrett, C.J.R., J.R. Keeley and D.A. Greenberg. 1978. Tidal mixing versus thermal stratification in the Bay of Fundy and the Gulf of Maine. Atmosphere-Ocean 16: 403-423.

Gordon, H.R. 1978. Removal of atmospheric effects from satellite imagery of the oceans. Appl. Optics 17: 1631-1636.

Gordon, H.R., D.K. Clark, J.W. Brown, O.B. Brown, R.H. Evans and W.W. Broenkow. 1983. Phytoplankton pigment concentrations in the Middle Atlantic Bight: comparison between ship determinations and Coastal Zone Color Scanner estimates. Appl. Optics 22: 20-36.

Greenberg, D.A. 1979. A Numerical Model Investigation of Tidal Phenomena in the Bay of Fundy and Gulf of Maine. Mar. Geodesy 2: 161-187.

Harrison, W.G. 1976. Nitrate metabolism of the red tide dinoflagellate Gonyaulax polyedra Stein. J. Exp. Mar. Biol. Ecol. 21: 199-209.

Holligan, P.M., M. Viollier, C. Dupouy and J. Aiken. 1983. Satellite studies on the distributions of chlorophyll and dinoflagellate blooms in the western English Channel. Cont. Shelf. Res. 2: 81-96.

Holligan, P.M., R.P. Harris, R.C. Newell, D.S. Harbour, R.N. Head, E.A.S. Linley, M.I. Lucas, P.R.G. Tranter and C.M. Weekley. 1984. Vertical distribution and partitioning of organic carbon in mixed, frontal, and stratified waters of the English Channel. Mar. Ecol. Prog. Ser. 14: 111-127.

Holligan, P.M., P.J.leB. Williams, D. Purdie and R.P. Harris. 1984. Photosynthesis, respiration and nitrogen supply of plankton populations in stratified, frontal and tidally mixed shelf waters. Mar. Ecol. Prog. Ser. 17: 201-213.

Holligan, P.M., W.M. Balch and C.M. Yentsch. 1984. The significance of a subsurface chlorophyll, nitrite and ammonium maxima in relation to the supply of nitrogen for phytoplankton growth in stratified waters of the Gulf of Maine. J. Mar. Res. 42: 1051-1073.

Hopkins, T.S. and N. Garfield. 1979. Gulf of Maine intermediate water. J. Mar. Res. 39: 103-139.

Hovis, W.A., C.K. Clark, F. Anderson, R.W. Austin, W.H. Wilson, E.T. Baker, D. Ball, H.R. Gordon, J.L. Mueller, S.Y. El Sayed, B. Sturm, R.C. Wrigley and C.S. Yentsch. 1980. Nimbus-7 coastal zone color scanner: system description and initial imagery. Science 210: 60-63.

Hurst, J.W. and C.M. Yentsch. 1981. Patterns of intoxification of shellfish in the Gulf of Maine coastal waters. Can. J. Fish. Aquat. 38: 152-156.

Incze, L.S. and C.M. Yentsch. 1979. Accumulation of algal biotoxins in mussels. Chapter 9. In: Mussel Culture in North America, pp. 223-246. R. Lutz (ed.). Elsevier, North Holland.

Karunasagar, I., H.S.V. Gowda, M. Subbaraj, M.N. Venugopal and I. Karunasagar. 1984. Outbreak of paralytic shellfish poisoning in Mangalore, west coast of India. Current Science 53: 247-249.

Koblentz-Mishke, O.J., V.V. Volkovinsky and J.G. Kabanova. 1970. Plankton Primary Production of the World Ocean. In: Scientific Exploration of the South Pacific. W.S. Wooster (ed.). Natl. Acad. Sci., Washington, D.C.

Legendre, L. 1981. Hydrodynamic control of marine phytoplankton production: The paradox of stability. In: Ecohydrodynamics, pp. 191-207. J.C.J. Nihoul (ed.). Elsevier, Amsterdam.

Margalef, R., M. Estrada and D. Blasco. 1979. Functional morphology of organisms involved in red tides, as adapted to decaying turbulence. In: Toxic Dinoflagellate Blooms. Proc. 2nd Intnl. Conf. on Toxic Dinoflagellate Blooms, pp. 89-94. D.L. Taylor and H.H. Seliger (eds.). Elsevier North Holland.

Miller, G.R. 1966. The flux of tidal energy out of the deep oceans. J. Geophys. Res. 71: 2485-2489.

Pingree, R.D. 1978. Mixing and stabilization of phytoplankton distributions on the Northwest European continental shelf. In: Spatial Pattern in Plankton Communities, J.H. Steele (ed.), pp. 181-200. Plenum Press, New York and London.

Pingree, R.D. and D.K. Griffiths. 1978. Tidal fronts on the shelf seas around the British Isles. J. Geo. Res. (Oceans & Atmospheres) 83: 4615-4622. Chapman Conference Special Session.

Pingree, R.D. and L. Maddock. 1979. Tidal flow around an island with a regularly sloping bottom topography. J. mar. biol. Ass. U.K. 59: 699-710.

Pingree, R.D., P.M. Holligan and G.T. Mardell. 1978. The effects of vertical stability on phytoplankton distributions in the summer on the northwest European Shelf. Deep-Sea Res. 25: 1011-1028.

Pingree, R.D., P.R. Pugh, P.M. Holligan and G.R. Forster. 1975. Summer phytoplankton blooms and red tides along tidal fronts in the approaches to the English Channel. Nature 258: 672-677.

Robinson, I.S. 1981. Tidal vorticity and residual circulation. Deep-Sea Res. 28A: 195-212.

Schantz, E.J. 1970. Algal toxins. In: Properties and Products of Algae. J.E. Zajic (ed.). Plenum, NY.

Selvin, R.C., C.M. Lewis, C.M. Yentsch and J.W. Hurst. 1984. Seasonal persistence of resting cyst toxity in the dinoflagellate Gonyaulax tamarensis var. excavata. Toxicon 2: 817-820.

Simpson, J.H. and D. Bowers. 1981. Models of stratification and frontal movement in shelf seas. Deep-Sea Res. 28A: 727-738.

Simpson, J.H. and J.R. Hunter. 1974. Fronts in the Irish Sea. Nature 250: 404-406.

Simpson, J.H., P.B. Tett, M.L. Argote-Espinoza, A. Edwards, K.L. Jones and G. Savidge. 1982. Mixing and phytoplankton growth around an island in a stratified sea. Cont. Shelf Res. 1: 15-31.

Tangen, K. 1979. Dinoflagellate blooms in Norwegian waters. In: Toxic Dinoflagellate Blooms, pp. 179-182. D.L. Taylor and H.H. Seliger (eds.). Elsevier/North Holland.

Taylor, F.J.R. 1984. In: Seafood Toxins. E. Ragelis (ed.). ACS Symposium Volume.

Thayer, P.E., J.W. Hurst, C.M. Lewis, R.C. Selvin and C.M. Yentsch. 1983. Distribution of resting cysts of Gonyaulax tamarensis var. excavata and shellfish toxicity. Can. J. Fish Aquat. Sci. 40: 1308-1314.

Townsend, D.W., C.M. Yentsch, C.E. Parker, W.M. Balch and E.D. True. 1983. An island mixing effect in the coastal Gulf of Maine. Helgol. Meeresunters 26: 348-356.

Tyler, M.A. and H.H. Seliger. 1980. Annual subsurface transport of a red tide dinoflagellate to its bloom area: water circulation patterns and organism distributions in the Chesapeake Bay. Limnol. Oceanogr. 23: 227-246.

Vermersch, J.A., R.C. Beardsley and W.S. Brown. 1979. Winter circulation in the Western Gulf of Maine: Part 2. Current and pressure observations. J. Phys. Oceanogr. 9: 768-784.

White, A. 1984. In: Seafood Toxins, E. Ragelis (ed.). ACS Symposium Volume.

Yentsch, C.M. 1984. Paralytic shellfish poisoning: An emerging perspective. In: Seafood Toxins, pp. 2-36. E. Ragelis (ed.). American Chemical Society Symposium Series No. 262.

Yentsch, C.M., B. Dale and J.W. Hurst. 1978. Coexistence of toxic and non-toxic dinoflagellates resembling Gonyaulax tamarensis in New England coastal waters (N.W. Atlantic). J. Phycol. 14: 330-332.

Yentsch, C.M., T.L. Cucci and D.A. Phinney. 1984. Flow cytometry and cell sorting: Problems and promises for biological ocean science research. In: Marine Phytoplankton and Productivity, Lecture Notes on Coastal and Estuarine Studies, pp. 141-155. O. Holm-Hansen, L. Bolis and R. Gilles (eds.). Springer-Verlag.

Yentsch, C.S. 1975. New England coastal waters - an infinite estuary. ACS Symposium Series No. 18. Marine Chemistry in the Coastal Environment. T.M Church (ed.), pp. 608-617.

Yentsch, C.S. and N. Garfield. 1981. Principal areas of vertical mixing in the waters of the Gulf of Maine, with reference to the total productivity of the area. In: Oceanography From Space, pp. 303-312. J.F.R. Gower (ed.). Plenum Press, NY.

DYNAMICS OF LARVAL HERRING (<u>CLUPEA</u> <u>HARENGUS</u> L.) PRODUCTION IN TIDALLY MIXED WATERS OF THE EASTERN COASTAL GULF OF MAINE

D.W. Townsend
Bigelow Laboratory for Ocean Sciences
McKown Point
W. Boothbay Harbor, ME 04575

J.J. Graham
Maine Department of Marine Resources
W. Boothbay Harbor, ME 04575

D.K. Stevenson
Department of Zoology
University of Maine
Orono ME 04469
and
Maine Department of Marine Resources
W. Boothbay Harbor, ME 04575

INTRODUCTION

The interactions of tidal mixing and plankton dynamics in shelf seas have been receiving increasing attention in recent years. It is now generally recognized that favorable conditions for phytoplankton growth, i.e., compromises between incident light levels, or upper mixed layer depth, and inorganic nutrient concentrations, are often achieved in shallow mixed areas and at fronts in deeper regions where thermally stratified and tidally mixed waters meet. Not surprisingly, it is becoming increasingly apparent that much of the overall biological productivity of the Gulf of Maine can be ascribed to such mixing processes (Yentsch and Garfield, 1981).

The importance of tidal mixing in the Gulf of Maine, where tides range from about 2 m in the west to >6 m in the east, has been recognized since the time of Henry Bigelow (1927), and is illustrated in Figures 1 and 2. Cooler surface water temperatures, which appear darker in satellite infrared images, usually represent those areas where tidal mixing is strong enough to overcome thermal stratifica-

Lecture Notes on Coastal and Estuarine Studies, Vol. 17
Tidal Mixing and Plankton Dynamics. Edited by J. Bowman, M. Yentsch and W. T. Peterson
© Springer-Verlag Berlin Heidelberg 1986

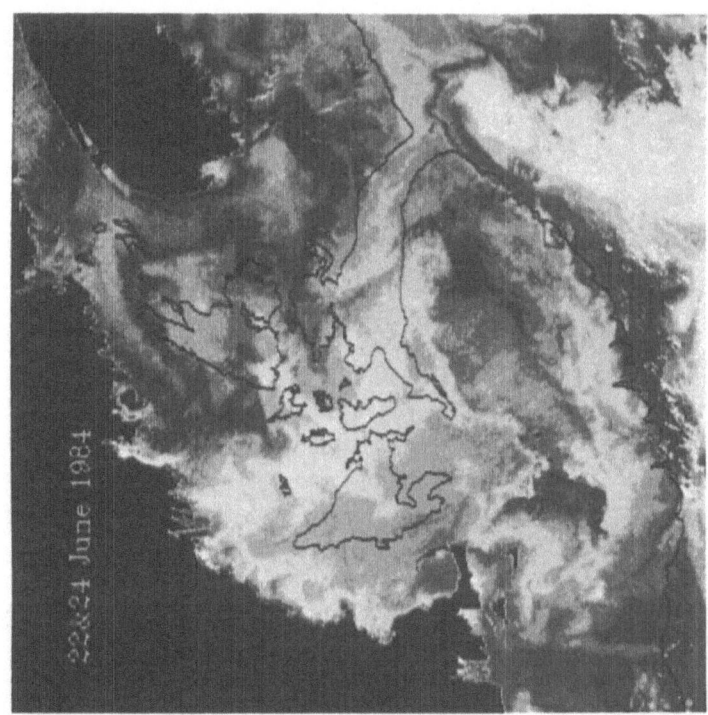

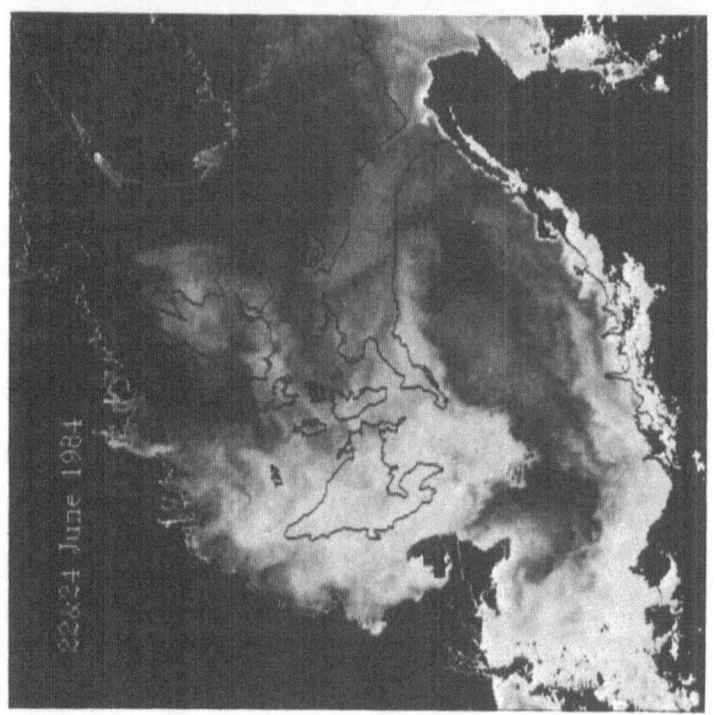

Figure 1. Composite satellite infrared image of the Gulf of Maine taken June 22 and 24, 1984. The panel on the right is a black and white photograph of a color enhancement and shows greater detail along the eastern Maine coast. The 200 m bottom contour is given.

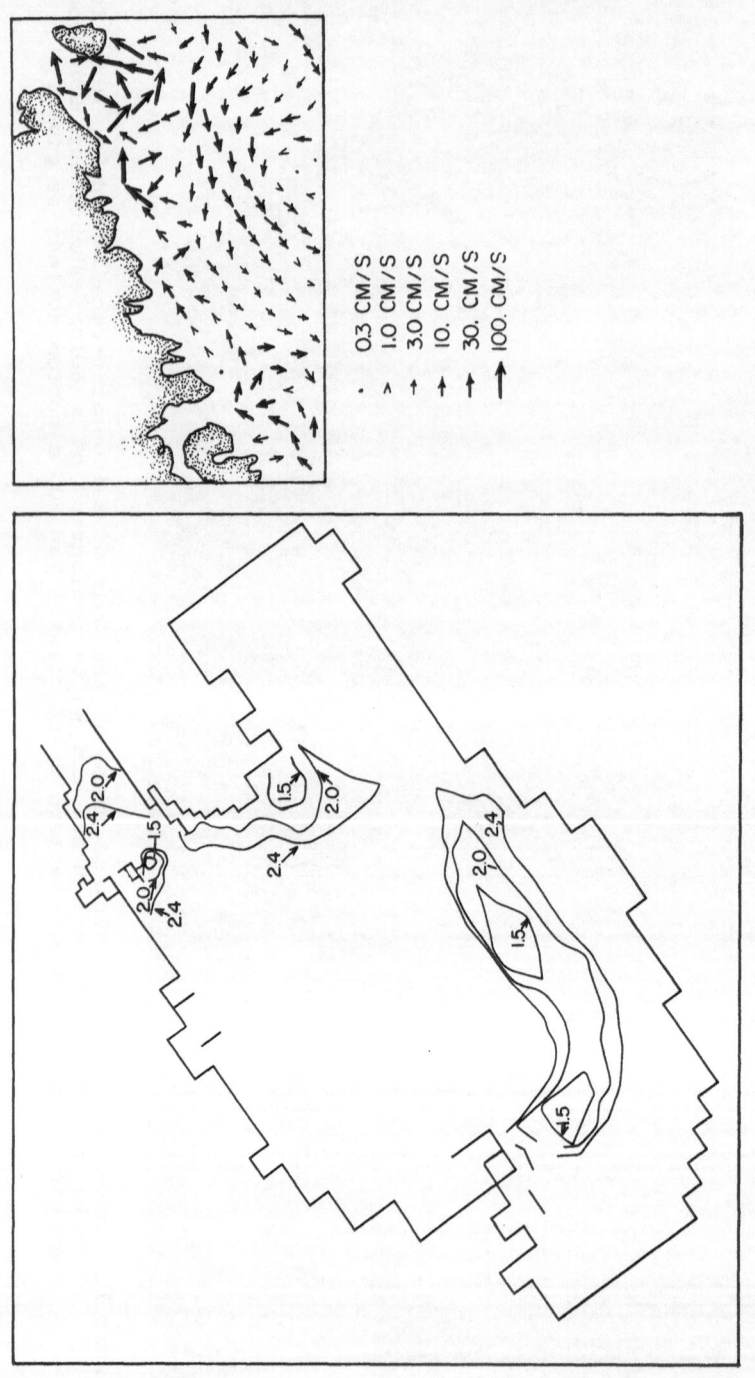

Figure 2. Contours of log (H/D) where H = water depth and D = the rate of tidal energy dissipation (from Garrett et al., 1978) for the Gulf of Maine-Georges Bank region, and model-produced residual currents driven by wind and tidal forcing (from Greenberg, 1983) off the coast of eastern Maine.

tion. These shallower areas of the Gulf include the southwest Nova
Scotian shelf, Georges Bank, the Bay of Fundy and its approaches, and
the coastal waters of New England. With a few exceptions, these
areas of cool surface water temperatures coincide with the tidally
mixed regions predicted by Garrett et al. (1978) using the Simpson
and Hunter (1974) stratification parameter, a ratio between water
column depth and tidal current speed. Since tides mix the water from
the bottom up, it is assumed that for a given tidal current speed and
value for bottom friction, mixing can be expected only out to a
certain depth; for the Gulf of Maine mixing by tides can be expected
only within the 50 to 60 m isobaths (Garrett et al., 1978; Yentsch
and Garfield, 1981). It appears unlikely, then, that tidal mixing
alone is responsible for maintaining the cooler surface water temperatures which extend beyond the 100 m isobath along the coast of
eastern Maine. Rather, Brooks (1985) has suggested that the cooler
waters along the Maine coast are advected from the tidally mixed
waters in the east as part of a cyclonic gyre over Jordan Basin. He
points out that, in keeping with accounts by earlier workers (Gran
and Braarud, 1935; Graham, 1970), these coastal waters depart the
coast east of Penobscot Bay producing the plume feature which extends
into the Gulf. We show in the discussion which follows, that these
mixing and advective processes may have significant consequences for
fish larvae.

Important herring spawning grounds are located in the tidally
mixed waters off the coast of eastern Maine and in the vicinity of
Grand Manan Island (Bigelow and Schroeder, 1953; Iles and Sinclair,
1982; Graham, 1982). Recently, it has been suggested that tidally
mixed areas in general serve as important herring spawning grounds
and retention areas for the larvae, and that the geographic size of
such areas determines the maximum size of a given adult spawning
population (Iles and Sinclair, 1982). Our work concerns itself with
the nature of larval herring production in the eastern coastal waters
of Maine, and how the physical and biological features of these
tidally mixed waters affect the distribution and abundance of herring
larvae and their zooplankton food.

MATERIALS AND METHODS

Temperature, zooplankton and ichthyoplankton were sampled at the stations shown in Figure 3a on two cruises on the Canadian research vessel J.L. Hart during 4-7 Oct and 29 Nov-3 Dec 1982. Temperatures were measured with reversing thermometers. The zooplankton was sampled using an 80-cm diameter 80-μm mesh net hauled vertically from near the bottom to the surface at about 1 m ·sec^{-1}. A 61-cm bongo net (Posgay and Marak, 1981), fitted with 505-μm mesh nets and digital flowmeters, was used to sample herring larvae. Samples were preserved in 5% buffered formalin. The bongo nets were towed in a double oblique manner to within 10-20 m of the bottom at a ship speed of 3.5 knots. A Boothbay Depressor trawl (Graham and Vaughan, 1966) was used at 14 of the 27 stations sampled on the 29 Nov-3 Dec cruise. It was used to test whether larger larvae might be caught more efficiently with the trawl; a comparison between the trawl and bongos showed no significant differences in the sizes of larvae caught with the two gears.

A second set of cruises followed in the fall of 1983 using the University of Maine's R/V Lee. On 13-15 Sept and 20-21 Sept, zooplankton and herring larvae were sampled at the stations shown in Figure 3b using the 80-cm 80-μm mesh net and 61-cm 505-μm mesh bongo net as described above. It was our intent in conducting two consecutive weekly surveys to document changes in the larval density pattern which would be indicative of transport along the coast from the spawning grounds in eastern Maine. For this reason samples were collected from Grand Manan Channel to Mt. Desert Island (Fig. 3). A small-scale survey of herring larvae was conducted on 27-28 September in the area shown in Figure 3b. The purpose of this was to determine larval distributions in the immediate vicinity of known egg beds in order to test whether or not small-scale larval density data could be used to locate egg beds and to examine dispersal of larvae from individual point sources rather than from an entire spawning ground. Sampling during the third survey was therefore concentrated in the Machias Bay area with stations located much closer together than in the other surveys.

258

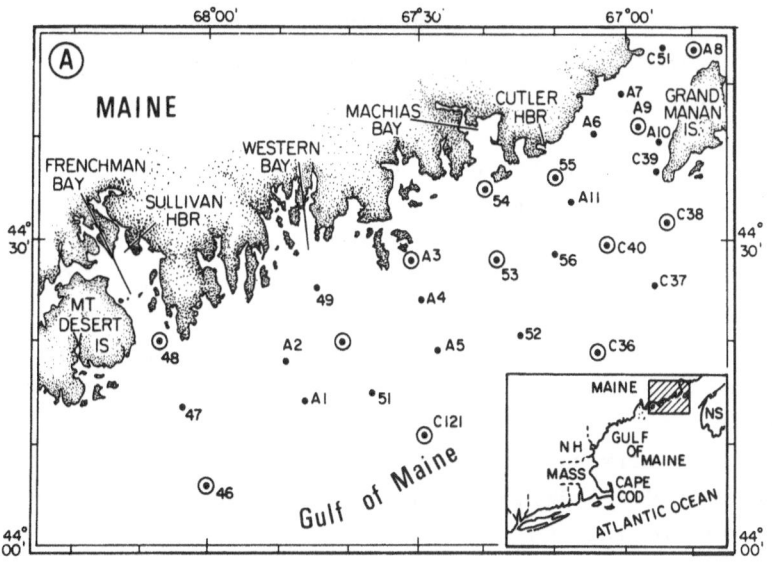

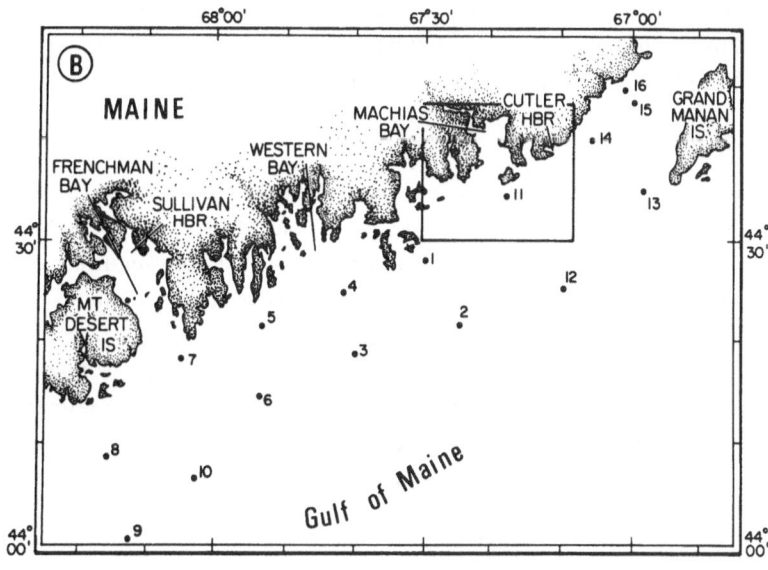

Figure 3. (A) Station locations for the 4-7 Oct and 29 Nov-3 Dec 1982
cruises. (Θ) = Sampling for zooplankton, herring larvae
and hydrography; (●) = sampling for herring larvae and
hydrography. Zooplankton were not sampled at stations 55
and A9 on the latter cruises. (B) Station locations for
the 13-15 Sept and 20-21 Sept 1983 cruises. Herring larvae
and zooplankton were sampled at each station; stations
13-16 were not sampled on the latter cruise. Station
locations for the 27 Sept 1983 small scale herring larvae
survey are shown by the enclosed box in B.

RESULTS

Egg Bed Locations

Herring in the Gulf of Maine spawn in late summer and fall and lay demersal and adhesive eggs in small, discrete patches, generally in depths of 10-100 m, along the coast and on offshore shoals and ledges (Bigelow and Schroeder, 1953; Sindermann, 1979); spawning occurs earlier in eastern Maine waters than anywhere to the west (Graham, 1982; Kelly and Stevenson, in press). Egg bed sites in the study area were determined by interviewing local fishermen who have observed herring eggs adhering to lobster pots. These interviews revealed that herring generally spawn in a number of discrete locations in the Grand Manan Channel and in the coastal waters near Machias, Little Machias, and Little Kennebec Bays (Fig. 4). These findings confirm published reports based on larval abundance data (Graham, 1982) which show that eastern Maine is an important spawning ground. Spawning also takes place in the vicinity of Grand Manan Island (Iles and Sinclair, 1982), although the number of adult spawning herring in that area has apparently diminished in recent years.

Three egg beds were located by lobster fishermen in 1983 (Fig. 4). The first was located south of Great Head near Cutler Harbor; eggs were first noticed in this location on 23-24 August. Two additional sites were located east and west of Cross Island on 8 and 14 September. All three sites were near the 40 m contour. Since lobster fishermen haul their pots every 3-4 days, the dates when eggs were first observed should have roughly corresponded to spawning dates. No attempts were made to locate egg beds in the Cutler-Machias area in 1982 or in the Western Bay area in either 1982 or 1983.

Hydrography

Our data support the general contention that the degree of vertical mixing along coastal Maine increases from west to east (Fig. 5, here using only the vertical distributions of temperature). This is in general agreement with Bigelow (1927) and the model predictions by Garrett et al. (1978) which showed greatest tidal mixing around Grand Manan Island.

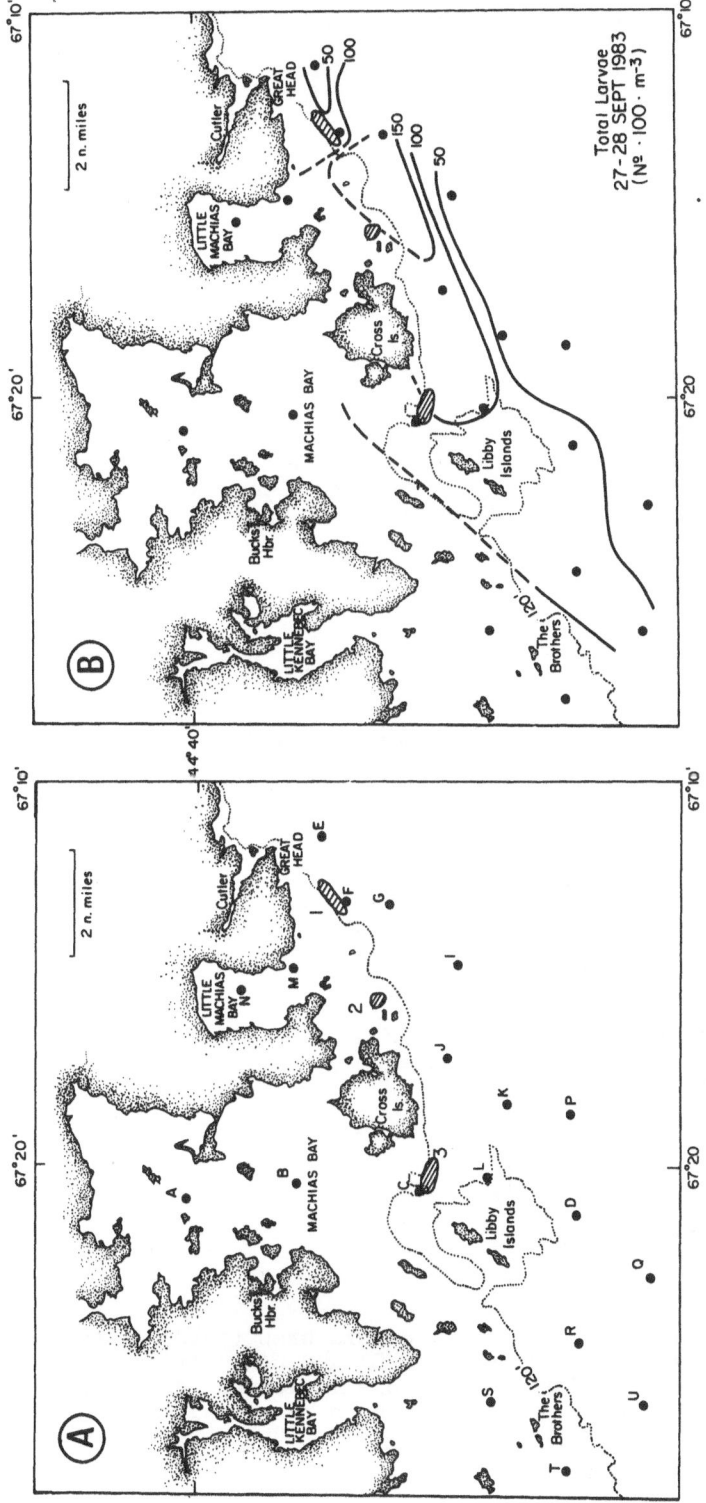

Figure 4. (A) Station locations for small scale survey of herring larvae 27-28 Sept 1983 and locations of egg beds (hatched). (B) Contours of densities of herring larvae (all sizes) for the 27-28 Sept survey given as a number of larvae per 100 m³.

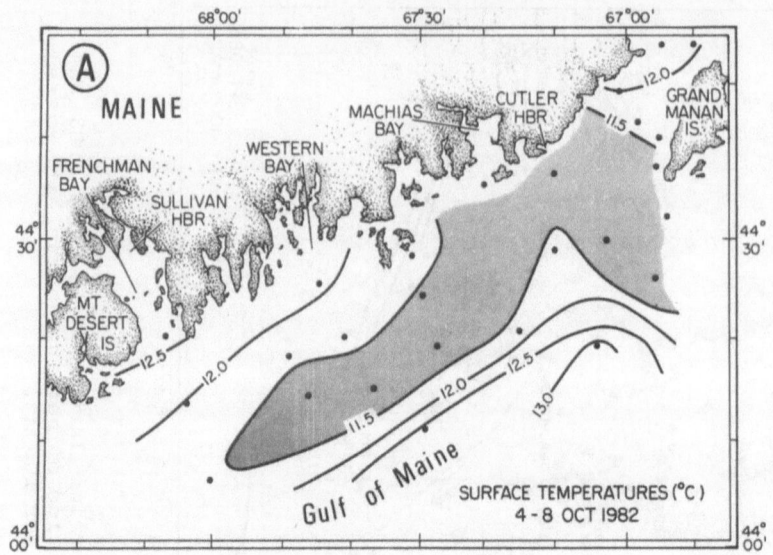

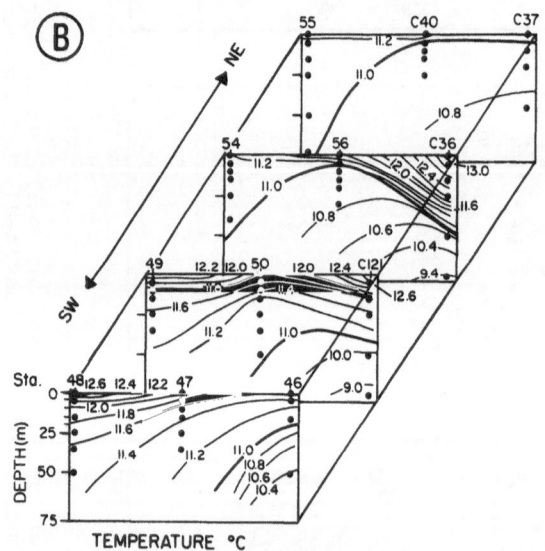

Figure 5. (A) Contours of surface temperatures (shaded area indicates surface temperatures <11.5°C) and, (B) vertical sections of temperatures for the stations indicated, during the 4-7 Oct 1982 cruise. All temperatures were fairly uniform (8.5-8.8°C) during the 29 Nov-3 Dec 1982 cruises and are not plotted.

The residual currents in these waters are quite complex. Bumpus and Lauzier (1965) showed using drift bottles that the general pattern of surface flow is from east to west along the coast, and that there may be an offshore component midway between Grand Manan Island and Mount Desert Island (Fig. 6). Greenberg's (1983) model-produced residual currents (Fig. 2), driven by tidal forcing and summer wind-stress (the latter from values given by Saunders (1977)) shows a series of complex eddies in the waters between Grand Manan Island and a point halfway to Mount Desert Island. These eddies could certainly provide a mechanism for retention of herring larvae as postulated by Iles and Sinclair (1982).

Our plots of the temperature structure (Fig. 5) suggest that the residual currents are probably more like those given by Bumpus and Lauzier (1965). Figure 5 shows that the cooler (<11.5^{o}C) tidally mixed waters just to the west of Grand Manan flow to the west and then move more offshore as a tongue of cool surface water. Such a flow was first discussed by Gran and Braarud (1935) and the tongue feature shows up nicely in the enhanced satellite infrared image (Fig. 1). As discussed earlier, Brooks (1985) has shown that this coastal flow is not due principally to tides but has a strong baro-clinic component and is part of a cyclonic gyre in the eastern Gulf. The distributions of herring larvae appear to be closely related to this coastal tongue.

Larval Herring Distributions

The distributions of larvae deduced from our 4-7 Oct 1982 cruise show two centers of abundance (Fig. 7). Figure 7b shows an area off Western Bay with a high catch rate of young larvae (<10 mm) suggesting the close proximity of a discrete spawning site. Figures 7 c and d show another center of abundance of older larvae (>10 mm) off Machias Bay. A third, less dense, group of older larvae occurred just off Grand Manan Island.

Densities of larvae detected on our 29 Nov to 3 Dec cruise were considerably depleted by mortality and drift (Figs. 8 a-d). We were unable to demonstrate transport of the abundance centers of larvae, owing to the time elapsed between the two cruises (ca. 2 mo). However, a series of samples from Sullivan Harbor (Fig. 3) collected

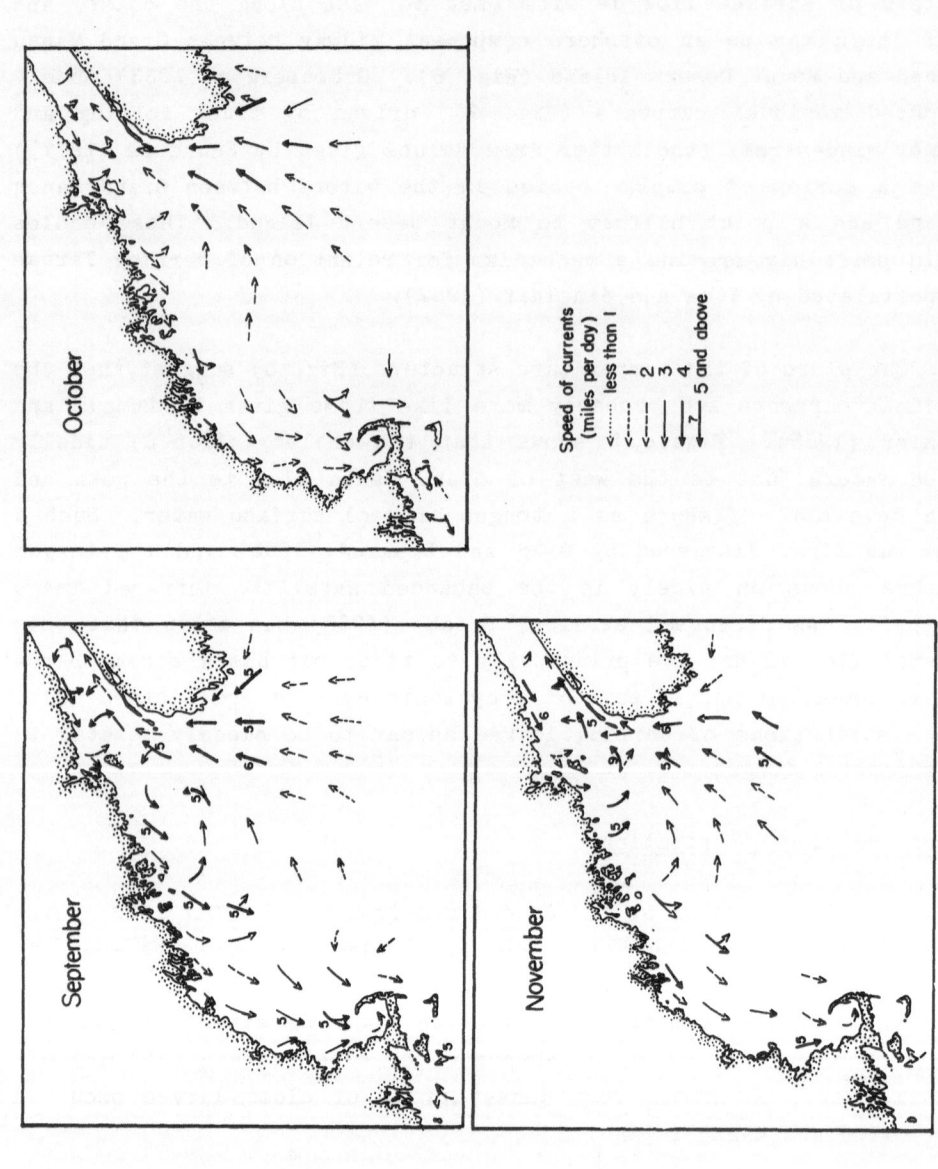

Figure 6. Inferred surface circulation from drift bottle returns (from Bumpus and Lauzier, 1965).

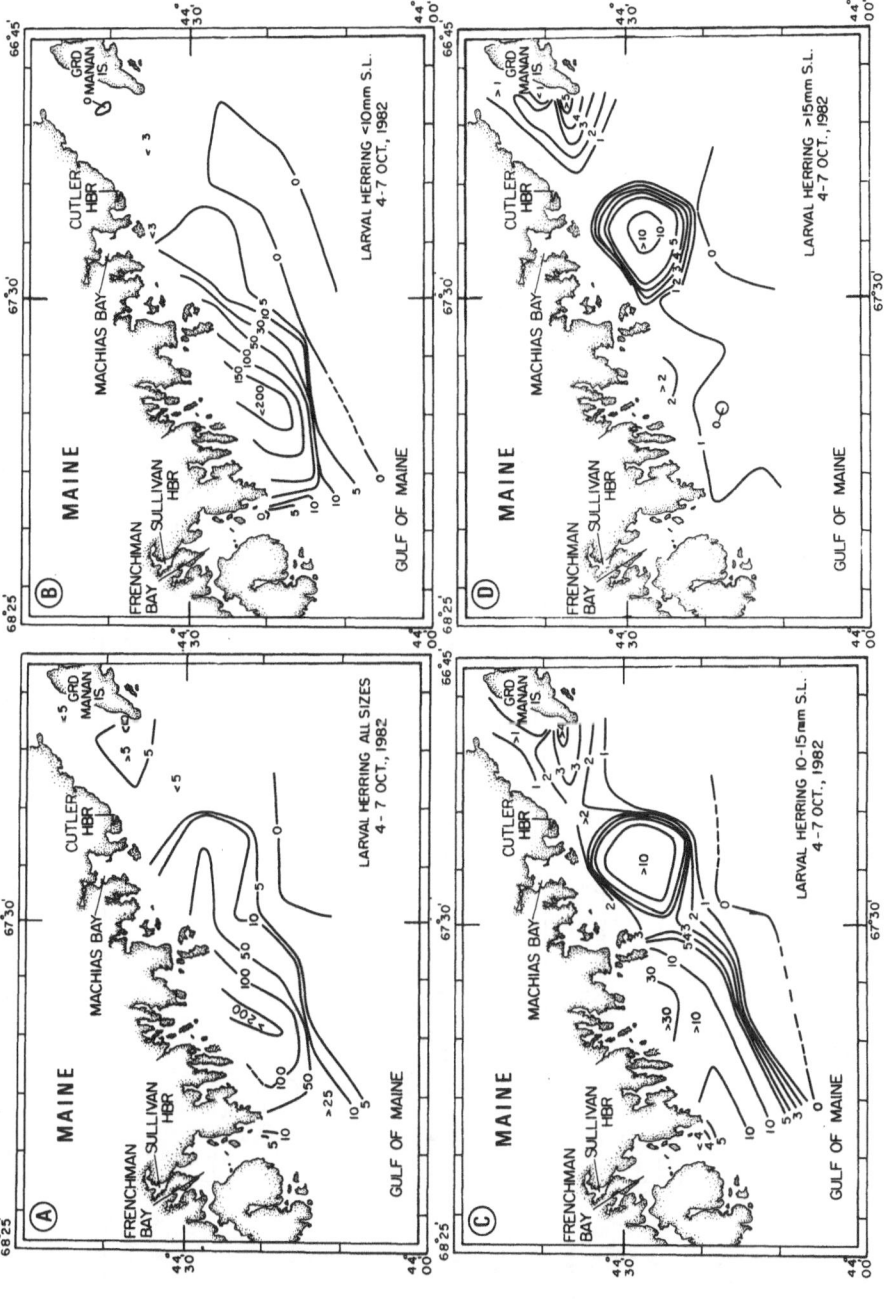

Figure 7. Distributions of herring larvae by size for the 4-7 Oct 1982 cruise. Contours are in numbers of larvae under a m² of sea surface.

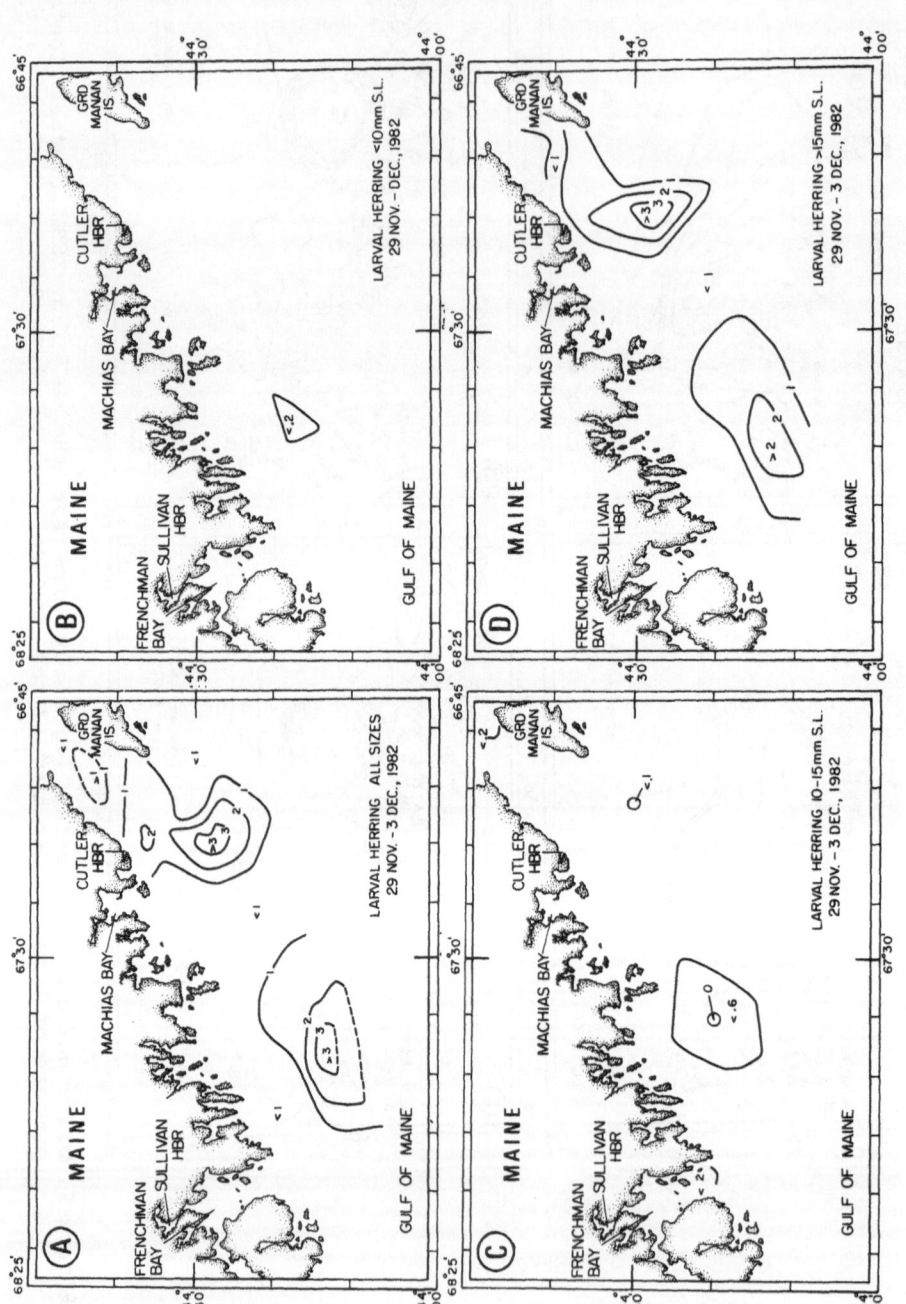

Figure 8. Distributions of herring larvae by size for the 29 Nov-3 Dec 1982 cruise. Contours are in numbers of larvae under a m² of sea surface.

each year by one of us (Graham) showed that larvae began to arrive in this inshore area in October, culminating in a small abundance peak (ca. $1.5 \cdot 100$ m^{-3}) in November. This also points out the fact that larval transport may include both active and passive components, since we have already noted that the coastal current turns offshore east of this inshore nursery area.

Our 1983 cruises were spaced one week apart, and afforded us the opportunity to examine more closely the advective transport and dispersion of larvae (Figs. 9 and 10). As we found in 1982, there were two centers of abundance of larvae, one off Machias Bay, representing recently hatched larvae, and a second center off Western Bay. This second center was comprised of larvae somewhat older (one to two weeks older) than the first, and may represent an aggregation rather than a spawning site. Transport of larvae to the west and somewhat offshore was evident between the two cruises. Such transport is in keeping with the residual current pattern discussed earlier. The dispersion and transport of larvae from the Machias Bay spawning area is reflected in the more even distribution of larvae shown by the second cruise (Fig. 10a). The results of the small scale sampling conducted on 27-28 Sept show more clearly the dispersion and transport of larvae to the west from the egg beds (Fig. 4). Given an incubation time of 10-15 days (Bigelow and Schroeder, 1953) and the fact that the majority of larvae collected in the 27-28 Sept. survey were still <10 mm, it seems probable that at least some of these larvae had hatched from the Cross Island egg beds. It was also evident from the results of this survey that larvae were initially being dispersed primarily along the coast, perhaps as part of the coastal tongue discussed above, rather than immediately into the bays and inlets (Fig. 4).

Zooplankton Distributions

The distributions of zooplankton on each cruise in 1982 and 1983 all showed the same general pattern: an increase in abundance from east to west (Figs. 11 and 12). On the 4-7 Oct 1982 cruise the abundances of naupliar and post naupliar stages of copepods increased from a low in Grand Manan channel to a high in the offing of Frenchman Bay (Fig. 11a and b). The same pattern held on the 29 Nov-3 Dec

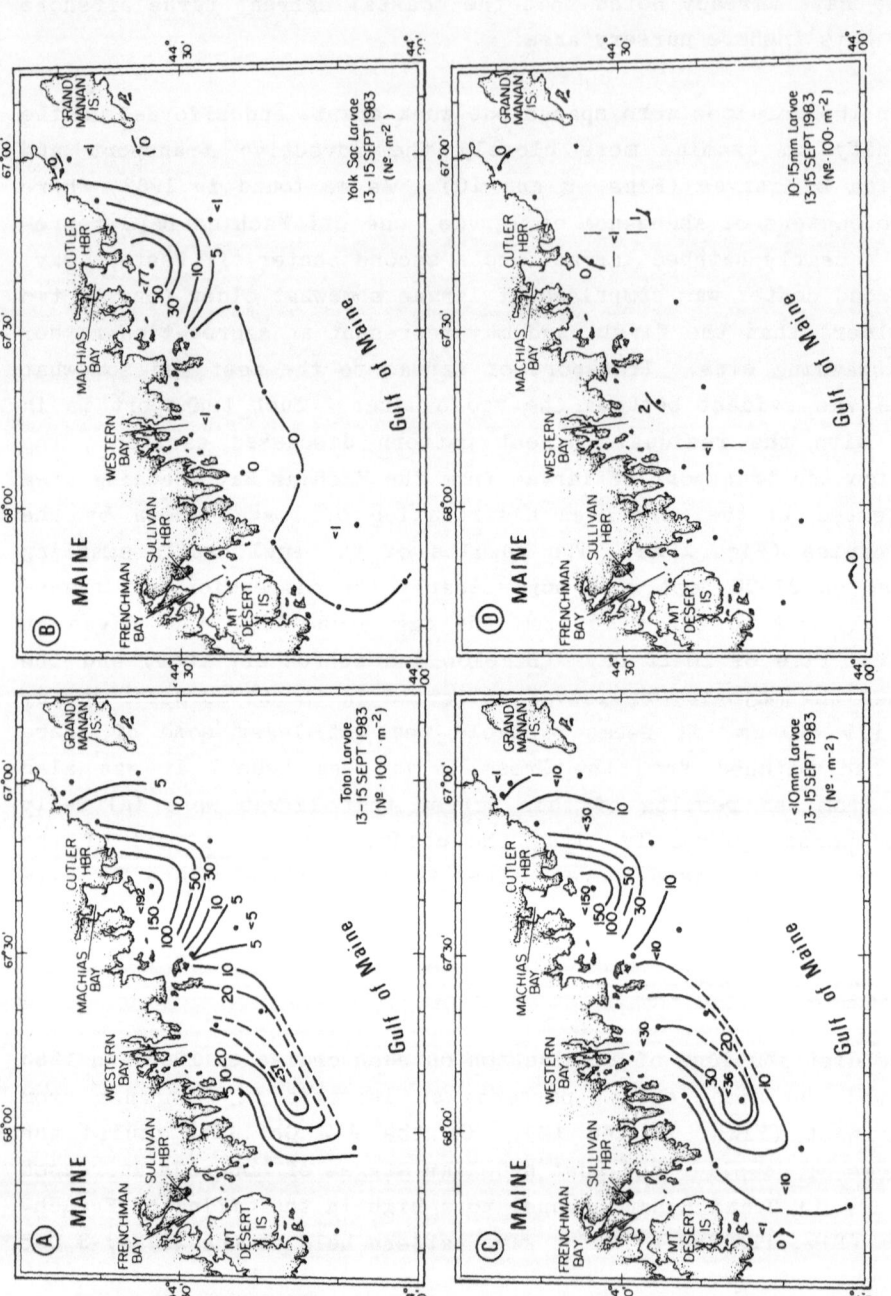

Figure 9. Distributions of herring larvae by size for the 13-15 Sept 1983 cruise. Contours are in numbers of larvae under a m² of sea surface.

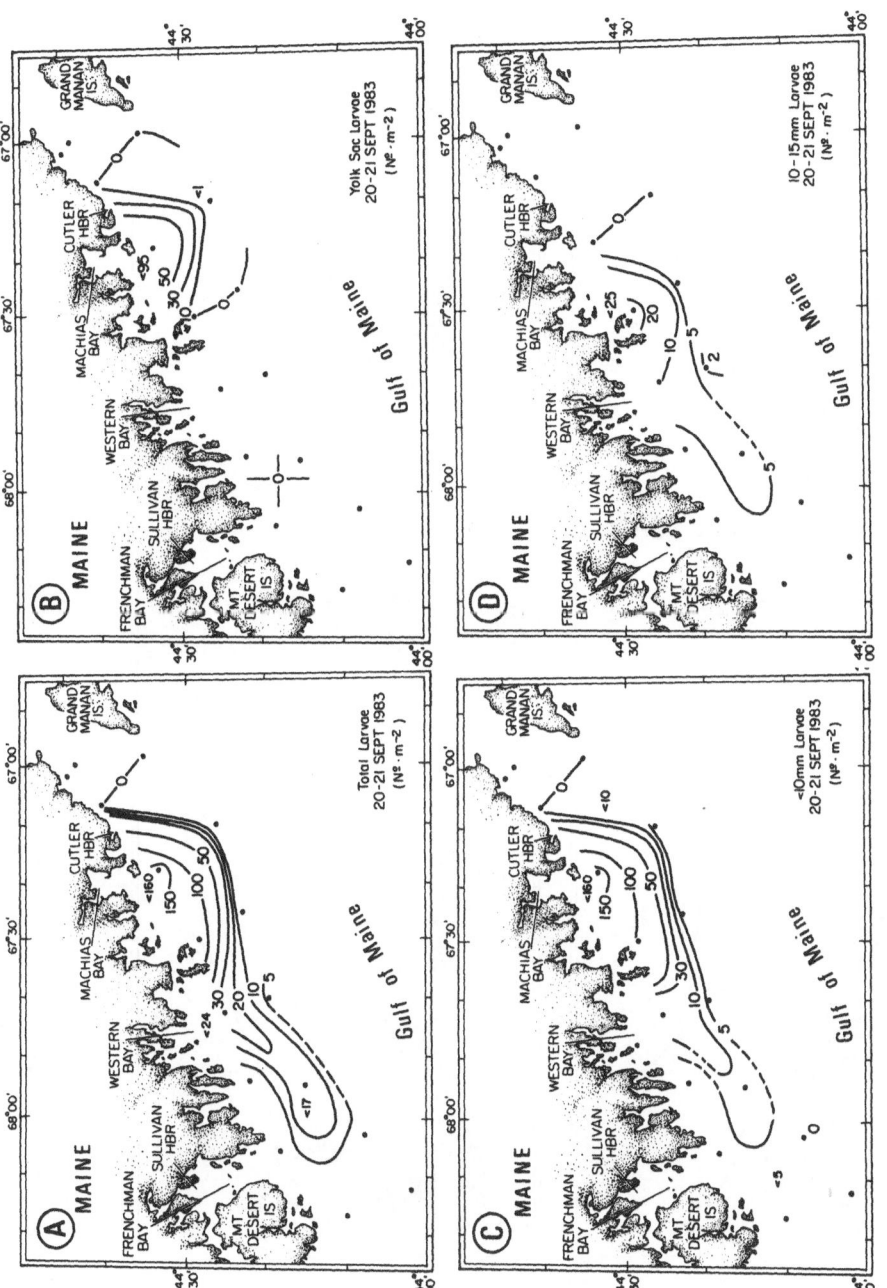

Figure 10. Distributions of herring larvae by size for the 20-21 Sept 1983 cruise. Contours are in numbers of larvae under a m² of sea surface.

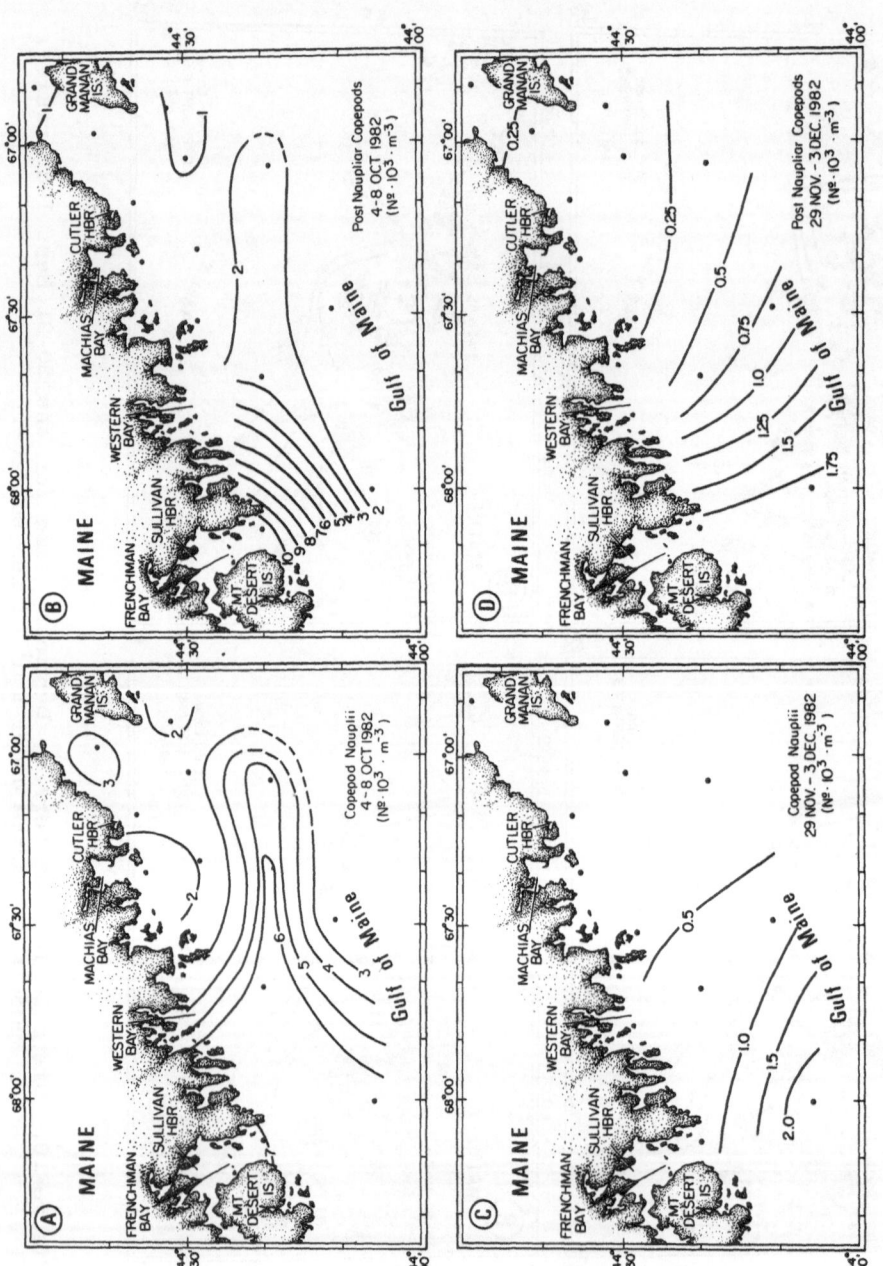

Figure 11. Distributions of naupliar and post-naupliar copepods for the two 1982 cruises. Contours are in thousands of animals per m³.

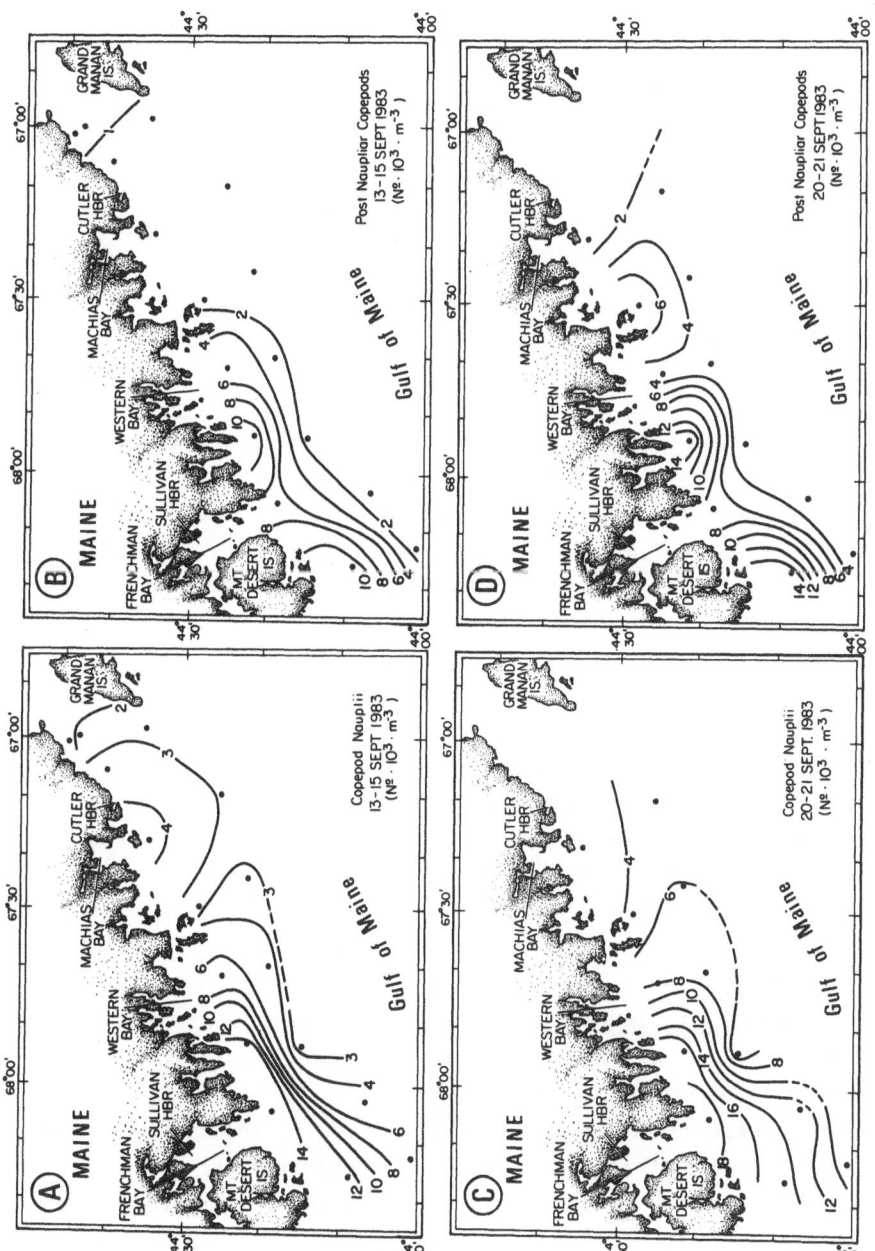

Figure 12. Distribution of naupliar and post-naupliar copepods for the two 1983 cruises. Contours are in thousands of animals per m³.

1982 cruise, but the overall densities were much reduced during this early winter sampling period (Fig. 11c and d) In September of 1983 the densities of copepod nauplii were greater than during October the previous year, yet the same pattern of highest densities off French-man Bay was again evident (Fig. 12a and c). The distributions of post naupliar copepods during the September 1983 cruises also displayed an increase in abundance from east to west, but did not show highest densities immediately off Frenchman Bay as did the nauplii (Fig. 11b and d). Rather, the densities of copepodid and adult stages increased from offshore to inshore where there were two areas of high concentration on either side of Frenchman Bay. The dominant species of copepods during each of the cruises included Pseudocalanus sp., Acartia longiremis, Microsetella norvegica, Oithona sp. and Centropages spp. A tintinnid Tintinnopsis sp. was also abundant on the 4-7 Oct 1982 cruise (Fig. 13), and had a maximum density ($789 \cdot m^{-3}$) at station 50 (Fig. 3). It was much less abundant on the second cruise that year ($<200 \cdot m^{-3}$ at 10 of the 11 stations). Tintinnopsis sp. was abundant on the second cruise in 1983 (20-21 Sept) but only at the offshore stations, reaching $28286 \cdot m^{-3}$ at station 10 (Fig. 13), while densities were less than $500 \cdot m^{-3}$ at all stations during the first cruise (13-15 Sept). A rotifer Synchaeta sp. was also more abundant on this second cruise and showed the same pattern as Tintinnopsis sp. in being more abundant at the offshore stations (Fig. 13).

Larval Feeding and Condition Factors

The distribution of zooplankton appears to be advantageous to those fish larvae which are transported to the west. After having hatched in the east, off Machias Bay, the larvae are transported into waters richer in zooplanktonic food. This may represent mixing with waters already high in zooplankton abundances or that the zooplankton propagate while being transported in the same water mass as the larvae. Even for those larvae which are transported offshore, the higher densities of Tintinnopsis and Synchaeta could serve as forage. Our preliminary examinations of larval gut contents from the 1982 cruises showed that Tintinnopsis was fed upon, although to a lesser extent than copepod nauplii and copepodites. The young larvae sampled in 1983 had a very low (ca. 10%) incidence of food in the gut, and we were unable to assess food preferences. However, we did evaluate relative condition factors (post yolk sac) of those larvae

272

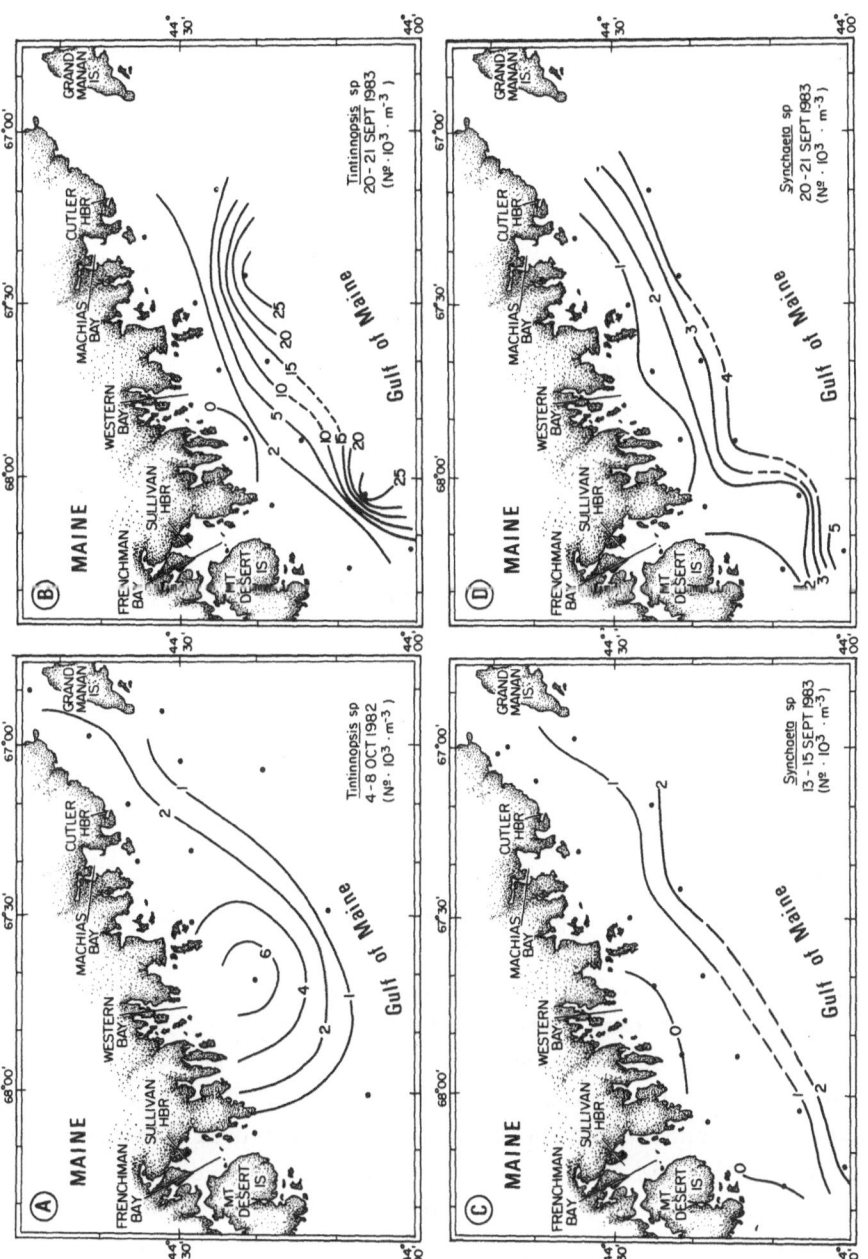

Figure 13. Distributions of Tintinnopsis sp. and Synchaeta sp. Contours are in thousands of animals per m³.

sampled in 1983 using the method given in Townsend (1983) in order to evaluate the relative physiological well-being of the larvae. Stated briefly, the condition factor incorporated the allometric equation W = aL^b where W = dry weight of a larva, L = total length and a and b are constants. All the data for that year were used to solve the allometric equation for the exponent, b, by using a geometric mean functional regression (Ricker, 1973). The condition factor for a larva is then K = W/L^b. These results are given in Figure 14, and demonstrate that recently hatched larvae sampled in the immediate vicinity of the egg beds had the highest condition factors, which we assume reflected good health after recently hatching and being in the transition period between relying on yolk sac reserves and exogenous feeding. As these larvae presumably are transported toward the west the condition factors drop considerably, but increase somewhat as they enter the zooplankton-richer waters to the west. Thus the relative condition factors are highest in the hatching area and in the area to the west which has greater concentrations of potential food organisms.

DISCUSSION

The results which we present here represent only the beginnings of an ongoing research effort to understand the dynamics of larval herring production in the eastern coastal Gulf of Maine, but already an interesting paradigm is emerging. The conceptual model presented by Iles and Sinclair (1982) suggests that Atlantic herring spawning sites are situated in tidally mixed waters to assure, among other reasons, the retention of larvae within the spawning area. They assumed that it would then follow, and the empirical evidence bore this out, that the size of a given adult herring spawning stock would be dictated by the size of the tidally mixed larval retention area. Our data from coastal waters do not fit this general pattern. Rather than retention within tidally mixed waters in the east, which represent only a relatively small geographic area (Fig. 2), we observed that the larvae were being transported to the west and into waters progressively more stratified. This apparently involved either selective tidal transport by the larvae or passive drift in the non-tidal residual flow, which is to be expected of recently mixed waters. Such a flow to the west and offshore was illustrated by our observations of water temperatures and larval distributions, and

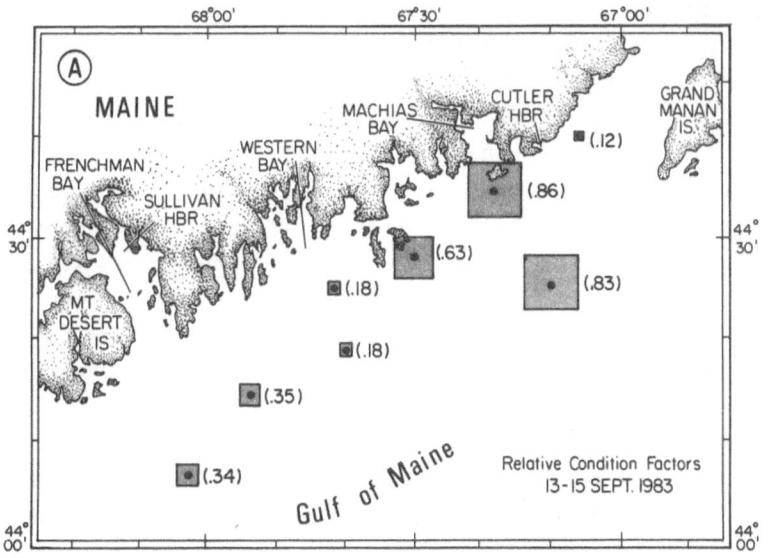

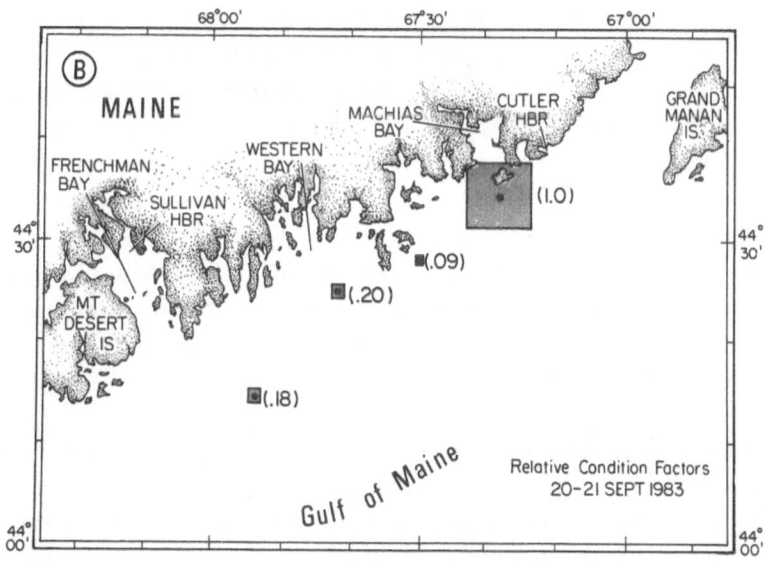

Figure 14. Relative condition factors for herring larvae collected
13-15 and 20-21 Sept 1983. The linear dimensions of each
box are relative to the condition factor measured at that
station (given in parentheses).

generally agrees with the patterns reported by Gran and Braarud (1935), Bumpus and Lauzier (1965), Graham (1970), and Brooks (1985). Further, we believe that it is more advantageous to the survival of the larvae to be transported away from the spawning site; the waters in the vicinity of the spawning site are low in potential zooplankton forage, but to the west food supplies are greater. For those larvae which are transported to western inshore areas, such as Frenchman Bay and Sullivan Harbor, an increased supply of zooplankton would then be available. Even for those larvae which are transported more offshore, there may be greater densities of food resources, in the form of tintinnids and rotifers, than available within the spawning area. Also, spawning in tidally well mixed waters, such as the eastern coastal Gulf of Maine where tidal excursions extend several kilometers, would result in effective dispersal and reduction of larval densities after hatching from the demersal eggs, consistent with the hypothesis that larval fish survival is density dependent (Shepherd and Cushing, 1980).

Since the demise of the Georges Bank herring spawning group (Anthony and Waring, 1980), one would expect there to be greater pressure on the remaining Gulf of Maine spawning populations to maintain the production of larvae necessary to support the fishery. Should Iles and Sinclair be correct in their assessment of the limitations imposed by the areal extent of the tidally mixed spawning areas, only a relatively small spawning stock could be supported by these waters. We suggest that the larval populations leave their restrictive "retention area/spawning site" less densely aggregated, as a result of tidal dispersion, and enter a transport system which provides additional nursery grounds.

In the case of the eastern coastal Gulf of Maine spawning area, it appears that the larvae are released "upstream" from food rich waters such as has been reported for tropical coral reef fishes (Johannes, 1978), and that the period of larval drift which ensues represents an important stage in their life history. It would appear that the herring larvae are drifting to the west along with recently mixed waters high in phytoplankton abundances (Yentsch and Garfield, 1981) and are ready to prey upon the early developing stages of zooplankton, which are drifting in the same water mass along with the larvae and which most likely propagate in response to the vertical mixing and increase in primary production.

Admittedly, the data we present are incomplete in that they deal only with the zooplankton and herring larvae, but we feel we have constructed a working hypothesis which can be tested. We will need to verify in the future our assumption that the injection of inorganic nutrients is enhanced by the intense vertical mixing in the waters of eastern Maine and those surrounding Grand Manan Island, and that these waters nurture a developing phytoplankton and zooplankton population as this water mass is advected downstream and to the west. This research is presently underway.

ACKNOWLEDGEMENTS

We wish to thank Dr. T.D. Iles and the Department of Fisheries and Oceans for shiptime on the J.L. Hart. We thank also Keith Sherman, Ken Abott, David Hodges, Kevin Kelly, Andrea Swiecicki, Maureen Maritato, Michael Dunn and Kerry Lyons for their valuable assistance. The satellite image analysis in Figure 1 was performed by the University of Rhode Island's Oceanographic Remote Sensing Laboratory. Image processing software was developed and has been maintained by O. Brown, R. Evans, J. Brown and A. Li at the University of Miami. This work was funded by grants from NOAA Sea Grant and the State of Maine. This is Bigelow Laboratory for Ocean Sciences contribution number 84013.

REFERENCES

Anthony, V.C. and G. Waring. 1980. The assessment and management of the Georges Bank herring fishery. Rapp. P.-v. Réun. Cons. int. Explor. Mer, 1977: 72-111.

Bigelow, H.B. 1927. Physical oceanography of the Gulf of Maine. Bull. U. S. Bur. Fish. 40: 1-509.

Bigelow, H.G. and W.C. Schroeder. 1953. Fishes of the Gulf of Maine. Fish. Bull. U.S. Fish. Wild. Serv. 53(74): 577 p.

Brooks, D.A. 1985. Vernal circulation in the Gulf of Maine. J. Geophys. Res. 90(C3): 4687-4705.

Bumpus, D.F. and L. Lauzier. 1965. Surface circulation on the continental shelf of eastern North America. Folio F. Serial Atlas of the Marine Environment. American Geograph. Soc., N.Y.

Garrett, C.J.R., J.R. Keeley and D.A. Greenberg. 1978. Tidal mixing versus stratification in the Bay of Fundy and Gulf of Maine. Atmosphere-Ocean 16: 403-423.

Graham, J.J. 1970. Coastal currents of the western Gulf of Maine. International Commission for the Northwest Atlantic Fisheries Res. Bull. No. 7: 19-31.

Graham, J.J. 1982. Production of larval herring, Clupea harengus, along the Maine coast, 1964-1978. J. Northw. Atl. Fish. Sci. 3: 63-85.

Graham, J.J. and G.B. Vaughan. 1966. A new depressor design. Limnol. Oceanogr. 11: 130-135.

Gran, H.H. and T. Braarud. 1935. A quantitative study of the phytoplankton in the Bay of Fundy waters. J. Biol. Bd. Can. 1: 279-467.

Greenberg, D.A. 1983. Modeling the mean barotropic circulation in the Bay of Fundy and Gulf of Maine. J. Phys. Oceanogr. 13: 886-904.

Iles, T.D. and M. Sinclair. 1982. Atlantic herring: Stock discreteness and abundance. Science 215: 627-633.

Johannes, R.E. 1978. Reproductive strategies of coastal marine fishes in the tropics. Envir. Biol. Fish. 3: 65-84.

Kelly, K.H. and D.K. Stevenson. In press. Fecundity of Atlantic herring (Clupea harengus L.) from three spawning locations in the western Gulf of Maine in 1969 and 1982. J. Northw. Atl. Fish. Sci. 6(2).

Posgay, J.A. and R.R. Marak. 1981. The MARMAP bongo zooplankton samples. J. Northw. Atl. Fish. Sci. 1: 91-99.

Ricker, W.E. 1973. Linear regressions in fishery research. J. Fish. Res. Bd. Can. 30: 409-434.

Saunders, P.M. 1977. Wind stress on the ocean over the eastern continental shelf of North America. J. Phys. Oceanogr. 7: 555-566.

Shepherd, J.G. and D.H. Cushing. 1980. A mechanism for density-dependent survival of larval fish as the basis of a stock-recruitment relationship. J. Cons. int. Explor. Mer 39: 160-167.

Simpson, J.H. and J.R. Hunter. 1974. Fronts in the Irish Sea. Nature 250: 404-406.

Sindermann, C.J. 1979. Status of Northwest Atlantic herring stocks of concern to the United States. Nat. Mar. Fish. Serv., Tech. Ser. Rep. 23: 449 p.

Townsend, D.W. 1983. The relations between larval fishes and zooplankton in two inshore areas of the Gulf of Maine. J. Plank. Res. 5: 145-173.

Yentsch, C.S. and N. Garfield. 1981. Principal areas of vertical mixing in the waters of the Gulf of Maine, with reference to the total productivity of the area. In: Oceanography From Space, pp. 303-312. J.F.R. Gower (ed.). Plenum, N.Y.

THE IMPORTANCE OF BATHYMETRY TO SEASONAL PLANKTON
BLOOMS IN HECATE STRAIT, B.C.

R.I. Perry*
Department of Oceanography
University of British Columbia
Vancouver, B.C. V6T 1W5 Canada

B.R. Dilke
P.O. Box 2085
Sidney, B.C., Canada

INTRODUCTION

The shallow sea tidal mixing model proposed by Simpson and Hunter (1974) is based on two competing physical processes: the tendency for a water column to stratify due to a surface buoyancy flux (resulting from heat or fresh water), and the tendency for mixing derived from the tide and wind to prevent this stratification. Given a constant buoyancy flux over an area, the resulting pattern of mixed and stratified water masses is due to variations in the tidal and wind energy dissipation rates. Both these rates depend on the water depth (Simpson et al., 1978), but tidal mixing will be more closely coupled with bathymetry because of the inverse relation between depth and current velocity. This model of contrasting mixing regimes has been used successfully to describe plankton dynamics during summer (e.g. Pingree, 1978; and other chapters in this volume) by representing nutrient and light conditions favorable for phyto-plankton growth.

In spring, the development of the seasonal thermocline is governed by the same physical processes, that is, stratification of surface waters due to an increasing buoyancy flux, which is opposed by tidal and wind mixing. The principal source of this increasing buoyancy is typically surface heating, while the tide should become more important as a source of mixing as the frequency of wind events decreases from winter to summer. Pingree (1975) concluded that tidally-induced turbulence exerted the principal control on the

*Present Address: Department of Fisheries and Oceans, Marine Fish Division, Biological Station, St. Andrews, N.B. EOG 2X0, Canada

Lecture Notes on Coastal and Estuarine Studies, Vol. 17
Tidal Mixing and Plankton Dynamics. Edited by J. Bowman, M. Yentsch and W. T. Peterson
© Springer-Verlag Berlin Heidelberg 1986

formation of the seasonal thermocline in the Celtic Sea, and that mixed and stratified regions could exist simultaneously, separated by a horizontal temperature gradient. James (1980) suggested that vertical temperature gradients initially occur only where tidal mixing is sufficiently weak, but that once formed, these gradients are sharpened by wind mixing. Assuming the efficiency of tidal mixing is constant seasonally, bathymetry has as important an influence on the formation of mixed and stratified waters during spring as it does in summer.

Sverdrup (1953) combined this concept of the seasonal shoaling of the surface mixed layer with the concept of the critical depth to predict conditions favorable for the occurrence of phytoplankton blooms in spring. Stratification would develop, and phytoplankton would be maintained within a more favorable light regime, once wind mixing had decreased to the extent that it no longer overlapped with tidal mixing along the bottom. Phytoplankton biomass will also increase, although much more slowly, in well-mixed waters with the general seasonal increase of light and temperature (Eppley, 1972).

It is in stratified waters of coastal temperature seas, therefore, that the pattern of a spring and fall bloom can be expected. In contrast, tidally well-mixed waters will show a gradual increase in production to a summer maximum (Le Fèvre et al., 1983). Boalch et al. (1978) have demonstrated maximal primary production by April in the central English Channel, leading to a diatom bloom (Holligan and Harbour, 1977), while maximum production in the southwestern English Channel did not occur until June. Reference to Pingree et al. (1978) shows this latter station was located in the well-mixed waters of the Ushant frontal system.

There is, however, a situation analogous to the Sverdrup (1953) model which does not depend on the formation of the seasonal thermocline for a bloom to occur. This is where an area is sufficiently shallow that it remains well-mixed year-round, yet the bottom acts as an effective limit to the depth of mixing. Although this region may have a higher energy dissipation rate, it may retain a mean light regime adequate for a bloom to occur.

It is apparent, therefore, that bathymetry across a given area can be a principal determinant of the pattern of seasonal plankton blooms. In this chapter, we examine the effects of marked bathy-metric variations on the development and characteristics of spring and summer plankton blooms across Hecate Strait, a wide coastal seaway in northern British Columbia. We demonstrate that bathymetry is the major feature influencing the pattern of blooms and their composition seasonally by, in effect, reducing the mixed layer in winter and spring, and maintaining the mixed layer during summer.

THE STUDY AREA

Hecate Strait, British Columbia, is a wide (100 km) strait separating the Queen Charlotte Islands from the mainland (Fig. 1). It is open to the south through Queen Charlotte Sound, and to the north through Dixon Entrance. Its most striking feature is the east-west variation of bathymetry, particularly in the northern half of the strait. A deep trough (over 100 m) runs along the eastern side parallel to the mainland coast, while the western side is dominated by a shallow bank with depths less than 30 m (Fig. 2).

Only general patterns of residual currents have been studied in the area. There is a northward-flowing current during winter, driven by prevailing winds, while winds weaken and currents become more variable during summer. One early study suggests a gyre over the shallow banks of northwestern Hecate Strait (Bell, 1963). A summer estuarine-type circulation is present in Dixon Entrance, driven by the Skeena and Nass Rivers, but its principal flow is northwest to the Pacific Ocean, rather than south into Hecate Strait. Some fresh water is added to Hecate Strait by runoff from the Mainland shore, although examination of salinity distributions presented in Dodimead (1980) suggests this effect is relatively restricted. Further details of the oceanography of the region are available in Thompson (1981).

SEASONAL PATTERNS

Spring Blooms

The distribution of plankton across Hecate Strait during spring (April 13-14, 1980) is presented in Fig. 3. Samples were collected from the seawater intake system (nominal depth 3 m) of a commercial oil tanker while underway, as part of a coastal ship-of-opportunity

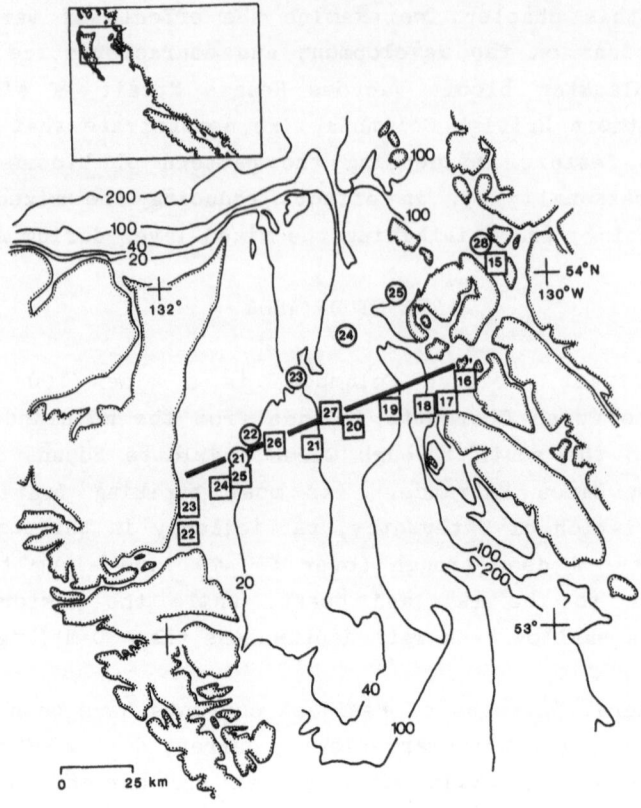

Figure 1. Hecate Strait bathymetry (depth in meters) and station locations. Numbers in square boxes represent station positions for Cruise 12 (February 1980); numbers in circles represent station positions for Cruise 13 (April 1980). Solid line across Hecate Strait indicates the continuous transect in August 1980.

program. Zooplankton were collected from the same depth using a high-speed sampler towed from the stern. This figure clearly indicates a plankton bloom on the shallow western shelf, which was not found on the eastern side of the strait. Although low in chlorophyll a, Station 21 adjacent to the Queen Charlotte Islands had the

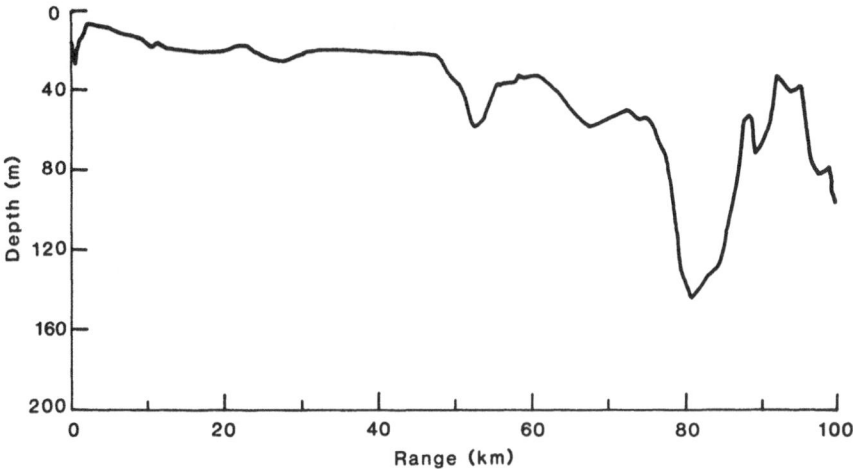

Figure 2. Bathymetric profile across Hecate Strait along the line of the continuous transect (Fig. 1).

highest phaeopigment (2.66 µg L^{-1}) and diatom concentrations on this cruise, suggesting a declining bloom. Zooplankton were four times more abundant (up to 244 m^{-3}), and showed a greater variety of species composition in the waters of the western shelf. Larval fish distributions reflected the same pattern, with greater diversity and over ten times greater abundance on the western side, as sampled by Mason et al. (1981) less than 1 week later.

During winter (February 3-4, 1980; Fig. 4), concentrations of chlorophyll a and diatoms were low across the strait while nutrients were high, typical of temperate waters under conditions of low insolation and frequent mixing by storms. However, chlorophyll was significantly ($p < 0.05$) different between western (mean 0.23 µg L^{-1}) and eastern (mean 0.12 µg L^{-1}) regions.

The distribution of phytoplankton biomass during spring is predicted by Sverdrup's (1953) critical depth model when it includes limitation of the surface mixed layer by bathymetry. Figure 5 indicates critical and mixed depths calculated separately for the eastern and western strait, and the relative importance of bathymetry. As described in Perry (1984), these depths represent monthly means for the period 1954 to 1971. In the western strait, the critical depth exceeded the depth of the bottom from March to

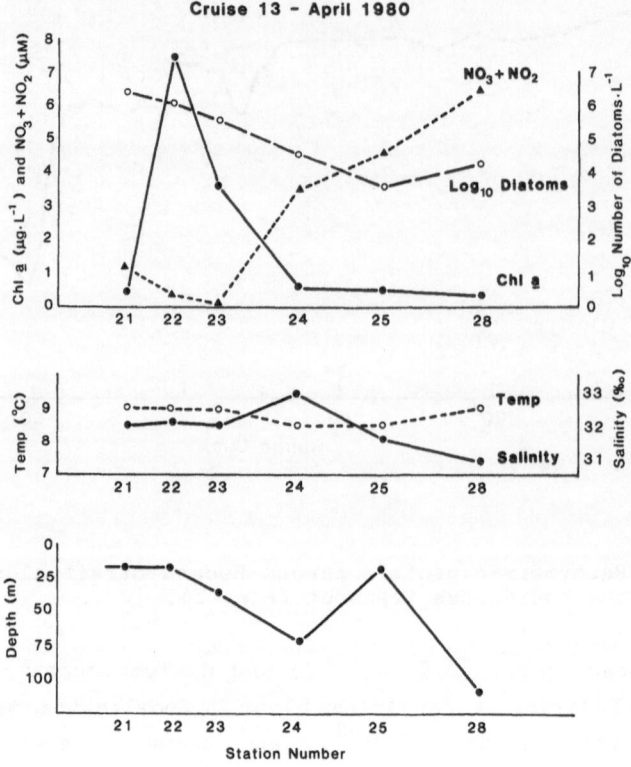

Figure 3. Discrete station data for Cruise 13 (13-14 April 1980) across Hecate Strait. Locations as in Fig. 1. Top: 3 m chlorophyll a and NO₃ + NO₂ concentrations, diatom abundance; Middle: temperature and salinity; Bottom: bottom depth at each station.

August-September, while the bottom of the eastern side was too deep to have a direct influence on the surface mixed layer. This suggests light and mixing conditions would be conducive for a bloom during March in the western region, as was observed, but not until May on the eastern side. A further indication this spring bloom is related to bathymetry is obtained by comparing their relative scales across the strait. According to Fig. 3, the high biomass region extended from the Queen Charlotte Islands eastward to Station 23 but not Station 24, representing a distance of 40 to 70 km. Figure 2

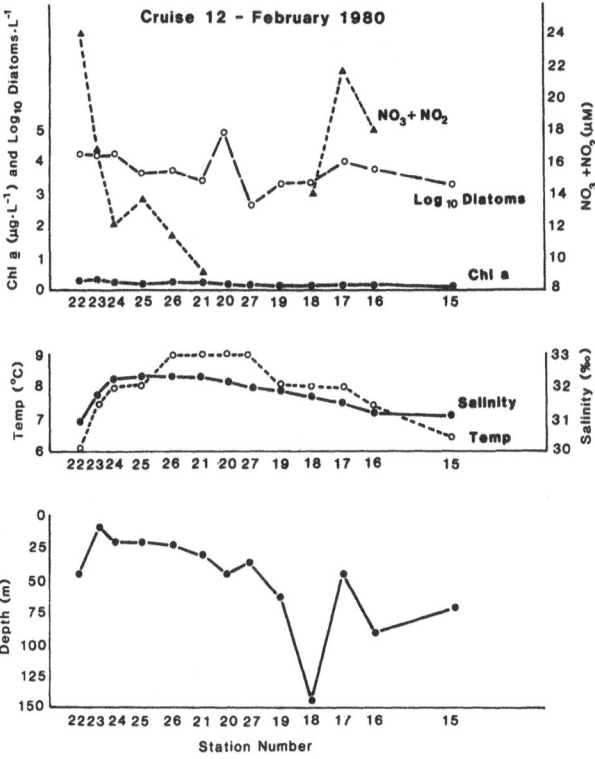

Figure 4. Discrete station data for Cruise 12 (3-5 February 1980) across Hecate Strait. Locations as in Fig. 1. Top: 3 m chlorophyll a and NO_3 + NO_2 concentrations, diatom abundance; Middle: temperature and salinity; Bottom: bottom depth at each station.

indicates the cross-strait dimension of the shallow western bank is similar at 55 to 60 km.

Bathymetry may also explain the differences of chlorophyll concentration across the strait during winter. As with initiation of the spring bloom, the shallow depth of the western shelf imposes a limit to the depth of the mixed layer, thereby increasing the mean light intensity to which cells will be exposed. This increase may not be enough to produce a bloom, but it could cause higher production than in the more deeply mixed waters of the eastern strait. A similar mechanism, involving limitation of the surface mixed layer

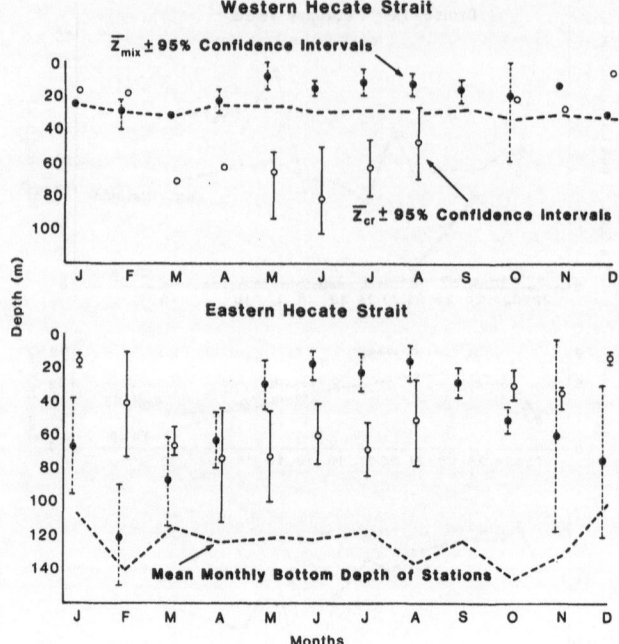

Figure 5. Hecate Strait critical depth - mixed depth model results. Horizontal dashed line represents the mean bottom depth of the set of stations sampled each month. Data are a composite from 1954 to 1971.

depth by a shelf-break front, was proposed by Fournier et al. (1979) to account for enhanced chlorophyll concentrations during winter off Nova Scotia.

Summer Blooms

The distribution of phytoplankton biomass across northern Hecate Strait during summer 1979 to 1980 was markedly different from that during spring. Perry et al. (1983) have described this system, and its relation to variations of tidal mixing induced by the bathymetry.

In their study, mixed and stratified water masses during summer were predicted using the Simpson-Hunter (1974) stratification parameter, derived from a numerical tidal model (Kinney et al., 1976). The current ellipses generated by this model (Fig. 6A) show a clear correspondence with bathymetry (Fig. 1), the largest velocities occurring in the shallowest waters. Figure 6B shows the distribution of this stratification parameter, predicting a mixed-transitional water mass on the western side, and a stratified water mass on the eastern side, using the critical value of 1.5 (Pingree, 1978). Slightly higher critical values of 2.0 to 2.5 (Fearnhead, 1975) and 2.5 to 3.0 (James, 1977) would extend the well-mixed region to the whole of the western shelf.

The existence of these water masses as predicted, was verified independently of the tidal model using a bulk stratification parameter. This parameter was calculated as the mean difference in density between the surface and a given depth. It clearly indicated the influence of the Skeena River, principally in northern Dixon Entrance, although fresh water input to the eastern side of Hecate Strait would serve to strengthen stratification and emphasize the differences with the mixed region. That this tidal mixing model does apply in areas where fresh water runoff occurs was demonstrated by Bowman and Esaias (1981) for Long Island Sound.

Perry et al. (1983) show the correspondence of plankton biomass with these vertical mixing characteristics. Water samples collected from the ship's intake at discrete stations in the mixed waters of the western strait (verified as such by XBT casts) typically had chlorophyll a concentrations less than 1 µg L^{-1} and diatom concentrations of 10^5 cells L^{-1}, while concentrations in stratified eastern waters were up to 5 µg L^{-1} and 10^6 L^{-1}, respectively. Nutrient concentrations across the strait were very low, less than 0.5 µM NO_3 + NO_2 for example, except in the vicinity of the Skeena River, where presumably entrainment of deep water into the surface layers occurred.

Small-scale variations of temperature and chlorophyll fluorescence between water masses were examined in August 1980 (Fig. 7). They were clearly related to bathymetry and vertical mixing characteristics, with higher temperature and lower fluorescence on the

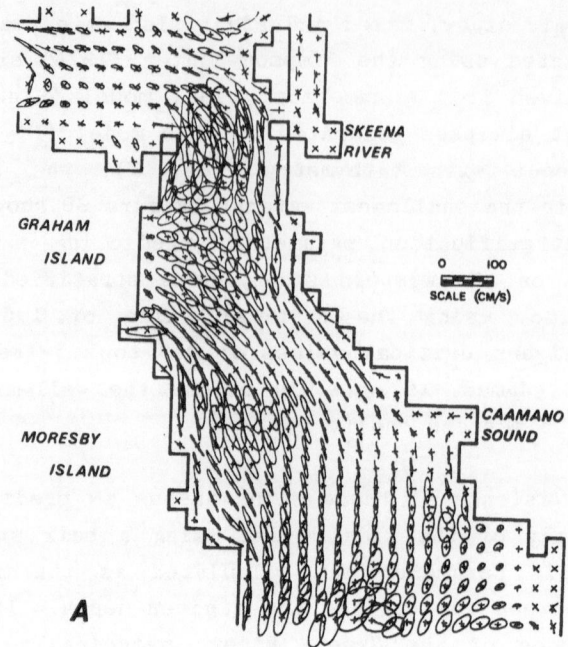

Figure 6. A. Tidal current ellipses computed from the oil spill drift
prediction model for Hecate Strait and Dixon Entrance.
Redrawn with permission after Kinney et al. (1976).

shallow mixed side of the strait.   The abrupt decrease in temperature
and increase of fluorescence occurred at a range of 40 km, coincident
with the edge of the western shelf and the onset of vertical stratification.

Comparison Between Seasons

The usual explanation for the low phytoplankton biomass on the
well-mixed side during summer is light limitation, as cells are mixed
to a greater depth than in adjacent stratified waters.   However, a
region that is sufficiently shallow to limit the depth of the surface
mixed layer during spring, thereby creating favorable conditions for
an early bloom, may also be too shallow for phytoplankton production
to be light limited during summer.   This appears to be the case in
Hecate Strait (Fig. 8), where there is no difference between the mean
mixed layer light intensities of both eastern and western sides, as

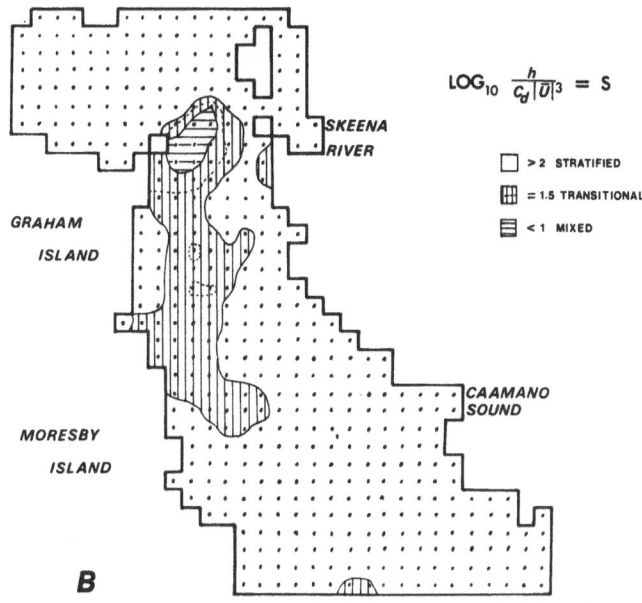

$$LOG_{10} \frac{h}{C_d|\overline{U}|^3} = S$$

☐ > 2 STRATIFIED

▦ = 1.5 TRANSITIONAL

▤ < 1 MIXED

SKEENA RIVER

GRAHAM ISLAND

CAAMANO SOUND

MORESBY ISLAND

B

Figure 6. B. Calculations of the Simpson-Hunter stratification parameter. Dots represent data locations (from Perry et al.,
1983).

calculated using the mixed depths of Fig. 5. The timing of the spring
bloom in the western strait predicted by the critical depth model is
also predicted by Fig. 8, using the mean light intensity suggested by
Gieskes and Kraay (1975) from studies in Dutch coastal waters.

Perry et al. (1983) suggested that the low phytoplankton biomass
on the mixed side was due to nutrient limitation. Assuming near-
surface nutrient concentrations represented those throughout the
water column, the mixing induced by the shallow bottom prevented the
accumulation of a deep nutrient pool during summer. In the deep
eastern region, however, nutrients accumulated below the thermocline
and were available to resupply the surface waters when mixed by local
storms. A similar situation, with phytoplankton production in well-
mixed waters limited by nutrients rather than light during summer,
was shown by Wafar et al. (1983) for a bay in the western English
Channel.

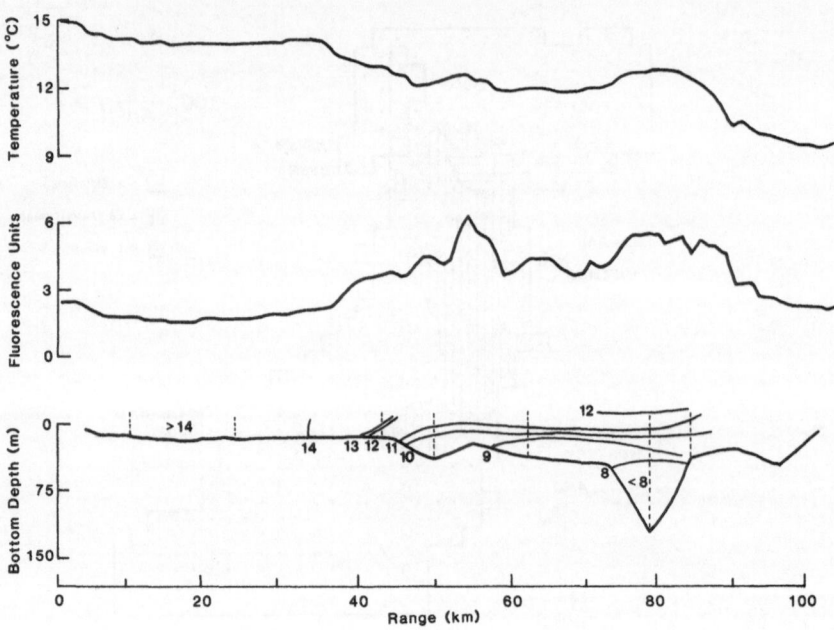

Figure 7. Cruise 15, August 1980. Continuous transect from the Queen
Charlotte Islands (on the left) to the mainland, indicated
on Fig. 1. Top: near-surface (3 m) temperature; Middle:
near-surface relative fluorescence; Bottom: vertical
temperature structure (^{o}C) from XBT profiles (vertical
dots) and the approximate bathymetry (from Perry et al.,
1983).

Phytoplankton taxonomic composition tracked very clearly the
seasonal switch of biomass between eastern and western Hecate Strait.
Figure 9 presents histograms of composition, calculated as (Mackas
and Sefton, 1982):

weighted % all samples = $(\Sigma^{N}_{j=1} B_{ij}) / (\Sigma^{P}_{i=1} \Sigma^{N}_{j=1} B_{ij}) * 100$

where N is the number of stations in a region for that cruise, P the
number of taxonomic groups, and B_{ij} the relative biomass index (abundance * relative size coefficient) of taxon i at station j. An
indication of the total biomass in each region is given in Fig. 9 as
the Normalized (per station in each area) Biomass Index.

Flagellates dominated the biomass across the strait during
winter, although diatom biomass was relatively greater over the
western shelf than the eastern trough. This is in accord with the
distribution of chlorophyll a during winter described above, and for
the same reason. Reid et al. (1983) found the concentration of

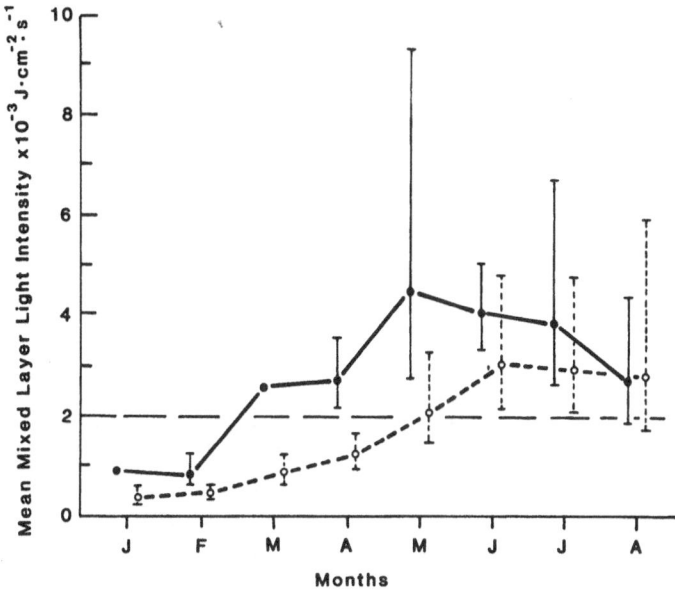

Figure 8. Monthly mean mixed layer light intensities for west (solid
line) and east (short dashed line) Hecate Strait. Vertical
bars represent 95% confidence intervals of the mean. Long
dashed line represents the light intensity suggested by
Gieskes and Kraay (1975) as the critical intensity for
spring phytoplankton blooms.

diatoms in the North Sea during winter increased due to shoaling of
the mixed layer, and Schnitzer (1979, quoted in Farmer et al., 1982)
observed that in well-mixed waters the ratio of 1% light depth to
maximum depth could be used to predict the dominant phytoplankton
group, i.e. microflagellates if <0.5, diatoms if >0.5. During
winter, this ratio will be greater for shallow western Hecate Strait
than for the eastern side.

The phytoplankton bloom during spring is clearly indicated, with
the normalized biomass index 34 times greater on the western shelf,
and obviously dominated by diatoms. By summer, the situation had
switched completely, with greater total biomass, particularly of
diatoms, on the eastern side. This is contrary to the distribution
across a typical tidal front, where higher diatom concentrations are
predicted in mixed waters due to light limitation, and higher flagel-

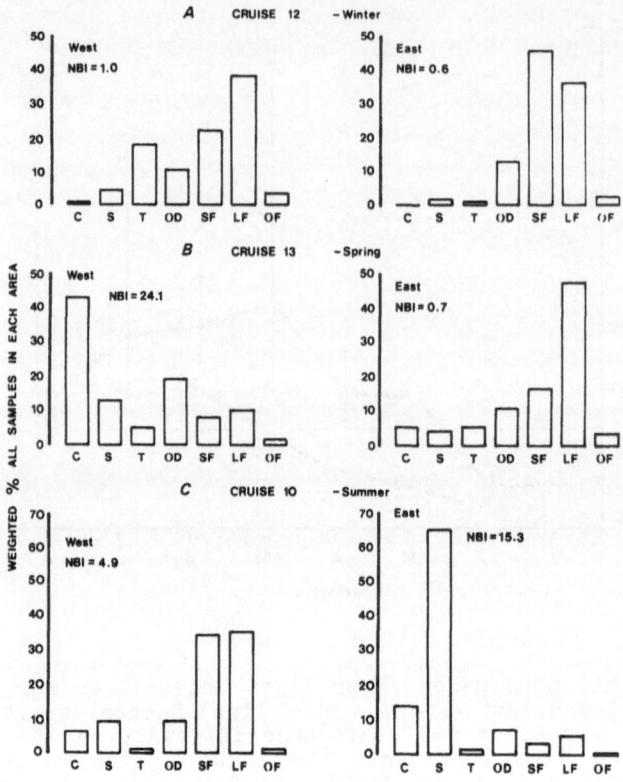

Figure 9. Phytoplankton taxonomic composition histograms for east and
west Hecate Strait during winter 1980 (A), spring 1980 (B)
and summer 1979 (C). NBI: Normalized (per station in each
area) Biomass Index (x 10^{-5}); C: <u>Chaetoceros</u> spp.; S:
<u>Skeletonema</u> <u>costatum</u>; T: <u>Thalassiosira</u> spp.; OD: other
diatoms; SF: small flagellates (<5 µm); LF: large flagellates (6-15 µm); OF: other flagellates (including
dinoflagellates).

late concentrations in stratified waters due to nutrient limitation
(Pingree <u>et</u> <u>al</u>., 1978). Since mean light conditions across the
strait in summer were similar, the pattern of phytoplankton composition reflects that expected from nutrient limitation, particularly as
flagellates are able to grow within a wider range of nutrients than
diatoms (Parsons <u>et</u> <u>al</u>., 1978).

DISCUSSION

Bathymetry exerts both direct and indirect control over the seasonal and spatial pattern of plankton blooms. Indirect control occurs by vertical mixing, which depends upon the depth of the water and the velocity of the tidal currents across the bottom. Vertical mixing determines the development of the seasonal thermocline, and therefore the distribution of mixed and stratified water masses during spring and summer. Each water mass has its own characteristic annual pattern of production and biomass accumulation. Those potentially stratified during summer have a typical temperate pattern, with the spring diatom bloom occurring in areas of weak tidal mixing (Pingree et al., 1976; Fasham et al., 1983). Mixed waters have an annual pattern more typical of high latitudes, with a summer peak of production and biomass (e.g. Wafar et al., 1983).

Direct control occurs when the critical depth exceeds the depth of the bottom, making the degree of vertical mixing and the formation of a thermocline unnecessary for the development of a spring bloom. This suggests three distinct regions could exist along a gradient of bathymetry (Fig. 10): well-mixed with a spring bloom, where the critical depth has exceeded the bottom; well-mixed, where light limitation prevents a spring bloom; and marginally stratified, where the bloom develops along with stratification. In Hecate Strait, with its simplified two-step pattern of bathymetry, the middle region has been reduced or eliminated.

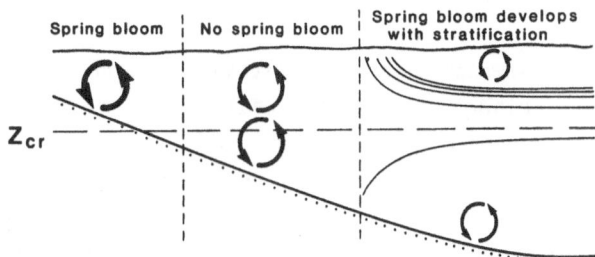

Figure 10. Representation of the interaction of mixing, the critical depth (Z_{cr}), and the bottom to produce three areas of spring bloom development along a gradient of bathymetry. Mixing occurs via the wind (near-surface arrows) and the tide (near-bottom arrows).

A consequence, however, of an early spring bloom in shallow mixed waters appears to be nutrient limitation during summer. Mixing will keep particulate matter and nutrients in suspension, and prevent the accumulation of a deep reserve pool. Advection of nutrients onto the shelf may be insufficient to meet production requirements, particularly if exchange is inhibited by a frontal system along the perimeter of the shelf. This implies that the annual production pattern of very shallow, well-mixed waters will be altered to resemble that of stratified waters, and vertical mixing becomes important only during summer when it prevents nutrient accumulation.

Bathymetry also influences the importance of nutrient regeneration from the benthos during summer. Pomroy et al. (1983) estimated that nutrient efflux from sand sediments in a Bristol Channel embayment satisfied the phytoplankton nitrogen demand when productivity was low, but met only 16% of demand during peak productivity in June. Although they have not been well studied, the benthic communities of Hecate Strait may vary considerably depending on tidal stress, as Wildish and Peer (1983) have indicated for the Bay of Fundy. If nutrients are released from these sediments, the impact is likely to be greater on the shallow side during summer where tidal mixing can distribute them effectively throughout the water column.

In situ regeneration from zooplankton is another potential source of nutrients to the mixed region, and has been suggested as the major source to well-mixed waters of Georges Bank (Schlitz and Cohen, 1984). Its potential has not been studied across Hecate Strait, although zooplankton biomass in near-surface waters was lowest in summer on the western shelf.

One further comment concerns bathymetry and the effect of wind mixing. Surface winds may be as effective as tidal currents at mixing the water column (Kullenberg, 1983), particularly in shallow waters without strong thermal gradients. Winds will be most effective during winter and spring, but this is also the period when production does not depend on the intensity of mixing in shallow waters. The importance of wind decreases towards summer with the decrease in frequency and intensity of storms, leaving the tide to supply the bulk of the energy for mixing the water column.

Bathymetry is therefore a dominant feature across Hecate Strait influencing the spatial distribution of plankton biomass and the seasonal occurrence of phytoplankton blooms. In shallow western Hecate Strait, the bottom limits the depth of the mixed layer in spring, and maintains the mixed layer in summer, preventing the accumulation of nutrients. Although occurring in each season, tidal mixing does not influence phytoplankton bloom dynamics across Hecate Strait in winter or spring.

ACKNOWLEDGMENTS

We wish to thank T.R. Parsons for initiating and coordinating this study, Imperial Oil of Canada, Ltd. for support and use of their vessel as a ship-of-opportunity, J.S. Parslow for many useful comments on early drafts of the manuscript and H.M. Dovey for drafting the figures. Figures 6 and 7 are reproduced with permission from the Canadian Journal of Fisheries and Aquatic Sciences. This study was supported by N.S.E.R.C. Strategic Grant No. G0068 to T.R. Parsons.

REFERENCES

Bell, W.H. 1963. Surface current studies in the Hecate Strait model. Fish. Res. Bd. Can. MS Rep. Ser. (Limnol. Oceanogr.) 159: 4 p.

Boalch, G.T., D.S. Harbour and E.I. Butler. 1978. Seasonal phytoplankton production in the western English Channel 1964-1974. J. Mar. Biol. Assoc. U.K. 58: 943-953.

Bowman, M.J. and W.E. Esaias. 1981. Fronts, stratification, and mixing in Long Island and Block Island Sounds. J. Geophys. Res. 86C: 4260-4264.

Dodimead, A.J. 1980. A general review of the oceanography of the Queen Charlotte Sound-Hecate Strait-Dixon Entrance region. Can. MS Rep. Fish. Aquat. Sci. 1574: 248 p.

Eppley, R.W. 1972. Temperature and phytoplankton growth in the sea. Fish. Bull. 70: 1063-1085.

Farmer, F.H., G.A. Vargo, C.A. Brown, Jr. and O. Jarrett, Jr. 1982. Spatial distributions of major phytoplankton community components in Narragansett Bay at the peak of the winter-spring bloom. J. Mar. Res. 40: 593-614.

Fasham, M.J.R., P.M. Holligan and P.R. Pugh. 1983. The spatial and temporal development of the spring phytoplankton bloom in the Celtic Sea, April 1979. Prog. Oceanogr. 12: 87-145.

Fearnhead, P.G. 1975. On the formation of fronts by tidal mixing around the British Isles. Deep-Sea Res. 22: 311-321.

Fournier, R.O., M. van Det, J.S. Wilson and N.B. Hargreaves. 1979. Influence of the shelf-break front off Nova Scotia on phytoplankton standing stock in winter. J. Fish. Res. Bd. Can. 36: 1228-1237.

Gieskes, W.W.C. and G.W. Kraay. 1975. The phytoplankton spring bloom in Dutch coastal waters of the North Sea. Neth. J. Sea Res. 9: 166-196.

Holligan, P.M. and D.S. Harbour. 1977. The vertical distribution and succession of phytoplankton in the western English Channel in 1975 and 1976. J. Mar. Biol. Assoc. U.K. 57: 1075-1093.

James, I.D. 1977. A model of the annual cycle of temperature in a frontal region of the Celtic Sea. Estuar. Coast. Mar. Sci. 5: 339-353.

James, I.D. 1980. Thermocline formation in the Celtic Sea. Estuar. Coast. Mar. Sci. 10: 597-607.

Kinney, P.J., J.C.H. Mungall, C.E. Abel, R.O. Reid, A.C. Vastano and R.E. Whitaker. 1976. Oil spill movement predictions. Kitimat Pipe Line Project. In: Kitimat Pipe Line Ltd. Termpol Submission re Marine Terminal at Kitimat, B.C., Vol. V, Appendix XII, Oil Spill Contingency Plan, Attachment 3. Vancouver, B.C.

Kullenberg, G. 1983. Mixing processes in the North Sea and aspects of their modelling. In: North Sea Dynamics, J. Sundermann and W. Lenz (eds.), p. 349-369. Springer-Verlag, Berlin.

Le Fèvre, J., M. Viollier, P. Le Corre, C. Dupouy and J.-R. Grall. 1983. Remote sensing observations of biological material by LANDSAT along a tidal thermal front and their relevancy to the available field data. Estuar. Coast. Shelf Sci. 16: 37-50.

Mackas, D.L. and H.H. Sefton. 1982. Plankton species assemblages off southern Vancouver Island: geographic pattern and temporal variability. J. Mar. Res. 40: 1173-1200.

Mason, J.C., O.D. Kennedy and A.C. Phillips. 1981. Canadian Pacific coast ichthyoplankton survey, 1980. Ichthyoplankton. Cruise four (April 15-23). Can. Data Rep. Fish. Aquat. Sci. 278: 80 p.

Parsons, T.R., P.J. Harrison and R. Waters. 1978. An experimental simulation of changes in diatom and flagellate blooms. J. Exp. Mar. Biol. Ecol. 32: 285-294.

Perry, R.I. 1984. Plankton blooms of the British Columbia northern shelf: seasonal distributions and mechanisms influencing their formation. Ph.D. Thesis, University of British Columbia, Vancouver, B.C. 198 p.

Perry, R.I., B.R. Dilke and T.R. Parsons. 1983. Tidal mixing and summer plankton distributions in Hecate Strait, British Columbia. Can. J. Fish. Aquat. Sci. 40: 871-887.

Pingree, R.D. 1975. The advance and retreat of the thermocline on the continental shelf. J. Mar. Biol. Assoc. U.K. 55: 965-974.

Pingree, R.D. 1978. Mixing and stabilization of phytoplankton distributions on the northwest European continental shelf. In: Spatial Pattern in Plankton Communities, J.H. Steele (ed), pp. 181-220. Plenum Press, NY.

Pingree, R.D., P.M. Holligan and G.T. Mardell. 1978. The effects of vertical stability on phytoplankton distributions in the summer on the northwest European shelf. Deep-Sea Res. 25: 1011-1028.

Pingree, R.D., P.M. Holligan, G.T. Mardell and R.N. Head. 1976. The influence of physical stability on spring, summer and autumn phytoplankton blooms in the Celtic Sea. J. Mar. Biol. Assoc. U.K. 56: 845-873.

Pomroy, A.J., I.R. Joint and K.R. Clarke. 1983. Benthic nutrient efflux in a shallow environment. Oecologia (Berlin) 60: 306-312.

Reid, P.C., H.G. Hunt and T.D. Jones. 1983. Exceptional blooms of diatoms associated with anomalous hydrographic conditions in the southern bight in early 1977. J. Plank. Res. 5: 755-765.

Schlitz, R.J. and E.B. Cohen. 1984. A nitrogen budget for the Gulf of Maine and Georges Bank. Biol. Oceanogr. 3: 203-222.

Schnitzer, M.B. 1979. Vertical stability and the distribution of phytoplankton in Long Island Sound. M.Sc. Thesis, SUNY, Stony Brook, NY. 108 p.

Simpson, J.H., C.M. Allen and N.C.G. Morris. 1978. Fronts on the continental shelf. J. Geophys. Res. 83C: 4607-4614.

Simpson, J.H. and J.R. Hunter. 1974. Fronts in the Irish Sea. Nature 250: 404-406.

Sverdrup, H.V. 1953. On conditions for the vernal blooming of phytoplankton. J. Cons. Inst. Explor. Mer 18: 287-295.

Thompson, R.E. 1981. Oceanography of the British Columbia coast. Can. Spec. Publ. Fish. Aquat. Sci. 56: 291 p.

Wafar, M.V.M., P. Le Corre and J.L. Birrien. 1983. Nutrients and primary production in permanently well-mixed temperate coastal waters. Estuar. Coast. Shelf Sci. 17: 431-446.

Wildish, D.J. and D. Peer. 1983. Tidal current speed and production of benthic macrofauna in the lower Bay of Fundy. Can. J. Fish. Aquat. Sci. 40(Suppl. 1): 309-321.

THE EFFECTS OF SEASONAL VARIATIONS IN STRATIFICATION
ON PLANKTON DYNAMICS IN LONG ISLAND SOUND

W.T. PETERSON
Marine Sciences Research Center
SUNY - Stony Brook
Stony Brook, NY 11794

INTRODUCTION

A considerable amount of physical and biological oceanographic
research has been carried out on tidal mixing in shallow seas and
estuaries, much of which is summarized in this volume. As is evident
from the table of contents, most of the biological studies have
focused on the relationships between mixing, light, nutrients and
phytoplankton. Very little work has been done on higher trophic
level organisms. This chapter attempts to show how herbivorous
copepods fit into the "Tidal Mixing Paradigm," using data from an
ongoing study of the population dynamics of zooplankton in Long
Island Sound, New York.

The Tidal Mixing Paradigm is summarized as follows: in shallow
seas and estuaries, phytoplankton is affected by temporal and spatial
variations in tidal mixing in several ways. First, tidal mixing is a
significant process in resupplying nutrients to the photic zone after
the spring bloom and during summer months thus enhancing primary
production during months when one would not normally expect high
productivity (Pingree et al., 1976). Second, spatial variations in
tidal mixing can lead to ecological zonation along a gradient
determined by levels of turbulence and water depths: a well-mixed
shallow water column nearshore can be separated from deeper offshore
stratified water by a marginally stratified transition (or frontal)
zone. Phytoplankton tend to be uniformly distributed with depth
within well mixed water columns but concentrated at or above the
pycnocline in stratified columns (Holligan and Harbour, 1977; Holligan et al., 1984a). Primary production tends to be highest in the
transition zone (Holligan et al., 1984b), especially during late
spring and summer months. Detritus may accumulate at the transition
zone and fuel a detritus-bacteria based food chain (Floodgate et al.,
1981) which may be especially important during times of low phytoplankton biomass, such as late autumn and winter.

Lecture Notes on Coastal and Estuarine Studies, Vol. 17
Tidal Mixing and Plankton Dynamics. Edited by J. Bowman, M. Yentsch and W. T. Peterson
© Springer-Verlag Berlin Heidelberg 1986

The phytoplankton community is characterized by a dominance of diatoms in the well-mixed and transitional zones, and dinoflagellates in the stratified waters (Holligan and Harbour, 1977), a feature which could influence the production of bacteria and zooplankton. Differences in phytoplankton abundance and species composition among the three ecological zones could directly affect copepod grazing, egg production and developmental rates. Indeed, these are probably the more important rate processes which must be studied when looking for relationships between mixing, phytoplankton and zooplankton. Ecological zonation along a tidal mixing gradient provides the zooplanktonologist with a unique situation in which to test specific hypotheses on the effects of variations in phytoplankton abundance and species composition on the dynamics (i.e., fecundity, grazing and growth rates) of copepod populations in the field. I suggest here that studies which focus on zooplankton dynamics (rather than zoo-plankton distribution and abundance) are what is needed to extend the Tidal Mixing Paradigm to include higher trophic level organisms.

Most of our research to date has focused on mixing and plankton dynamics in Long Island Sound on a seasonal time scale at a few fixed locations. Spatial variability has not been investigated in any detail for logistical reasons. The bulk of this paper is, therefore, a description of how changes in the annual plankton production cycle are related to the annual cycle of stratification in Long Island Sound. Data from two years, 1982 and 1983, are discussed. Three topics are treated:

1. the seasonal cycles of water column stratification and plankton abundances;

2. the effects of seasonal changes in stratification
   on the vertical distribution and taxonomic composition of the phytoplankton, and

3. the combined effects of stratification and
   changes in phytoplankton species composition on
   copepod fecundity.

The chapter concludes with a discussion of some suggestions for future research. It is hoped that the analyses presented here will yield some insights into how both temporal and spatial variations in hydrodynamic processes affect biological productivity, especially of second and third trophic level organisms.

METHODS

Hydrography and plankton abundances were studied from samples collected at two stations in central Long Island Sound on a weekly basis from 1982 through 1985. Figure 1 is a chart showing station locations. One is a shallow water shore station (indicated by a star) and the other a deep water station (station 4; 37 m water depth), representative of the deep central basin of the Sound. In addition, two other stations, stations 2 and 7, both in 20 m of water, were each studied for a one year period, in 1982 and 1983.

Temperature, salinity and conductivity were measured with a Beckman RS-5 induction salinometer. Water samples for nitrate, chlorophyll and zooplankton were collected from discrete depths with the aid of a Jabsco pump (flow rate 10 ℓ min^{-1}) and a 1.9 cm i.d. hose. From each sampling depth, a 100 ml sample was filtered through a GF/C filter for later analysis of chlorophyll, and the filtrate was frozen for later analysis of nitrate concentration. A 3 ℓ sample was filtered through a 64 μm Nitex screen, rinsed into a sample bottle and preserved with 5% formalin for later enumeration of the copepod eggs and other smaller zooplankton. Macrozooplankton was sampled with a 1/2 m 202 μm net towed vertically through the entire water column. During May 1981, the zooplankton was sampled from discrete depths with the aid of a Pacer centrifugal pump (flow rate 250 ℓ min^{-1}) and 5 cm i.d. hose.

In the laboratory, nitrate and chlorophyll were analyzed by standard methods (Strickland and Parsons, 1972). For the zooplankton pump samples collected in 1982 and 1983, the entire sample was counted, yielding a count of 100 to 800 animals per sample. For the 1981 pump samples and the plankton net samples, duplicate 1 ml piston pipette samples were withdrawn and counted. Copepods were identified to life cycle stages and species. The data on egg abundance (from the pump samples) were converted to in situ rate of egg production using the egg-ratio method:

$$F = B \ / \ A \ D$$

where F is fecundity in units of eggs female^{-1} day^{-1}, B is egg abundance in number liter^{-1}, A is female abundance in number liter^{-1}, and D the egg development time at the observed in situ water temperature.

RESULTS

Distribution of the Stratification Index. Considerable spatial variability in tidal mixing is found in Long Island Sound (Bowman and Esaias, 1981). Figure 1 shows the distribution of the h/u^3 tidal mixing-stratification index (S) in Long Island and Block Island Sounds. Areas where S > 1.5 tend to be less well mixed than areas where S < 1.5. (S = log h/u^3 where h = depth in meters and u the amplitude of the tidal current in m s^{-1}; Simpson and Hunter, 1974.) Regions of relatively high mixing are found along the nearshore margins of the central Sound, east of 73°15'W, and throughout the easternmost portions of the Sound, east of 72°30'W longitude. Intense mixing is seen around two headlands, Eaton's Neck (73°25'W) and Crane Neck (73°10'W), and at the eastern entrance to the Sound (72°15'W). Turbulent tidally-driven mixing is sufficiently weak throughout the deeper parts of the central Sound that the water column becomes stratified from May through September. Similarly, mixing in the western portions of the Sound are weak so the water column tends to be more highly stratified there than anywhere else in the Sound.

TEMPORAL VARIABILITY

Water Column Stability. As well as the spatial variability discussed above, Long Island Sound is characterized by temporal variability in water column stratification. The seasonal cycle of air temperature, coupled with seasonal changes in runoff, contribute to a well-defined cycle in the buoyancy of surface waters of the Sound. This produces the dominant annual signal in temporal variability in stratification. Superimposed on the seasonal cycle of stratification is shorter time scale variability associated with wind events and spring-neap differences in tidal mixing. Little is known about the effects of either of these latter two forces on mixing processes in Long Island Sound, other than the work of Schnitzer (1979) who found only modest onshore-offshore variations in the locations of the frontal zone as a function of spring-neap tides, during two cruises in July 1978.

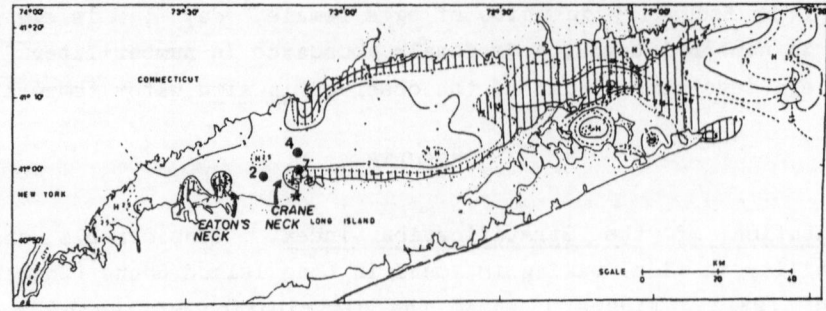

Figure 1. A chart of Long Island Sound showing station locations and
          the distribution of the h/u³ parameter. Stations are
          indicated by a star (shore station) and closed circles
          (stations 2,4, and 7). The hatched areas are regions which
          tend to have relatively high mixing rates (redrawn from
          Bowman and Esaias, 1981).

Figure 2 shows the typical changes in the stability of the water
column that occur at stations 2, 4, and 7 over an annual cycle.
During each year of this study, the water column was weakly stratified during winter and early spring, began to stratify significantly
at the end of April, reached maximum stability in mid-June, destratified briefly in mid-August due to intense wind events, then
completely destratified in late September.

Increases in stability are due to increased input of both heat
and freshwater during the spring and summer (Figure 3). The lack of
any great change in stability from mid-June through mid-August
suggests that the rate of buoyancy input to surface waters is
balanced by downward cross-pycnocline transfer of heat and freshwater
similar to that described by Pingree and Pennycuick (1976) for the
English Channel.

Nitrate and Phytoplankton. The seasonal cycle of nitrate and chlorophyll for the year 1982 is shown in Figure 4. Nitrate concentrations
were highest during the winter months, but decreased from 20 µM to
0.5 µM during the first major phytoplankton bloom of the year, in

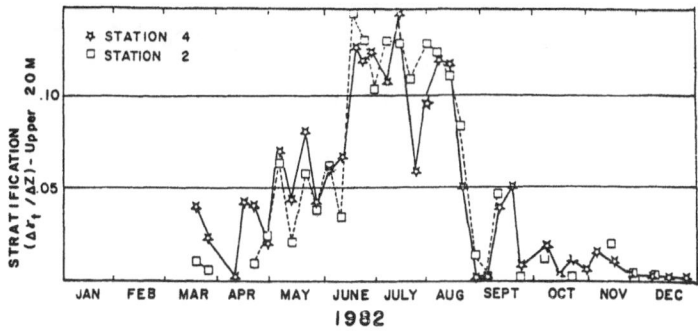

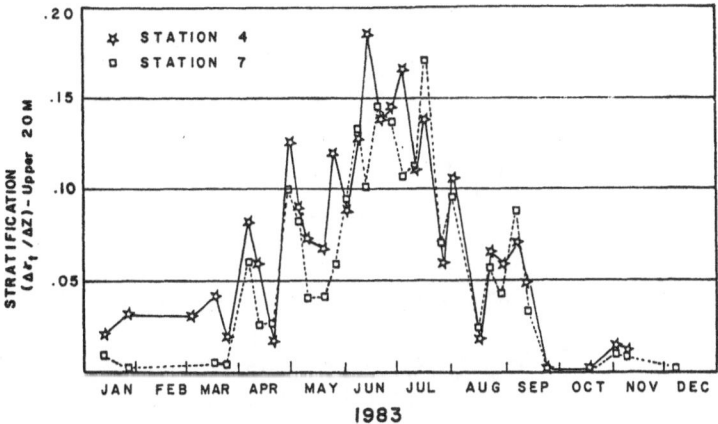

Figure 2. Seasonal changes in water column stability during 1982 and 1983.

late February. After the bloom, surface nitrate concentrations remained low, between 0.1 and 0.5 µM through the spring and summer. Surface concentrations did not increase until the Sound destratified in mid-September. Bottom water nitrate concentrations began to increase in early May coincident with the onset of seasonal stratification (Figure 5). The rate of change of nitrate concentration in bottom waters was approximately linear, 0.036 µM per day, and maximum concentrations (approximately 3µM) were reached in early July. Thus, a reservoir of nutrient-rich water exists in Long Island Sound which, if exchanged across the pycnocline by tidal mixing, could enhance primary productivity in late spring and early summer (if phytoplankton growth is nutrient-limited).

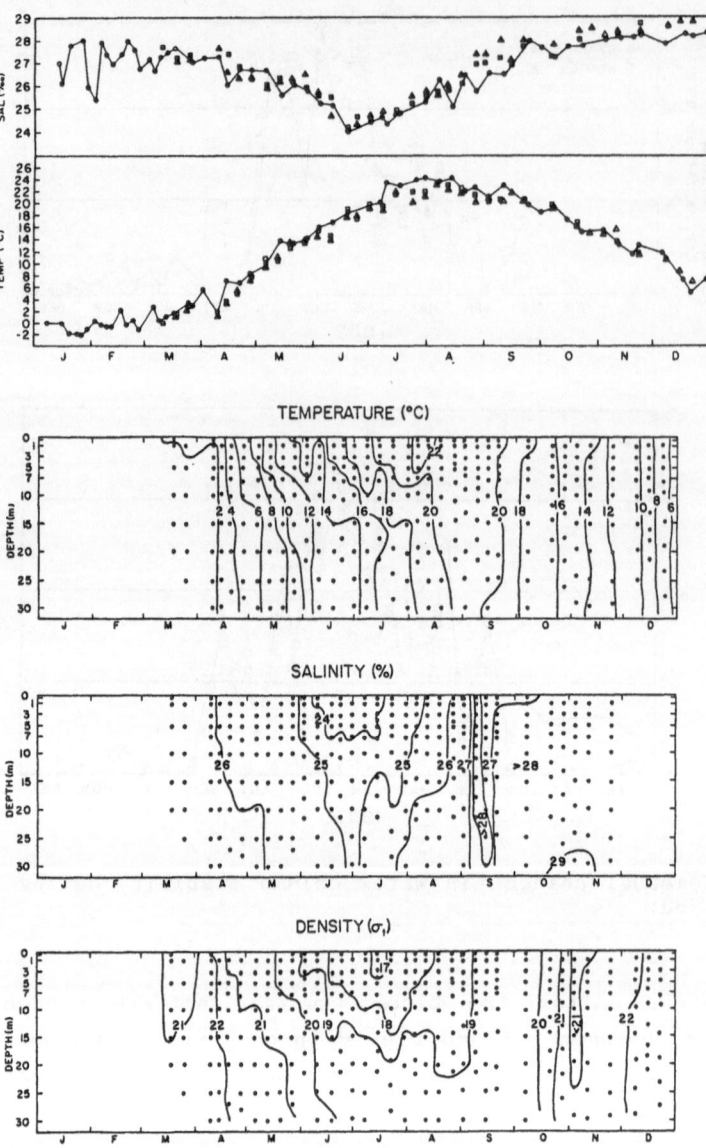

Figure 3. The annual cycle of surface temperature and salinity at the shore station (solid line) and offshore stations 2 (squares) and 4 (triangles) during 1982 (UPPER). Time-depth plot of temperature, salinity and density at the offshore station 4 in 1982 (LOWER).

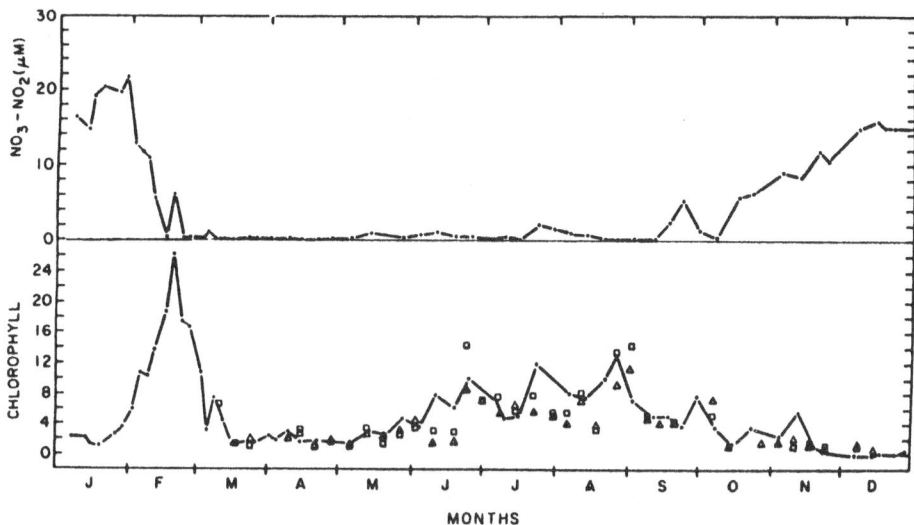

Figure 4. The annual cycle of nitrate and chlorophyll at the surface at the shore station (solid line) and offshore station 2 (squares) and station 4 (triangles) during 1982.

The annual "spring" bloom was initiated in early February of 1982 (Figure 4) and reached a peak biomass of 26 μg chl ℓ^{-1} on 19 February 1982 (at the shore station) then abruptly declined to 4 μg ℓ^{-1}. In 1983, the bloom was initiated in late February, reached 18 μg chl ℓ^{-1} on 1 March then remained at this concentration or as high as 22 μg ℓ^{-1} before rapidly declining on 25 March. During both years, the bloom terminated when nitrate-nitrogen supply was exhausted and the bloom dissipated by sinking out of the water column (Figure 6). Decaying phytoplankton in March/April along with copepod fecal pellets in April/May/June may be the primary source of organic nitrogen which through microbial activity (including nitrification), leads to the observed steady increase in bottom water nitrate concentration.

After the spring bloom, chlorophyll concentration remained relatively low and fairly constant from mid-March through mid-June in 1982 (Figure 7). During this three-month period, the biomass of phytoplankton may have been low because either growth was limited by nutrient flux (ammonium regeneration by zooplankton and cross-pycno-

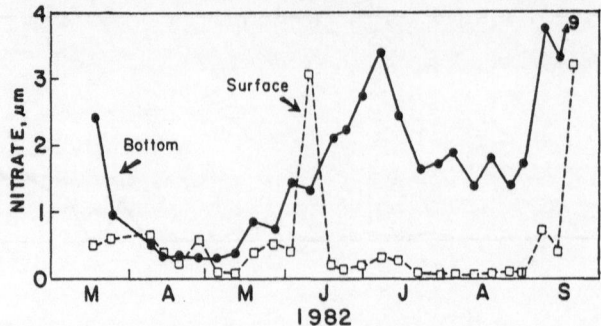

Figure 5. Concentration of nitrate in surface and bottom waters at station 4 during 1982. The "spike" in surface concentration on 9 June resulted from complete mixing of the water column during a violent wind storm on 7-8 June 1982.

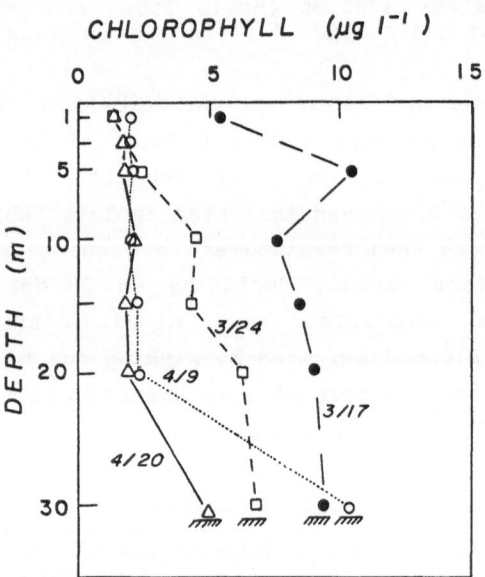

Figure 6. Vertical distribution of chlorophyll during and immediately after the spring bloom in 1982.

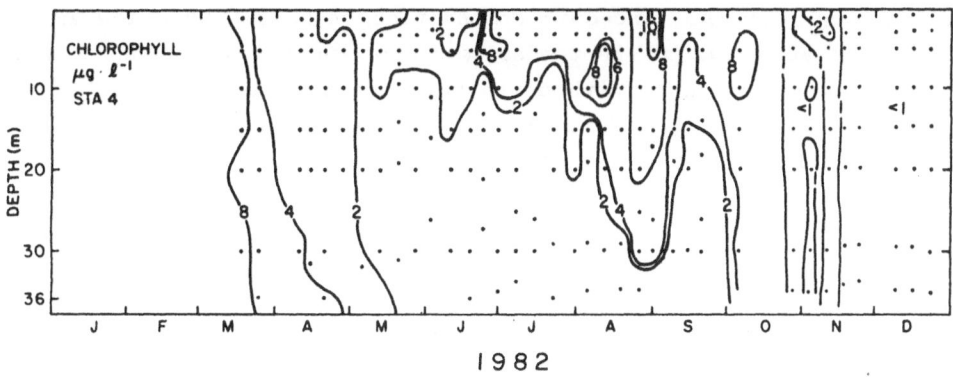

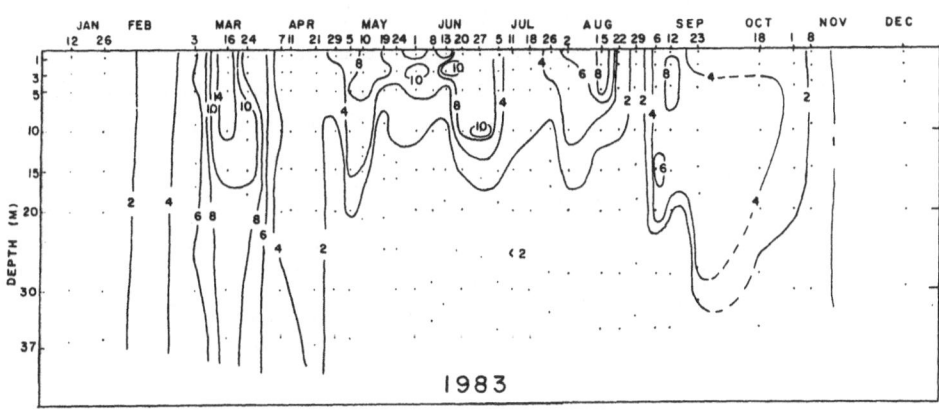

Figure 7. Time-depth variations in chlorophyll concentration at the offshore station 4 during 1982 (upper and 1983 (lower).

cline mixing of deep nutrient-rich water), or the biomass was reduced by zooplankton grazing. In June of each year, phytoplankton abundance began to increase (following a reduction in zooplankton abundance, see Figure 9). Blooms were initiated aperiodically from July through September, probably because of reduced grazing pressure during the annual ctenophore invasion in a manner similar to that described for Narragansett Bay by Deason and Smayda (1982). During autumn of both years, chlorophyll concentrations usually ranged from 2-4 µg ℓ^{-1}.

The onset of stratification in April had two notable effects on the phytoplankton: (1) the vertical distribution of chlorophyll was modified and (2) the taxonomic composition changed. When the water column was well mixed (September through April), chlorophyll was distributed homogenously with depth but when the water column became stratified, chlorophyll was abundant only within the upper 10 m of the water column (Figure 7). This change in the pattern of vertical distribution occurred when the stratification exceeded 0.05 σt units m^{-1}, thus for Long Island Sound, this physical stratification threshold defines the onset of biological stratification.

Associated with the change in water column stability was a change in taxonomic composition of phytoplankton, from chain-forming diatoms (March-April) to a dominance of dinoflagellates and other small flagellates, from May through August (Figure 8). After destratification in late summer, diatoms again dominated the phytoplankton biomass. The major diatom species in Long Island Sound are Skeletonema costatum and Thalassiosira nordenskioldii during the spring bloom, Thalassionema nitzschoides, several Thalassiosira species and Rhizosolenia delicatulum later in spring, and S. costatum, T. nitzschoides, Ditylum brightwellii, Coscinodiscus, Leptocylindricus danicus and Thalassiosira pseudonana in late summer. The dinoflagellate genera Prorocentrum, Peridinium, Gyrodinium and Exuviella are important during the stratified season, from May through August (Conover, 1956; Riley, 1967; Capriullo and Carpenter, 1983). Cryptomonads (3-5 μm diameter) dominate cell counts in late-spring and early summer.

Copepods. Copepods were not affected by changes in stratification in the same way as were the phytoplankton. There were no changes in abundance or species composition, and only very slight changes in vertical distribution coincident with the onset of seasonal stratification (Peterson and Ausubel, 1984). Changes in abundance and species composition were seen over the annual cycle but these were associated with the seasonal cycle of temperature rather than with the stratification cycle. This information is briefly summarized below:

Two copepod assemblages occur during the course of the year (Figure 9), a boreal assemblage from February through July dominated by Temora longicornis, Acartia hudsonica, and Pseudocalanus sp., and

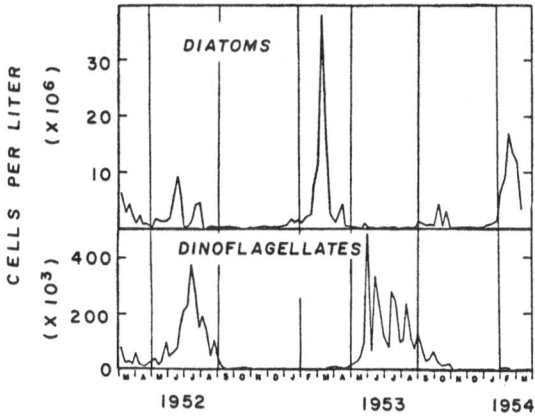

Figure 8. Abundance of diatoms and dinoflagellates in central Long Island Sound during 1952-1954 (redrawn from S. Conover, 1956).

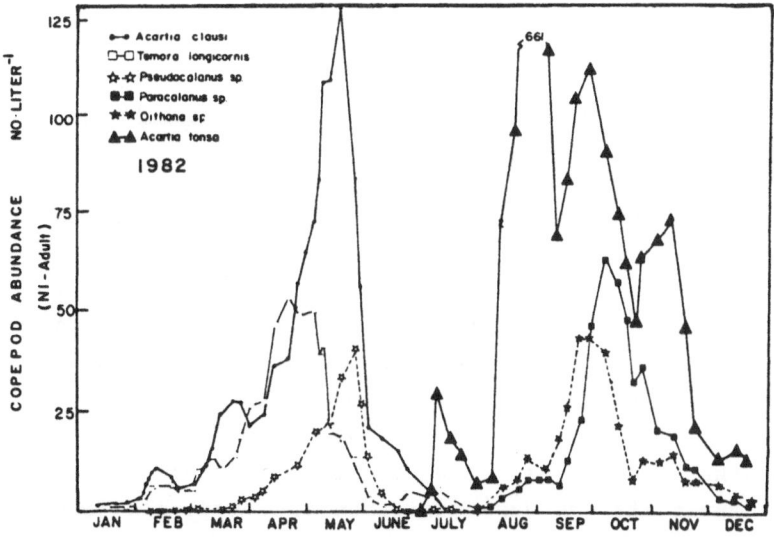

Figure 9. Seasonal cycle of copepod abundances at the shore station in Long Island Sound during 1982.

a "warm water" assemblage dominated by <u>Acartia</u> <u>tonsa</u>, <u>Paracalanus</u> <u>crassirostris</u> and <u>Oithona</u> <u>similis</u>. The boreal forms dominate the spring and early summer zooplankton until temperatures within the upper mixed layer exceed 19°C, at which time they become replaced by the warm water forms. Population numbers of the summer species remain high through October or November, then begin to decline when temperatures fall below 15°C. A few adult individuals persist through most months of the year, so during the winter, all six ecologically dominant copepod species are found in Long Island Sound. Only two of them begin to increase in numbers with the initiation of the spring bloom in February, <u>T</u>. <u>longicornis</u> and <u>A</u>. <u>hudsonica</u>. This suggests that even though summer species do manage to survive over a temperature range from below 0° to 25°C, they are not able to produce eggs at cool temperatures.

The onset of seasonal stratification was correlated with changes in the <u>fecundity</u> of the boreal copepods. During both years of this study, maximum egg production by female <u>Temora</u> <u>longicornis</u> occurred during and after the spring bloom, in March (Fig. 10), but ceased altogether in late April and May after the onset of stratification. Rates were about 30 eggs female^{-1} day^{-1} in early spring, nearly zero in May, and about 10 eggs female^{-1} day^{-1} in June. During 1982, females of <u>Acartia</u> <u>hudsonica</u> and <u>Pseudocalanus</u> produced their greatest number of eggs after stratification, during the interval 27 April-19 May, and again 18-30 June. During 1983, <u>Acartia</u> <u>hudsonica</u> were most fecund from 19 May-20 June, and <u>Pseudocalanus</u> from 24 May-27 June.

These observations on temporal variability in copepod fecundity have led to three hypotheses:

(1) Changes in phytoplankton species composition associated with changes in seasonal stratification might be controlling copepod fecundity,

(2) the three boreal copepods which co-occur during March through June partition the available food resource in different ways, and

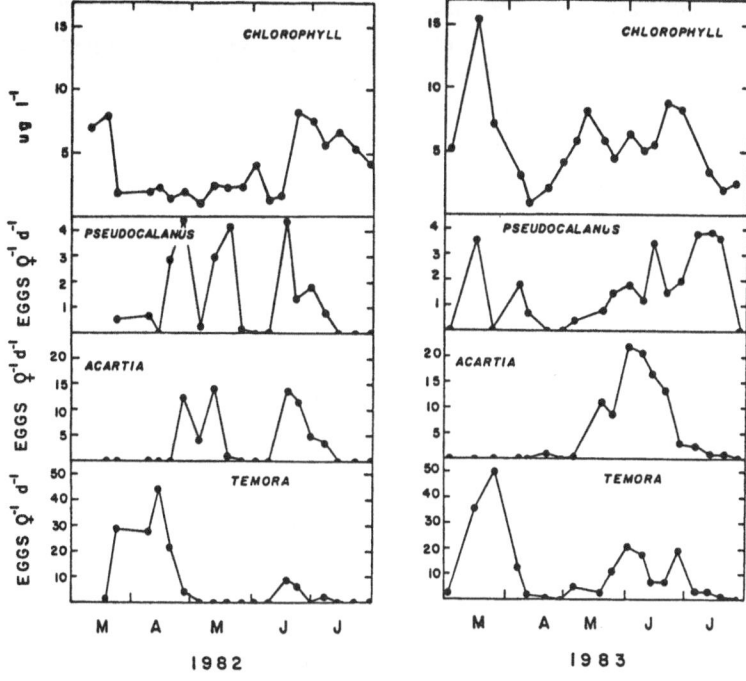

Figure 10. Relationship between phytoplankton biomass and the in situ
egg production rate for Pseudocalanus sp., Acartia hudsonica and Temora longicornis in 1982 and 1983 at the
offshore station 4.

(3) egg production by boreal copepods in Long Island Sound is
     food-limited, since egg production by any one species
     occurred only for a few weeks during the six month growth
     season (February-July).

FUTURE RESEARCH NEEDS

There are three general results of most field studies of tidal
mixing in shallow seas and estuaries and phytoplankton. First,
phytoplankton biomass is usually greatest in shoal well-illuminated
mixed water columns or in frontal zones. Second, in mixed water
columns, phytoplankton biomass is evenly distributed with depth,
whereas in stratified water columns, phytoplankton is usually
abundant only at or near the pycnocline or within the upper mixed
layer. Third, diatoms dominate in mixed regions, dinoflagellates in

stratified regions. These patterns seem to recur both in systems characterized by temporal (seasonal) as well as spatial (zonal) variations in hydrography (Bowman et al., 1981; Holligan and Harbour, 1977; Holligan et al., 1984a).

Little is known about how mixing processes affect zooplankton. Three general questions have been studied to a varying extent for phytoplankton and these are recast and discussed below with respect to zooplankton:

1. Are there any relationships between spatial variations in water column stratification and the biomass and species composition of zooplankton?

2. Is the vertical distribution of zooplankton controlled by physical mixing alone?

3. Are there any effects of temporal and spatial differences in phytoplankton abundance and/or species composition on fecundity, grazing or growth of zooplankton?

Zooplankton Species Composition and Abundance. In most (if not all) temperate latitude continental shelf pelagic ecosystems, ecological zonation of zooplankton species is seen in an inshore-offshore direction. Among the copepods, typically Acartia and Centropages species occur nearshore and a Metridia and Calanus species offshore. Also, cladocerans and larvae of benthic invertebrates are invariably abundant only within a few kilometers of the coast. Three other common neritic copepods, Pseudocalanus, Paracalanus parvus and Oithona similis tend to be broadly distributed over much of the shelf. For examples of these distribution patterns see Peterson et al. (1979), Oregon coast; Smith and Vidal (1984), Bering Sea shelf; Judkins et al. (1980), New York Bight; Bigelow (1924), Gulf of Maine; Holligan (1981) and Holligan et al. (1984b), northwest European shelf.

The relative importance of physical factors such as advection and tidal mixing vs. biological factors such as phytoplankton abundance, predation and behavior in maintaining the zonal patterns of zooplankton abundance is not known. Since different species with different life history strategies dominate in inshore vs. offshore zones, it is certainly inappropriate to discuss differences in "total zooplankton" abundance or biomass between zones. Rather, one must study spatial gradients in abundance at the species level. In order to determine the relative importance of physical vs. biological factors in maintaining zonal gradients, descriptions of the flow field need to be coupled with descriptions of the vertical distribution of various zooplankton species and life cycle stages. Of particular importance is an understanding of how cross-frontal mixing between a tidally-mixed and stratified water column may modify zonal distribution patterns of zooplankton. Also, a determination of the biological factors which control mortality and fecundity schedules are needed to understand how these variables control zooplankton abundance and population dynamics.

Direct Effects of Mixing on Vertical Distribution. The vertical distribution of copepods in tidally-mixed vs. stratified water columns has been studied by Turner and Dagg (1983), Lough (1984), Holligan et al. (1984b), Townsend (this volume) and Peterson (unpublished data). Turner and Dagg showed that the vertical distribution of most of the abundant copepod species in the New York Bight and Georges Bank was markedly different in stratified compared to isothermal regions. During their sole cruise in October 1978, they reported that in the stratified waters south of Long Island, Centropages typicus and Paracalanus parvus were found mostly above the thermocline, but in the mixed waters over Georges Bank, they were evenly distributed with depth. Oithona similis and Pseudocalanus sp. could be abundant either above or below the thermocline but were evenly distributed with depth over Georges Bank.

Lough (1984) studied the distribution and abundance of larval haddock and cod, and their zooplanktonic prey over Georges Bank in April and May 1981. He found that "prey" (defined to be copepods of <0.45 mm in width) ranged in abundance from 5-10 liter^{-1} in late April to 20-30 liter^{-1} in late May, and were evenly distributed with depth in the well-mixed parts of the Bank. Abundances were highest at the stratified stations, reaching 60 liter^{-1} just above the thermocline.

Townsend (this volume) found a similar result for the Gulf of Maine during autumn months: lowest copepod nauplii abundances were found in mixed waters, highest in stratified zones.

Holligan et al. (1984a) studied the vertical distribution of phytoplankton, bacteria, protozoans, microzooplankton and macrozooplankton at three stations representative of the stratified, frontal and mixed regions of the English Channel in July 1981. They too found that copepod eggs and nauplii and most copepod species were aggregated above the thermocline at the stratified station, but were evenly distributed with depth at the mixed station.

During our routine plankton sampling in Long Island Sound, the vertical distribution of boreal copepod in mixed and stratified waters was studied on several dates. On three of four dates in May 1981, Acartia hudsonica and Pseudocalanus sp. were most abundant between 5-10 m and 15-20 m, respectively, at the offshore station but were evenly distributed with depth at the inshore mixed station (Figure 11). The same patterns were seen on four of six dates in May-June 1983 (not shown). The other dominant copepod, Temora longicornis, was most abundant in the bottom waters of both stations, not surprising because this species prefers deep water by day and vertically migrates to the surface by night.

In conclusion, all evidence collected to date suggests first, that copepod abundances are certainly no greater (perhaps even less) in mixed as compared to stratified regions, and second, that both micro- and macrozooplankton tend to be evenly distributed with depth in shallow mixed water columns, but sort themselves out into different vertically-segregated zones in stratified water columns.

Effects of Temporal Variations in Mixing on Copepod Physiological Processes. One of the key results of our work in Long Island Sound is that temporal variations in copepod fecundity were correlated with changes in phytoplankton species composition (from diatoms to dinoflagellates), which in turn were associated with the onset of seasonal stratification. If these variations in fecundity also occur spatially, then this may partially explain ecological zonation of species. If, for example, an "offshore" species (which is acclimated to feeding on dinoflagellates) finds itself "inshore" where diatoms dominate, and if this individual is unable to produce

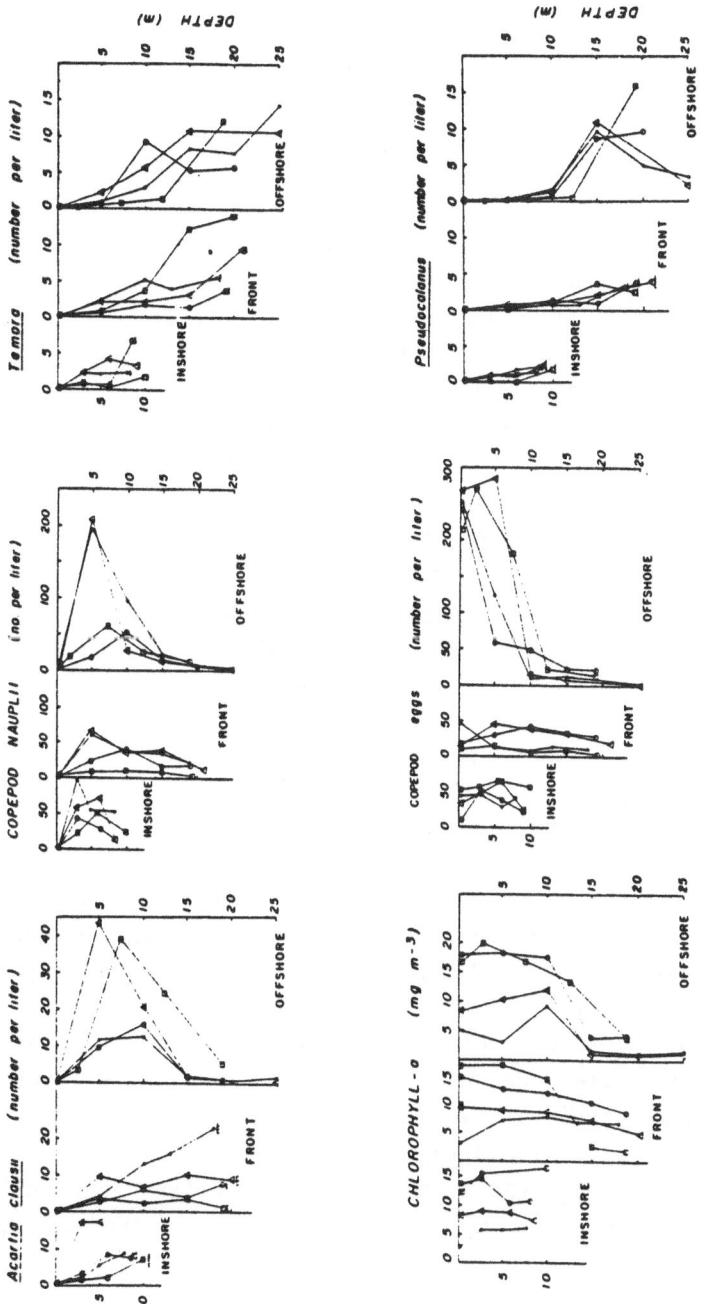

Figure 11. The vertical distribution of copepods in Long Island Sound in May 1981 at an offshore station (Station 4), frontal station (Station 7) and an inshore station (1 km inshore of station 7; see Figure 1). Squares for samples collected on 19 May, circles 20 May, triangles 22 May and filled circles 25 May 1981.

eggs on a diatom diet, a population of such individuals would not develop. Both Grice et al. (1981) and Oviatt (1981), who studied plankton dynamics in experimental mesocosms, reported that zooplankton biomass was far greater in stratified than in mixed mesocosms. This could be a result of differential fecundity and growth rates arising from observed differences in phytoplankton taxonomic composition (flagellates vs. diatoms) in the two experimental systems.

So little is known of the effects of phytoplankton species on zooplankton growth, fecundity and feeding that generalizations cannot be made. This does seem to be a fertile area of research however. Hitchcock (1982) has clearly shown that diatoms and dinoflagellates differ greatly in their nutritional value: diatoms contain half the caloric value of a dinoflagellate on a volume basis, and have less protein, carbohydrate and lipid on a per cell basis. Very little is known of how food quality may affect zooplankton physiological processes. But, very specific dietary requirements have been shown for the tintinnid Favella which requires dinoflagellates for growth (Stoecker et al., 1983; Gold, 1970), and for the copepod Calanus pacificus which grew faster and reached a larger terminal size in culture when fed the dinoflagellate Gymnodinium splendens rather than diatoms (Paffenhofer, 1976). Morey-Gaines (1979) reported a greater rate of egg production by Acartia tonsa and Paracalanus parvus when fed a diet of the dinoflagellate Gonyaulax polyedra than when fed diatoms such as Leptocylindricus danicus and Chaetoceros curvisetus.

Also, recent work by Price et al. (1983) and others have shown that some copepods have very sophisticated feeding behaviors and are capable of manipulating particles which may be ingested or rejected on an individual basis. Such selective feeding behavior may very well be linked to day-to-day nutritional needs. For example, a ripe female which is producing eggs may require a higher daily intake of lipids (or lipid precursors) than immature individuals, because eggs have a different ratio of lipid to biomass than a whole animal. In this regard, Prahl et al. (1984) have shown that the lipid content of Calanus eggs is predominantly fatty acids whereas that of whole animals is mostly fatty acids and sterols.

Nutritional quality and species composition of phytoplankton (rather than phytoplankton concentration), may be the two most important variables in setting egg production, grazing and growth rates of herbivorous copepods. Coastal environments which have zonal gradients in mixing are ideal places for studying this problem because on any given day, one could collect seawater containing natural particulate assemblages dominated by either diatoms or dinoflagellates from mixed and stratified environments, and then study the effects of these different natural assemblages on fecundity, grazing and growth of individuals in _in situ_ incubations.

Techniques exist for the rapid, short-term measurement of egg production, grazing and growth rates. Fecundity can be investigated using bottle incubation techniques, over 6 to 24 hour intervals. In this method, a small number of females are added to one liter bottles filled with 64 μm filtered natural seawater and incubated for one day or less. At the end of the experimental period, the contents are filtered out onto a fine-mesh Nitex screen, and the number of eggs and females counted. This simple technique has been successfully used to describe the _in situ_ rates of egg production of a number of coastal copepods including _Centropages typicus_ (Dagg, 1978), _Acartia clausii_ (Uye, 1982), _Acrocalanus inermis_ (Kimmerer, 1984), _Calanus pacificus_ (Runge, 1985), _Pseudocalanus_ sp. (Dagg et al., 1984), and _Acartia tonsa_ (Beckman, 1985).

Feeding activity can be quickly assayed by using the gut fluorescence technique, and feeding rates can be calculated from empirically derived data on gut evacuation rate as a function of temperature (Dagg and Wyman, 1983). Growth rate of copepods may in some cases be estimated from measurements of molting rates of field-collected animals held in containers for one or two days. This notion was first tested by Burkhill and Kendall (1982). However, Miller et al. (1984) have shown that the technique may not always give unbiased results because there may be a diel cycle in copepod molting.

In conclusion, a number of simple techniques exist that allow the researcher to easily study the indirect effects of spatial variations in hydrodynamic processes on second trophic level organisms. Field measurements of fecundity, feeding and growth could be supplemented with controlled laboratory experiments which look at the effects of various dinoflagellate and diatom clones on copepod fecundity (cf. Checkley, 1980), feeding and growth. Until such field measurements and laboratory experiments are done, we really cannot proceed very efficiently towards an understanding of the importance of hydrodynamic processes and changes in phyto-plankton species composition and food quality on zooplankton population dynamics.

Additional data on the abundance and species composition of zooplankton in mixed, frontal and stratified zones are certainly needed, but studies which focus on the relationships between the physiological rate processes and hydrodynamics should be given the highest priority (following recommendations of Legendre and Demers, 1984).

REFERENCES

Beckman, B. 1985. Egg production by the copepod Acartia tonsa in Long Island Sound. M.S. Thesis, SUNY, Stony Brook.

Bigelow, H. 1924. Plankton of the offshore waters of the Gulf of Maine. Bull. U.S. Bur. Fisheries 60: 1-509.

Bowman, M.J. and W.E. Esaias. 1981. Fronts, stratification and mixing in Long Island and Block Island Sounds. J. Geophys. Res. 86: 4260-4264.

Bowman, M.J., W.E. Esaias and M.B. Schnitzer. 1981. Tidal stirring and the distribution of phytoplankton in Long Island and Block Island Sounds. J. Mar. Res. 39: 587-598.

Burkhill, P.H. and T.F. Kendall. 1982. Production of the copepod Eurytemora affinis in the Bristol Channel. Mar. Ecol. Prog. Ser. 7: 21-31.

Capriullo, G.M. and E.J. Carpenter. 1983. Abundance, species composition and feeding impact of tintinnid micro-zooplankton in central Long Island Sound. Mar. Ecol. Prog. Ser. 10: 277-288.

Checkley, D.M. 1980. The egg production of a marine planktonic copepod in relation to its food supply: laboratory studies. Limnol. Oceanogr. 25: 430-446.

Conover, S.A.M. 1956. Oceanography of Long Island Sound, 1952-1954. IV. Phytoplankton. Bull. Bingham. Oceanogr. Coll. 15: 62-112.

Dagg, M.J. 1978. Estimated, in situ rates of egg production for the copepod Centropages typicus (Kroyer) in the New York Bight. J. Exp. Mar. Biol. Ecol. 43: 183-196.

Dagg, M.J. and K.D. Wyman. 1983. Natural ingestion rates of the copepods Neocalanus plumchrus and N. cristatus calculated from gut contents. Mar. Ecol. Prog. Ser. 13: 37-46.

Dagg, M.J., M.E. Clarke, T. Nishiyama and S.L. Smith. 1984. Production and standing stock of copepod nauplii, food items for larvae of the walleye polluck Theragra chalcogramma in the southeastern Bering Sea. Mar. Ecol. Prog. Ser. 19: 7-16.

Deason, E.E. and T.J. Smayda. 1982. Ctenophore-zooplankton-phytoplankton interactions in Narragansett Bay, Rhode Island, USA, during 1972-1977. J. Plank. Res. 4: 203-217.

Floodgate, G.D., G.E. Fogg, D.A. Jones, K. Lochte and C.M. Turley. 1981. Microbiological and zooplankton activity at a front in Liverpool Bay. Nature 290: 133-136.

Gold, K. 1970. Cultivation of marine ciliates (Tintinnida) and heterotrophic flagellates. Helgol. wiss. Meeresunters. 20: 264-271.

Grice, G.D., R.P. Harris, M.R. Reeve, J.F. Heinbokel and C.O. Davis. 1981. Large scale enclosed water column ecosystems: an overview of Foodweb I, the final CEPEX experiment. J. Mar. Biol. Ass. U.K. 60: 401-414.

Hitchcock, G.L. 1982. A comparative study of the size-dependent organic composition of marine diatoms and dinoflagellates. J. Plank. Res. 4: 363-377.

Holligan, P.M. 1981. Biological implications of fronts on the northwest European continental shelf. Phil. Trans. R. Soc. Lond. A. 302: 547-562.

Holligan, P.M. and D.S. Harbour. 1977. The vertical distribution and succession of phytoplankton in the western English Channel in 1975 and 1976. J. Mar. Biol. Ass. U.K. 57: 1075-1093.

Holligan, P.M., R.P. Harris, R.C. Newell, D.S. Harbour, R.N. Head, E.A.S. Linley, M.I. Lucas, P.R.G. Tranter and C.M. Weekley. 1984a. Vertical distribution and partitioning of organic carbon in mixed, frontal and stratified waters of the English Channel. Mar. Ecol. Prog. Ser. 14: 111-127.

Holligan, P.M., P.J.leB. Williams, D. Purdie, and R.P. Harris. 1984b. Photosynthesis, respiration and N supply of plankton populations in stratified, frontal and tidally mixed shelf waters. Nature 17: 201-213.

Judkins, D.C., C.D. Wirick and W.E. Esaias. 1980. Composition, abundance, and distribution of zooplankton in the New York Bight, September 1974-September 1975. Fish. Bull. U.S. 77: 669-683.

Kimmerer, W.J. 1984. Spatial and temporal variability in egg production rates of the calanoid copepod Acrocalanus inermis. Mar. Biol. 78: 165-169.

Legendre, L. and S. Demers. 1984. Towards dynamic biological oceanography. Can. J. Fish. Aquat. Sci. 41: 2-199.

Lough, R.G. 1984. Larval fish trophodynamic studies on Georges Bank: Sampling strategy and initial results. Flodevigen rapportser. 1: 395-434.

Miller, C.B., M.E. Huntley and E.R. Brooks. 1984. Post-collection molting rates of planktonic, marine copepods: Measurement, applications and problems. Limnol. Oceanogr. 29: 1274-1289.

Morey-Gaines, G. 1979. The ecological role of red tides in the Los Angeles-Long Beach Harbor food web. In: Toxic Dinoflagellate Blooms. D.L. Taylor and H.H. Seliger (eds.), pp. 315-320. Elsevier/North Holland.

Oviatt, C.A. 1981. Effects of different mixing schedules on phytoplankton, zooplankton and nutrients in marine microcosms. Mar. Ecol. Prog. Ser. 4: 57-67.

Paffenhofer, G.-A. 1976. Feeding, growth, and food conversion of the marine planktonic copepod Calanus helgolandicus. Limnol. Oceanogr. 21: 39-50.

Peterson, W.T., C.B. Miller and A. Hutchinson. 1979. Zonation and maintenance of copepod populations in the Oregon upwelling zone. Deep-Sea Res. 26: 467-494.

Peterson, W.T. and S.J. Ausubel. 1984. Diets and selective feeding by larvae of Atlantic mackerel Scomber scombrus on zooplankton. Mar. Ecol. Prog. Ser. 17: 65-75.

Pingree, R.D. and L. Pennycuick. 1976. Transfer of heat, freshwater and nutrients through the seasonal thermocline. J. Mar. Biol. Ass. U.K. 55: 261-274.

Pingree, R.D., P.M. Holligan, G.T. Mardell and R.N. Head. 1976. The influence of physical stability on spring, summer and autumn phytoplankton blooms in the Celtic Sea. J. Mar. Biol. Ass. U.K. 56: 845-873.

Prahl, F.G., G. Eglinton, E.D.S. Corner, S.C.M. O'Hara and T.E.V. Forsberg. 1984. Changes in plant lipids during passage through the gut of Calanus. J. Mar. Biol. Assoc. U.K. 64: 317-334.

Price, H.J., G.-A. Paffenhofer and J.R. Strickler. 1983. Modes of cell capture in calanoid copepods. Limnol. Oceanogr. 28: 116-123.

Riley, G.A. 1967. The plankton of estuaries. In: Estuaries, pp. 313-326, G.H. Luaff (ed.). American Assoc. Advancement of Science Publ. No. 83.

Runge, J.A. 1985. Relationships of egg production of Calanus pacificus to seasonal changes in phytoplankton availability in Puget Sound, Washington. Limnol. Oceanogr. 30: 382-396.

Schnitzer, M.B. 1979. Vertical stability and the distribution of phytoplankton in Long Island Sound. M.S. Thesis, SUNY, Stony Brook.

Simpson, J.H. and J.R. Hunter. 1974. Fronts in the Irish Sea. Nature 250: 404-406.

Smith, S.L. and J. Vidal. 1984. Spatial and temporal effects of salinity, temperature and chlorophyll of the communities of zooplankton in the southeast Bering Sea. J. Mar. Res. 42: 221-257.

Stoecker, D., L.H. Davis and A. Provan. 1983. Growth of *Favella* sp. (Ciliata: Tintinnina) and other microzooplankters in cages incubated *in situ* and comparison to growth *in vitro*. Mar. Biol. 75: 293-302.

Strickland, J.D.H. and T.R. Parsons. 1972. A practical handbook of seawater analysis. Bull. Fish. Res. Bd. Can. 167: 1-310.

Turner, J.T. and M.J. Dagg. 1983. Vertical distributions of continental shelf zooplankton in stratified and isothermal waters. Biol. Oceanogr. 3: 1-40.

Uye, S. 1982. Fecundity studies of neritic calanoid copepods *Acartia clausi* Giesbrecht and *A. Steueri* Smirnov: A simple empirical model of daily egg production. J. Exp. Mar. Biol. Ecol. 50: 255-271.

OBSERVATIONS OF THE STRUCTURE OF CHLOROPHYLL A

IN CENTRAL LONG ISLAND SOUND

R.E. Wilson and A. Okubo

Marine Sciences Research Center

SUNY

Stony Brook, NY 11794

and

W.E. Esaias

Goddard Space Flight Center

Greenbelt Road

Greenbelt, MD 20771

INTRODUCTION

During the week of 20 June 1977 we conducted a week-long chlorophyll mapping experiment in central Long Island Sound. The objective
of this experiment was to examine the horizontal distribution of
chlorophyll a within the upper water column and to infer the relative
importance of biological and physical processes in producing structure of different length scales. The impetus for this experiment was
provided by the investigations of Platt et al. (1970), Platt (1972),
Denman and Platt (1975) and Fasham and Pugh (1976).

Each cruise involved continuous underway sampling along horizontal transects at a depth of 1 m for in vivo fluorescence, salinity
and temperature, and simultaneously at a depth of approximately 7.5 m

Lecture Notes on Coastal and Estuarine Studies, Vol. 17
Tidal Mixing and Plankton Dynamics. Edited by J. Bowman, M. Yentsch and W. T. Peterson
© Springer-Verlag Berlin Heidelberg 1986

for _in vivo_ fluorescence and temperature. For sampling at 1 m, water was pumped through the ship's hull and directed to the various instruments; the instruments used were a Turner Design[R] Model 10-005R flow through fluorometer and a Plessey 6600T Thermosalinograph[R]. For sampling at depth we used an _in situ_ towed fluorometer system.

OBSERVATIONS

The horizontal distribution of chlorophyll _a_ at 1 m depth for 22 June (Fig. 1A) shows that riverine nutrient sources represent a major source of variability at long wavelength (of the order 10 km) for chlorophyll structure in the central Sound. Within our study region three major riverine sources can be identified along the Connecticut shore: the Pequonnock River, the Housatonic River and the Quinnipiac River; the Nissequogue River on Long Island apparently represents another important nutrient source.

The accompanying distributions of temperature (Fig. 2A,B) and salinity (Fig. 3) at 1 m depth are complex. It is important to note that these distributions are not synoptic and that the mapping survey took more than 6 hours to complete. There is evidence for significant lateral temperature structure across the Sound and for the intrusion of low salinity water from the Connecticut into the central basin, but the relationship between this structure and the distribution of chlorophyll _a_ is not clear.

Continuous records for fluorescence, temperature, salinity and sigma-t at 1 m depth (Fig. 4) along two transects across the Sound show an abundance of small scale structure. The distributions are characterized by discontinuities in fluorescence and physical water properties. Major discontinuities occurring in the central basin

323

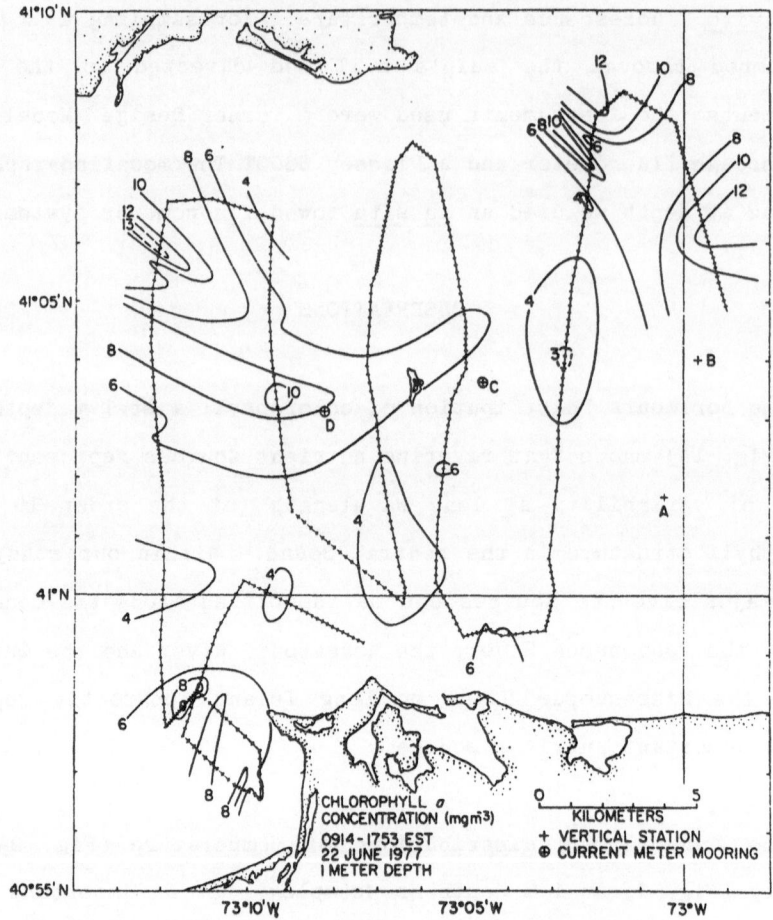

Figure 1. A. Horizontal distributions of chlorophyll a at 1 m depth
in central Long Island Sound on 22 June 1977.

represent a change in sigma-t of approximately 1 unit. The sharpness
of the discontinuities and the magnitude of the density change across
them suggests that advective processes are important in maintaining
this structure. Coherency between fluctuations in fluorescence and
fluctuations in temperature and salinity at 1 m is not high even at
shorter wavelengths. Discontinuities in fluorescence do, however,
coincide with discontinuities in temperature and salinity, and the
region of high fluorescence in the central basin does appear to be
associated with a particular water mass delineated by two
discontinuities.

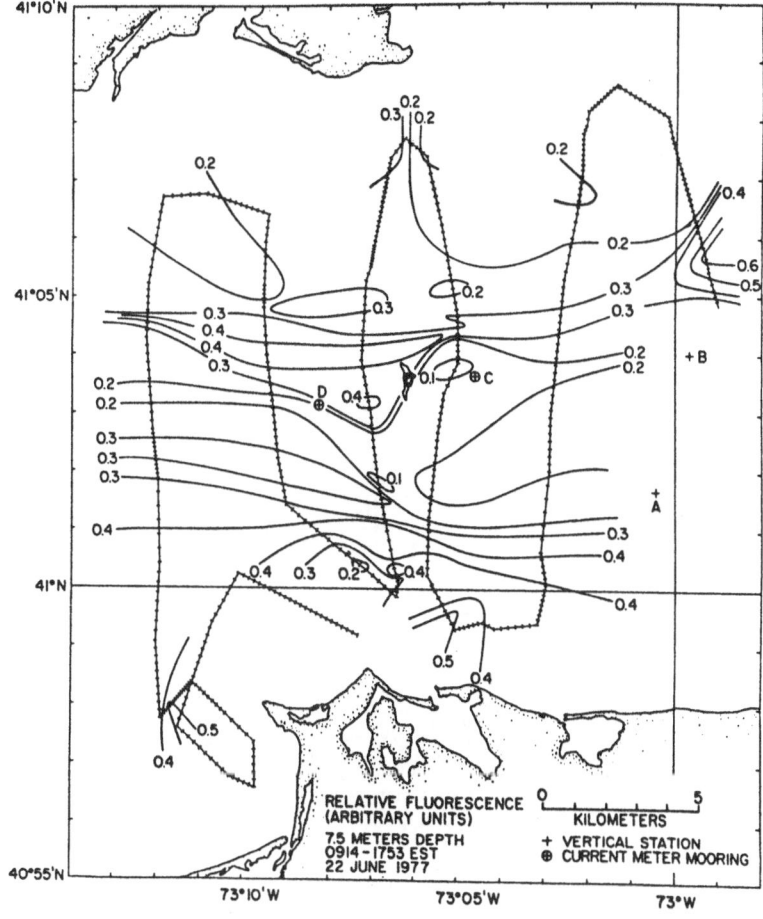

Figure 1. B. Horizontal distributions of relative fluorescence at 7.5 m depth in central Long Island Sound on 22 June 1977.

The coherency between fluctuations in fluorescence at 1 m and 7.5 m along the transects is low (Fig. 5); the same is true for the coherency between fluctuations in temperature at 1 and 7.5 m. Figure 5 does show very high coherency between fluctuations in fluorescence and temperature at 7.5 m depth at wavelengths from a few hundred meters to several kilometers along the two transects suggesting that at those wavelengths chlorophyll at depth behaves nearly as a passive scalar. It should be noted that although the fluorescence records in Fig. 5 are from fluorometers on the same relative scale, the fluorometers are essentially uncalibrated relative to one another.

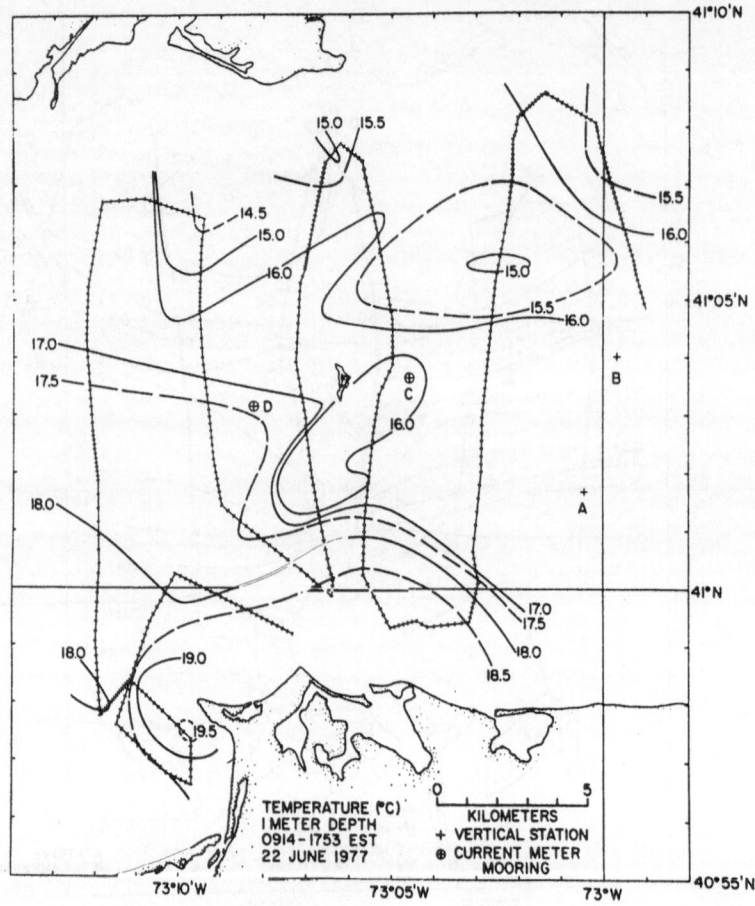

Figure 2. A. Horizontal distributions of temperature at 1 m depth in central Long Island Sound on 22 June 1977.

DISCUSSION

Some insight into the advective and mixing processes responsible for the small scale structure encountered along horizontal transects (Fig. 4) is provided by the temperature-salinity diagrams in Figure 6. The diagrams for the eastern and western transects were constructed from approximately 280 and 150 data points, respectively, sampled at 15 s intervals at 1 m depth along the transects; the ship speed along the transects was approximately 7 knots.

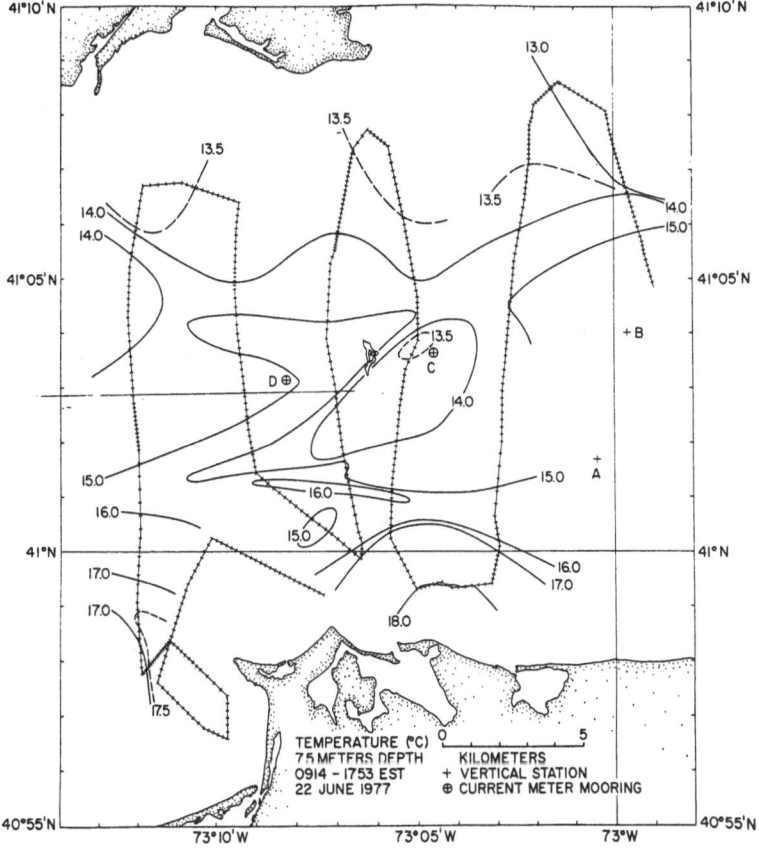

Figure 2. B. Horizontal distributions of temperature at 7.5 m depth in central Long Island Sound on 22 June 1977.

Both diagrams exhibit a number of separate water masses formed by the mixing of distinct water types with particular temperature and salinity characteristics. Data points in the lower part of each diagram represent a water mass produced by the mixing of high temperature and relatively high salinity found shoreward of the stationary tidal front found off Long Island (Bowman and Esaias, 1977; Bowman et al., 1981) with lower temperature and slightly more saline water found seaward of the front. Fluorescence levels within this water mass are some of the lowest encountered along the transects. A second water mass is formed by the mixing of water with temperature

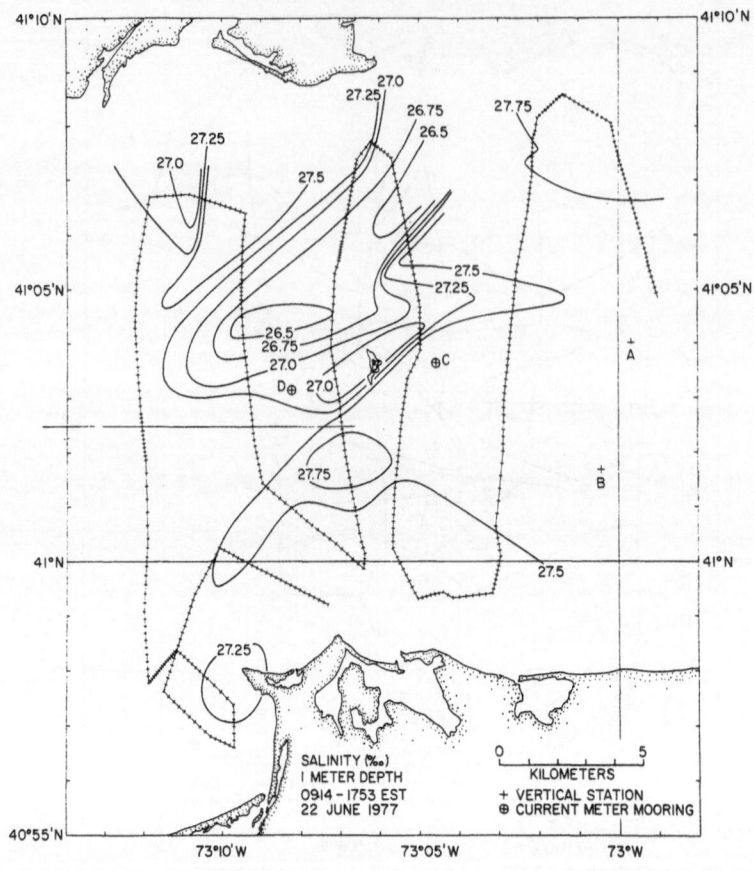

Figure 3. Horizontal distributions of salinity at 1 m depth in central Long Island Sound on 22 June 1977.

(16.5-17.2°C) and relatively low salinity (26.8-27.0 °/oo) with water of the central Sound (15.0-15.8°C, 27.4-27.5 °/oo). Fluorescence levels within this water mass are the highest encountered along the transects. A third water mass with moderately high fluorescence levels is formed by the mixing of even lower salinity water with water of the central Sound. It is shown most clearly on the diagram for the eastern transect.

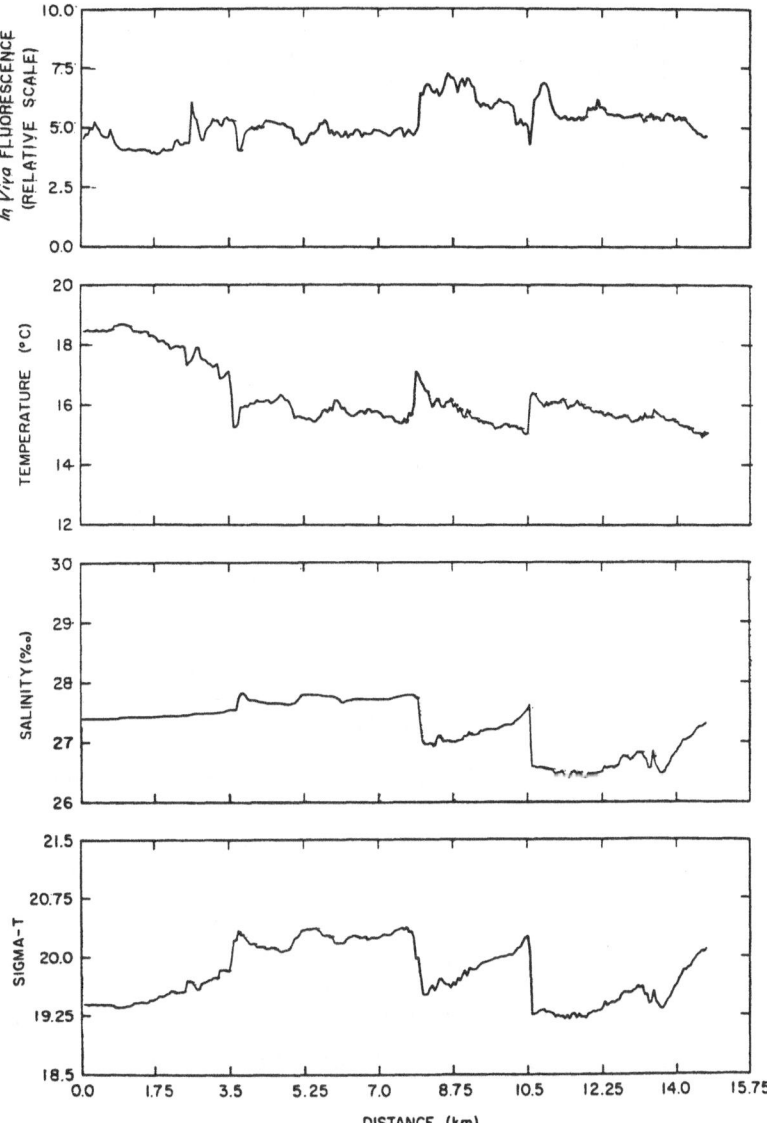

Figure 4. Records of in vivo fluorescence, temperature, salinity and sigma-t at 1 m depth along transects on 22 June (Figures 1-3) from approximately 40°59'N, 73°04'W to 41°07'N, 73°05'W (left) and from approximately 41°06'N, 73°07'W to 41°02'N, 73°07'W (right).

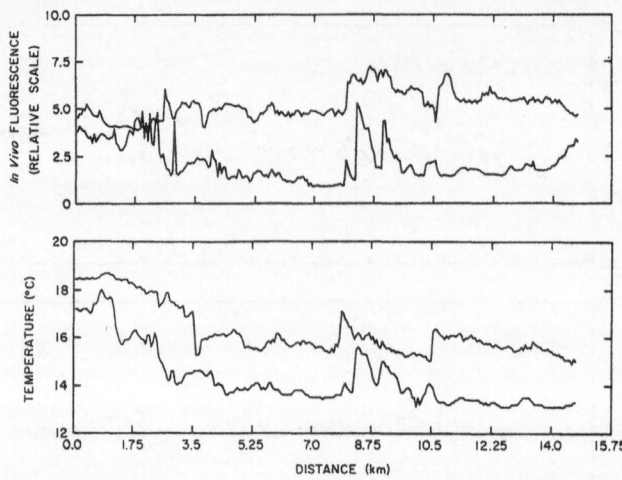

Figure 5. Records of <u>in vivo</u> fluorescence at 1 m and 7.5 m depth and temperature at 1 m and 7.5 m depth along transects referenced in Figure 4.

It is enlightening to examine more carefully the horizontal distributions of temperature and fluorescence within the second water mass with the highest fluorescence levels. It was encountered along the western transect between approximately 1.5 and 3 km (Fig. 4) and along the eastern transect between approximately 7.5 and 10.5 km (Fig. 4). It is bounded to the south by the sharp front and to the north by a somewhat less pronounced discontinuity. The horizontal distributions of temperature and fluorescence at 1 m depth within this water mass measured at each of the transects are plotted in Figure 7.

It is possible to examine some of the characteristics of chlorophyll growth and turbulent mixing within this somewhat isolated and bounded water mass by comparing the observed distributions in Figure 7 to those predicted by a simple reaction-diffusion model. Consider

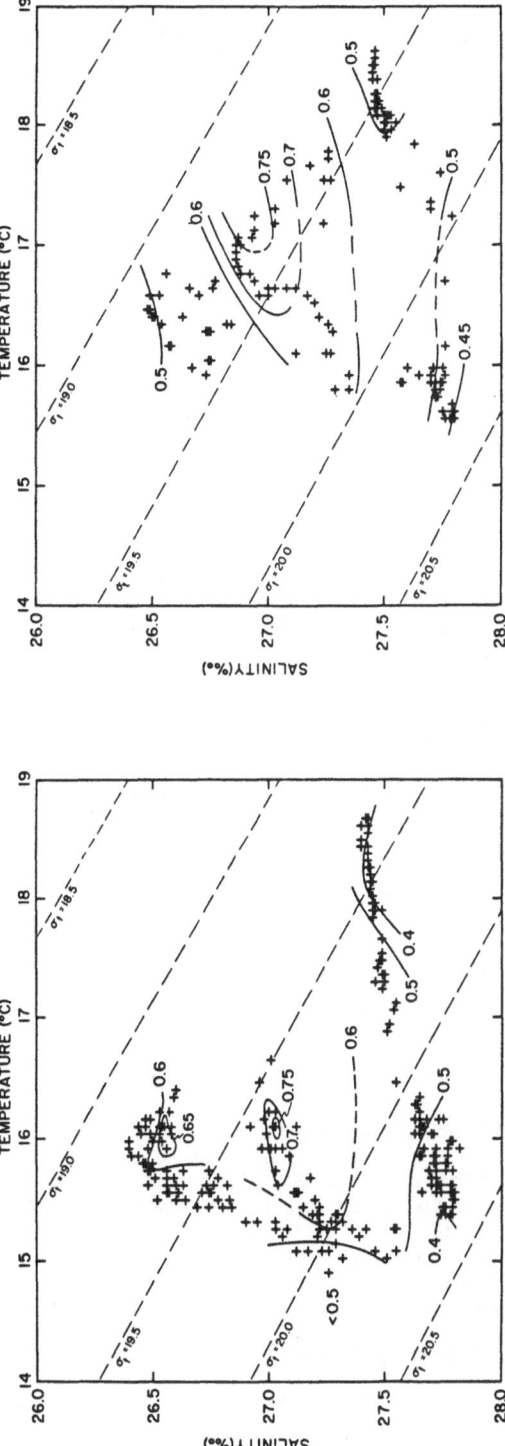

Figure 6. Temperature-salinity diagrams constructed from observations at 1 m depth along transects referenced in Figure 4. The diagram on the left is for the eastern transect and that on the right is for the western transect. Contours represent isolines of in vivo fluorescence on a relative scale.

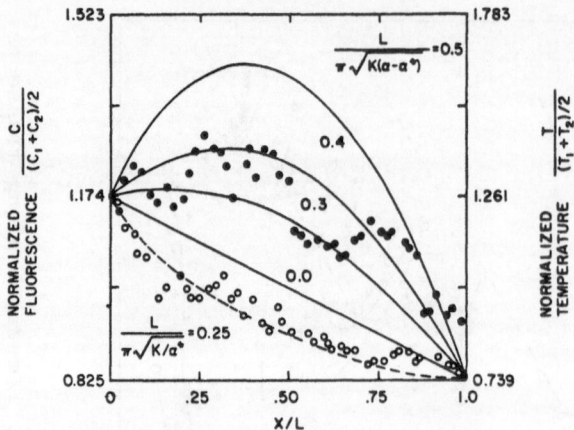

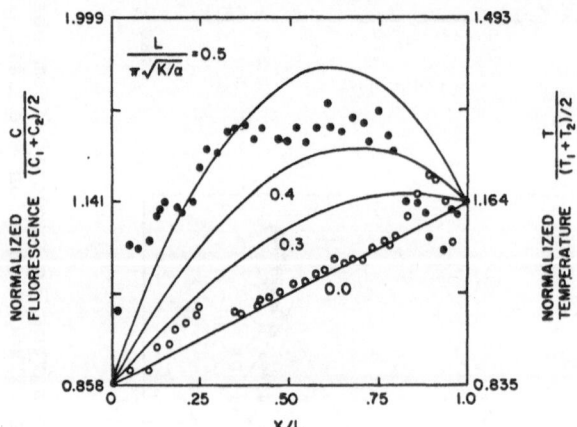

Figure 7. Comparisons of observed horizontal fluorescence and temperature structure (closed and open circles, respectively) with that predicted by a simple reaction-diffusion model. Fluorescence and temperature are normalized by the mean of the boundary values (see text). The diagram on the left is for the eastern transect and that on the right is for the western transect.

first a steady state balance between horizontal diffusion and first order growth $K \dfrac{d^2 C}{dx^2} + \alpha C = 0$ where C is chlorophyll concentration, x is the horizontal coordinate, K is a constant horizontal eddy diffusivity and α is a constant growth rate. Appropriate boundary conditions would be $C = C_1$ at $x = 0$ and $C = C_2$ at $x = L$ where L is the horizontal extent of the water mass. If we normalize x by L: $x = x'L$, and concentration C by $(C_1+C_2)/2$: $C = C'(C_1+C_2)/2$, we obtain $\dfrac{d^2C'}{dx'^2} + \pi^2(\dfrac{L}{L_c})^2 C' = 0$ where $L_c = \pi\sqrt{K/\alpha}$ with boundary conditions $C' = C_1' = \dfrac{C_1}{(C_1+C_2)/2}$ at $x' = 0$ and $C'' = C_2' = \dfrac{C_2}{(C_1+C_2)/2}$ at $x' = 1$.

This has the solution

$$C'(x') = C_1' \frac{\sin\pi \frac{L}{L_c}(1-x')}{\sin\pi \frac{L}{L_c}} + C_2' \frac{\sin\pi \frac{L}{L_c}x'}{\sin\pi \frac{L}{L_c}} \qquad (1)$$

It is advantageous to normalize concentration by $(C_1+C_2)/2$ because this allows the <u>total excess chlorophyll</u> within the water mass: $\int_0^1 C'dx' - (C_1'+C_2')/2$ to be expressed as a function of $\dfrac{L}{L_c}$ <u>only</u>. Regardless of the normalization employed, however, the concentration $C'(x')$ will depend on <u>both</u> C_1 and C_2.

Figure 7 provides a comparison of the structure represented by the solution (1) with $\dfrac{L}{L_c}$ as a parameter with the fluorescence and temperature structure within the water mass encountered along the western transect. The solution (1) appears to represent the major features in the fluorescence structure for $\dfrac{L}{L_c}$ between 0.4 and 0.5 and the temperature structure for $\dfrac{L}{L_c} = 0.0$.

Horizontal temperature structure within the water mass encountered along the eastern transect (Fig. 7) is more complex than that encountered along the western transect. The reduced vertical stratification along the eastern transect (Fig. 5) suggests that vertical as well as horizontal mixing is important. It is possible to simulate the effect of vertical mixing on the horizontal temperature structure at 1 m depth as a simple first order reaction (see, for example, Ichiye, 1952). The equation for temperature T would be $K\dfrac{d^2T}{dx^2} - \alpha^*T = 0$ or when normalized $\dfrac{d^2T'}{dx^2} - \pi^2(\dfrac{L}{L_c})^2\, T' = 0$ where $L_c^* = \pi\sqrt{K/\alpha^*}$. This equation has the solution

$$T'(x') = T_1'\,\frac{\sinh\pi\,\dfrac{L}{L_c^*}\,(1-x')}{\sinh\pi\,\dfrac{L}{L_c^*}} + T_2'\,\frac{\sinh\pi\,\dfrac{L}{L_c^*}\,x'}{\sinh\pi\,\dfrac{L}{L_c^*}} \qquad (2)$$

For $\dfrac{L}{L_c^*} = 0.25$ the solution (2) adequately represents the temperature structure along the eastern transect (Fig. 7).

The reduced vertical chlorophyll structure along the eastern transect provides further evidence for the importance of vertical mixing (Fig. 5); an appropriate equation for the horizontal chlorophyll distribution would be in this case $K\dfrac{d^2C}{dx^2} + (\alpha-\alpha^*)C = 0$ or when normalized $\dfrac{d^2C'}{dx'^2} + \pi^2\,((\dfrac{L}{L_c})^2 - (\dfrac{L}{L_c^*})^2\,)C' = 0$ where $(\dfrac{L}{L_c})^2 - (\dfrac{L}{L_c^*})^2 = \dfrac{L^2}{\pi^2\,K/(\alpha-\alpha^*)}$. This equation has a solution which is of the same form as (1) but with $\dfrac{L}{L_c}$ replaced by $((\dfrac{L}{L_c})^2 - (\dfrac{L}{L_c^*})^2)^{\frac{1}{2}}$. The solution represents the major features of the fluorescence structure along the eastern transect (Fig. 7) for $\dfrac{L}{\pi\sqrt{K/(\alpha-\alpha^*)}}$ between 0.3 and 0.4 . For a value of $\dfrac{L}{L_c^*} = 0.25$ determined from the distribution of temperature this requires that $\dfrac{L}{L_c}$ be between 0.39 and 0.47 which is consistent with the estimates for $\dfrac{L}{L_c}$ obtained from the western transect.

We can assess the validity of our steady state approximation by considering the behavior of the solution to the time dependent reaction-diffusion problem: $\dfrac{dC}{dt} = K\dfrac{d^2C}{dx^2} + \alpha C$ with boundary conditions $C = C_1$ at $x = 0$ and $C = C_2$ at $x = L$ and initial conditions $C = f(x)$

at t = 0. This has solution (Carslaw and Jaeger, 1959; p. 102)

$$C(x,t) = \frac{2}{L} \sum_{m=1}^{\infty} e^{\alpha(1-n^2 (\frac{L_c}{L})^2)t} \sin n\pi \frac{x}{L} \int_{o}^{L} f(x') \sin \frac{n\pi x'}{L} dx' +$$

$$\frac{2}{\pi} (\frac{L_c}{L})^2 \sum_{n=1}^{\infty} n [C_1 \frac{1-e^{(1-n^2 (\frac{L_c}{L})^2)t}}{(n^2(\frac{L_c}{L})^2 - 1)} - (-1)^n C_2 \frac{1-e^{\alpha(1-n^2 (\frac{L_c}{L})^2)t}}{(n^2(\frac{L_c}{L})^2 - 1)}]$$

$$\sin \frac{n\pi x}{L} \tag{3}$$

where again $L_c = \pi\sqrt{K/\alpha}$. The solution (3) will tend to a stable steady state solution given by (1) only if $\alpha(1-n^2(\frac{L_c}{L})^2) < 0$ which requires that $L < L_c$. Our earlier analyses suggested, in fact, that $\frac{L}{L_c}$ was of the order 0.45. The solution (3) also shows that the longest characteristic time scale to reach steady state is $\frac{1}{\alpha|1-(\frac{L_c}{L})^2|}$. In our case for $\frac{L}{L_c}$ of the order 0.45 this becomes approximately $\frac{1}{4\alpha}$; the time scale to reach a steady state distribution is ¼ the time scale for phytoplankton growth and the steady state distribution represented by (1) should be approached relatively quickly.

We remark that $L_c = \pi\sqrt{K/\alpha}$ is identified as a critical size of water mass in which phytoplankton bloom can be maintained (Kierstead and Slobodkin, 1953). If the horizontal extent of the water mass L is larger than L_c, the growth process dominates horizontal mixing and the chlorophyll structure cannot remain in steady state. If, on the other hand, L is smaller than L_c, horizontal mixing checks the growth process and, combined with the boundary conditions, the chlorophyll structure can reach steady state.

ACKNOWLEDGEMENT

This work was supported by the National Science Foundation under grant OCE-7610730.

REFERENCES

Bowman, M.J. and W.E. Esaias. 1977. Fronts, jets, and phytoplankton patchiness. In: Bottom Turbulence, J. Nihoul (ed.). Proc. 8th Internatl. Colloq. Ocean Hydrodynamics, Elsevier Oceanography Series, 19: 255-265.

Bowman, M.J., W.E. Esaias and M.B. Schnitzer. 1981. Tidal stirring and the distribution of phytoplankton in Long Island and Block Island Sounds. J. Mar. Res. 39: 587-603.

Carslaw, H.S. and J.C. Jaeger. 1959. Conduction of Heat in Solids. Oxford Univ. Press, London. 496 pp.

Denman, K.L. 1976. Covariability of chlorophyll and temperature in the sea. Deep-Sea Res. 23: 539-550.

Denman, K.L. and T. Platt. 1975. Coherences in the horizontal distributions of phytoplankton and temperature in the upper ocean. Mémoires de la Société royale des sciences de Liège 7: 19-30.

Fasham, M.J.R. and P.R. Pugh. 1976. Observations on the horizontal coherence of chlorophyll a and temperature. Deep-Sea Res. 23: 527-538.

Kierstead, H. and L.B. Slobodkin. 1953. The size of water masses containing plankton blooms. J. Mar. Res. 12: 141-147.

Ichiye, T. 1952. On the use of T-S diagrams in shallow water. Memoirs of the Kobe Marine Observatory 10: 25-45.

Platt, T. 1972. The feasibility of mapping the chlorophyll distribution in the Gulf of St. Lawrence. Fish. Res. Bd. Can., Tech. Rept. 332, 8 pp.

Platt, T., L.M. Dickie and R.W. Trites. 1970. Spatial heterogeneity of phytoplankton in a near-shore environment. J. Fish. Res. Bd. Can. 27: 1453-1473.

PHYTOPLANKTON PATCHINESS IN THE CENTRAL LONG ISLAND SOUND

A. Okubo and R.E. Wilson
Marine Sciences Research Center
State University of New York
Stony Brook, NY 11794

INTRODUCTION

As shown in the previous study (Wilson et al., 1985), patchiness in chlorophyll a is observed in the northern and southern coastal regions of Long Island Sound. This patchiness is associated with riverine fronts (northern part) or with tidal streaming (southern part) (Bowman and Esaias, 1977). On the other hand, the distribution of chlorophyll a in the central region of Long Island Sound is fairly uniform, which suggests that a critical size of patchiness in the region should be much larger than one kilometer (Fig. 1).

We moored two Endeco Model 105 current meters on a single taut wire in the center of the study region (marked by $^+$C in Fig. 1) to measure the vertical shear in horizontal currents. The absolute values of the components of current shear along the principal axes of the flow are plotted as a function of time in Fig. 2. The component of shear across the Sound exhibited fluctuations which were significantly greater than in amplitude and lower frequency than those in the component along the Sound. The spectra of the fluctuations in the shear (Fig. 3) show that the dominant peak occurs at the semi-diurnal frequency for the component across the Sound.

In the following model we will demonstrate that the vertical shear in the across-Sound current produces a strong shear effect on the phytoplankton distribution, resulting in a critical patch size which is of the order of 10 km rather than 1 km.

Critical Size Problem for Phytoplankton Growth in an Oscillatory Current with Shear

The central region of Long Island Sound is characteristic of a strong tidal flow superposed in a weak mean flow. Horizontal currents in the sea are usually associated with shears. The vertical shear

Lecture Notes on Coastal and Estuarine Studies, Vol. 17
Tidal Mixing and Plankton Dynamics. Edited by J. Bowman, M. Yentsch and W. T. Peterson
© Springer-Verlag Berlin Heidelberg 1986

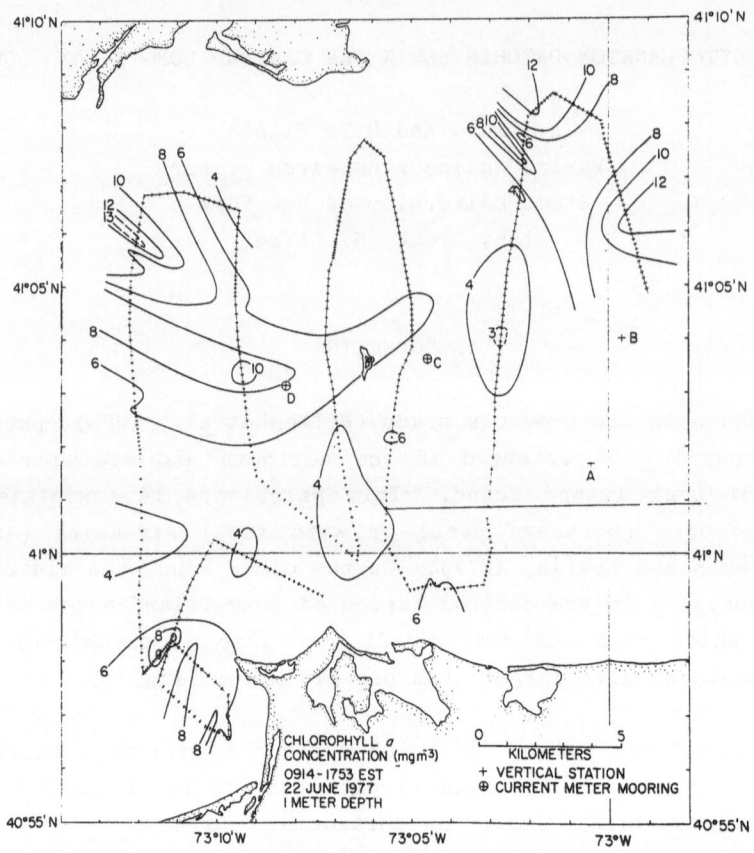

Figure 1. Horizontal distribution of chlorophyll a at 1 m depth in
central Long Island Sound on 22 June 1977.

effective to horizontal mixing is considered to be of primary importance compared with the horizontal shear.

An important role for the current shear in contributing to horizontal dispersion (the "shear effect") has been well recognized
(Fischer et al., 1979). Thus a patch of phytoplankton in the sea
will be subject to the shear effect, which produces an effective
diffusion in the horizontal direction. If the patch is surrounded by
ambient water, physiologically unsuitable for the growth of plankton,
a part of the population in the patch is continuously lost to the
surrounding water owing to horizontal diffusion. Unless the reproduction of organisms within the patch counterbalances the diffusion
loss, this plankton patch would cease to exist.

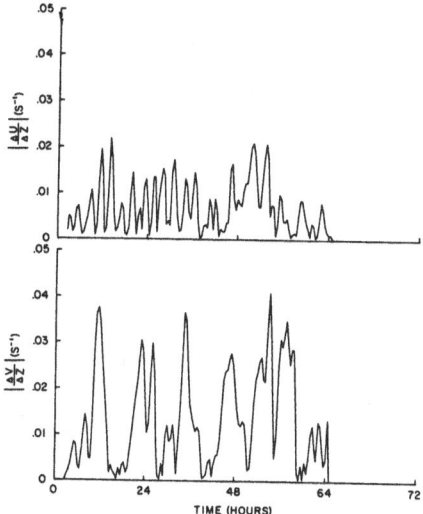

Figure 2. Absolute value of components of vertical current shear along principal flow axis 87.5°T (upper figure) and normal to principal flow axis 357.5°T (lower figure) for the period 0840 24 October to 1220 27 October.

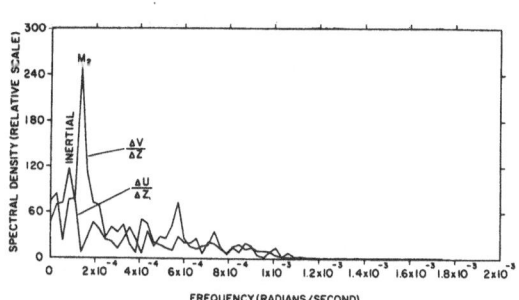

Figure 3. Spectra of the fluctuations in both components of vertical current shear (Figure 2).

Now the loss of organisms due to diffusion occurs through the boundary of the patch, and hence its rate is proportional to surface area. On the other hand, reproduction takes place locally within the patch, and hence its rate is proportional to volume. Thus there exists a minimum critical size for the patch, or simply critical size, below which no increase of plankton population is possible (Okubo, 1980). We now estimate this critical size in the central Long Island Sound.

Model for Critical Size

We consider a mass of water assumed to be biologically bounded both vertically and horizontally, i.e. the environmental water outside the mass of water is totally unsuitable for phytoplankton growth. Also assumed are exponential growth in plankton population in the water mass, vertical uniform but time dependent shear in the current, and constant diffusivities in the horizontal and vertical directions (Wilson and Okubo, 1978).

The basic equation for the problem is thus written as

$$\frac{\partial S}{\partial t} + \Omega(t)z \frac{\partial S}{\partial x} = K_h \nabla_1^2 S + K_z \frac{\partial^2 S}{\partial z^2} + \alpha S \qquad (1)$$

where   t = time;

   x,y = horizontal axes;

   z = vertical axis;

$S(x,y,z,t)$ = phytoplankton concentration;

   $\Omega(t)$ = time dependent vertical shear;

   $K_h K_z$ = horizontal and vertical diffusivities;

   α = specific growth rate; and

   $\nabla_1^2 = \dfrac{\partial^2}{\partial x^2} + \dfrac{\partial^2}{\partial y^2}$

Equation (1) is subject to the following initial and boundary conditions:

At t=0; S = f(x,y,z) for $0 < x < L_x$, $0 < y < L_y$, $0 < z < H$ (2)

At x = 0, L_x; S = 0   (3)

At y = 0, L_y; S = 0   (4)

At z = 0; $K_z \, \partial S/\partial z = 0$ (5)

At z = H; S = 0   (6)

Because of the shear term in (1), the problem is somewhat difficult to treat analytically. We will instead take the following approximated method. First, find the Green's function of (1) in an infinite domain, evaluate the effective dispersion coefficients in the principal axes of diffusion, and use these coefficients in a generalized diffusion-reaction equation for obtaining the critical size of water mass.

Let $S_I(x,y,z,t)$ be the Green's function, which obeys the following equation:

$$\frac{\partial S_I}{\partial t} + \Omega(t)z \frac{\partial S_I}{\partial x} = K_h \nabla_I^2 S_I + K_z \frac{\partial^2 S_I}{\partial z^2} + \alpha S_I \qquad (7)$$

Subject to

$$\text{At } t = 0, \ S_I = \delta(x)\delta(y)\delta(z) \qquad (8)$$

After lengthy, but somewhat straightforward calculations, the solution is obtained in terms of the principal axes of diffusion (X,Y,Z) as

$$S_I (X,Y,Z,t) = \frac{e^{\alpha t}}{(2\pi)^{3/2} \ \sigma_x \sigma_y \sigma_z} \cdot \exp - \left\{ \frac{X^2}{2\sigma_x^2} + \frac{Y^2}{2\sigma_y^2} + \frac{Z^2}{2\sigma_z^2} \right\} \qquad (9)$$

where

$$\sigma_x^2 = \tfrac{1}{2} [(\sigma_1^2 + \sigma_2^2) + \{(\sigma_1^2 - \sigma_2^2)^2 + 4\sigma_1^2\sigma_2^2\lambda^2\}^{\frac{1}{2}}] \qquad (10)$$

$$\sigma_y^2 = 2 K_h t \qquad (11)$$

$$\sigma_z^2 = \tfrac{1}{2} [(\sigma_1^2 + \sigma_2^2) - \{(\sigma_1^2 - \sigma_2^2)^2 + 4\sigma_1^2\sigma_2^2\lambda^2\}^{\frac{1}{2}}] \qquad (12)$$

$$\sigma_1^2 = 2K_h t \ \{1 + \varepsilon \ (t^{-1}G + \phi^2 - 2t\phi F)\} \qquad (13)$$

$$\sigma_2^2 = 2 K_z t \qquad (14)$$

$$\varepsilon = K_z/K_h \qquad (15)$$

$$\lambda^2 = \varepsilon \ (\phi - t^{-1}F)^2 \ / \ \{1 + \varepsilon \ (t^{-1}G + \phi^2 - 2t\phi F)\} \qquad (16)$$

$$\phi(t) = \int_0^t \Omega(t')dt' \qquad (17)$$

$$F(t) = \int_0^t \phi(t')dt' \qquad (18)$$

$$G(t) = \int_0^t \phi^2(t')dt' \qquad (19)$$

The angle θ between the coordinate axes (x,z) and the principal axes of diffusion (X,Z) is given by

$$\theta = \tfrac{1}{2}\tan^{-1} \{2\sigma_1\sigma_2\lambda \ / \ (\sigma_1^2 - \sigma_2^2)\} \qquad (20)$$

From now on we consider a case of an oscillatory shear with a tidal frequency ω; $\Omega(t) = \Omega_1 \sin\omega t$. We then obtain

$$\sigma_1^2 = 2K_h t \left[1 + \varepsilon \frac{\Omega_1^2}{\omega^2} \left\{1 + \frac{\cos 2\omega t}{2} - \frac{3}{4} \frac{\sin 2\omega t}{\omega t}\right\}\right] \tag{21}$$

$$\sigma_2^2 = 2K_z t \tag{22}$$

$$\lambda^2 = \varepsilon \frac{\Omega_1^2}{\omega^2} \left(\cos\omega t - \frac{\sin\omega t}{\omega t}\right)^2 \Big/ \left[1 + \varepsilon \frac{\Omega_1^2}{\omega^2} \left\{1 + \frac{\cos 2\omega t}{2} - \frac{3}{4} \frac{\sin 2\omega t}{\omega t}\right\}\right] \tag{23}$$

As we show later, for Long Island Sound

$$\varepsilon \sim 0(10^{-3}) \tag{24}$$

$$\frac{\Omega_1^2}{\omega^2} \sim 0(10^4 - 10^6) \tag{25}$$

so that we may approximate (10), (11), (12), and (20) by

$$\sigma_x^2 = 2K_h \varepsilon \frac{\Omega_1^2}{\omega^2} t \left[1 + \frac{\omega^2}{\Omega_1^2} \left\{\varepsilon^{-1} + \frac{1}{4} \frac{\cos^2\omega t}{1 + \frac{1}{2}\cos 2\omega t}\right\}\right] \tag{26}$$

$$\sigma_y^2 = 2K_h t \tag{27}$$

$$\sigma_z^2 = 2K_z t \left(1 - \frac{1}{4} \frac{\cos^2\omega t}{1 + \frac{1}{2}\cos 2\omega t}\right) \tag{28}$$

$$\theta \sim \frac{1}{2} \tan^{-1} \{0(\omega/\Omega_1)\} \tag{29}$$

Therefore, the principal axes of diffusion asymptotically coincide with the coordinate axes, and effective diffusivities in the principal axes are given by

$$A_x(t) = \frac{1}{2} d\sigma_x^2/dt = K_h \varepsilon \frac{\Omega_1^2}{\omega^2} \left[1 + \frac{\omega^2}{\Omega_1^2}\left\{\varepsilon^{-1} + \frac{1}{4} \frac{\cos^2\omega t}{1 + \frac{1}{2}\cos 2\omega t}\right\}\right] \tag{30}$$

$$A_y(t) = \frac{1}{2} d\sigma_y^2/dt = K_h \tag{31}$$

$$A_z(t) = \frac{1}{2} d\sigma_z^2/dt = K_z \left(1 - \frac{1}{4} \frac{\cos^2\omega t}{1 + \frac{1}{2}\cos 2\omega t}\right) \tag{32}$$

Since a diffusing patch in the shearing current tends to line up in the direction of the coordinate system (x,y,z), our discussion for the critical size will be restricted to the asymptotic stage. The solution for the Green's function (8) implies that asymptotically the distribution of phytoplankton in the shearing current is obeyed by the following diffusion equation

$$\frac{\partial S}{\partial t} = A_x(t) \frac{\partial^2 S}{\partial x^2} + A_y(t) \frac{\partial^2 S}{\partial y^2} + A_z(t) \frac{\partial^2 S}{\partial z^2} + \alpha S \qquad (33)$$

Equation (33) is subject to the initial and boundary conditions mentioned before.

Let us assume the solution

$$S = \sum_{k=1}^{\infty} \sum_{m=1}^{\infty} \sum_{n=0}^{\infty} \Psi_{kmn}(t) \sin(^{k\pi x}/L_x) \sin(^{m\pi y}/L_y) \cos(\frac{(2n+1)\pi z}{2H}) \qquad (34)$$

which satisfies the boundary conditions automatically. Substituting (34) into (33) we have

$$\frac{d\Psi_{kmn}}{\Psi_{kmn} dt} = -\{(\frac{k^2\pi^2}{L_x^2}) A_x(t) + (\frac{m^2\pi^2}{L_y^2}) A_y(t) + (\frac{(2n+1)^2\pi^2}{4H^2}) A_z(t) - \alpha\} \qquad (35)$$

which can be integrated into

$$\Psi_{kmn}(t) = \Psi_{kmn}(0) \exp[\alpha t - (\frac{k^2\pi^2}{L_x^2}) \int_0^t A_x(t^1)dt^1 - (\frac{m^2\pi^2}{L_y^2}) \int_0^t$$

$$A_y(t^1)dt^1 - (\frac{(2n+1)^2\pi^2}{4H^2}) \int_0^t A_z(t^1)dt^1] \qquad (36)$$

where $\Psi_{kmn}(0)$ is determined from the initial condition:

$$\psi_{kmn}(0) = \frac{8}{L_x L_y H} \int_0^{L_x} dx \int_0^{L_y} dy \int_0^H dz \, f(x,y,z)\sin(\frac{k\pi x}{L_x})\sin(\frac{m\pi y}{L_y})$$

$$\cos(\frac{(2n+1)\pi z}{2H}) \qquad (37)$$

Substitution of (36) with (37) into (34) yields the solution

$$S(x,y,z,t) = \frac{8}{L_x L_y H} \sum_{k=1}^{\infty} \sum_{m=1}^{\infty} \sum_{n=0}^{\infty} \int_0^{L_x} dx \int_0^{L_y} dy \int_0^{L_z} dz \; f(x,y,z)$$

$$\sin(\frac{k\pi x}{L_x}) \sin(\frac{m\pi y}{L_y}) \cos(\frac{(2n+1)\pi z}{2H}) \exp[\alpha t - (\frac{k^2 \pi^2}{L_x^2}) \int_0^t A_x(t^1)dt^1 -$$

$$(\frac{m^2 \pi^2}{L_y^2}) \int_0^t A_y(t^1)dt^1 - (\frac{(2n+1)^2 \pi^2}{4H^2}) \int_0^t A_z(t^1)dt^1] \tag{38}$$

Thus the time-dependent exponential factor in (38) determines the asymptotic behavior of the phytoplankton patch. Since the periodic components in A's make insignificant contributions to the long-term integral, the exponential factors becomes asymptotically

$$\exp[\alpha - (\frac{k^2 \pi^2}{L_x^2}) \{K_h (\varepsilon \frac{\Omega_1^2}{\omega^2} + 1)\} - (\frac{m^2 \pi^2}{L_y^2}) K_h - (\frac{(2n+1)^2 \pi^2}{4H^2}) K_z]^t$$

$$\tag{39}$$

The organism population will then be unable to maintain itself against diffusion unless there is at least one term in the exponential factor for which

$$\alpha - (\frac{k^2 \pi^2}{L_x^2}) \{K_h (\varepsilon \frac{\Omega_1^2}{\omega^2} + 1)\} - (\frac{m^2 \pi^2}{L_y^2}) K_h - (\frac{(2n+1)^2 \pi^2}{4H^2}) K_z \geq 0$$

$$\tag{40}$$

Since the smallest values of k and m are one and zero for n , a necessary condition for maintenance of the population is

$$\alpha \geq (\frac{\pi^2}{L_x^2}) \{K_h (\varepsilon \frac{\Omega_1^2}{\omega^2}) + 1\} + (\frac{\pi^2}{L_y^2}) K_h + (\frac{\pi^2}{4H^2}) K_z \tag{41}$$

Note that if $\Omega_1 = 0$, i.e. no shear, (41) is reduced to

$$\alpha \geq (\frac{\pi^2}{L_x^2}) K_h \tag{42}$$

for one-dimensional case.

From (42) the critical patch size L_c is obtained by

$$L_c = \pi (K_h / \alpha)^{\frac{1}{2}} \qquad (43)$$

which is the formula given by Kierstead and Slobodkin (1953) and Skellam (1951).

In the central Long Island Sound, $K_h \sim 10^4$ cm^2/sec and $\alpha \sim 10^{-5}$/sec (1 div/day), which gives L_c of the order of km. Without shear in the tidal current the patchiness as small as a kilometer would therefore exist. In the presence of shear, however, the first term on the right-hand side of (41) plays a crucial role in eliminating all but a large-scale patchiness. Consider the critical size $(L_x)_c$ which is given rise to by the balance between the growth rate and longitudinal diffusion rate, i.e.

$$(L_x)_c = \pi \left\{ \frac{K_n (\varepsilon \Omega_1^2 / \omega^2 + 1)}{\alpha} \right\}^{\frac{1}{2}} \qquad (44)$$

From Figures 2 and 3, the amplitude of shear in the across flow and the dominant frequency of oscillation are estimated to be $\Omega_1 = 4.10^{-2}$/sec and $\omega = 1.4 \ 10^{-4}$/sec, so that $\Omega_1^2 / \omega^2 \sim 8 \cdot 10^4 = 0(10^5)$. The horizontal and vertical diffusivities are assumed to be $K_h \sim 10^4$ cm^2/sec, $K_z \sim 10$ cm^2/sec, respectively, so that $\varepsilon = 0(10^{-3})$. The phytoplankton growth rate is taken to be $\alpha = 0(10^{-5}$/sec), i.e. 1 div/day. Using these values in (44), we obtain

$$(L_x)_c \sim 10 \text{ km} \qquad (45)$$

Thus the critical patch size in the across Sound direction is about 10 km, which is one order of magnitude larger than that of (43).

ACKNOWLEDGEMENTS

This work was supported by the National Science Foundation under Grant OCE-7610730.

REFERENCES

Bowman, M.J. and W.E. Esaias. 1977. Fronts, jets, and phytoplankton patchiness. In: Bottom Turbulence, pp. 255-268. J. Nihoul (ed.). Elsevier Oceanography Series 19, NY.

Fischer, H.B., E.J. List, R.C.Y. Koh, J. Imberger and N.H. Brooks. 1979. Mixing in Inland and Coastal Waters. Academic Press, NY. 483 p.

Kierstead, H. and L.B. Slobodkin. 1953. The size of water masses containing plankton bloom. J. Mar. Res. 12: 141-147.

Okubo, A. 1980. Diffusion and ecological problems: mathematical models. Springer-Verlag, Berlin and New York. 254 p.

Skellam, J.G. 1951. Random dispersal in theoretical populations. Biometrika 38: 196-218.

Wilson, R.E. and A. Okubo. 1978. Longitudinal dispersion in a partially mixed estuary. J. Mar. Res. 36: 427-447.

Wilson, R.E., A. Okubo and W.E. Esaias. 1985. Observations of the structure of chlorophyll a in central Long Island Sound. This volume.

EMPIRICAL MODELS OF STRATIFICATION VARIATION
IN THE YORK RIVER ESTUARY, VIRGINIA, USA

D. Hayward, L.W. Haas, J.D. Boon, III, K.L. Webb and K.D. Friedland
Virginia Institute of Marine Sciences
Gloucester Point, VA 23062

INTRODUCTION

The influence of tides on estuarine processes has long been of interest to estuarine hydrographers and biologists. Recently it was shown that moderately stratified subestuaries of the lower Chesapeake Bay undergo periodic destratification (i.e. surface to bottom salinity differences less than 1 per mille) which correlate with the occurrence of high spring tides (Haas, 1977). In the York River, destratification events were reported to persist for up to three days and result in near vertical homogeneity. Subsequent observations indicated that destratification is preceded by the influx of less saline water into the mouth of the River, an occurrence suggested as instrumental in the subsequent destratification processes (Hayward et al., 1982).

To date, studies of the spring tide associated destratification phenomenon in the York River have related salinity difference with tidal height and predicted tides. Measurements of salinity difference have lacked either sufficient frequency (i.e. >=0.25 per day) or duration (i.e. >=1 year) to properly address the potential effect of non-predictable and seasonal environmental factors such as fresh water flow and wind on the rapid destratification process in this estuary. The availability of a longer term set of observations of sufficient frequency has permitted us to analyze the effects of several factors on the process of vertical mixing in the lower York River. Factors examined include observed and predicted tidal ranges and heights and nontidal sea level at Gloucester Point and Hampton Roads, water temperature, wind speed and direction, and fresh water discharge of the York and Rappahannock Rivers.

This work was conducted as part of a study of variations in phytoplankton population and productivity in the lower York River. The purposes of the models were primarily to provide a basis for interpolation of values for salinity difference between samples and

Lecture Notes on Coastal and Estuarine Studies, Vol. 17
Tidal Mixing and Plankton Dynamics. Edited by J. Bowman, M. Yentsch and W. T. Peterson
© Springer-Verlag Berlin Heidelberg 1986

estimate SMLD for the extent of the study, and secondarily to aid in predicting stratification conditions for future studies.

METHODS

From February 12, 1982 to April 4, 1983, water samples were collected at 1,3,5, and 12 m depths from a pier at the United States Coast Guard Reserve Training Center at Yorktown, Virginia (Fig. 1). This pier provided land based access to both surface and subhalocline water in the lower York River throughout the entire year. Samples from the top three depths were collected with a closing water bottle while samples from 12 m were collected with a hand vacuum pump attached to a plastic tube with the inlet located approximately one meter above the bottom. Samples were stored in glass bottles until analyzed with a Beckmen Induction Salinometer, Model RS-7B, which provides a precision of 0.01 per mille. All samples were collected at slack before ebb and sampling frequency varied from 0.29 to 1.0 per day and included 157 observations over a period of 434 days. The station was also occupied on 13 occasions from 31 August to 30 September 1983 to collect data for a verification study. Wind speed and direction were taken from the Local Climatological Data, Monthly Summary, for Norfolk, Virginia, International Airport, 45 km from the sample site (Fig. 1). Resultant direction and resultant speed were used (NOAA, 1982, 1983). Fresh water river flow data were supplied by the U.S. Geological Survey in Richmond, Virginia, and were from gauging stations on the Pamunkey and Mattaponi Rivers, primary tributaries of the York River, about 120 km upstream of the sampling site, and the Rappahannock River near Fredericksburg, Virginia (Fig. 1). The Rappahannock River flow was included because of the possibility that the influx of less saline water to the Chesapeake Bay near the mouth of the York River might influence stratification within the York River (Hayward et al., 1982). Both the Pamunkey and Mattaponi records contained sequences of anomalous or missing data. In order to generate a complete record for York River flow, estimates of Mattaponi flow were made from Pamunkey flow and vice versa to fill in. The regression model explained 60% of the variation in 318 daily mean flows. Estimates were necessary for 116 of 868 daily mean flows.

Observed daily nontidal sea levels and tidal ranges were from the tidal record collected at the Virginia Institute of Marine Science (VIMS) located 1.9 km from the sampling site (Fig. 1). Tidal

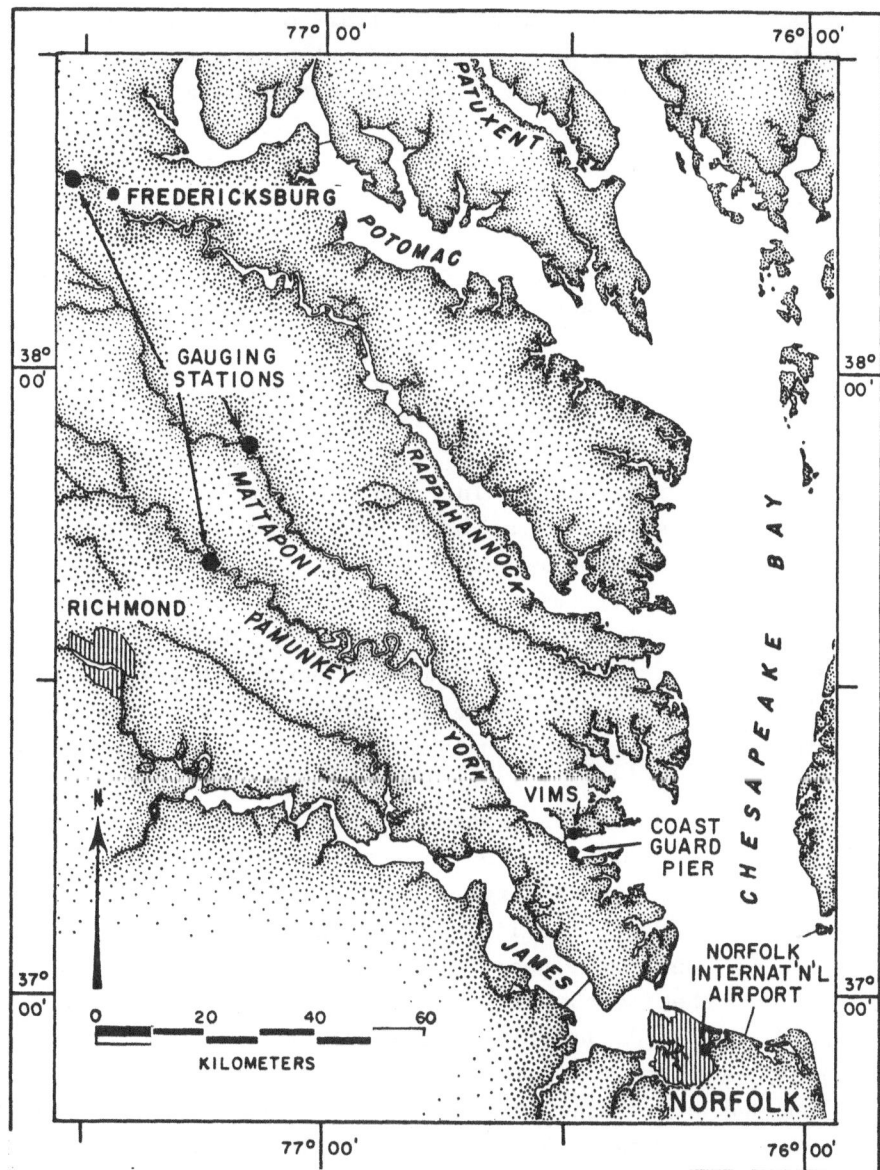

Figure 1. Regional map showing locations of: A. Sampling Station; B. Tide gauge station; C. Norfolk International Airport; and flow gauges for D. Pamunky, E. Mattaponi and F. Rappahannock Rivers.

heights were recorded at 6 minute intervals on a Fischer-Porter tide gauge located at the end of the VIMS pier at Gloucester Point. After the records are edited for anomalous and missing data, hourly heights were produced using a 5 point moving average with a one hour step interval, effectively removing high frequency "noise" while leaving tidal frequencies at virtually full response. Accuracy was maintained through frequent time and height comparisons between recorded data and external reference readings and was normally better than ± 0.02 m. Predicted tidal heights were calculated using procedures described in Boon and Kiley (1978). Daily nontidal sea levels were calculated from predicted and observed hourly heights using the filter proposed by Doodson and Warburg (1941), centered on 1200 hours of each day.

The dependent variables for the correlation and regression analyses were measures of salinity stratification calculated as either (1) the difference between the mean of the salinities at 1,3, and 5 m and the salinity at 12 m or (2) the difference between the salinity at 1 m and the salinity at 12 m. The first was more strongly correlated and unless otherwise noted will be the value intended when salinity difference is indicated. The data (Fig. 2) clearly indicate strong fortnightly as well as seasonal variation.

All aspects of the tidal signal used as factors in the model were assessed on a daily basis to correspond with the comparable values for wind and fresh water flow. Mean daily tidal range was calculated by summing two flood tide ranges (low height to high height) and two ebb tide ranges (high height to low height) and dividing by four. When necessary, immediately preceding or immediately following low or high heights from the previous or following day, respectively, were used to make a full complement of ranges. Mean daily flood and ebb ranges were calculated using the appropriate two of those four. Extreme daily ranges were selected as the greatest of the ranges in a given category. If only one range in a category was completely within a given day, that range was used. Mean and extreme daily high and low tidal heights were calculated or selected, respectively, from the appropriate heights when two were available. If only one occurred, that one was selected. The resulting tidal factors were lagged 0 to 12 days for correlation analysis. Unweighted moving averages of 2 to 7 days lagged 0 to 5 days were also subjected to correlation analysis.

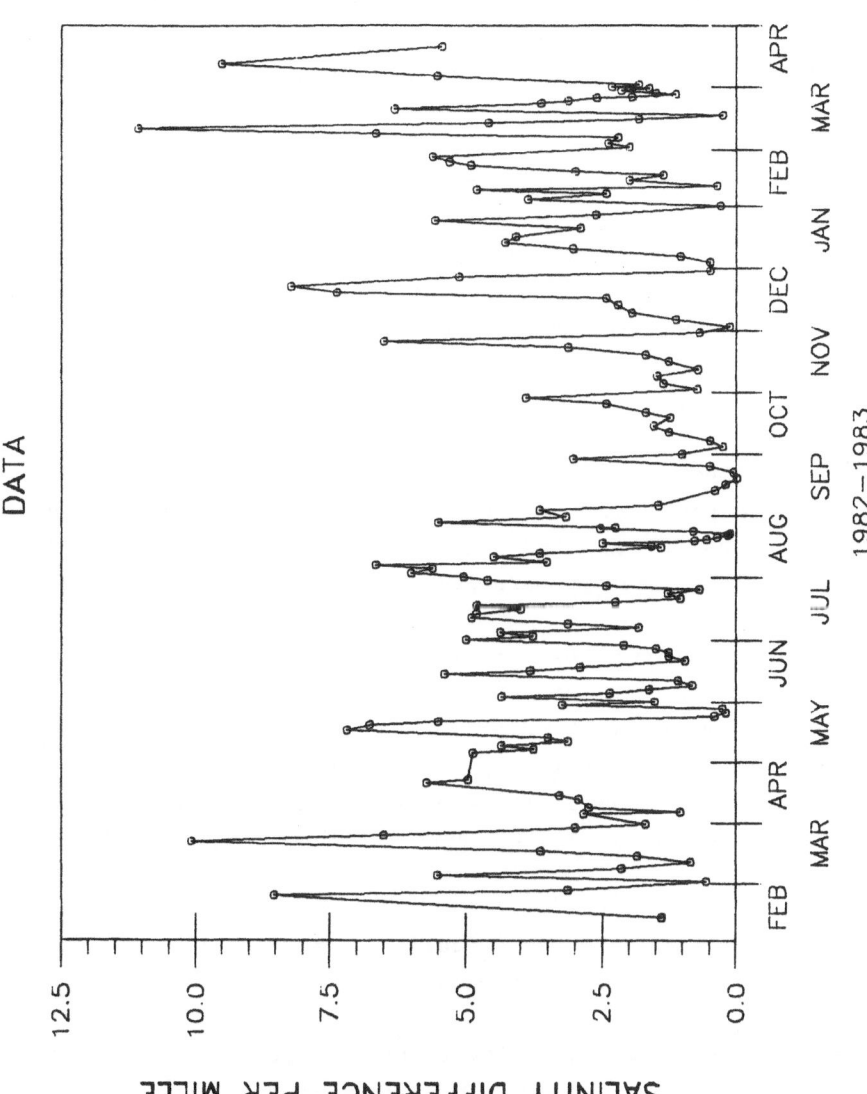

Figure 2. Line graph of salinity differences per mille from February 1982 to April 1983. Circles indicate data points.

Because of the possibility that wind induced large scale water movement in the Chesapeake Bay near the river mouth, as well as direct stress of wind axial to the river, might influence stratification within the river (Hayward et al., 1982), calculated stress of wind from nonaxial as well as axial directions was examined for relation to stratification variation. A directional wind stress for each day and each direction was calculated by weighting the square of the resultant daily wind speed by a directional coefficient based on the vector component of the resultant direction in the direction of interest raised to the fourth power. Powers 1 through 6 were tested; 4 was found to give the best model fit. This directional coefficient weighted winds at the direction of interest by a factor of 16 and wind 180° from the direction of interest by 0. The directional weighting was appropriate because of the asymmetry of geomorphology of the river basin, the Chesapeake Bay and their connections and the assumed corresponding asymmetry of fetch. Directional coefficients were calculated at 15 degree intervals from 0 degrees as well as axially to the lower and upper river basins. The axial directions of the upper river basin were considered to be 140° and 320°, and the lower basin to be 83° and 263° after Kiley (1980). The resulting weighted wind stress factors were lagged 0 to 9 days for analysis.

Correlation analysis of fresh water flow was performed with flows lagged 0 to 30 days for the York River and 0 to 80 days for the Rappahannock River, as well as with flows categorized by value and lagged. The results suggested that lag was related to flow volume logarithmically, therefore fresh water flow factors used in the model were generated using variable lags calculated by the following algorithm:

$$\text{Variable Lag} = \text{Maximum Lag} - \text{Scaling Factor} * \text{Log 10}$$
$$(\text{Fresh Water Discharge m}^3 \text{ s}^{-1}) \tag{1}$$

The Maximum Lag vs. Scaling factor space was systematically searched for optimum correlation with the residuals from a model containing tidal and wind stress terms. A variable lag was calculated from the flow for each calendar day and added to that day to determine an effective date for that flow. All flows for each effective date were summed and a five-day moving average was calculated to smooth the resulting signal. Descriptive statistics for the unlagged and lagged fresh water flows and other factors used in modeling are presented in Table 1.

Initial correlation analyses and multiple regression analyses were performed using SPSS procedures (Nie et al., 1975; Hull and Nie, 1981). The search for optimum correlation between calculated effective fresh water flow and salinity difference was a grid search programmed by the senior author. The fit of SMLD to salinity difference was calculated using the gradient-expansion algorithm for a least-squares fit of a non-linear function found in Bevington (1969).

Salinity difference is limited at the low extreme by zero and essentially unlimited in the upper extreme; tidal range as well as nontidal sea level were negatively correlated with salinity difference but have very similar limit characteristics. That is, exponentially increasing values of salinity difference were associated with predictor values approaching zero almost asymptotically. In order to overcome this numerical sticking point, these tidal factors were scaled and complemented, i.e.

$$\text{predictor} = 1 - (\text{factor value} / \text{scaling constant}) \qquad (2)$$

where the scaling constants were close to the maximum values for these factors found in the data and were estimated by iterative approximation. The predictors thereby generated tend to keep the high order transforms very close to zero over a broad region. This provides a good approximation of the nearly asymptotic zero limit for salinity difference while producing very large values in association with very small tidal ranges.

The process for generating the regression model included 1) initially setting high (P = 0.2) probability levels and presenting range and sea level factors to the SPSS stepwise selection procedure. Terms were raised to higher powers as long as the fit was improved at that probability level. 2) Using the terms selected, and adding the wind terms, the probability level was lowered (P = 0.1) and the backward elimination procedure was used. 3) Finally, the power to which each of the range and sea level terms was raised was varied and changes which improved fit were kept.

Table 1. Descriptive Statistics for Factors Included in Modeling.

Factor	Maximum	Minimum	Mean
Salinity difference	11.06	0.01	2.929
Rappahannock River flow, unlagged	515.4	4.2	42.2
Rappahannock River flow, variably lagged	256.2	3.8	40.4
York River flow, unlagged	318.5	3.2	49.3
York River flow, variably lagged	200.0	1.1	43.1
Observed mean sea level	1.62	0.46	0.889
Predicted mean sea level	0.40	0.16	0.357
Observed daily extreme tidal range	1.20	0.45	0.816
Predicted daily extreme tidal range	1.13	0.49	0.795
Temperature in surface mixed layer	28.8	2.9	16.2
Resultant wind speed	46.0	1.0	14.6

Units are as follows: Salinity difference, per mille; Fresh water flows, $m^3 s^{-1}$; Sea level, meters above mean low water; Tidal range, meters; Temperature, $^{\circ}C$; Wind speed, kmh^{-1}. Tidal observations are for Gloucester Point.

RESULTS

The results of the initial correlation analyses are presented in Table 2. Aspects of tidal range and height were examined. Lagged daily values were generally correlated at about -0.60, with a lag of around 2-3 days. Notable exceptions are observed mean and extreme high tide height values from Gloucester Point (GP.OMHH and GP.OXHH) which correlated best at 0 and 1 days, respectively, with much lower correlations of -0.26. Apparently the noise from non-astronomical factors strongly masked any predictive capacity in this signal. Also of interest is the positive correlation with predicted extreme low tide height at Hampton Roads (HR.PXLH). Low tides are associated with high tidal ranges which are associated with lower salinity differences.

The best correlations are generally with unweighted moving averages 7 days long, centered on day 3. Shorter spans of 4 and 3 days are centered on days 2-3. The strongest associations overall are seen in the extreme observed and predicted ranges at Gloucester Point (GP.OSR and GP.PXR) with 7 day averages centered on day 3, giving correlations of -0.675 and -0.672, respectively. These factors were therefore selected for use in the regression models. The predicted tidal range signal for the period of the study is shown in Fig. 3. Observed tidal factors produced models which fit the data much less well than those constructed with predicted values under the same constraints of significance (Table 3), and so were not used in generating the models combining wind factors with tides.

The association of the salinity difference with tidal range was found to be non-linear. Models constructed from simple lagged extreme tidal range and nontidal sea level were associated with central values fairly well, producing R^2 values of about 0.50, but tended to severely underpredict the large positive extremes (Table 3). The use of higher order transforms of lagged range and nontidal sea level improved the overall fit by more than 15% and particularly improved the estimates of the high extremes while slightly degrading the low extreme estimates (Table 3). Models constructed with high order transforms of the scaled and complemented predictors produced an additional improvement of about 20% in the fit and very good prediction of the highest salinity differences (Table 3).

Table 2. Best simple correlations of salinity differences with non-averaged and unweighted moving averages of various tidal aspects lagged by various amounts.

Tidal aspect	Salinity Difference I (N=157) Non-Averaged		Salinity Difference I (N=157) Unweighted Average			Salinity Difference II (N=147) Non-Averaged		Salinity Difference II (N=147) Unweighted Average		
	Lag	Corr	Lag	Aver	Corr	Lag	Corr	Lag	Aver	Corr
GP.OMR	3	-0.603	0	7	-0.655	3	-0.598	0	7	-0.646
GP.PMR	3	-0.637	0	7	-0.655	2	-0.616	1	4	-0.623
HR.PMR	3	-0.632	0	7	-0.651	3	-0.612	1	4	-0.627
GP.OXR	4	-0.601	0	7	-0.675	1	-0.541	0	7	-0.648
GP.PXR	3	-0.641	0	7	-0.671	3	-0.619	0	7	-0.648
HR.PXR	3	-0.622	0	7	-.0648	3	-0.600	0	7	-0.625
GP.OMFR	2	-0.592	0	7	-0.662	2	-0.610	0	7	-0.654
GP.PMFR	3	-0.633	0	7	-0.653	3	-0.612	0	7	-0.630
HR.PMFR	3	-0.631	0	7	-0.650	3	-0.611	0	7	-0.625
GP.OXFR	4	-0.622	0	7	-0.675	2	-0.574	0	7	-0.654
GP.PXFR	2	-0.561	0	7	-0.656	2	-0.611	0	7	-0.637
HR.PXFR	4	-0.613	0	7	-0.654	2	-0.600	0	7	-0.629
GP.OMER	3	-0.588	0	7	-0.642	3	-0.588	0	7	-0.630
GP.PMER	3	-0.641	0	7	-0.656	3	-0.619	1	4	-0.635
HR.PMER	3	-0.633	0	7	-0.651	3	-0.614	1	4	-0.630
GP.OXER	4	-0.594	0	7	-0.658	3	-0.574	0	7	-0.632
GP.PXER	3	-0.646	0	7	-0.670	3	-0.626	0	7	-0.647
HR.PXER	3	-0.628	0	7	-0.656	3	-0.613	0	7	-0.639

Table 2. Continued.

Tidal aspect	Salinity Difference I (N=157)					Salinity Difference II (N=147)				
	Non-Averaged		Unweighted Average			Non-Averaged		Unweighted Average		
	Lag	Corr	Lag	Aver	Corr	Lag	Corr	Lag	Aver	Corr
GP.OMHH	1	-0.216	0	7	-0.231	0	-0.253	0	7	-0.246
GP.PMHH	3	-0.615	2	3	-0.613	2	-0.601	1	4	-0.598
HR.PMHH	2	-0.654	1	4	-0.656	2	-0.634	1	4	-0.630
GP.OXHH	1	-0.205	0	7	-0.247	1	-0.266	0	7	-0.258
GP.PXHH	3	-0.626	2	3	-0.626	3	-0.614	1	4	-0.613
HR.PXHH	3	-0.648	1	4	-0.662	2	-0.627	1	4	-0.643
HR.PXLH	4	0.221	3	2	0.219	3	0.216	2	3	0.212

Salinity difference I = salinity at 12 m; mean of salinity at 1,3, and 5 m.

Salinity difference II = salinity at 12 m; salinity at 1 m.

Tidal aspect legend: GP = Gloucester Point, HR = Hampton Roads, O = Observed, P = Predicted, M = Daily mean, X = Daily extreme, F = Flood, E = Ebb, H = High or Height, L = Low, R = Range, i.e. tidal height difference between a low and a subsequent high or vice versa as appropriate.

Consecutive averages are unweighted moving averages ending on the day designated as Lag and including the number of days designated as Aver, e.g. those designated (0|7) indicate an average of the 7 consecutive days ending with the day on which the salinity measurements were taken, (2|3) indicates a 3 day average ending 2 days before the salinity measurement.

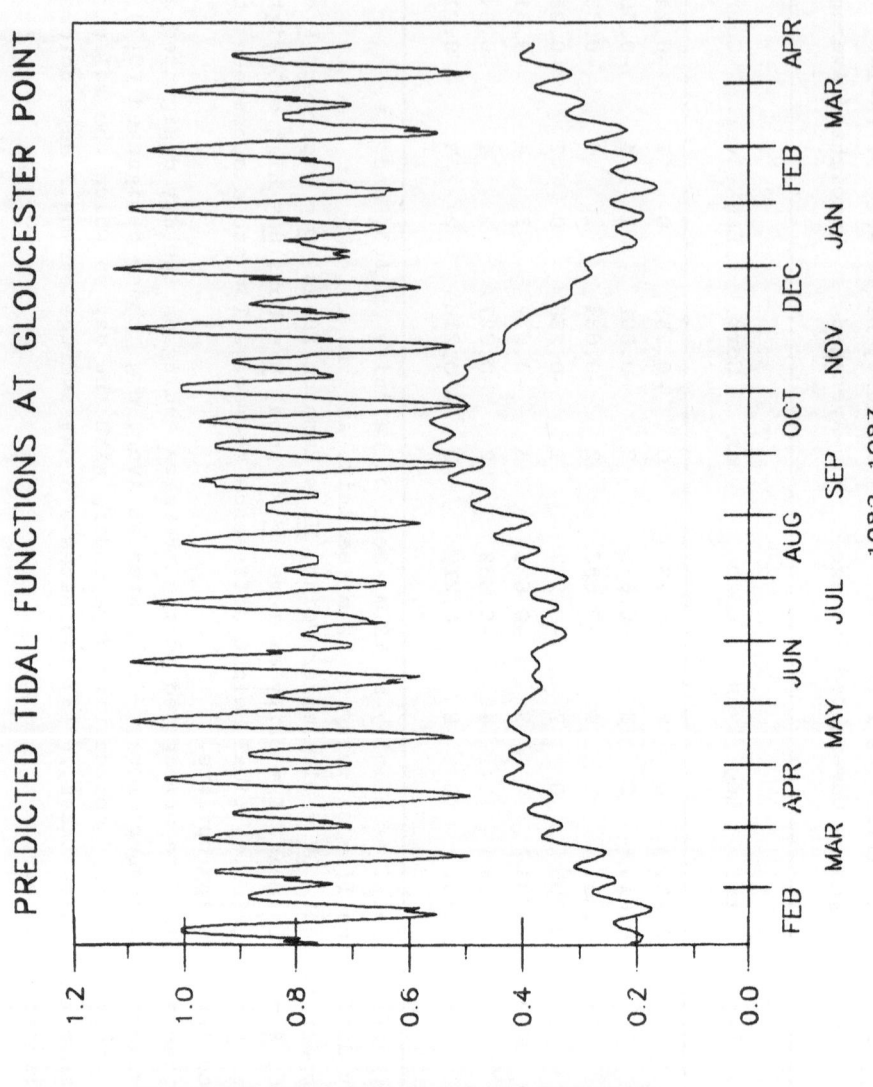

Figure 3. Predicted tidal functions at Gloucester Point. The upper line represents daily extreme tidal range. The lower line represents daily nontidal sea level.

Table 3. Comparison of the results of various models.

| Model Description | Predicted Salinity Difference | | R^2 | RMSE |
	Maximum	Minimum		
Predicted Tidal Aspect Models:				
Range only model:				
Complemented, high order	7.69	0.41	0.50	1.55
Range and mean sea level models:				
Uncomplemented, first order	6.53	-0.51	0.51	1.55
Uncomplemented, high order	7.26	-0.72	0.59	1.43
Complemented, high order	10.24	-0.61	0.67	1.26
Range, mean sea level, and wind:				
Complemented, high order	11.09	-1.01	0.80	0.99
Observed Tidal Aspect Model:				
Range and mean sea level:				
Complemented, high order	10.03	-0.03	0.59	1.46

A model composed exclusively of predicted tidal range factors is associated with 50% of the variability (Table 3) and reproduces the dominant fortnightly pattern of variation in salinity difference (Fig. 4a,c). It does not, however, reflect the seasonal variations well. Possible candidate predictors for this seasonal variation were water temperature and mean sea level (Fig. 3). Water temperature was found to be essentially unrelated to salinity difference, while the inclusion of nontidal sea level terms produced a 15% improvement in model fit and brought the predicted seasonal variation into very close accord with the observed values (Table 3, Fig. 4b,d).

Wind stress factors which correlated most strongly with salinity differences were those axial to the upper and lower York River basins and from the North and Northeast. Wind blowing axially down the upper basin lagged 0-2 days is positively correlated with salinity difference possible indicating increased flow of surface water into the lower river. This effect was found by Kiley (1980) in the correlation of wind stress and current meter data. Wind from the North, Northeast, axial to the lower river, or blowing up the upper river generally show negative correlations with salinity difference when lagged 0-2 days; this indicates direct mixing and possibly the enhanced transport of fresher water into the river from the region of the Chesapeake Bay just northeast of the river mouth, consequently reducing or reversing the horizontal salinity gradient in the York River (Hayward et al., 1982). Three-day lags show some reversals in correlation, possibly indicating the presence of some rebound phenomenon, i.e. the river emptying after a wind driven filling episode or vice versa.

The addition of wind stress terms enabled the model to make much better predictions of some obvious anomalies. For example, in March 1983 (Fig. 5a,b), where the residual is reduced from -4.59 to 0.50 per mille by the addition of the effect of 40 km h^{-1} northeast wind event which occurred on the previous day. The inclusion of wind stress terms improved the overall fit by 20%.

In spite of efforts to relate fresh water flow to salinity difference, no significant correlation could be found. The best correlation for York River flow was found with a maximum lag of 24.18 days and a scaling factor of 4.12. The longest lag generated from

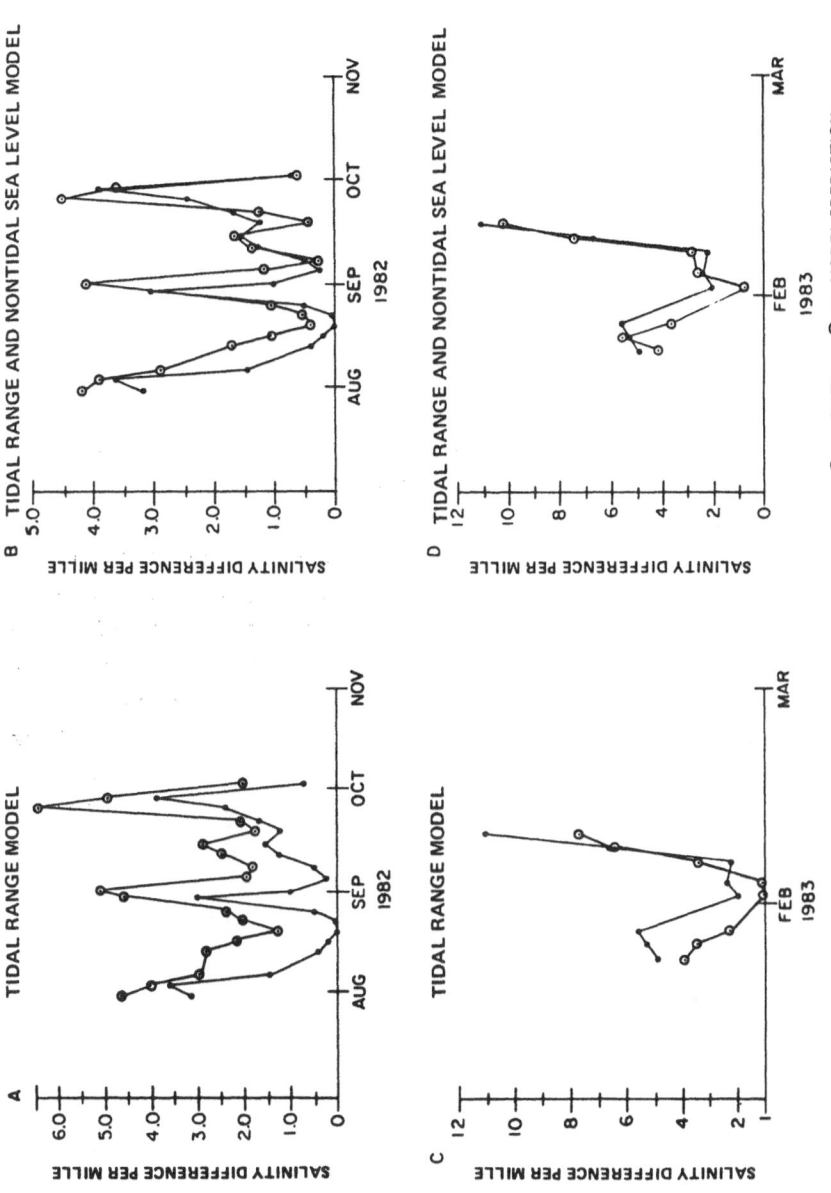

Figure 4. Comparison of model predictions using only tidal range factors with predictions using both tidal range and nontidal sea level factors. The top figures, 4A and 4B, illustrate the improvement for the fall season while the bottom figures, 4C and 4D, illustrate the improvement for spring, demonstrating the seasonal influence of nontidal sea level.

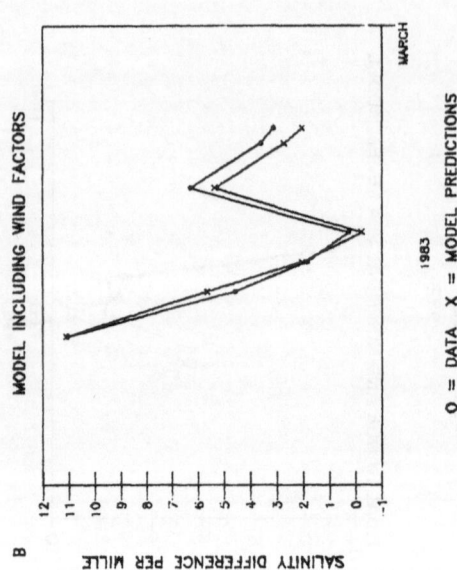

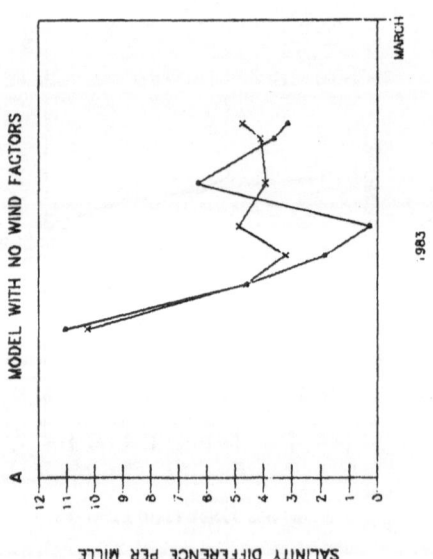

Figure 5. Comparison of model predictions using only tidal factors with those using both tidal and wind factors. The improvement in fit demonstrates the influence of wind in cases which cannot be predicted by tidal factors.

the data was 22 days, the shortest was 14 days, and the lag of the mean value was 17 days. The variation in salinity difference associated with York River fresh water flow as <0.1% and the significance of the association was P > 0.8; therefore, it was not included as a term in the complete model. The best correlation for the Rappahannock River flow was found with a maximum lag of 55.01 days and a scaling factor of 7.32. The longest lag generated from the data was 50 days, the shortest was 35 days, and the lag of the mean value was 43 days. The variation in salinity difference associated with Rappahannock River fresh water flow was <0.25% and the significance of the association was P > 0.2, and it also was not included in the final model. This lack of association in spite of flow variations over two orders of magnitude may be due to the low absolute magnitude of the flow. Hydrodynamic model studies of the James River (C. Cerco, pers. comm.) indicate that the maximum flows recorded in the York River system during this study would not be large enough to cause changes in salinity difference sufficient to produce significant association.

The final model (Fig. 6) includes 25 terms: 7 tidal range terms, 8 mean sea level terms, and 10 wind stress terms. The tidal terms are all individually significant at $P < 0.02$, and the wind terms at $P < 0.05$. An F statistic for the model was calculated, $F(4.07, 77, 79)$, $P < 1.0e-8$. Coefficients for the model are presented in Table 4. The form of the model is

$$SD(d) = C + \sum_i a_i \ R(d-L1_i)^{b_i} + \sum_j c_j \ M(d-L2_j)^{e_j} +$$
$$\sum_k f_k \ V(d-L3_k)^2 \ 1 + \cos D(d-L3_k)^4 \qquad (3)$$

where

$SD(d)$ = Salinity difference on the day of interest.

C = Constant.

a_i = Regression coefficient for the ith range constituent.

$R(d-L1_i)$ = Scaled and complemented predicted extreme tidal range on the day $L1_i$ days before the day of interest, i.e.
$$1 - \frac{Range}{0.98}$$

$L1_i$ = Lag for the ith range constituent.

b_i = Power to which the ith range constituent is raised.

c_j = Regression coefficient for the jth sea level constituent.

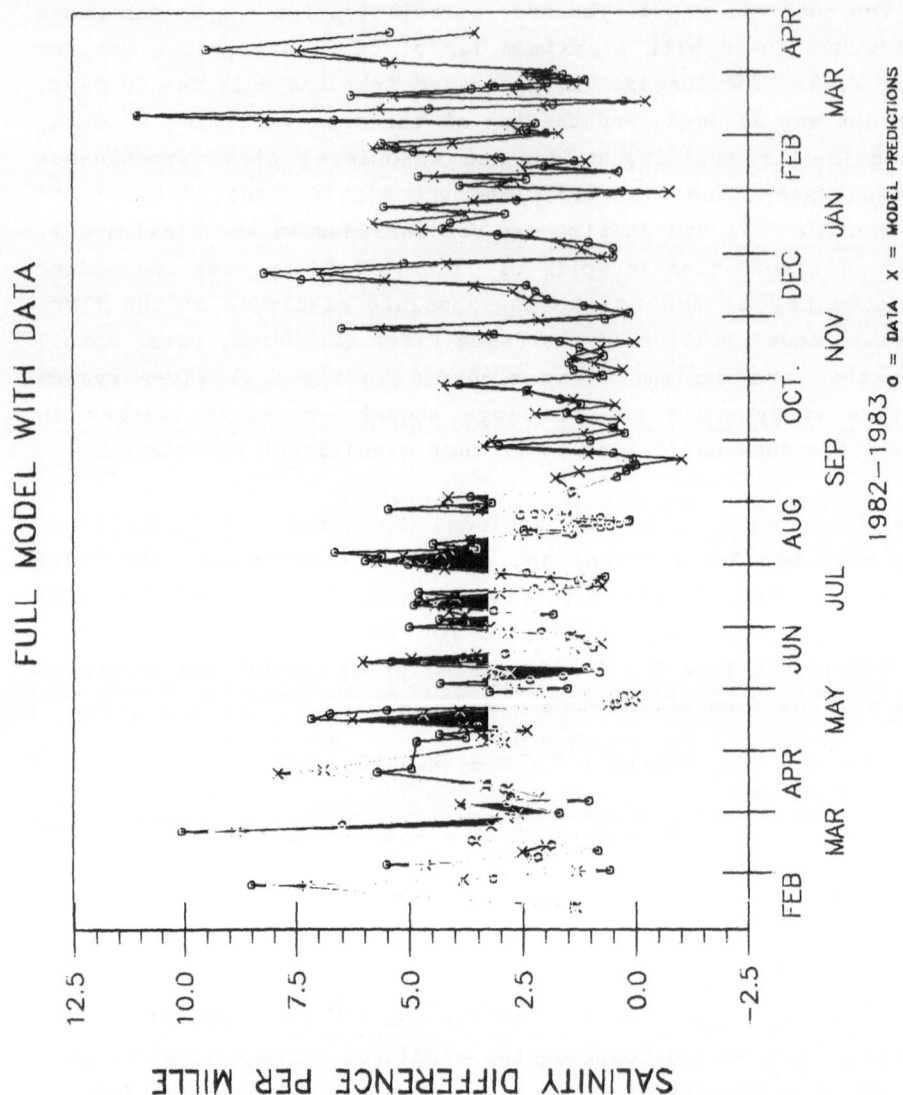

Figure 6. Comparison of all data points with corresponding values predicted by the full model.

Table 4. Coefficients for terms included in the full model.

Range Terms:				Significance:
i	$L1_i$	b_i	a_i	
1	5	1	7.59704	0.0000
2	0	1	10.27152	0.0000
3	0	3	-59.63675	0.0000
4	1	5	609.55842	0.0001
5	6	19	0.1139 E+07	0.0045
6	1	9	-5411.37775	0.0049
7	2	4	-66.21355	0.0165

Sea Level Terms:				
j	$L2_j$	e_j	c_j	
1	4	3	257.01694	0.0000
2	5	3	-413.37466	0.0000
3	6	1	5.62696	0.0000
4	6	3	160.48822	0.0000
5	0	7	-295.14167	0.0000
6	2	7	-885.06131	0.0004
7	1	6	721.60371	0.0014
8	4	6	173.99482	0.0072

Wind Stress* Terms:				
k	$D(d-L3_k)$	$L3_k$	f_k	
1	83°	1	-0.23072 E-03	0.0000
2	45°	3	-0.60460 E-03	0.0007
3	320°	0	0.21325 E-03	0.0018
4	0°	0	-0.14609 E-03	0.0029
5	0°	3	0.40860 E-03	0.0036
6	263°	2	-0.16598 E-03	0.0038
7	83°	2	-0.44177 E-03	0.0042
8	0°	2	-0.10743 E-03	0.0280
9	320°	3	-0.20834 E-03	0.0346
10	320°	2	0.1557 E-03	0.0477
Constant			-0.15819	0.5198

*Wind stress coefficients assume wind speed in km h^{-1}.

$M(d-L2_j)$ = Scaled and complemented predicted mean sea level on the day $L2_j$ days before the day of interest, i.e.

$$1 - \frac{Level}{0.98}$$

$L2_j$ = Lag for the jth sea level constituent.

e_j = Power to which the jth sea level constituent is raised.

f_k = Regression coefficient for the kth wind stress constituent.

$V(d-L3_k)$ = Resultant wind speed on the day $L3_k$ days before the day of interest.

$D(d-L3_k)$ = Resultant wind direction on the day $L3_k$ days before the day of interest.

$L3_k$ = Lag for the kth wind stress constituent.

Salinity difference is valuable to biologists as an indirect indicator of turbulent mixing above the halocline. Another parameter of interest which might be estimated from salinity difference is the depth to which that mixing extends. We have defined SMLD as the first depth at which a salinity gradient of 0.2 per mille per 0.1 m is encountered. Using a set of data limited to those days for which we have salinity profiles with 0.1 m depth resolution (N = 11), we modeled SMLD as a function of salinity difference. The following equation accounts for 81% of the variation in SMLD at a significance level of P < 0.1 (Fig. 7).

$$SMLD = exp(3.0666 - 0.6064 * \text{salinity difference}^{0.6528}) \qquad (4)$$

For a salinity difference of 0 per mille this function predicts a maximum mixing depth of 21.47 m which is approximately the maximum depth of the lower York River. For the mean salinity difference found in this study, 2.929 per mille, it predicts a SMLD of 6.32 m.

CONCLUSIONS

While neither of these models can be considered causal, the strong associations which have been shown demonstrate their broad usefulness. As the major portion of variation can be predicted from astronomical tide predictions, salinity difference, and secondarily, SMLD, can be predicted for the indefinite future. This predictive capability will be helpful for water quality managers concerned with predicting pollution impact as well as for experimentally-oriented

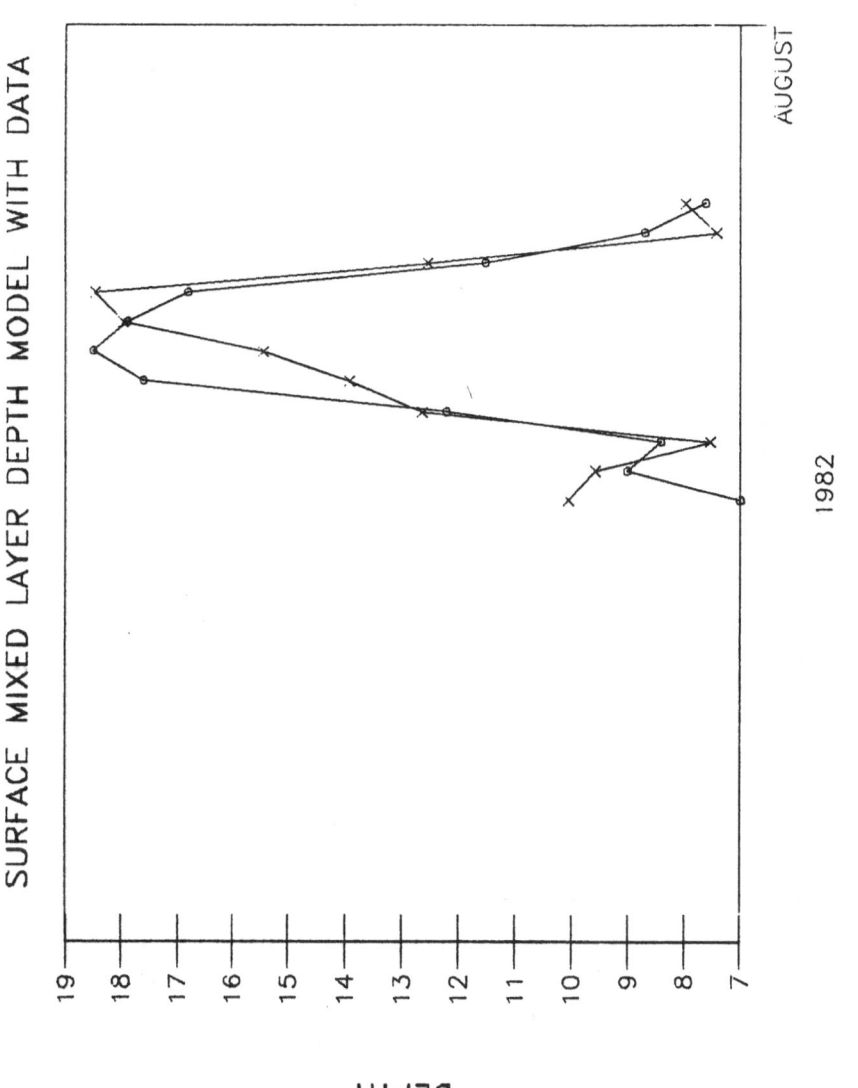

Figure 7. Comparison of surface mixed layer depth model predictions with data values.

estuarine scientists concerned with surface mixed planktonic phenomena. With the addition of wind stress factors, even more precise
hindcasting can be accomplished which will enable managers to assess
major changes in the mixing pattern.

ACKNOWLEDGEMENTS

Contribution No. 1256 from the Virginia Institute of Marine
Science, College of William and Mary. This research was supported in
part by NSF Grant OCE8026030 (to K.L. Webb). We thank Donald Wright,
Evon Ruzecki and the editors for critically reviewing the manuscript.

REFERENCES

Bevington, P.R. 1969. Data Reduction and Error Analysis for the
    Physical Sciences. McGraw-Hill, NY. 336 p.

Boon, J.D. III and K.P. Kiley. 1978. Harmonic analysis and tidal
    prediction by the method of least squares, a user's manual.
    Special Rept. No. 186 in Appl. Mar. Sci. and Ocean. Engn. of the
    Virginia Inst. Mar. Sci.

Doodson, A.T. and H.D. Warburg. 1941. Admiralty Manual of Tides.
    His Majesty's Stationery Office, London, pp. 110-112.

Haas, L.W. 1977. The effect of the spring-neap tidal cycle on the
    vertical salinity structure of the James, York and Rappahannock
    Rivers, Virginia, U.S.A. Estuar. Coast. Mar. Sci. 5: 485-496.

Hayward, D.M., C.S. Welch and L.W. Haas. 1982. York River destrati-
    fication: an estuary-subestuary interaction. Science 216:
    1413-1414.

Hull, C.H. and N.H. Nie. 1981. SPSS Update 7-9. McGraw-Hill, NY.

Kiley, K.P. 1980. The relationship between wind and current in the
    York River Estuary, Virginia, April 1973. M.A. Thesis, College
    of William and Mary, 195 pp.

Nie, N.H., C.H. Hull, J.G. Jenkins, K. Steinbrenner, D.H. Bent.
    1975. Statistical Package for the Social Sciences. McGraw-
    Hill, NY.

NOAA. 1982, 1983. Local climatological data, monthly summary for
    Norfolk, VA. National Climatic Data Center, Ashville, NC.

TEMPORAL AND SPATIAL SEQUENCING OF DESTRATIFICATION
IN A COASTAL PLAIN ESTUARY*

E.P. Ruzecki and D.A. Evans
Virginia Institute of Marine Science
School of Marine Science
College of William and Mary
Gloucester Point, VA 23062

INTRODUCTION

Periodic destratification of the York River estuary of Virginia was noted by Haas (1977) to closely follow the fortnightly neap/spring cycling of tides. Continued investigations of this tributary to Chesapeake Bay prompted Hayward, Welch and Haas (1982) to formulate a conceptual model (henceforth referenced as the HWH model) of destratification-restratification based on limited observations in the lower York River. The HWH model progresses through the following steps:

1) Relatively fresh water is advected into the York River from Chesapeake Bay as a result of tidal current phase relationships and increased tidal excursion during spring tides.

2) This reduces (and may reverse) the horizontal pressure gradient which drives normal estuarine circulation thereby limiting the importation of relatively saltier bottom water from the bay.

3) Normal mixing processes enhanced by stronger currents during spring tides and the limited availability of saltier bottom water combine to produce a well-mixed water column.

4) Restratification of the water column starts when reduced post-spring tidal currents fail to advect fresher water into the York from the bay.

*Contribution No. 1255 from the Virginia Institute of Marine Science

Lecture Notes on Coastal and Estuarine Studies, Vol. 17
Tidal Mixing and Plankton Dynamics. Edited by J. Bowman, M. Yentsch and W. T. Peterson
© Springer-Verlag Berlin Heidelberg 1986

5) Reestablishment of the two layered estuarine circulation occurs which supplies higher salinity bottom water from the bay.

The hypothetical HWH model was tested, in part, by observing surface markers in the Chesapeake Bay Hydraulic Model of the U.S. Army Corps of Engineers, Stevensville, Maryland. The test showed advection into the mouth of the York from up-bay during spring tides but movement of markers through only 15 to 30 percent of the same distance during neap tide. To further test the HWH model, we extensively measured the salinity structure across Chesapeake Bay and along the axis of the York River during an eleven-day period bracketing spring tides in August 1982 and analyzed drogued buoy data taken in the hypothesized freshwater source region during the associated spring and neap tides. Our interest was in both the temporal/spatial variation of the destratification-restratification along the axis of the York River and across Chesapeake Bay and the intrusion of freshened water from the bay into the river.

Results of these measurements support the HWH model and are consistent with the hypothesis that destratification-restratification of more landward portions of the York River may be controlled by conditions other than those explicit in the model. We propose a companion model which may be applicable to broader regions than those of estuary-subestuary combinations discussed in the HWH model.

York River-Chesapeake Bay Sampling

Daily CTD casts were made at stations across Chesapeake Bay and along the axis of the York River from 16 to 26 August 1982 (Fig. 1). York River stations (Y1 to Y15) were sampled at predicted slack water before ebb tide (SBE), and on 17, 19 and 21 August, at local slack water before flood tide (SBF). Eight additional stations were occupied in the vicinity of New Point Comfort (stations A-H, Fig. 1) at approximately local SBF on 18 August, the day of maximum predicted tidal heights.

On 19 and 26 August drogued buoys were released along a line extending eastward from New Point Comfort to mid bay (Fig. 1, Stations 1-5) to coincide with local SBF during spring and neap tides. Buoys were tracked for half a tidal cycle and periodic surface salinity samples were taken at their locations.

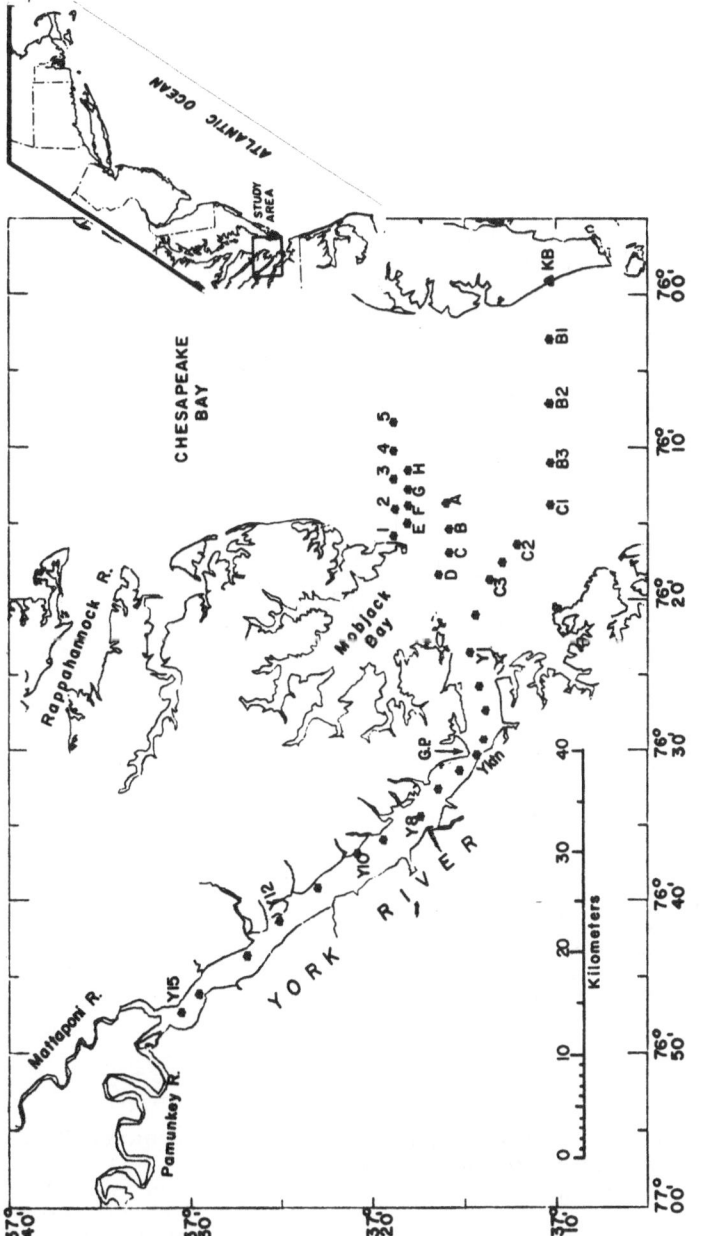

Figure 1. Location of stations occupied in Chesapeake Bay and the York River. Identified features are Gloucester Point (G.P.), Yorktown (Yktn) and Kiptopeke Beach (KB). New Point Comfort is at the tip of the peninsula separating Mobjack Bay from Chesapeake Bay.

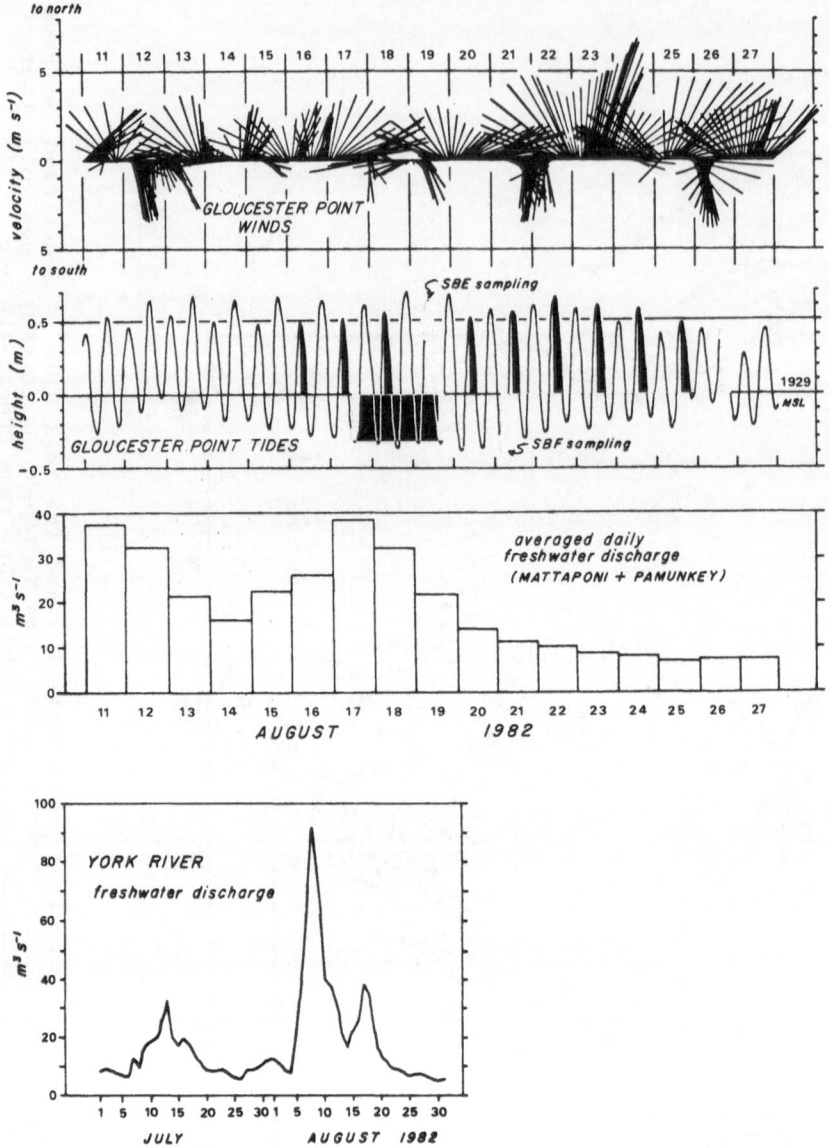

Figure 2. Variations of wind and tidal height at Gloucester Point and
streamflow prior to and during the study.

Variations of wind and tidal height at Gloucester Point (GP in
Fig. 1) and streamflow prior to and during the study are shown in
Figure 2.   Streamflow information was obtained from the U.S. Geolo-

gical Survey, Richmond, VA and is shown as daily sums for gauging stations on the primary tributaries to the York - the Mattaponi and Pamunkey Rivers, where gauging stations are 88 and 118 km from the head of the York, respectively.

RESULTS

Where applicable to a one or two dimensional interpretation, results are presented in a coordinate system where, in general, east is to the right, distance increases upstream, and depth increases downward. In figures showing isohalines, contour spacing is 1.0 unit with half unit isohalines occasionally shown as dashed lines.

Daily Salinity Distributions

Salinity distributions for the cross-bay and York River axis sampling at SBE and for the York River axis at SBF are shown in Figure 3. Isohaline patterns for these figures suggest the study area consisted of three primary regions (U, D and B designated by diagonal and vertical hatching: $\setminus$, $/$, and $|$, respectively) and two intermediate regions (I and M designated by cross hatching). Within the York River:

> Region U is a shallow, upstream region, 20 to 50 km from the river mouth (Stations Y9-Y15), where thalweg depths are on the order of 10 m and isohalines had a generally sigmoid shape with longitudinal slopes on the order of 12×10^{-4}.

> Region D is the deep region between the river mouth (Station Y1 at km 0) and the Yorktown/Gloucester Point constriction at km 10 (Station Y5) where, during SBE, isohalines were generally horizontal with longitudinal slopes on the order of 5×10^{-4}.

> Region I is an intermediate region between U and D, 10 to 20 km from the mouth (Stations Y6-Y8) where isohaline structure was similar to that in Region D at SBE and Region U at SBF.

The two regions bayward of the mouth of the York River are:

> Region B which extends across the main stem of Chesapeake Bay 12 to 35 km east of the mouth of the York (Stations B1-C2). In

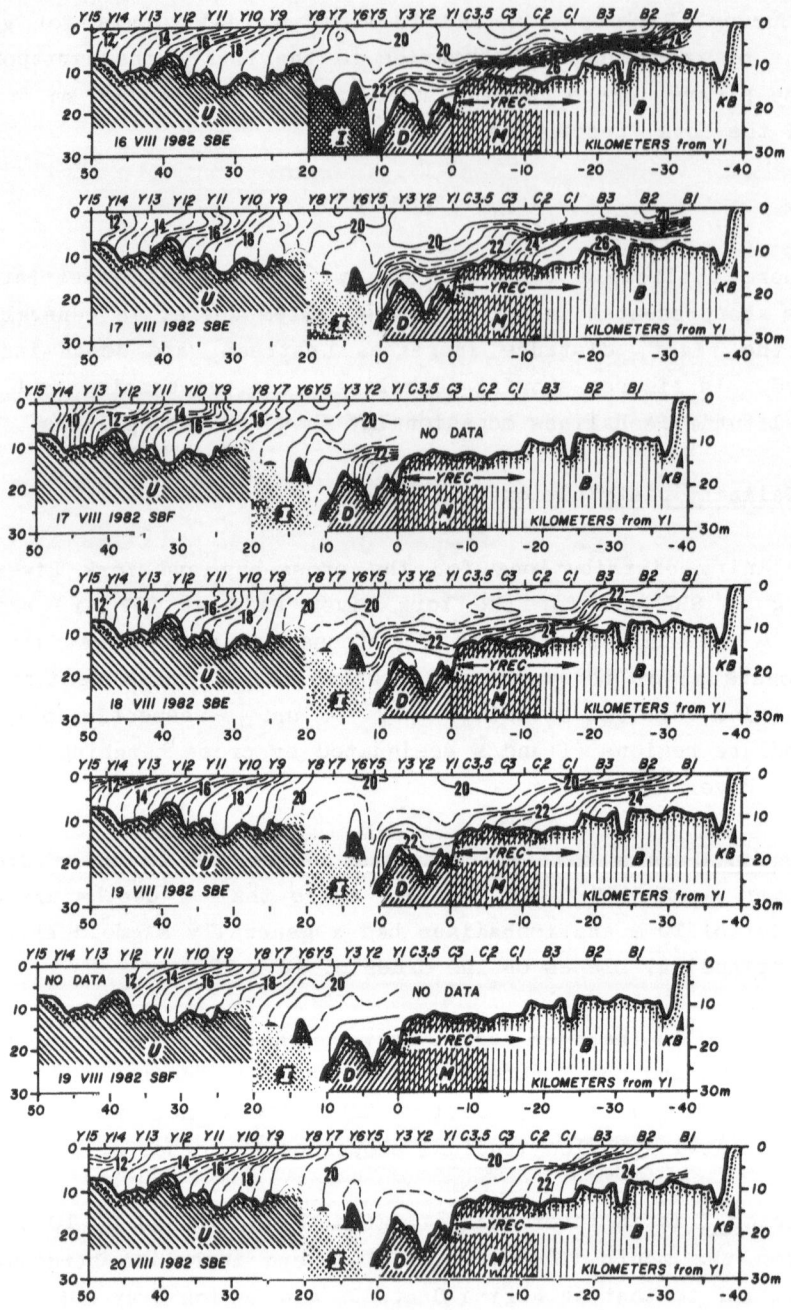

Figure 3. Salinities between Kiptopeke Beach (KB) and West Point (km 50) at SBE and SBF in the York River. SBF panels are offset by the approximate distance of a tidal excursion in the upper river.

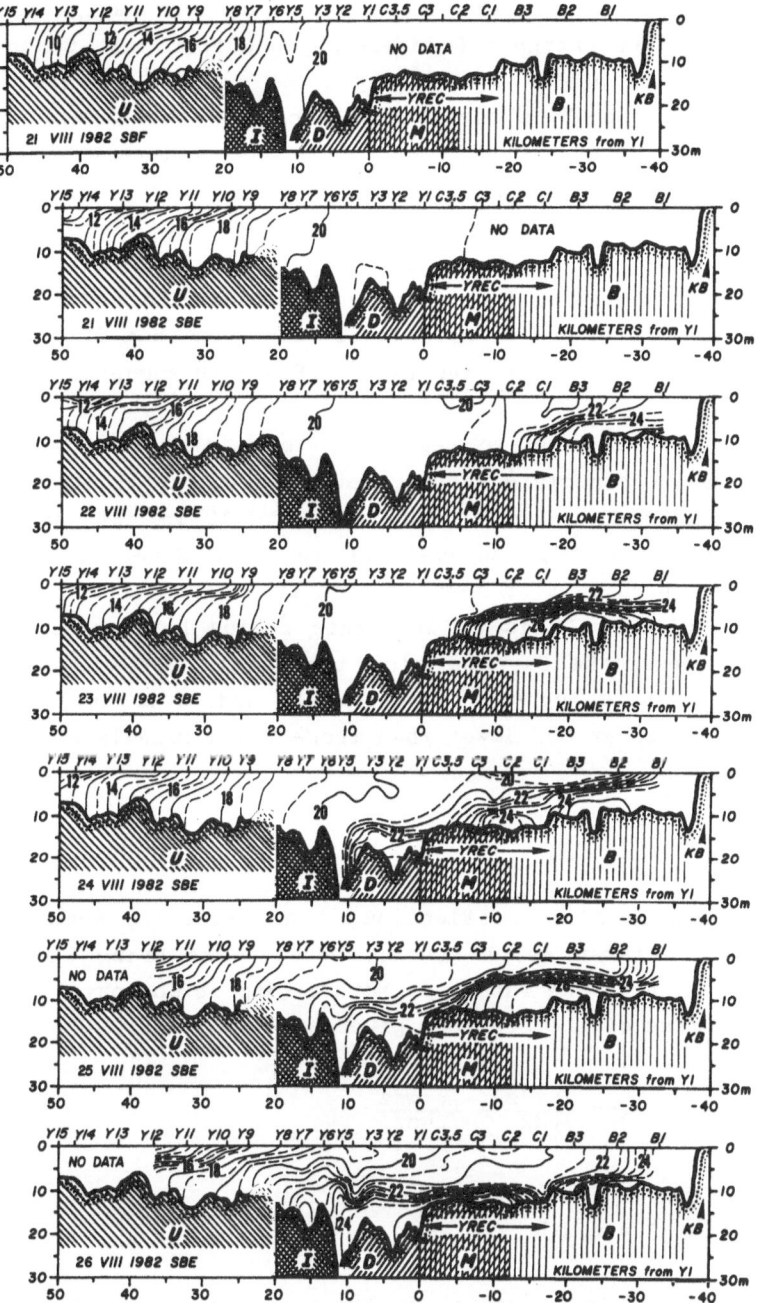

Figure 3. Continued. Salinities between Kiptopeke Beach (KB) and West Point (km 50) at SBE and SBF in the York River. SBF panels are offset by the approximate distance of a tidal excursion in the upper river.

this region, with the exception of dredged navigation channels (York River Entrance Channel-YREC and York Spit Channel at -24 km), water depths are on the order of 10 m. This region was characterized by a persistent halocline which varied in maximum gradient from 4.2 salinity units per m (measured over 1.0 m intervals) on 16 August (Station B3) to 1.5 per m on 19 August (Station C1). During the study, salinities in Region B varied from a maximum of 28 at bottom on 16 August (Station B3) to less than 20 at the surface (Station B2, 17 August and Station C2, 24 August) while bottom salinities were less than 25 on the days of predicted maximum spring tides (18 and 19 August).

Region M, similar to Region I, is an intermediate region with isohaline structure varying between that of Regions B and D. It extends from the York River mouth to 12 km into the bay (Stations Y1 to C2) and contains the York River Entrance Channel along which saltier bottom water from Chesapeake Bay enters the York River. It is an interface region between the York River and Chesapeake Bay and, for the first half of the study and later (16-20 August, as well as 22 and 24 August), had surface salinities at SBE lower than those found immediately upstream in Region D.

Stability and Mixing

The degree of stratification of an estuarine segment at slack water can be taken as the energy required to bring about complete vertical mixing. The result of such mixing is an upward shift in the center of mass of the section, given by

$$\zeta = \frac{\int zwdz}{\int wdz} - \frac{\int \rho zwdz}{\int \rho wdz} \tag{1}$$

where $\rho = \rho(z)$ = density
   $w = w(z)$ = width

Change in the potential energy per unit volume of the section, assuming unit axial length, Δx, and lateral homogeneity, is

$$P = \frac{\Delta PE}{\Delta \, Vol} = g\zeta \int \frac{\rho w \Delta x dz}{Vol} \tag{2}$$

or, substituting for ζ and writing Vol $= \int \Delta xwdz$ gives:

$$P = \frac{g}{\int wdz} \quad \frac{\int \rho wdz \quad \int zwdz}{\int wdz} - \int \rho wzdz \qquad (3)$$

(In Regions D, I and U, w(z) was used. In Regions B and M, w(z) was assigned a value of 1 m.) P differs from the frequently used stability parameter $\delta S/S_o$ (Hansen and Rattray, 1966) in that it 1) represents the static, slack water condition, 2) accounts for the depth dependence of width of a section and 3) varies with the depth and strength of the pycnocline. Comparisons of P with $\delta S/S_o$ show that, in general, when $\delta S/S_o > 0.1$ indicating stratified conditions $P > 20$ Joules m^{-3}. Distribution of P from Kiptopeke Beach to West Point for each sampling period is shown in Fig. 4.

The close association between estuarine destratification and spring tides prompted us to examine the changes in available tidally attributable kinetic energy which could be converted to turbulent energy and used for mixing. Following the arguments of Simpson and Hunter (1974), we determined the mean rate of tidal energy dissipation per unit volume for each lunar day prior to sampling from:

$$\frac{dE}{dt} = \frac{4k}{3\pi h} \rho \, \overline{u_o^3} \qquad (4)$$

where k = a constant of the assumed quadratic friction law (estimated

as 2×10^{-3} after Csanady, 1982, p. 12)

h = water depth taken as mean cross-sectional water depth in Regions D, I and U and total water depth in Regions B and M

$\overline{u_o^3} = \frac{1}{4} \Sigma \, u_p^3$ where u_p is the predicted maximum flood and ebb tidal current for the previous two tidal cycles

$\rho \approx 10^{-3}$ kg m^{-3}

The total kinetic energy generated between sampling periods is then

$$M = \frac{dE}{dt} \Delta t \qquad (5)$$

where Δt is taken as 9×10^4s.

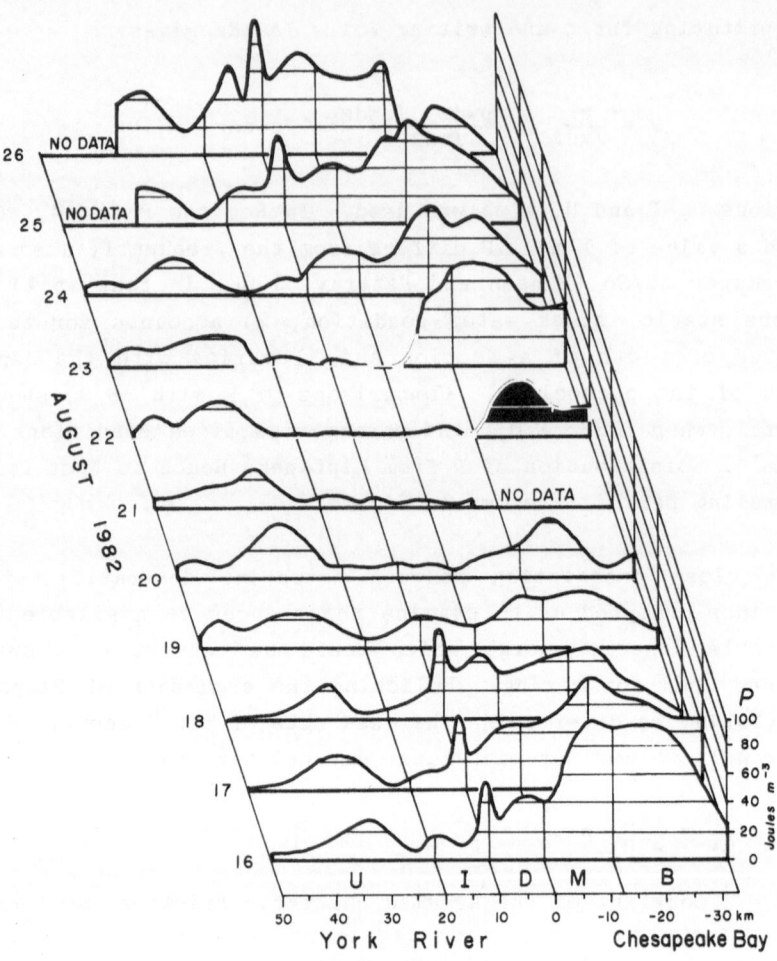

Figure 4. Daily longitudinal variations of stratification (as Joules m^{-3} required to destratify water) at SBE between stations B_1 (-32 km) and Y15 (50 km).

Simpson and Hunter (1974) use h/u^3 as a mixing parameter and show that for continental shelf regions, a value of $h/u^3 \simeq 55$ is found in frontal regions which separate stratified from well-mixed shelf waters. Taken over a tidal cycle and assuming $\rho \simeq 10^3$ kg/m^3 and $k = 2 \times 10^{-3}$, we find $h/u^3 \simeq 55$ is equivalent to $M \simeq 1.4 \times 10^3$ Joules m^{-3}. We identify this as M_m, the energy unit per volume required to maintain a destratified condition in opposition to a

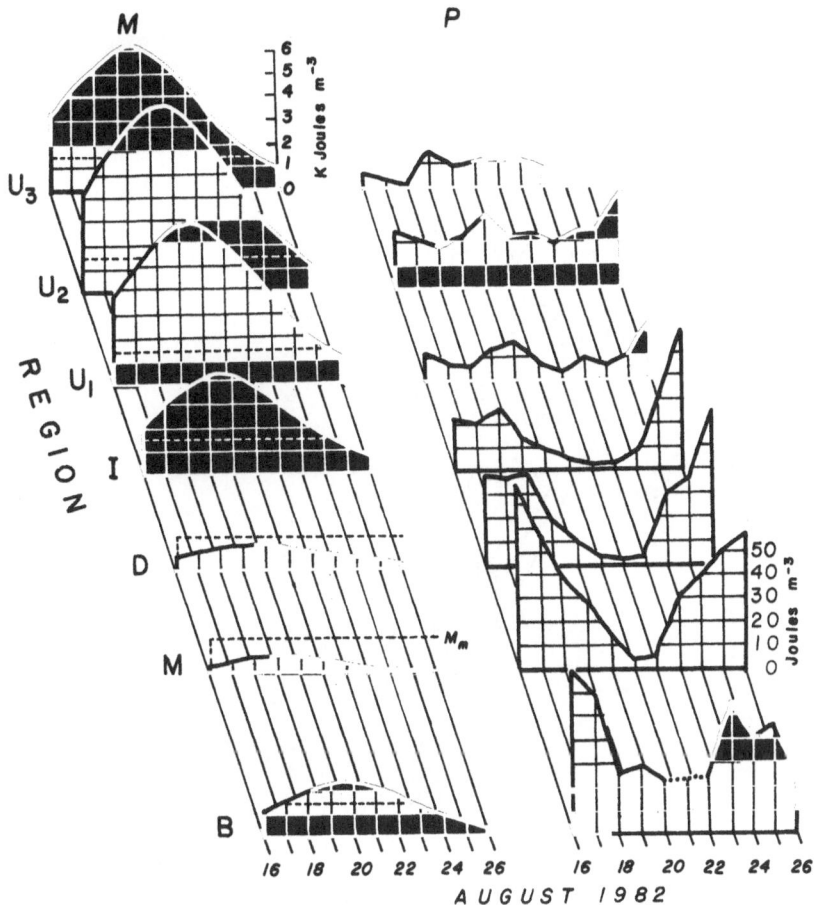

Figure 5. Daily variations of average available tidal kinetic energy
(M as K Joules m^{-3}) and energy required to destratify water
(P as Joules m^{-3}) for regions and subregions. Dashed line
(M_m) indicates kinetic energy required to maintain
destratification.

given set of conditions which, if allowed to persist, would result in
a stratified water column.

The temporal sequence of spatially averaged values of P and M
along with M_m are shown in Figure 5 for Regions B, M, D and I and for
three subregions of Region U sampled:

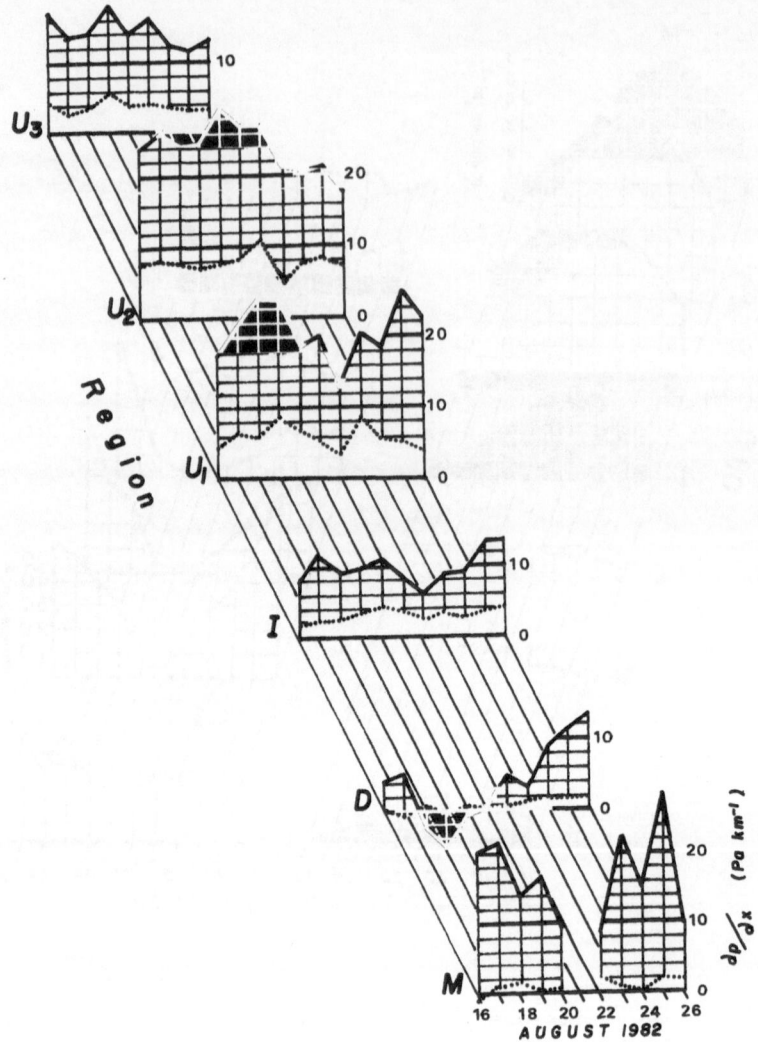

Figure 6. Daily variations in near surface (solid lines) and near
bottom (dotted lines) baroclinic component of the horizontal pressure gradient along the York River at SBE. Positive
values indicate greater pressure downstream. Near-bottom
∂p/∂x was calculated at the following depths in each
region: M-12m, D-18m, I-10m, U_1-12m, U_2-12m, U_3-7m.

U_1 = km 20 to km 29 where P < 20 Joules m^{-3} except for the final
day of the study

U_2 = km 29 to km 39.5 where P > 20 during the study, and

U_3 = km 39.5 to km 50 where P < 20 when this subregion was sampled

Horizontal Pressure Gradient

The HWH model hypothesizes a reduction or possible reversal of the horizontal pressure gradient in the Lower York River as a consequence of importation of fresher water from Chesapeake Bay. To investigate this hypothesis, the daily near-surface (3 m) and near-bottom baroclinic portion of $\delta p/\delta x$ was determined from our SBE density data along Regions M, D, I, U_1, U_2 and U_3. Our results (Fig. 6) support the HWH hypothesis in that they show a reversal of $\delta p/\delta x$ at 18 m in Region D on 18 and 19 August. Positive values of $\delta p/\delta x$ in Figure 6 indicate higher pressure downstream. Near-surface baroclinic pressure gradients were of the same sense and we hypothesize that they are in opposition to the barotropic pressure gradients along the river.

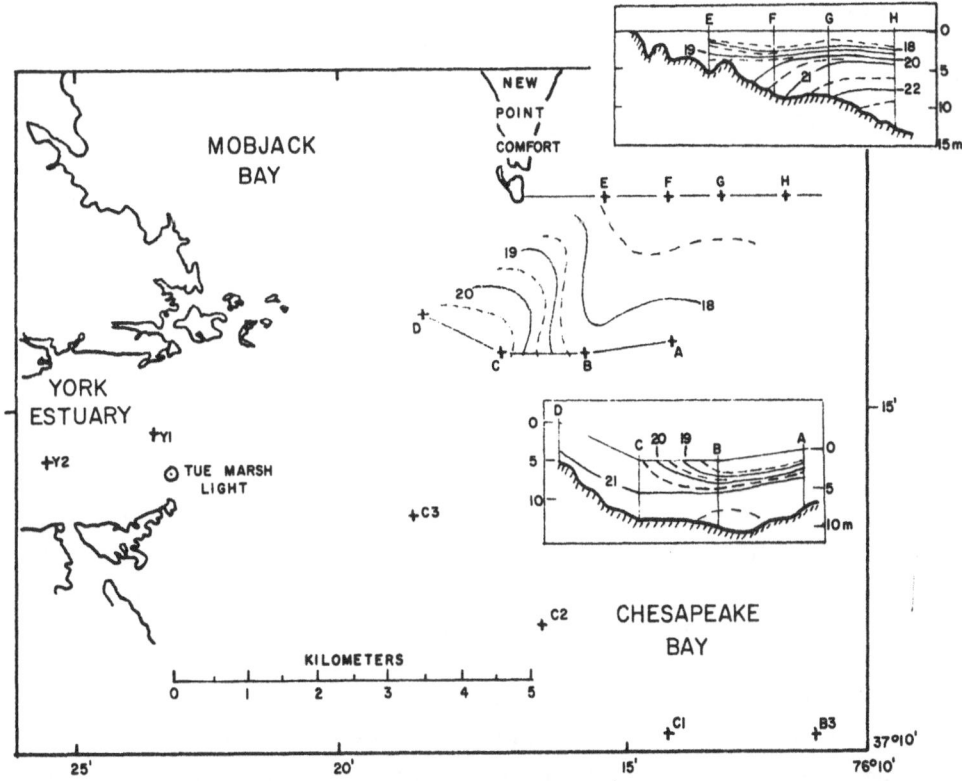

Figure 7. Salinity distribution near New Point Comfort on 18 August 1982. Insets show vertical distributions along east-west sections.

Salinity off New Point Comfort

Results of CTD casts at stations A through H in the vicinity of
New Point Comfort show a surface intrusion of low salinity water was
directed southwest from Chesapeake Bay towards the York River (Fig.
7). Low salinity water was confined to a 2 m thick lens with minimum
salinities less than 17.2 at station H (insets, Fig. 7). Underlying
water had maximum salinity at station H (greater than 22.7 at 11.7
m). Salinities at 4 m in this region averaged 20.8 with lowest
values at station E (19.6) and greatest values at station B (21.2)
suggesting the core of fresher water extended southward over the
shoal, nearshore region on the western side of Chesapeake Bay. CTD
stations off New Point Comfort were occupied between 1500 and 1700 h
on 18 August when predicted tidal currents indicated conditions in
Chesapeake and Mobjack Bays were at the end of the ebbing tide, while
near the mouth of the York (Station Y2), currents were between SBF and
maximum flood.

Currents off New Point Comfort

Five buoys, drogued at 1 m, were released from locations east of
New Point Comfort on 19 August and tracked for eight hours. The
experiment was repeated on 26 August (0400 to 1300 h) with three
buoys. Hourly buoy positions and surface salinities are plotted in
Figure 8.

Each drogued buoy experiment was conducted during the ebb
portion of a tidal cycle to compare water motions during spring (19
August; Fig. 8a) and neap tides (26 August; Fig. 8b). During each
experiment, buoy tracks show anticyclonic curvature with bouys either
ending in or tending towards Mobjack Bay. Within Chesapeake Bay,
buoy movement towards the South (seaward) was approximately 50%
greater during spring tide than during neap tide.

Anticyclonic curvature of buoy paths at the end of ebbing tide
is most likely the result of tidal current phase difference between
Chesapeake Bay and its tributaries, however, topographic steering by
York Spit (5 m isobath shown in Fig. 8) and wind forcing could also
be important (see Fig. 2).

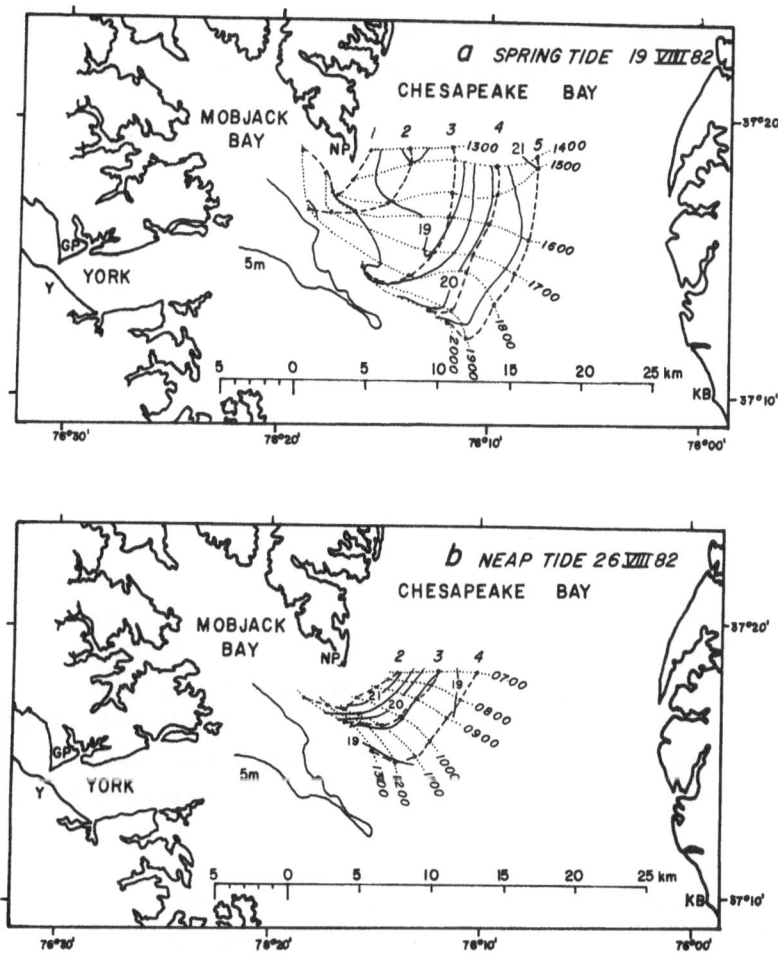

Figure 8. Surface salinities (solid lines) and drogued buoy paths
(dashed lines) during ebbing portion of tidal cycle off New
Point Comfort (NP) during spring tides (a) and neap tides
(b).

DISCUSSION

Periodic destratification of water in the York River appears to
be caused by three processes acting on separate portions of the
estuary. Destratification of the lower estuary, Region D, follows
the process outlined in the HWH model which is specific to conditions
of estuary-subestuary interactions such as those of the lower Chesapeake Bay system. We hypothesize that destratification in the

central portion of the York River (Region I and Subregion U_1) results from a more general condition where increased tidal kinetic energy during spring tides provides sufficient turbulent mixing to overcome stratification maintained by gravitational circulation. This condition may be enhanced by basin geometry. Destratification in the upstream portion of the York River (Subregion U_3) does not follow the orderly pattern observed at other locations (i.e. Region I) and we hypothesize this reach of the estuary response to riverflow modulated by increased local sea level during spring tides whereas persistent stratification in Subregion U_2 is a consequence of conditions in adjacent subregions.

The Lower Estuary and Adjacent Portions of Chesapeake Bay: Regions M and D

Salinity distributions near New Point Comfort on 18 August (Fig. 7) and drogued buoy movements on 19 August (Fig. 8a) indicate freshened water from more northerly portions of Chesapeake Bay was directed into Mobjack Bay during the time of maximum spring tides. Salinities at drogued buoy positions on 26 August (Fig. 8b) and cross-bay salinities (Fig. 3) show freshened water tended more towards the central to eastern portion of Chesapeake Bay during the period of neap tide. Although the spring tide buoy experiment results and associated salinities do not strictly conform to the HWH model, cross-bay salinity sections (Fig. 3) show lowered salinities from surface to 5 m in Region M on 16, 17, 19 and 20 August. We take this as an indication that fresher Chesapeake Bay water was entering Mobjack Bay on the flood tide and left Mobjack Bay on the following ebb to enter the York River on the subsequent flood tide. Both advection and vertical mixing were increased during spring tides as a result of stronger tidal currents. Increased vertical mixing in Region M is suggested from the salinity distributions shown in Fig. 3 and time-depth isohaline variations at individual stations in Regions M and D (Fig. 9). However, minimal tidal energy is available for mixing in this region (see Fig. 5) and we suggest that water advected into Region M was previously mixed elsewhere - possibly in Mobjack Bay. Reversal of the near bottom horizontal pressure gradient in the lower York coincided with maximum spring tides (Fig. 6) and the entry of high salinity bottom water from Chesapeake Bay was blocked (Fig. 3). This condition allowed the lower York to completely destratify within five days of maximum spring tides (23 August, Fig. 3). Restra-

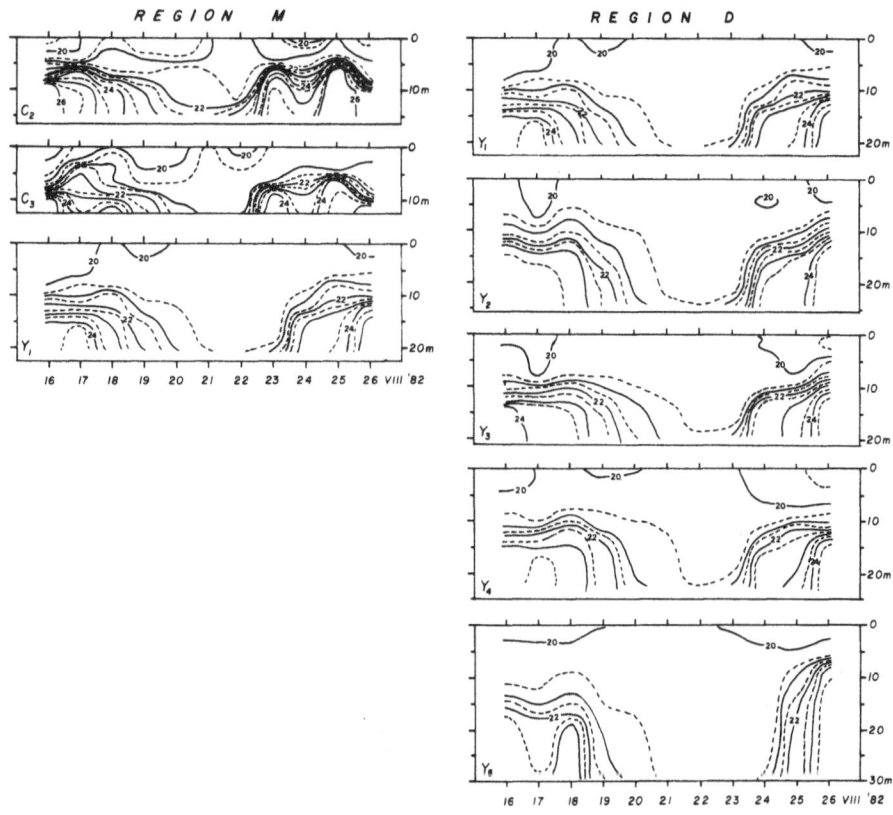

Figure 9. Time-depth variation of salinity at stations in Regions M
and D at SBE.

tification occurred abruptly the following day with the influx of
high salinity water from Chesapeake Bay. All phases of the HWH model
are therefore verified by the present study.

The Central Estuary: Region I and Subregion U_1

Our analysis shows that upstream of the Yorktown/Gloucester
Point constriction, a marked decrease in stratification was associated with an increase in available tidal energy (Figs. 4 and 5).
We offer the following conceptual model to explain the conditions
observed:

1) In the seaward portion of the central estuary (Region I) the destratification process begins when tidal mixing, driven by stronger currents, increases sufficiently to permanently erode the pycnocline established by gravitational circulation. Some critical value of the tidal mixing parameter, M, which is determined by the existing density gradient, must be attained to initiate the erosive process. Once pycnocline erosion is permanent (i.e. constant and persistent gravitational circulation is insufficient to maintain the initial stratified condition) reduced tidal currents are sufficient to continue the erosion towards complete destratification.

2) The process is augmented spatially in the landward direction by a reduction in a) basin cross sectional area, b) maximum depth along the thalweg, and c) cross sectionally averaged depth, all of which tend to increase tidal energy per unit bottom area. Temporal augmentation is effected by reduced apparent streamflow resulting from temporary upstream storage of freshwater during periods of increasing tidal amplitude and reduced gravitational circulation at the mouth of the estuary.

3) The rapid decrease in both maximum depth and cross sectionally averaged depth occurs in Region I and provides a barrier to upstream extension of higher salinity bottom water. The relatively steep bottom slope also serves to move bottom water upward during the flood portion of a tidal cycle which results in a) more uniform surface to bottom salinities, and b) a separation of lower salinity surface water into upstream and downstream segments at SBE.

4) Reduction of tidal currents diminishes mixing in Region I and allows gravitational circulation to reestablish stratified conditions. Stratification is also enhanced by the post-spring tide release of previously stored freshwater and by strengthening of gravitational circulation at the mouth of the estuary.

Formulation of this hypothetical model is based on the results of the present study, field observations presented by Hayward et al. (1982) and basin geometry shown in Figure 10.

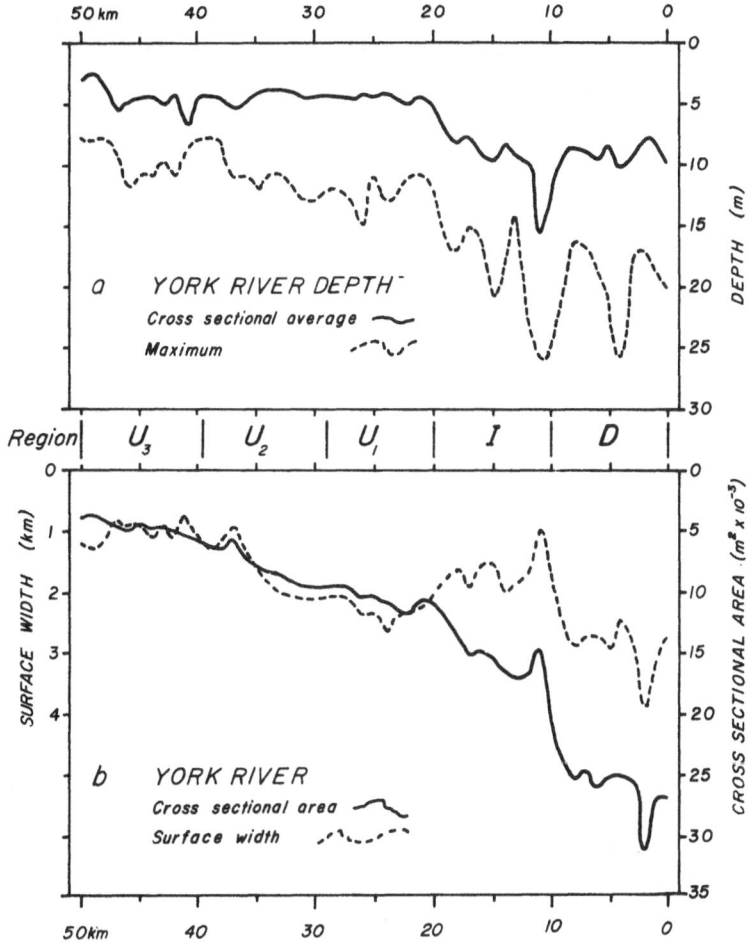

Figure 10. Longitudinal variations of depth (a) and cross-sectional
area, surface width (b) of the York River.

The Upper Estuary: Subregions U_2 and U_3

Streamflow modulation during the spring-neap cycle of tides
has been explained and demonstrated for macrotidal estuaries (Allen
et al., 1980). We suggest similar conditions occur in the York
River, a microtidal estuary, causing destratification in Subregion U_3
prior to and during the time of maximum spring tides (16, 17, 18
August; Fig. 3) and increased stratification immediately following
spring tides (19, 20 August; Fig. 3). This hypothesis is, in part,

supported by examination of tide records from Elsing Green in the Pamunkey River (Fig. 1) 200 km from the mouth of the York. These records (Hyer et al., 1971) show average values of local mean sea level increase by 15 cm during the period between mean and spring tides and then diminish to mean tide values during the three days following spring tides. During mean tides, 3×10^7 m^2 of wetlands in the Pamunkey and Mattaponi drainage basins are regularly flooded (G. Silberhorn, pers. comm.). A 15 cm increase in mean water level over this area during the week prior to spring tides could be accomplished by a steady streamflow of 7.4 m^3 s^{-1}. Release of this stored water during the three days following spring tides would be equivalent to augmenting a steady streamflow by 17.4 m^3 s^{-1}. To examine possible streamflow modulation during our study we lagged measured streamflow (Fig. 2) using the empirical relationship developed for York River flows by Hayward et al. (1985). Their formula: lag in days = 24.18 - 4.12 $\log_{10}$ flow in m^3 s^{-1} was based on gauged discharge and salinities near station Y4. Our interest, however, is in effective freshwater discharge at the head of the York and we use historical records (unpublished data) of near-surface currents to estimate time of travel between stations Y15 and Y4 as 9 days thus modifying the above equation to:

$$lag = 15.18 - 4.12 \log_{10} flow$$

Lagged flow data was then smoothed using a five point weighted average. We next assumed freshwater storage and release rates to be Gausian and approximated the modulated flow at the head of the York as shown in the upper panel of Figure 11. Comparing this with time-depth salinity variations at stations Y3, Y14 and Y15 (remainder of Fig. 11) shows depressed surface salinities are closely related to modulated lagged flows in further support of our hypothesis.

Finally, we hypothesize that persistent stratification in Subregion U_2 is the consequence of 'normal' gravitational circulation generated by high salinity destratified conditions in Subregion U_1, and low salinity destratified conditions in Subregion U_3. This hypothesis is, in part, supported by the persistent, relatively strong, horizontal pressure gradient across Subregion U_2 (Fig. 6).

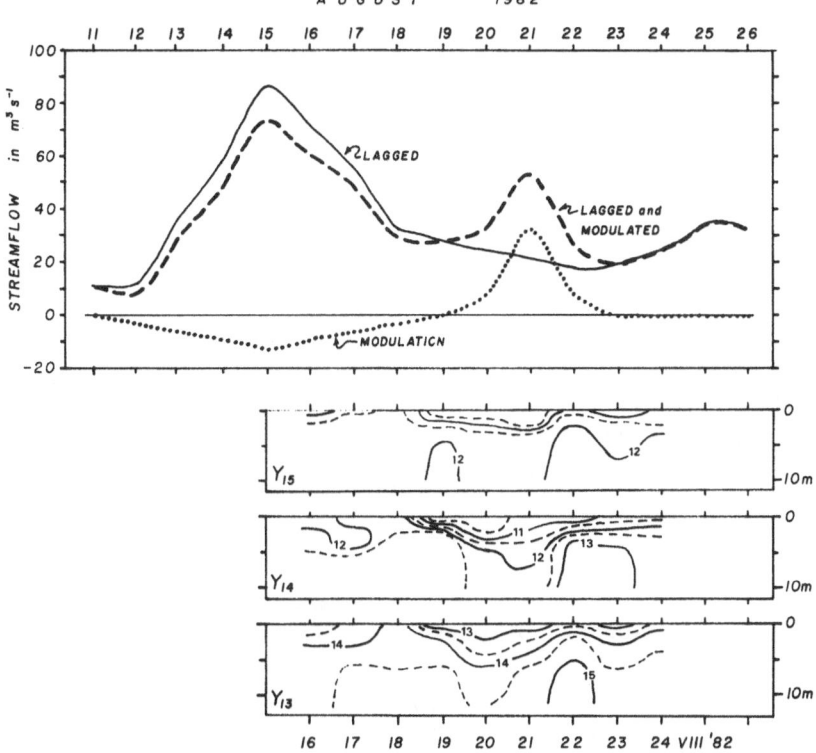

Figure 11.   Lagged York River flows, spring tide modulation and their combination (upper panel) and time-depth plots of isohalines at stations Y_{15}, Y_{14} and Y_{13}.

SUMMARY AND CONCLUSIONS

The fortnightly cycle of stratification and destratification in the York River is attributable to effects of the neap-spring cycling of tides. This interpretation is in conflict with the often made assumption of steady state conditions when circulation and mixing within coastal plain estuaries are considered. The conceptual model of destratification proposed by Hayward et al. (1982) is supported for the lower portion of the York which demonstrates estuary-subestuary interaction. Destratification of this portion of the estuary is highly dependent on mixing and advection in the adjacent portion of Chesapeake Bay. Cyclic destratification in the central portion of the York estuary is hypothesized to be the result of increased tidal mixing and basin geometry while apparent destratification of the upstream portion of the estuary is attributed to riverflow modulation during the spring-neap cycling of tides.

ACKNOWLEDGEMENTS

We are indebted to C.S. Welch for the results of his drogued buoy experiments off New Point Comfort. Several of our colleagues provided assistance during many hours of discussion and we acknowledge the particular interest of J. Brubaker and L. Haas in this respect. We appreciate the assistance of L. Huzzey, M.Z. Moustafa, T. Brooks, M.O. Green and D.S. Fenstermacher in the field, and G. Shaw in data processing. Comments on the manuscript by J.D. Boon, III, L. Haas, K.L. Webb, C.S. Welch, L.D. Wright and other unknown reviewers were most helpful. Finally, we wish to thank S. Crossley for manuscript preparation and N. Courtney for help with the figures. Ship time was graciously provided by L. Haas under National Science Foundation Grant OCE-8110396.

REFERENCES

Allen, G.P., J.C. Salomon, P. Bassoullet, Y. DuPenhoat and C. DeGrandpre. 1980. Effects of tides on mixing and suspended sediment transport in macrotidal estuaries. Sed. Geol. 26: 69-90.

Csanady, G.T. 1982. Circulation in the Coastal Ocean. R. Reidel, Dordrecht, Holland. 279 p.

Haas, L.W. 1977. The effect of the spring-neap tidal cycle on the vertical salinity structure of the James, York and Rappahannock Rivers, Virginia, U.S.A. Estuar. Coastal Mar. Sci. 5(4): 485-486.

Hansen, D.V. and M. Rattray, Jr. 1966. New dimensions in estuary classifications. Limnol. Oceanogr. 11(3): 319-326.

Hayward, D., L.W. Haas, J.D. Boon, III, K.L. Webb and K.D. Friedland. 1985. A regression model of neap-spring tidally associated stratification variation in the York River Estuary, Virginia. (in this volume)

Hayward, D., C.S. Welch and L.W. Haas. 1982. York River destratification: an estuary-subestuary interaction. Science 216: 1413.

Hyer, P.V., C.S. Fang, E.P. Ruzecki and W.J. Hargis, Jr. 1971. Hydrography and hydrodynamics of Virginia estuaries, II. Studies of the distribution of salinity and dissolved oxygen in the upper York system. Spec. Rept. in Appl. Mar. Sci. Ocean Eng. No. 13, VA Inst. Mar. Sci., Gloucester Pt., VA.

NOAA. 1981. Tidal current tables 1982, Atlantic Coast of North America. U.S. Dept. Commerce, NOAA, National Ocean Survey.

Simpson, J.H. and J.R. Hunter. 1974. Fronts in the Irish Sea. Nature 250: 404-406.

EFFECTS OF TIDAL MIXING ON THE PLANKTON AND BENTHOS
OF ESTUARINE REGIONS OF THE BAY OF FUNDY

G.R. Daborn
Department of Biology
Acadia University
Wolfville, Nova Scotia B0P 1X0

INTRODUCTION

In coastal waters, vertical mixing induced by wind or tidal currents is often associated with relatively high primary production resulting from nutrient recirculation. This general conceptual model is now well recognized (cf. Mann, 1982; Valiela, 1984; and other papers in this volume), and is variously invoked to explain many phenomena of oceans, coastal waters and lakes, from the initiation of seasonal blooms to the spatial distribution of areas of higher and lower production.

In shallow waters, benthic remineralization of nutrients is an important factor increasing nutrient availability, which stimulates phytoplankton production where vertical mixing occurs. Thus, the benthic and pelagic communities are tightly coupled (Hargrave, 1973, 1975; Rowe et al., 1975). Where stratification occurs, however, productivity of surface waters tends to be lower because of persistent losses of nutrients from the photic zone (Pingree, 1978; Pingree et al., 1975; Garrett and Loucks, 1976), unless the stratified area is coupled closely with a frontal zone system where vertical mixing occurs (Fournier et al., 1979, 1984).

In many estuarine areas also, studies of benthic remineralization rates indicate that vertical mixing does enhance surface primary production: in Narragansett Bay, for example, 40-50% of nitrogen and phosphorus utilized by phytoplankton during the year is derived from recycling by the benthic community (Nixon, 1981). There is, however, no a priori reason to expect the coastal mixing model to apply to estuaries, particularly those with moderate to high flow ratios or receiving large quantities of allochthonous organic input. On the contrary, the continuous supply of "new" nutrients through river input could compensate for removal of nutrients by phytoplankton. Furthermore, in a stratified system phytoplankton cells remain in or

Lecture Notes on Coastal and Estuarine Studies, Vol. 17
Tidal Mixing and Plankton Dynamics. Edited by J. Bowman, M. Yentsch and W. T. Peterson
© Springer-Verlag Berlin Heidelberg 1986

near the photic zone (although translocated downstream), rather than mixed out of it. Thus, far from stratification being associated with low primary production as in some fjords (Matthews and Heimdal, 1980), estuaries and fjords that are influenced by nutrient-yielding boundary conditions may well be most productive where vertical mixing is least. Narragansett Bay is again an example: a marked halocline is only found in the upper Bay and river. This is the area of origin of the winter bloom which then spreads throughout the Bay (Kremer and Nixon, 1978). Another consequence of continuous nutrient supply to an estuary is the prolongation of net primary production throughout much of the year, instead of being largely restricted to brief spring and fall blooms initiated by turnover events.

Recent studies in the Bay of Fundy system suggest that where tidal range is high, and vertical mixing influences the whole of a relatively shallow estuarine water column, the results are quite the reverse of those predicted by the coastal model. Phytoplankton and zooplankton stocks are generally low, and production is modest, despite abundant and well distributed nutrients. Benthic populations on the other hand are moderately diverse and prolific. Where stratification occurs, planktonic production is relatively high, and benthic populations impoverished. These observations are explicable if vertical mixing of shallow water columns results either in removal of phytoplankton from the photic zone, or in extensive grazing on phytoplankton by benthic suspension feeders, so that benthos and zooplankton are in direct competition for a limited food source (e.g. Riley, 1972; Wolff et al., 1975; Matthews and Heimdal, 1980). In areas of extreme tidal currents, benthic communities are eliminated by scouring, whereas zooplankton populations are abundant and relatively productive, even where suspended sediment levels are very high (Daborn, 1984).

The hypothesis presented here is that where tidal mixing pro-duces homogeneity of the water column of shallow estuaries, phyto-plankton and zooplankton standing crops will be impoverished at least partly because of extensive grazing on phytoplankton by benthic suspension feeders. Conversely, stratification leads to higher planktonic populations and limited benthic ones.

The Bay of Fundy System

The Bay of Fundy System (Fig. 1) is a large, complex estuarine system noted mostly for its high tidal range which increases from 3-5 m at the confluence of the Bay and the Gulf of Maine to 12-16 m in the innermost headwaters. Tides are semi-diurnal, with moderate inequality. Associated with their movements are proportionately strong currents (<4 m·sec^{-1}) that generate considerable turbulence (Greenberg, 1984). Stratification occurs in the outer Bay during summer months, but is generally eliminated during fall and winter by meteorological conditions (Bailey, 1954; Garrett, 1977; Garrett et al., 1978). Inside a line joining the Saint John Estuary and the Annapolis Estuary (and corresponding roughly to the 100 m depth contour) the water is vertically mixed all year.

The inner regions are macrotidal estuaries with extensive intertidal areas and moderate to high turbidity (Table 1). Except for the Annapolis and Kennebecasis Estuaries which are stratified, water depths are shallow and the water column is vertically homogeneous for most of the year (Amos and Joice, 1977; Amos and Asprey, 1979, 1981; Daborn et al., 1982; Brown, 1984; Greenberg, 1984; Keizer et al., 1984). Tidal currents in inner regions may be so strong that the bottom is either scoured to bedrock or covered by mobile sand waves (Swift and McMullen, 1968; Amos and Joice, 1977; Middleton, 1977; Long, 1979). Because of tidal turbulence, suspended particulate matter (SPM) varies from <10 to >12 mg·ℓ^{-1}, and consequently light penetration ranges from 10+ m in Chignecto Bay and Minas Basin, to a few millimeters in the Cornwallis Estuary and inner regions of Cumberland Basin, Cobequid Bay and Shepody Bay.

The Annapolis Estuary is in marked contrast to the inner bays of the Fundy system. Tides in the Annapolis Basin range from 4 to 9 m, but the water is moderately clear of suspended particulate matter, and the depth of the photic zone varies seasonally between 8 and 16 m (Daborn et al., 1982). In 1960 a rockfill tidal dam was constructed at Annapolis Royal to provide control of water levels in the Annapolis River, and to prevent flooding of reclaimed marshland when high spring tides coincide with high river runoff. As a consequence, part of the previously homogeneous macrotidal estuary was converted into a stratified, microtidal (<1 m range) headpond. In the headpond and lower river, stratification is fairly stable during the warmer

Table 1. Characteristics of estuarine areas of the Bay of Fundy system (from various sources listed in text).

	Tidal range (m)	Depth range (m)	Water column	Suspended particulate matter (mg·ℓ^{-1})
Outer Bay of Fundy	3-5	80-200	Stratified in summer	<10
Inner Bay of Fundy	5-7	40-100	Mixed	<20
Passamaquoddy Bay	4-8	0-75	Mixed	-
Chignecto Bay	8-12	0-80	Mixed	5-50
Minas Basin	8-16	0-120	Mixed	5-50
Cobequid Bay	11-16	0-40	Mixed	50-1,000
Cumberland Basin	9-13	0-40	Mixed	50-1,500
Shepody Bay	9-13	0-45	Mixed	50-250
Cornwallis Estuary	9-13	0-15	Mixed	50-12,000
Annapolis Estuary				
– Basin	7-9	10-50	Mixed	<20
– Headpond	<1	<25	Stratified	<30
Saint John Estuary	4-9	5-60	Mixed	-
Kennebecasis Estuary	<1	5-60	Stratified	-

394

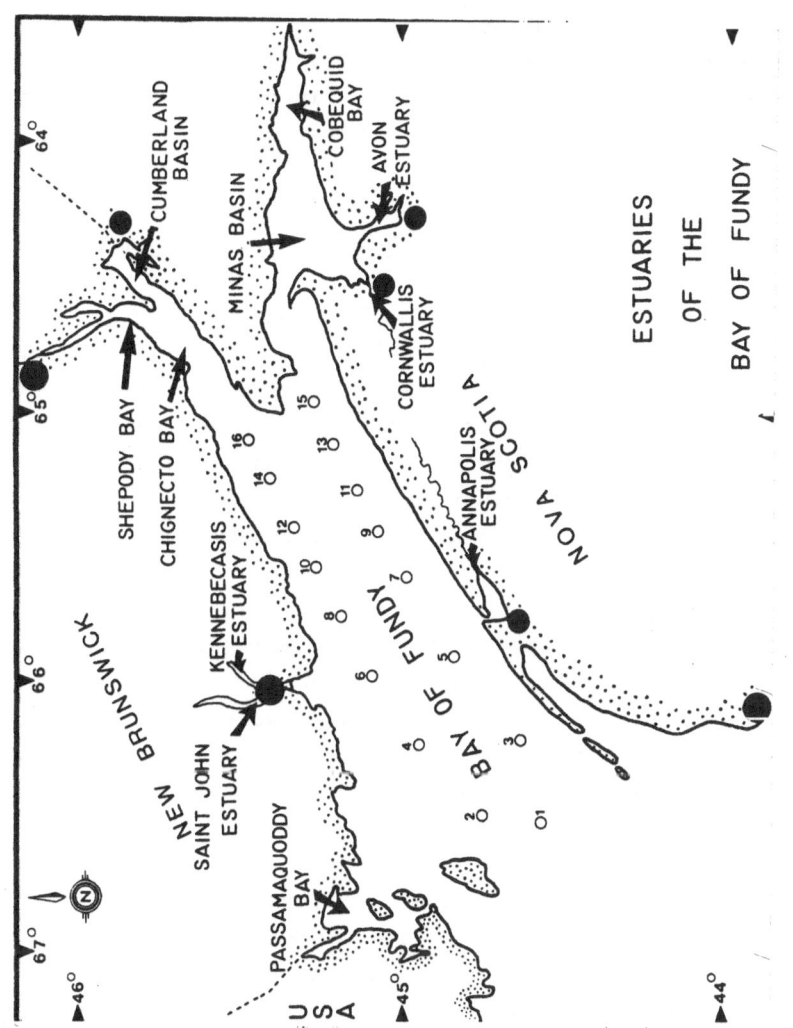

Figure 1. The Bay of Fundy system. O 1-16 transect stations.

months, with low salinity surface waters (<5 ^{o}/oo) overlying a deep salt wedge (23-30 ^{o}/oo). SPM levels are low throughout the system, but light penetration in the headpond and river is strongly restricted by high humic acid concentrations in the river water (Daborn et al., 1982).

Relationship Between Tidal Mixing and Pelagic Primary Production

According to the estuary tidal mixing hypothesis stated above, complete water column mixing should lead to diminished phytoplankton standing crop as benthic suspension feeders graze them down. Unfortunately, very few measurements of primary production have been made in the main Bay of Fundy, although the composition of the flora in the outer Bay has been the subject of several investigations (Gran, 1933; Davidson, 1934; Gran and Braarud, 1935; Lakshminarayana, 1983). During four seasonal cruises to the headwaters of the Bay, measurements of chlorophyll concentrations were made at 16 stations (cf. Fig. 1); maximum average biomass (as mg·m^{-3} averaged for the photic zone) occurred in summer at the mouth of the Bay (Fig. 2). During February and March cruises, chlorophyll levels were approximately the same throughout the Bay (Fig. 2), whereas in summer concentrations were lowest at stations 7-12, about 90-130 km from the mouth of the Bay but somewhat higher both at outer and inner stations. In general, during the summer months, when thermal stratification effects in the Bay are maximal (Garrett et al., 1978) stations less than 100 m deep exhibit sharply reduced chlorophyll levels. Therefore, when chlorophyll levels are plotted as a function of water column stability (as $\Delta\sigma_t$ calculated for the full water column), it is evident that higher chlorophyll values are found in regions showing the greatest tendency to stratify (Fig. 3).

During the same cruises, chlorophyll measurements were also made at stations in Chignecto Bay, Shepody Bay and Cumberland Basin (Prouse et al., 1984). Average chlorophyll biomass in Cumberland and Shepody remained similar throughout the year at 1.3-1.8 mg·m^{-3}, considerably higher than the winter values throughout the Bay, and the summer values of stations 7-13 (Fig. 2). High values of chlorophyll have been found in the most turbid estuarine regions of the system, and this has been attributed partly to resuspension of intertidal benthic diatoms (Prouse et al., 1984), and partly to the reduction in breakdown rate of chlorophyll when kept in the dark

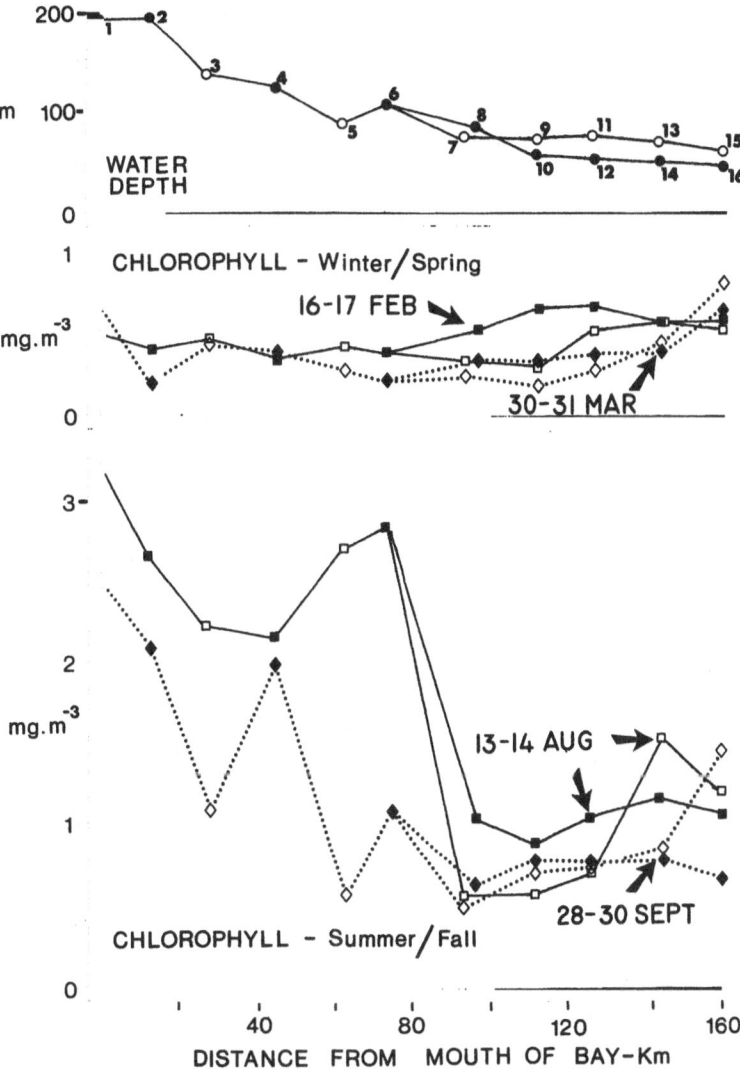

Figure 2. Average photic zone chlorophyll concentrations in the Bay of Fundy 1979-1980. Open symbols - stations 3-15 along Nova Scotia shore; closed symbols - stations 2-16 along New Brunswick shore (cf. Fig. 1).

(Daborn et al., 1982; Brown, 1984). Minimum values of chlorophyll seem therefore to be associated with intermediate values of tidal mixing in the Bay of Fundy system.

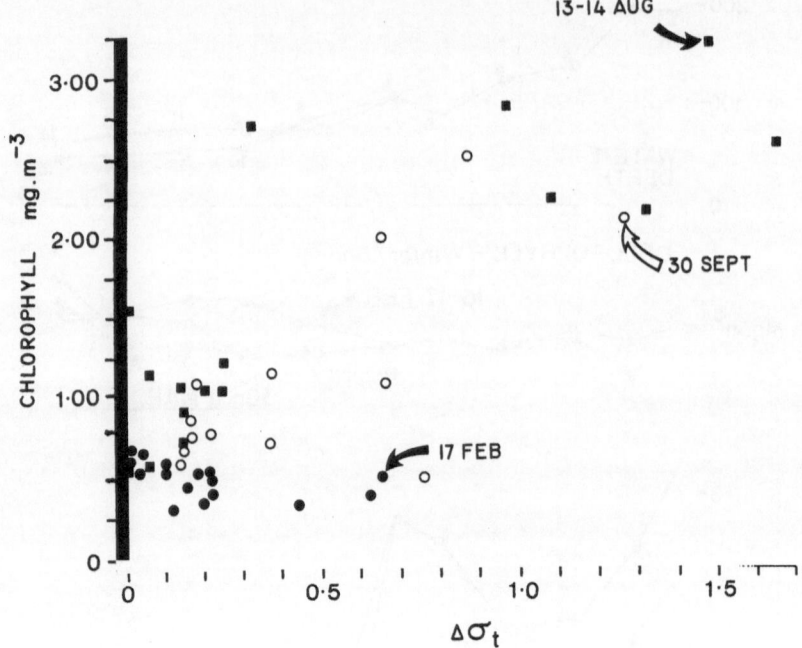

Figure 3. Relationship between photic zone chlorophyll concentrations and water column stability ($\Delta\sigma_t$) in the Bay of Fundy, 1979-80. ■ - 13-14 Aug 1979; O - 30 Sept 1979; ● - 17 Feb 1980.

The obvious decline in phytoplankton biomass with greater vertical mixing in the Bay of Fundy, coupled with a complementary decline in zooplankton abundance (Jermolajev, 1958) and the generally low levels of fish production in the inner Bay, induced Huntsman (1952) to conclude that the more turbid regions of the system would be even poorer in phytoplankton. Zooplankton abundances, however, tend to increase in more turbid regions, with a distinct change in species composition (Roff, 1983; Daborn, 1984). It would appear that the enhanced zooplankton biomass of the most turbid waters is supported by food resources other than active phytoplankton (Brown, 1984; Daborn, 1984).

Relationship Between Tidal Mixing and Benthic Production

In a series of studies, Wildish and his co-authors (Wildish, 1977, 1983; Peer et al., 1980; Wildish and Kristmanson, 1979; Wildish and Peer, 1983) have investigated the benthic fauna in the Bay of

Fundy. Wildish has concluded that tidal currents are the major determinant of benthic distribution and production in the lower Bay, through their control of sediment dynamics, and of settlement, feeding and growth of benthic animals. Maximum tidal current velocities are generally inversely correlated with water depth; thus, up to a limiting velocity (c. 100 cm·sec^{-1}) when bottom deposits become excessively coarse or scoured, the proportion of suspension feeding benthic organisms is also directly related to current speed (Wildish and Peer, 1983). The most important benthic organism in the Bay is the horse mussel, Modiolus modiolus, which, with the infaunal brachiopod Terebratulina septentrionalis produces >88% of total benthic secondary production.

According to the estuary tidal mixing model described above, maximal benthic populations should occur where water column mixing allows the benthos maximum access to phytoplankton biomass. This is quite clearly shown by the distribution of suspension feeder production in the Bay of Fundy (Fig. 4): maximal benthic production of >1000 g wet wt·m^{-2}yr^{-1} occurs in a band approximately 75-110 km from the mouth of the Bay. In deeper waters, production drops to 25-500 g wet wt·m^{-2}yr^{-1}, and is largely due to Terebratulina septentrionalis. In shallower waters the suspension feeding component declines somewhat, despite the higher current velocities, and total biomass and production of the benthic community decline also. These data have been interpreted by Wildish as support for the trophic group mutual exclusion hypothesis that identifies current velocity as the limiting factor of species composition and productivity (Wildish, 1977; Wildish and Peer, 1983).

It is important to note the coincidence between the peak of suspension feeding community and the sharp declines in chlorophyll concentrations midway up the Bay of Fundy. As implied by the coastal mixing model, the peak of benthic production probably results from the transition from stratified to mixed conditions at this point in the Bay. Vertical mixing of the water column conveys food from the photic zone directly to the benthos. Vertical mixing becomes even greater further up the Bay, yet we observe both a decline in benthic production and a rise in photic zone chlorophyll.

These observations are conformable with the hypothesis that in sufficiently shallow well-mixed water, suspension feeding benthos

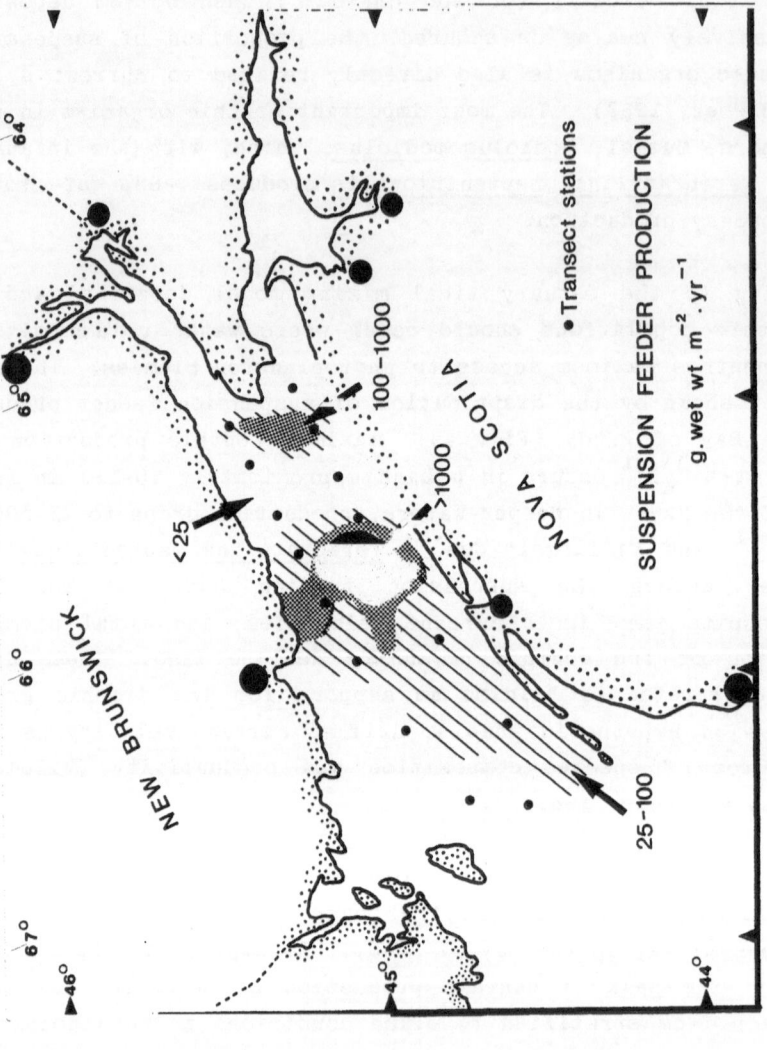

Figure 4. Production of macrobenthic suspension feeders in the Bay of Fundy. (After Wildish and Peer, 1983).

will effectively graze on phytoplankton biomass. Direct measurements of grazing or filtering rates are not available for _Modiolus_ in the Bay of Fundy. Schulte (1975), however, records filtration rates of 1.7-2.5 $\ell \cdot hr^{-1}$ per individual _Modiolus_ (75 mm length). In the area covered by stations 7-10 (Fig. 1), Wildish and Peer found an average of 140 _Modiolus_ per m^2, with an average biomass of 1650 g wet wt$\cdot m^{-2}$. Such a population could filter 6720 $\ell \cdot day^{-1}$, equivalent to nearly 10% of the water column. Clearly, this is insufficient to control phytoplankton populations in the region completely, but allows for a significant predation rate by benthic consumers upon primary producers that, coupled with relatively low local production because of turbidity and removal from the photic zone, could cause the low chlorophyll levels observed in this area.

In the intertidal area, where suspension feeding _Mytilus edulis_ are common, Frechette (1984) was unable to detect any correlation between filtration rate of the mussels and pigment concentrations except within 50 cm of the bed.

Plankton and Benthos in the Inner Bays

Very little research has been carried out on the subtidal faunas of Chignecto Bay, Shepody Bay, Cumberland Basin and Minas Basin. In the relatively clearer waters of Chignecto Bay and Minas Basin, bottom deposits tend to be coarse and hard (Long, 1979; Amos and Asprey, 1981), and, in the latter at least, are colonized by a well-developed suspension feeding community including _Modiolus_ and _Flustra_ (Macfarlane, pers. comm.). In the most turbid regions, however, the fauna becomes completely impoverished as current speeds and turbulence yield a mobile bottom of sand waves. Phytoplankton production is very low in Minas Basin and Chignecto Bay, generally <20 $g \cdot C \cdot m^{-2} \cdot yr^{-1}$ (Prouse et al., 1984), although recent measurements in the Southern Bight suggest that daily production values of greater than 250 $mg \cdot C \cdot m^{-2}$ may be common during the summer months (Brylinsky, pers. comm.). Despite these limitations, zooplankton populations increase with increasing SPM levels, and thus reach high values in turbid estuaries such as Cumberland Basin and the Cornwallis Estuary, where both subtidal benthos and living phytoplankton are non-existent (Daborn, 1984). These populations are apparently sustained by feeding on the large quantities of particulate organic matter derived from local saltmarshes and benthic diatoms (Prouse et al., 1984) and

their associated microflora (Cammen, 1984). Thus, extreme tidal
mixing removes benthic suspension feeders, and apparently creates
suitable conditions for zooplankton populations.

Relationships Between Tidal Mixing and Plankton and Benthos of the
Annapolis Estuary

Relatively more information is available on the Annapolis
Estuary (Fig. 5) than other tributary estuaries of the Fundy system.
With low SPM levels in the Basin (Table 1), a consequence of the
erosion-resistant substrates of the valley, light penetration is
generally good, and the photic zone extends between 8 and 16 m.
Phytoplankton abundance and chlorophyll levels are generally low,
however, the latter being <5 mg·m^{-3} except on rare occasions (Fig.
6). By contrast, light penetration is much more limited in the
Headpond and lower River because of high humic and tannic acids in

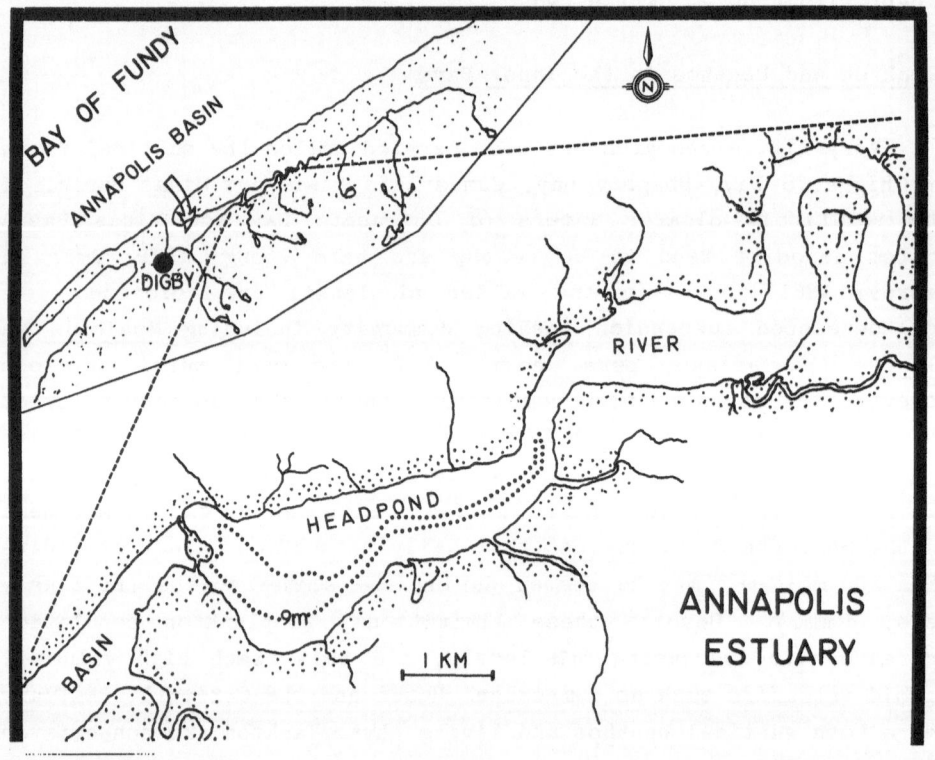

Figure 5. The Annapolis Estuary system.

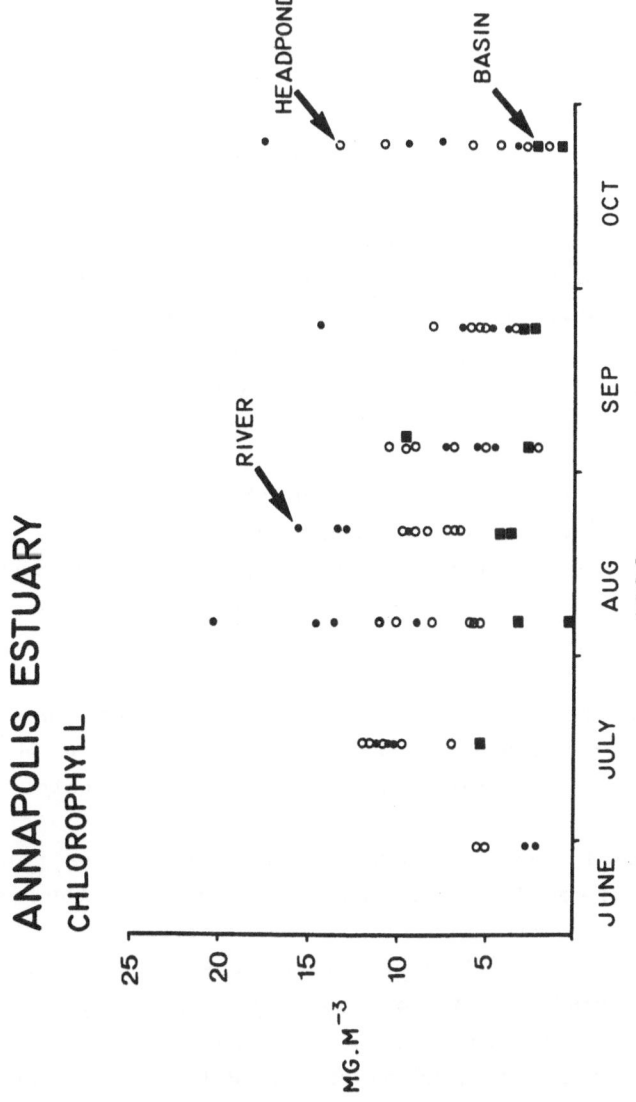

Figure 6. Seasonal variations in photic zone chlorophyll concentrations in the Annapolis Estuary, 1982. ■ - Basin; ○ - Headpond; ● - River stations.

the river waters, so that the photic zone extends only to the pycnocline, approximately 3 m below the surface (Daborn et al., 1979,
1982). Despite the limited light, chlorophyll concentrations are
high (<10 mg·m^{-3}) in the Headpond, and become even higher (<21 mg·
m^{-3}; exceptionally to 45 mg·m^{-3}) in the most strongly stratified
portions of the lower River. Rates of primary production up to 1.5 g
C·m^{-3}·day^{-1} have been measured in surface layers of the Headpond, but
with such rapid attenuation there is no indication of photoinhibition
at the surface, unlike the results for Minas Basin (Brylinsky, pers.
comm.). The 1% light level corresponds roughly to the pycnocline
depth.

Zooplankton populations are also much greater in the Headpond
than in the Annapolis Basin, and steadily increase in surface waters
with distance upstream from the Causeway (Daborn et al., 1982).
Thus, the highest zooplankton densities occur in the most stably
stratified portions of the estuary, where phytoplankton abundance is
also greatest.

Benthic populations show precisely the reverse relationship
throughout the estuary. Below the causeway a diverse and abundant
benthic fauna is found (Figs. 7 and 8) dominated by suspension
feeding forms such as Mytilus edulis, Mya arenaria and Balanus
balanoides (Daborn et al., 1982). Distribution of most individual
species is clearly linked to specific suitable substrates, which in
turn are determined by current velocities. As a consequence, abundance and biomass of the benthic community is greater in the intertidal zone than in the extensively scoured subtidal channels.

In the Headpond, however, substrate type varies progressively
with distance upstream from the Causeway. Near to the Causeway
bottom deposits are indicative of strong current flows: rocky
substrates colonized by a Mytilus-Balanus association, and sandy
deposits being occupied by a Mya-Mytilus-Polydora-Corophium association. Finer sediments prevail upstream, eventually giving way to a
deposit-feeding infaunal association dominated by Nephtys incisa and
Scoloplos acutus.

Although substrate is undoubtedly the major proximal determinant
of the benthic community (Fig. 7) there are indications that turbulence is an important factor. Suspension feeders such as Mytilus

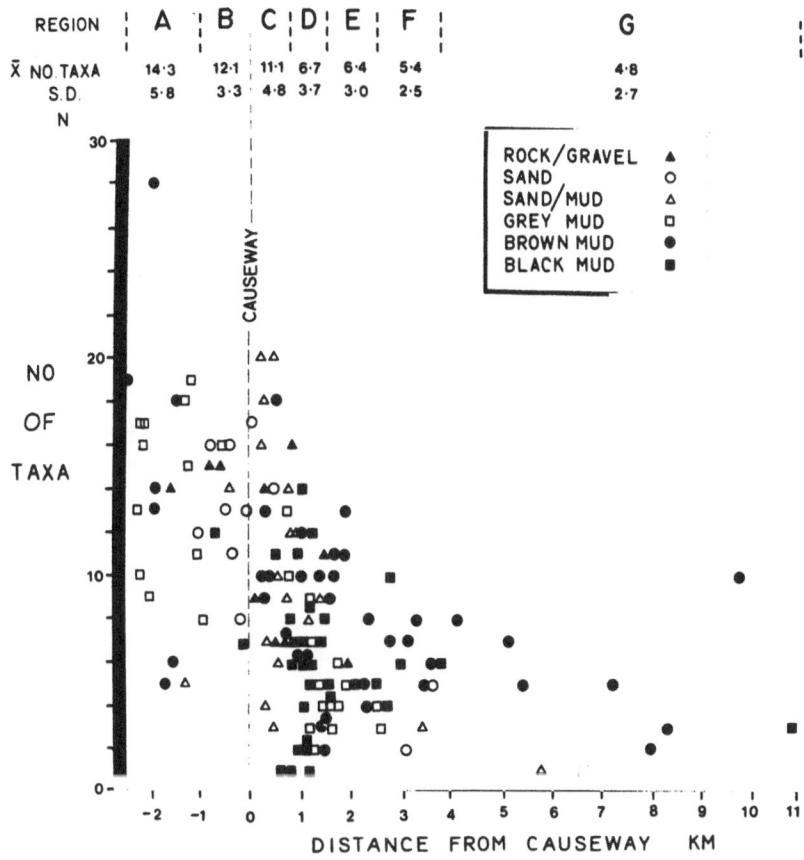

Figure 7. Relationships between benthic diversity (as Number of Taxa), substrate type and distance from Causeway in the Annapolis Estuary.

occur on soft mud deposits in near-shore localities where wave action maintains water circulation, but rarely on firm substrates in the stratified headpond.   In addition, the suspension feeding Cerasto- derma pinnulatum occurs in soft deposits over much of the Headpond, but steadily diminishes in abundance with distance upstream from the Causeway.   In general, benthic abundance decreases with distance upstream (Fig. 8), but since the dominant organisms also are smaller forms, biomass decreases even more sharply.

These results have been summarized in the conceptual model of the Annapolis Estuary shown in Fig. 9.   According to this interpre-

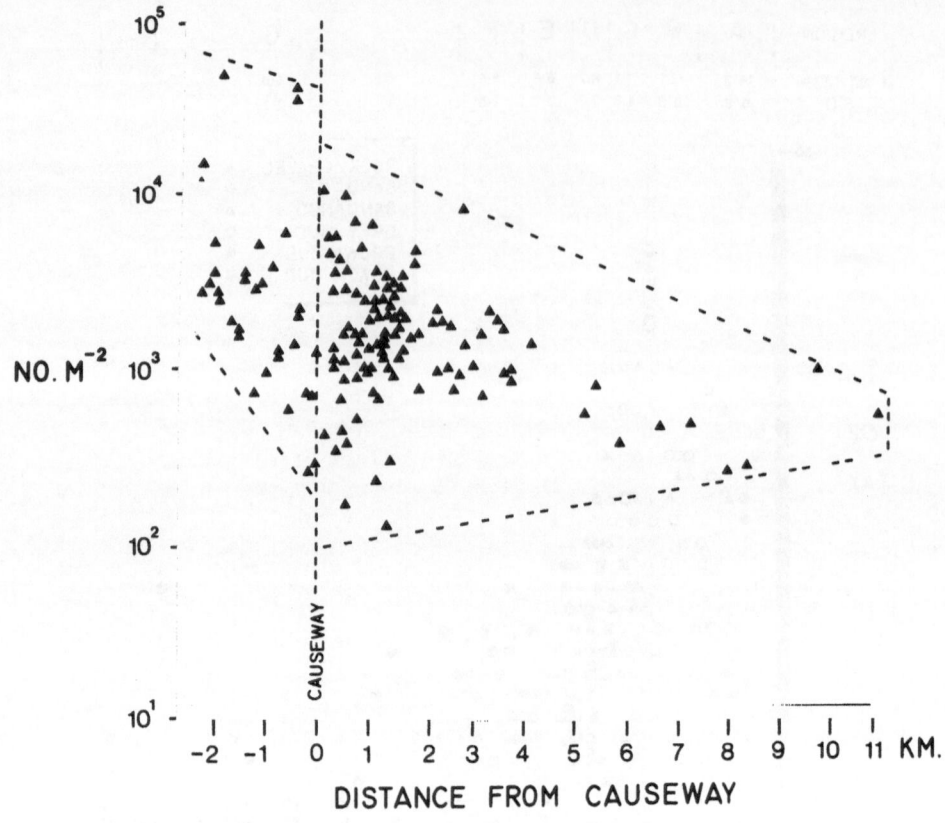

Figure 8. Relationship between abundance of benthic macroinverte-
brates and distance from the Causeway in the Annapolis
Estuary.

tation, extensive vertical mixing in the macrotidal Basin leads to an
abundant and diverse benthic community, the specific distributions of
which depend on substrate type, which is determined by tidal current
velocities. Despite good light penetration and apparently abundant
nutrients, phytoplankton and zooplankton standing crops are low,
presumably as a result of grazing on phytoplankton by suspension-
feeders. The Causeway itself constitutes a major, but not complete,
inhibitor of tidal movements. Tidal water enters the Headpond during
the flood, causing mixing in the immediate vicinity of the Causeway,
but with much less energy the heavier Basin waters tend to sink below
the river water, generating limited upstream currents in the underlying salt wedge. Benthic populations are associated with finer
sediments and diminish in diversity and abundance with distance

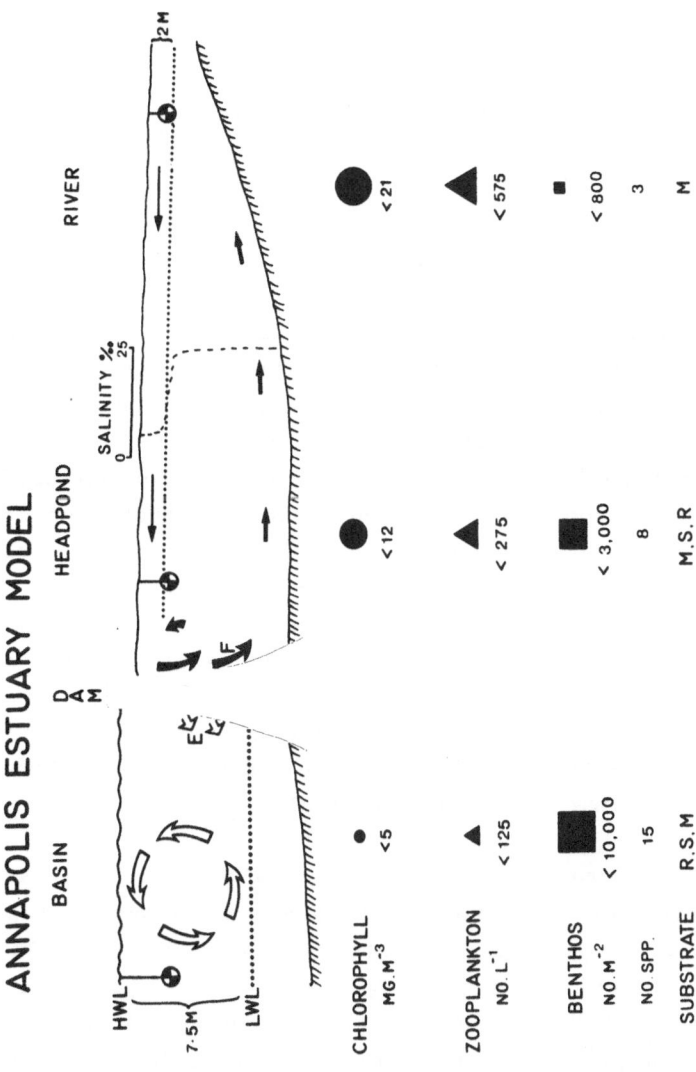

Figure 9. The Annapolis Estuary model. ⊕ Secchi disc depth. Dominant substrate types: R – rock/gravel; S – sand; M – mud.

upstream. In contrast, phytoplankton and zooplankton populations become extremely abundant in surface waters, despite the low light penetration. Because of persistent outflow at the surface, and the strong stratification, the products of primary production are translocated downstream and do not descend to feed the benthos until they have reached the mixed zone near the Causeway. Thus, the great abundance of zooplankters upstream, despite continued losses through river outflow, may be seen as a consequence of the lack of competition with suspension feeders on the bottom. In addition, planktivorous and omnivorous fish feed extensively on photic zone zooplankton in the Annapolis Headpond, whereas the same species feed primarily on benthos in Minas Basin (Redden, pers. comm.; Gilmurray and Daborn, 1981).

DISCUSSION

A number of observations made on estuarine regions of the Bay of Fundy thus appear to be incompatible with the standard mixing model applied to coastal waters, and more conformable with the proposals of Riley (1972), Wolff et al. (1975) and Matthews and Heimdal (1980) that vertical mixing may lead to direct competition for phytoplankton between the pelagic zooplankton and the benthos. Under such circumstances increasing vertical mixing enhances the tight coupling between phytoplankton and benthos (Rowe, 1971; Hargrave, 1973, 1975, 1980) with consequent reduction in zooplankton biomass. Where vertical mixing encompasses the whole of the water column, the system may become dominated by benthic suspension feeders because of their superior filtering capability. Quite obviously, pelagic planktivores should diminish or switch to feeding on the benthos, as has been seen in several cases (Gilmurray and Daborn, 1981; Scully, 1983).

In estuaries where nutrient input from the river is relatively high, the coastal model does not apply, since it is predicated upon the importance of recycling nutrients from the benthos. In such conditions, the phytoplankton and zooplankton appear to be favoured by any degree of stratification, developing abundance and standing crops in proportion to the stability of the water column. Conversely, without direct access to the phytoplankton, benthic populations are lower, and the remainder of the pelagic community should be modified accordingly.

Evidence for the superior competitive ability of benthic suspension feeders is readily obtainable from many other studies of feeding
rates (e.g. Tenore and Dunstan, 1973; Shulte, 1975; Wolff et al.,
1975). In applying the GEMBASE ecosystem model to predict the
consequences of a Severn tidal barrage, Radford (1981) found that
benthic suspension feeders were most successful at exploiting the
mixed regions behind the barrage, and were ultimately responsible for
reducing the phytoplankton bloom conditions that prevailed for the
first years following closure of the dam. This dominance of benthic
suspension feeders was attributed to their superior filtering capabilities, coupled with maintenance of high biomass levels all year
round, in contrast to the greatly varying, multivoltine zooplankton
populations.

The Bay of Fundy results, do, however, suggest that enhanced
benthic production with increased tidal mixing will only occur up to
limiting values of current velocity. In effect there is a "window"
of tidal currents that permits benthic dominance of the system. At
greater velocities, as are commonly encountered in the innermost
portions of the Fundy system, tidal scour coupled with high suspended
sediment levels estimates benthic suspension feeders. The system
once again becomes dominated by estuarine zooplankton populations,
which flourish in the absence of visual predators and benthic compe-
titors (Daborn, 1984). Paradoxically, these pelagic consumers are
supported primarily by producers associated with deposited sediments
- the saltmarsh macrophytes and benthic diatoms of the intertidal
zone (Prouse et al., 1984; Peer.et al., 1980).

While as yet unverified, this model of the influence of tidal
mixing on estuarine plankton and benthos populations provides a more
appropriate conceptual framework for the study of estuarine systems.
It conforms with Valiela's dictum that in coastal waters relatively
few dynamic processes determine most aspects of community structure,
productivity, diversity and succession (Valiela, 1984). Furthermore,
it offers a convenient quantitative mechanism for linking together
the physical and biological submodels upon which current simulation
models of estuarine ecosystems are based. In estuaries, modelling
biological processes is only as good as the physical modelling
allows, and coupling of the two is often a most difficult problem.

ACKNOWLEDGEMENTS

I am grateful to many people for providing unpublished data and for discussions upon which this paper has been based, in particular: M. Brylinsky, A. Redden, G. Brown, R. Gregory, and R. McFarlane. The mistakes are my own. It is a pleasure also to acknowledge support provided for our work on the Bay of Fundy system through grant A9679 from the Natural Sciences and Engineering Research Council, and grants from the Canadian National Sportmen's Fund and the Nova Scotia Tidal Power Corporation.

REFERENCES

Amos, C.L. and K.W. Asprey. 1979. Geophysical and sedimentary studies in the Chignecto Bay system, Bay of Fundy - a progress report. In: Current Research, Part B, Geological Survey of Canada Paper 79-18, pp. 245-252.

Amos, C.L. and K.W. Asprey. 1981. An interpretation of oceanographic and sediment data from the upper Bay of Fundy. Bedford Institute of Oceanography, Report Series BI-R-81-15, Dartmouth, N.S. 143 pp.

Amos, C.L. and G.H.E. Joice. 1977. The sediment budget of the Minas Basin, Bay of Fundy, N.S. Bedford Institute of Oceanography, Data Series BI-D-77-3, Dartmouth, N.S. 411 pp.

Bailey, W.B. 1954. Seasonal variations in the hydrographic conditions of the Bay of Fundy. Manuscript Report (Biological Series). Fish. Res. Bd. Can. No. 21. 56 pp.

Brown, G.S. 1984. The zooplankton of a turbid macrotidal estuary. Unpublished M.Sc. Thesis, Dept. Biology, Acadia Univ., Wolfville, N.S. 233 pp.

Cammen, L.M. 1984. Microbial ecology of the Bay of Fundy. In: Update on the Marine Environmental Consequences of Tidal Power Development in the Upper Reaches of the Bay of Fundy. D.C. Gordon and M.J. Dadswell (eds.). Can. Tech. Rep. Fish. Aquat. Sci. 1256: 115-133.

Daborn, G.R. 1984. Zooplankton studies in the upper Bay of Fundy since 1976. In: Update on the Marine Environmental Consequences of Tidal Power Development in the Upper Reaches of the Bay of Fundy. D.C. Gordon and M.J. Dadswell (eds.). Can. Tech. Rep. Fish. Aquat. Sci. 1256: 135-162.

Daborn, G.R., R.R.G. Williams, J.S. Boates and P.S. Smith. 1979. Limnology of the Annapolis River and Estuary: I. Physical and chemical features. Proc. N.S. Inst. Sci. 29: 153-172.

Daborn, G.R., A.M. Redden and R.S. Gregory. 1982. Ecological studies of the Annapolis Estuary 1981-82. Publ. No. 29, Acadia Univ. Inst., Wolfville, N.S. 80 pp.

Davidson, V.M. 1934. Fluctuations in the abundance of planktonic diatoms in the Passamaquoddy Region, New Brunswick, from 1924 to 1931. Contrib. Can. Biol. Fish. 8: 359-407.

Fournier, R.O., R. Ernst, N.B. Hargreaves, M. Van Det and D. Douglas. 1984. Variability of chlorophyll a off southwestern Nova Scotia in late fall and its relationship to water column stability. Can. J. Fish. Aquat. Sci. 41: 1730-1738.

Fournier, R.O., M. Van Det, J.S. Wilson and N.B. Hargreaves. 1979. Influence of the shelf-break front off Nova Scotia on phytoplankton standing stock in winter. J. Fish. Res. Bd. Can. 36: 1228-1237.

Frechette, M. 1984. The benthic-pelagic interface: mussels, phytoplankton, and the benthic boundary layer. In: Biology of the Sediment-Water Interface: Report of the St. Andrews Biological Station's 75th Anniversary Benthic Workshop. D.J. Wildish (ed.). Can. Tech. Rep. Fish. Aquat. Sci. 1263: 17-18.

Garrett, C.J.R. 1977. Tidal influences on the physical oceanography of the Bay of Fundy and Gulf of Maine. In: Fundy Tidal Power and the Environment, pp. 101-115. G.R. Daborn (ed.). Publ. No. 28, Acadia Univ. Inst., Wolfville, N.S.

Garrett, C.J.R. and R.H. Loucks. 1976. Upwelling along the Yarmouth shore of Nova Scotia. J. Fish. Res. Bd. Can. 33: 116-127.

Garrett, C.J.R., K.R. Keeley and D.A. Greenberg. 1978. Tidal mixing versus thermal stratification in the Bay of Fundy and Gulf of Maine. Atmosphere-Ocean 16: 403-423.

Gilmurray, M.C. and G.R. Daborn. 1981. Feeding relationships of the Atlantic silverside, Menidia menidia, in the Minas Basin, Bay of Fundy. Mar. Biol. Ecol. Prog. Ser. 6: 231-235.

Gran, H.H. 1933. Studies on the biology and chemistry of the Gulf of Maine. II. Distribution of phytoplankton in August 1932. Biol. Bull. 64: 159-182.

Gran, H.H. and T. Braarud. 1935. A quantitative study of the phytoplankton in the Bay of Fundy and the Gulf of Maine (including observations on hydrography, chemistry and turbidity). J. Biol. Bd. Can. 1: 279-467.

Greenberg, D.A. 1984. A review of the physical oceanography of the Bay of Fundy. In: Update on the Environmental Consequences of Tidal Power Development in the Upper Reaches of the Bay of Fundy. D.C. Gordon, Jr. and M.J. Dadswell (eds.). Can. Tech. Rep. Fish. Aquat. Sci. 1256: 9-30.

Hargrave, B.T. 1973. Coupling carbon flow through some pelagic and benthic communities. J. Fish. Res. Bd. Can. 30: 1317-1326.

Hargrave, B.T. 1975. The importance of total and mixed-layer depth in the supply of organic material to bottom communities. Symp. Biol. Hung. 15: 157-165.

Hargrave, B.T. 1980. Factors affecting the flux of organic matter to sediments in a marine bay. In: Marine Benthic Dynamics, pp. 243-263. K.R. Tenore and B.C. Coull (eds.). Univ. S. Carolina Press, Columbia, S.C.

Huntsman, A.G. 1952. The production of life in the Bay of Fundy. Trans. R. Soc. Can. Series 3, Section 5: 15-38.

Keizer, P.D., D.C. Gordon, Jr. and E.R. Hayes. 1984. A brief overview of recent chemical research in the Bay of Fundy. In: Update on the Marine Environmental Consequences of Tidal Power Development in the Upper Reaches of the Bay of Fundy. D.C. Gordon, Jr. and M.J. Dadswell (eds.). Can. Tech. Rep. Fish. Aquat. Sci. 1256: 45-53.

Kremer, J.N. and S.W. Nixon. 1978. A Coastal Marine Ecosystem. Springer-Verlag, NY. 217 pp.

Jermolajev, J.S.S. 1958. Zooplankton of the inner Bay of Fundy. J. Fish. Res. Bd. Can. 15: 1219-1228.

Lakshminarayana, J.S.S. 1983. Phytoplankton of the Quoddy Region. In: Marine and Coastal Systems of the Quoddy Region, New Brunswick. M.L.H. Thomas (ed.). Can. Spec. Publ. Fish. Aquat. Sci. 64: 176-192.

Long, B.F.N. 1979. The nature of bottom sediments in the Minas Basin System. Bedford Inst. Oceanogr. Data Series B1-D-79-4. 31 pp.

Mann, K.H. 1982. Ecology of Coastal Waters. Studies in Ecology, Vol. 8. Univ. Calif. Press, Berkeley. 322 pp.

Matthews, J.B.L. and B.R. Heimdal. 1980. Pelagic productivity and food chains in fjord systems. In: Fjord Oceanography, pp. 377-398. H.J. Freeland, D.M. Farmer and C.D. Levings (eds.). NATO Conf. Series IV4, Plenum Press, NY.

Middleton, G.V. 1977. The sediment regime of the Bay of Fundy. In: Fundy Tidal Power and the Environment, pp. 125-130. G.R. Daborn (ed.). Publ. No. 28, Acadia Univ. Inst., Wolfville, N.S.

Nixon, S.W. 1981. Remineralization and nutrient cycling in coastal marine ecosystems. In: Estuaries and Nutrients, pp. 111-138. B.J. Neilson and L.E. Cronin (eds.). Humana Press, NJ.

Peer, D., D.J. Wildish, A.J. Wilson, J. Hines and M.J. Dadswell. 1980. Sublittoral macro-infauna of the lower Bay of Fundy. Can. Tech. Rep. Fish. Aquat. Sci. No. 981. 74 pp.

Pingree, R.D. 1978. Mixing and stabilization of phytoplankton distributions on the northwest European continental shelf. In: Spatial Pattern in Plankton Communities, pp. 181-200. J.H. Steele (ed.). NATO Conf. Series IV3, Plenum Press, NY.

Pingree, R.D., P.R. Pugh, P.M. Holligan and G.R. Forster. 1975. Summer phytoplankton blooms and red tides along tidal fronts in the approaches to the English Channel. Nature 258: 672-677.

Prouse, N.J., D.C. Gordon, Jr., B.T. Hargrave, C.J. Bird, J. McLachlan, J.S.S. Lakshminarayana, J. Sita Devi and M.L.H. Thomas. 1984. Primary production:organic matter supply to ecosystems in the Bay of Fundy. In: Update on the Marine Environmental Consequences of Tidal Power Development in the Upper Reaches of the Bay of Fundy. Can. Tech. Rep. Fish. Aquat. Sci. No. 1256: 65-95.

Radford, P.J. 1981. Modelling the impact of a tidal power scheme upon the Severn Estuary Ecosystem. Proc. Int. Symp. Energy and Ecological Modelling. Louisville, KY, pp. 235-247.

Riley, G.A. 1972. Patterns of production in marine ecosystems. In: Ecosystem Structure and Function, pp. 91-110. J.A. Wiens (ed.). Oregon State Univ. Press, Corvallis.

Roff, J.C. 1983. Microzooplankton of the Quoddy Region. In: Marine and Coastal Systems of the Quoddy Region, New Brunswick. M.L.H. Thomas (ed.). Can. Spec. Publ. Fish. Aquat. Sci. 64: 201-214.

Rowe, G.T. 1971. Benthic biomass and surface productivity. In: Fertility of the Sea, pp. 441-454. J.D. Costlow (ed.). Gordon and Breach Sci. Publ., NY.

Rowe, G.T., C.H. Clifford, K.L. Smith and P.L. Hamilton. 1975. Benthic nutrient regeneration and its coupling to primary productivity in coastal waters. Nature 255: 215-227.

Schulte, E.H. 1975. Influence of algal concentration and temperature on the filtration rate of Mytilus edulis. Mar. Biol. 30: 331-341.

Scully, B. 1983. The utilization of an intertidal salt-marsh-mudflat system as a nursery area by smooth flounder, Liopsetta putnami (Gill). Unpublished M.Sc. Thesis, Dept. Biology, Acadia Univ., Wolfville, N.S. 183 pp.

Swift, D.J.P. and R.M. McMullen. 1968. Preliminary studies of intertidal sand bodies in the Minas Basin, Bay of Fundy, Nova Scotia. Can. J. Earth Sci. 5: 175-183.

Tenore, K.R. and W.M. Dunstan. 1973. Comparison of feeding and biodeposition of three bivalves at different food levels. Mar. Biol. 21: 190-195.

Valiela, I. 1984. Marine Ecological Processes. Springer-Verlag, NY. 546 pp.

Wildish, D.J. 1977. The marine and estuarine sublittoral benthos of the Bay of Fundy and Gulf of Maine. In: Fundy Tidal Power and the Environment, pp. 160-163. G.R. Daborn (ed.). Publ. No. 28, Acadia Univ. Inst., Wolfville, N.S.

Wildish, C.J. 1983. Sublittoral sedimentary substrates. In: Marine and Coastal Systems of the Quoddy Region, New Brunswick. M.L.H. Thomas (ed.). Can. Spec. Publ. Fish. Aquat. Sci. 64: 140-155.

Wildish, D.J. and D.D. Kristmanson. 1979. Tidal energy and sublittoral macrobenthic animals in estuaries. J. Fish. Res. Bd. Can. 36: 1197-1206.

Wildish, D.J. and D. Peer. 1983. Tidal current speed and production of benthic macrofauna in the lower Bay of Fundy. Can. J. Fish. Aquat. Sci. 40(Suppl. 1): 309-321.

Wolff, W.J., F. Vegter, H.G. Mulder and T. Meijs. 1975. The production of benthic animals in relation to the phytoplankton production. Observations in the saline Lake Grevelingen, The Netherlands. 10th European Symposium on Marine Biology, Ostend, Belgium. Vol. 2: 653-672.

REAL-TIME CHARACTERIZATION OF INDIVIDUAL MARINE PARTICLES
AT SEA: FLOW CYTOMETRY

C.M. Yentsch, T.L. Cucci, D.A. Phinney and J.A. Topinka
Jane J. MacIsaac Flow Cytometer/Sorting Facility
Bigelow Laboratory for Ocean Sciences
McKown Point
W. Boothbay Harbor, ME   04575 USA

INTRODUCTION

Oceanographers have always wrestled with the time-space dilemma. The enormity of the oceans and the manifold physical, chemical and biological processes compound to make 'sampling' in the oceanic realm an omni-present problem. For the most part, the oceanographer has been wed to the research vessel. This has heavily restricted and frequently determined experimental design. It is unlikely that measurements which are widely spaced in time and space will provide further fruitful information on the rate of processes in the oceans.

Ocean biologists need instrumentation that at least matches the capability of data sampling and handling of their chemical and physical colleagues. We have found optical techniques to be among the msot useful to achieve this goal.

Visible Light and Life are Closely Related

A multi-user Flow Cytometer/Sorting Facility, named in honor of the late Jane J. MacIsaac, has been established at Bigelow Laboratory for Ocean Sciences, West Boothbay Harbor, Maine 04575, through resources of the National Science Foundation and the Office of Naval Research. It is open to use by the entire oceanographic community. The focus is realization and utilization of the potential of optical instruments to characterize water masses and the organisms contained therein. Special attention is paid toward individual particle analysis.

Flow cytometric (FCM) instrumentation permits simultaneous multiple parameter analysis of individual particles at rapid rates

Lecture Notes on Coastal and Estuarine Studies, Vol. 17
Tidal Mixing and Plankton Dynamics. Edited by J. Bowman, M. Yentsch and W. T. Peterson
© Springer-Verlag Berlin Heidelberg 1986

$(10^3$ per second). Living aquatic plankton in the 1–200 μm size range can be characterized by cell volume, pigment fluorescence, dye fluorescence and light scattering properties per cell. Resulting data are informative in the analysis of subpopulations by cell volume (termed allometry) and pigment groups (termed ataxonomic analysis). This instrumentation permits the testing of hypotheses previously intractable by either traditional bulk analyses or size fractionation analyses and aids biological oceanographers in attempts to make real-time observations and assessments of plankton communities.

Traditionally, interdisciplinary studies have often been undertaken where small scale physical and chemical observations of aquatic features are made. Biologists involved in these studies are, for the most part, left with large numbers of filtered bulk samples to analyze. Some measurements, such as chlorophyll fluorescence, can be made in a continuous fashion (Lorenzen, 1966), however whole population assemblages are considered with no information as to individuals or groups which constitute the population.

The purpose of this chapter is to provide background information on innovative application of individual particle analysis in the oceanic environment. This includes Coulter Counter Spectra and Flow Cytometry. Great strides are also being made in image analysis, however these are not covered in this chapter.

A. Coulter Counter Spectra

Automated individual particle analysis is not new. Coulter Counter spectra, or so called Sheldon spectra (Sheldon and Parsons, 1967; Sheldon et al., 1972), have been gathered in aquatic systems for nearly 2 decades. Each particle, whose volume is estimated by the change in resistivity of an electrolyte across an orifice due to displacement by the particle, is allotted a "count" in one of 256 channels. The spectra have been analyzed in a variety of formats with a significant advantage being the observation of the kinetics of discrete subpopulations which are often present in the marine environment, as well as grazing effects (Sheldon, pers. com.).

Comparison of Coulter data and microscopic counts for unialgal cultures (Maloney et al., 1962) have found good agreement. However, in natural freshwater populations estimates of phytoplankton cell

number, volume and size distribution are often poor (Arvola, 1984).
Our intercomparison has resulted in a few important generalities:

  1) Large sized cells (20-45 μm); 1:1 correlation

  2) Medium sized cells (8-20 μm); reasonable correlation

  3) Small sized cells (2-8 μm); poor correlation, presumably
     due to detritus

  4) Pico-sized cells (<2 μm); no information.

Sheldon reports greater contamination from detritus in the 8-20 μm
size fraction (pers. comm.).

    Several realizations have emerged from these studies. First, it
is impossible to differentiate live vs. dead cells with electronic
particle counters. Second, detritus adds a severe problem as it is a
routine component in natural samples. Third, many pico-sized cells
were overlooked until the advent of epifluorescence microscopy and
the pioneering work of Waterbury et al. (1979) and Sieburth et al.
(1979).

B. Flow Cytometry

    Flow cytometry is a technique of measuring individual cells/
particles one-by-one in a fluid flow (Kruth, 1982; Melamed, et al.,
1979; and Trask et al., 1982). Many parameters can be measured.
Instruments are in existence which measure fluorescence, forward
angle light scatter, 90° light scatter, particle volume, and sonic
transmittance.

    The advance of flow cytometers as a general class of instruments
over previously existing techniques include 1) rapid analysis of
individual particles, 2) simultaneous multiparameter analysis and 3)
ability to sort subpopulations (Horan and Wheeless, 1977). Thus,
instead of a single particle volume measurement, we can now combine
this signal with fluorescence. If the particle is 10 μm in equivalent spherical diameter and it has fluorescence emission at >650 nm
when excited with 488 nm light, we can now state quite assuredly that
it is chlorophyll containing, therefore, phytoplankton. If the

particle has no fluorescence emission under the above conditions, then we can only speculate as to whether it is a) sediment, b) micro-heterotroph, c) detritus, d) other. With the addition of a fluorescent stain for proteinaceous material - a positively stained particle would lead us to favor the interpretation as b, a microheterotroph. An idealized simple scheme has been presented by Yentsch et al. (1983a). With sorting capabilities, a flow cytometer can then physically sort types of cells (phytoplankton, microheterotrophs, etc.) for collection and further analysis/verification.

The use of fluorescence is precisely where the gains have been made in recent years with the instrumentation described in this chapter. Criteria can be established which are specific to living phytoplankton. Figure 1 shows the autofluorescent structures of procaryote (predominately phycoerythrin) and eucaryote (chlorophyll) cells. Within these cells, structures can also be stained with fluorescent probes to further differentiate cell populations. Combining yet another parameter, light scatter, gives even further confidence in interpretation. Forward angle light scatter has some relevance to cell size (Spinrad and Yentsch, submitted), and 90° light scatter can be an index of surface texture and intracellular organization. We are thus approaching the real-time assessment of oceanic particles utilizing flow cytometers, of which several types exist, optimized for specific measurements. In doing so, we accept that some detail must be sacrificed in order to gain rapid real-time analysis.

Instrumentation

Commercial flow cytometers generally combine light scatter, fluorescence and particle volume measurements. Manufacturers in the U.S. include Becton Dickinson, Sunnyvale, CA; Coulter Electronics, Hialeah, FL; and Ortho Diagnostic Systems, Inc., Westwood, MA. Additionally, there are some European manufacturers. A summary of many features are covered by Van Dilla and Mendelsohn (1977) and more recently by Shapiro (1985).

Due to the high costs of commercial instruments, many investigators are constructing their own flow cytometers using the basic components of an epifluorescence microscope plus a channel analyzer and photomultiplier tubes. One such unit is described by Olson et

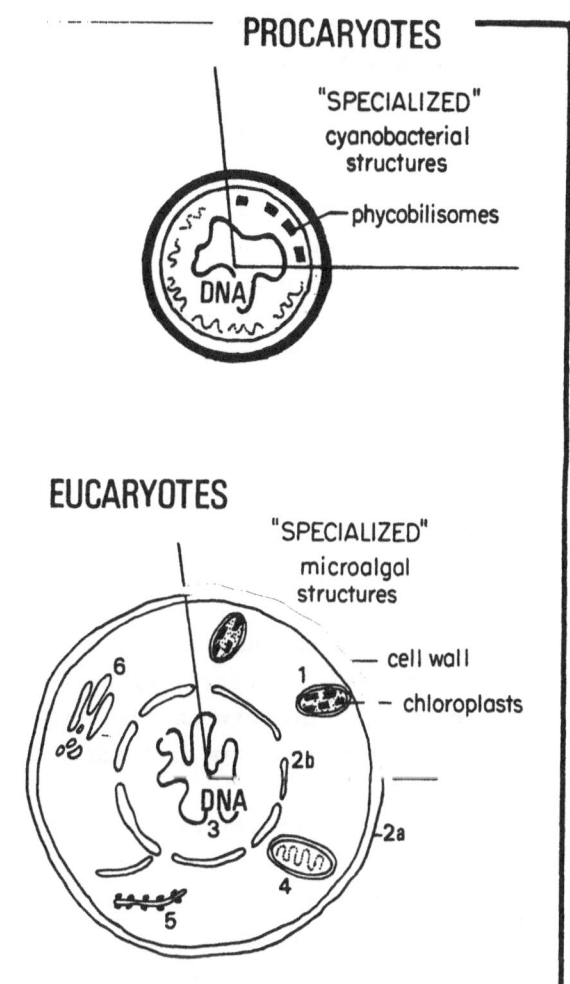

Figure 1. Schematic representation of procaryote and eucaryote cells depicting specialized features exploited by flow cytometry. Pigments are in the phycobilisomes (site of phycoerythrin fluorescence in procaryotes) of cyanobacteria, and the cell wall composition and chloroplasts (site of chlorophyll fluorescence) in other microalgae. Additionally, cell membranes (2a,b) and DNA (3) can be stained. There are fluorescent stains which emphasize the mitochondria (4). The ribosomes (5) and golgi apparatus (6) have not been focused upon, however, there are protein stains which highlight the ribosomes (5), the site of protein production.

al. (1983). There are, of course, limitations with such fabricated units, but if they can fulfill the needs of the research questions at hand, rapid automated measurements can be achieved for only a minor investment above that of an epifluorescence microscope.

One limitation imposed by the commercial instruments made by Ortho Diagnostics should be mentioned. The sample is injected into the laminar flow via a syringe pump which lacks flexibility for non-routine exploratory studies. Introduction of fixed sample volumes requires optimum cell concentrations of 10^5-10^6 cells·ml^{-1}. As illustrated in Figure 2, unconcentrated natural seawater samples (1-100μm size fraction) contain approximate cell concentrations (cells·ml^{-1}) of bacteria (10^6) and cyanobacteria (10^5) suitable for analysis. However, autotrophic phytoplankton (10^2-10^4 cells·ml^{-1}) and phagotrophic flagellates (10^3 cells·ml^{-1}) may need to be concentrated.

The Coulter EPICS V and the Becton Dickinson FACS Analyzer are both used in our laboratory to characterize marine phytoplankton populations. Many similarities between the instruments are to be noted, however certain differences restrict their usefulness.

The fluorescence activated cytometer system FACS Analyzer (Becton-Dickinson, Sunnyvale, CA) is a benchtop flow cytometer (photo, Figure 3). It utilizes a mercury arc lamp as its excitation source, which is as epi-illumination (Fig. 4). It has the capability of measuring 4 parameters simultaneously: impedence (Coulter) volume, two colors of fluorescence and 90° light scatter. The major mercury lines available are 405, 436 and 546 nm, with somewhat less output power at 365 and 488 nm. Output power is fixed (constant) and the instrument has no sorting capabilities. It is compact and rugged, therefore easily taken to sea.

The Coulter EPICS V (manufactured by Coulter Electronics, Hialeah, FL,) is a free-standing flow cytometer (photo, Figure 5) with one or two Coherent (Palo Alto, CA) Model 90-5 5 watt argon ion lasers as its excitation source (Fig. 6). Major lines are 488 and 514 nm (1500 and 2000 mW maximum output, respectively) with less output available at 365, 457, 465, 476, 496 and 502 nm (<600 mW). It has the capability to measure 6 parameters simultaneously (currently restricted in our laboratory to 4 by the number of detectors):

MICRO-FOOD WEB

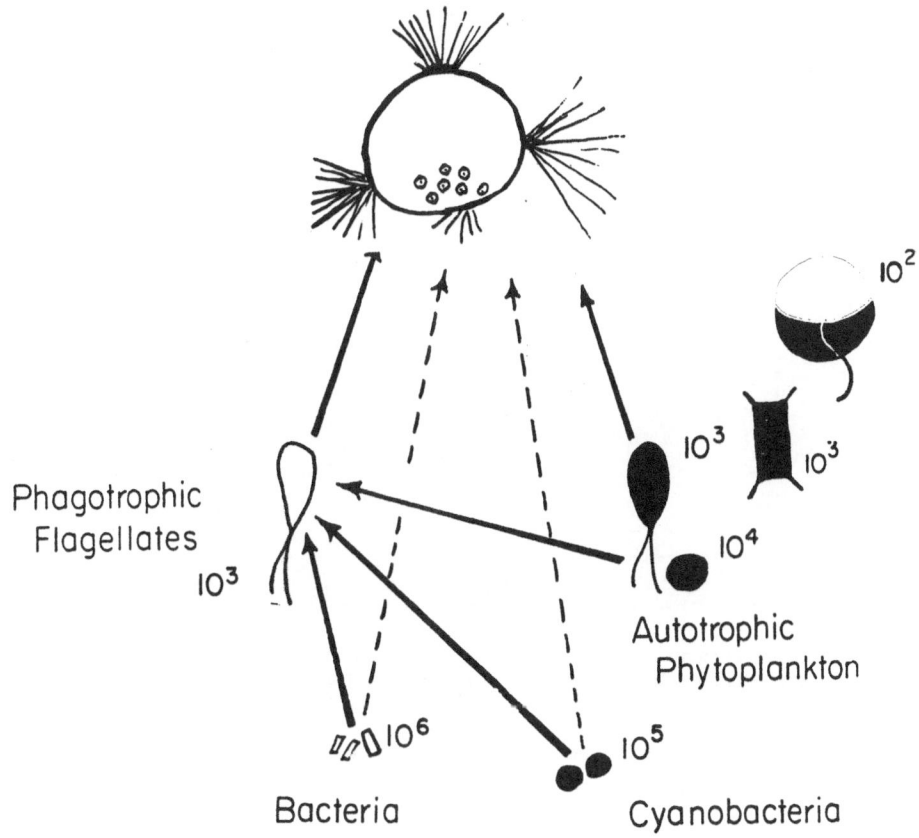

Phagotrophic
Flagellates

10^3

10^2

10^3

10^3

10^4

Autotrophic
Phytoplankton

10^6

Bacteria

10^5

Cyanobacteria

APPROXIMATE CELLS $\cdot$ ml^{-1}

Figure 2. Schematic representation (adapted from M. Landry, pers. comm. PRPOOS Program, University of Washington) depicting approximate numbers of cells on a per ml basis and trophic pathways. Blackened cells signify ability to characterize by autofluorescence.

Figure 3. Photograph of FACS Analyzer, bench top flow cytometer with epi-illumination from a mercury lamp.

forward angle light scatter (related to cell size), two colors of fluorescence and 90° light scatter. Output power from the laser is fully adjustable and sorting is a routine exercise. (For a more complete description of sorting, see Yentsch et al., 1983b). Simultaneous Coulter volume is not currently available for the EPICS. It has been taken to sea (Olson et al., submitted), however its physical size as well as power and cooling water requirements for the laser deem consideration. The Epics C is a somewhat smaller, single laser sorting system.

In both the FACS and EPICS instruments, sample delivery systems pass cells in single file through the focal point of the excitation source. Light from fluorescing phytoplankton is collected by a lens and delivered to photomultiplier tubes (PMT). A dichroic filter, placed at 45° to the incident light, splits the fluorescent light with short wavelengths going to the 'green' PMT and longer wavelengths to the 'red' PMT. Additional filters placed directly in front of the PMT's select the final wavelengths of fluorescence

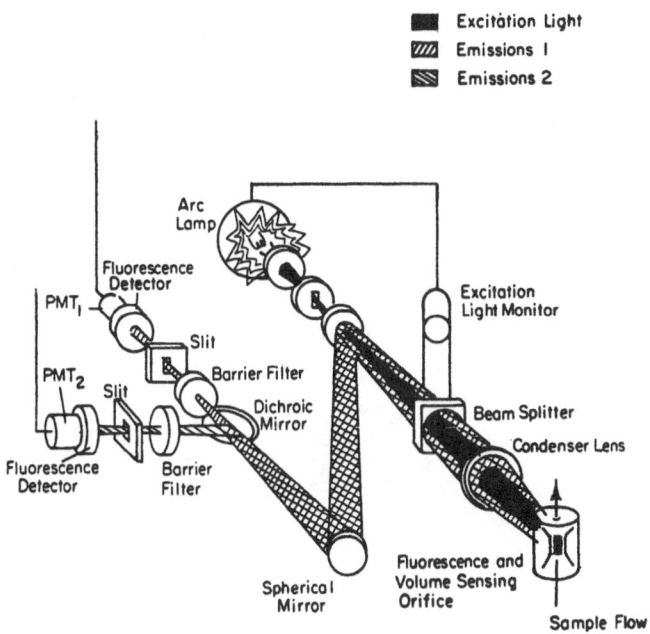

Excitation Light
Emissions I
Emissions 2

Figure 4. Schematic diagram of FACS Analyzer. Excitation light beam is solid and emission light is cross-hatched. There are four sensors, 2 PMTs for fluorescence, side scatter light detector, and volume sensor (independent of light source).

to be measured. Chlorophyll fluorescence emission is detected at wavelengths 650-700 nm and phycoerythrin fluorescence emission detected between 560-590 nm. Comparisons of the FACS and EPICS are summarized in Table I.

Each instrument has a dedicated high speed data acquisition and graphics display computer. Analog signals from each detector are digitized and treated in one of three ways.

1) List mode data acquisition, which means that every characteristic is identified and kept discrete with the cell identification, maintains all parameter data for each cell such that a complete fingerprint is established. The advantage is that data can be exhaustively re-analyzed with no loss of information on individual cells. The disadvantage is that much disc storage space is required.

Table I.

		FACS Analyzer	EPICS V
1.	size requirements	5'x 3'x2' (L X W X H) bench top	12'x 3'x 6' floor space free standing
2.	light source	mercury arc lamp standard power air cooled lamp fixed lamp output epi-illumination focused on orifice	5W argon ion laser 3 phase power 2.2 gal/min cooling water adjustable laser output
3.	cell sizing parameter	impedence volume simultaneous measurement performs 1-100 μm range	forward angle light scatter (FALS) $1-10^{\circ}$ or $1-19^{\circ}$ Coulter volume option is not a simultaneous measurement
4.	light scatter	90° (side scatter)	FALS plus 90°
5.	dual wavelength excitation	not available	need second laser
6.	simultaneous dual wavelength emission (ataxonomic analysis)	not currently available	available
7.	sorting capability	no	yes
8.	ease of data handling	good; fast	fair on instrument dedicated computer, better on independent extended analysis system
9.	total cost	< 80K	< 200K

Figure 5. Photograph of EPICS V flow cytometer/sorter. The excitation is achieved with an argon-ion laser with "infinitely" controllable light intensity.

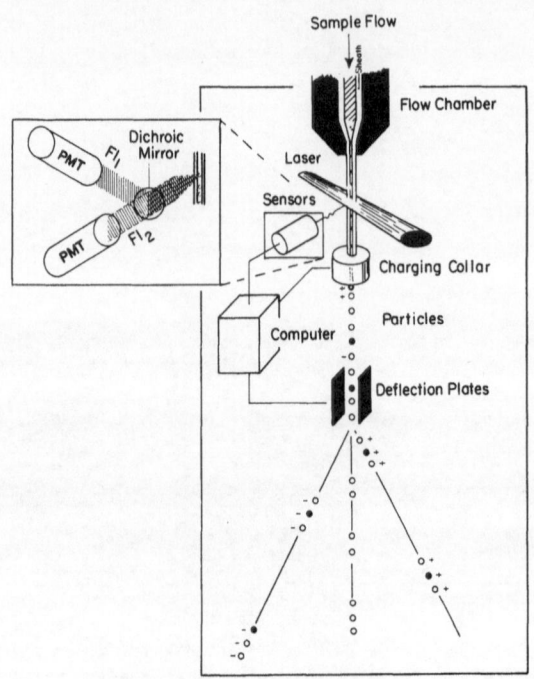

Figure 6. Schematic diagram of EPICS V flow cytometer and cell
sorter. There are four sensors, 2 PMTs for fluorescence, a
photodiode for forward angle light scatter (less obscura-
tion bar angle) and 90° light scatter can be collected by a
third PMT.

2) Single parameter histograms place digitized values (or
events) of one parameter into one of 256 channels of a linear or 3
decade log histogram. Channels represent relative units increasing
to the right. Once data are deposited in channels, they are
irretrievable, thus the identity of information on any particular
cell is lost.

3) Bivariate (dual parameter) histograms place data in one
channel of a 64x64 channel, x-y grid corresponding to the relative
units of two parameters. Again, the identity of information is
irretrievable once placed in the histogram.

"Gates" can be established on one or more parameters which
specify the criteria or range of channels containing cells of
interest. The computer will not recognize a cell that does not meet

the criteria (falls outside the proper channels) of a gated parameter, such that the information for that cell will not be placed in
a histogram. Gates can be used when re-analyzing list mode data to
establish subpopulations exhibiting specific attributes. "Sort
windows" are similar to gates, specifying the criteria of cells to be
sorted.

Instrument standards are used to assure stability of the flow
cytometer during an experiment. Fluorescent beads (microspheres) are
used to initially align the instrument for optimum performance.
Changes in excitation power, optical alignment and sample delivery
system flow characteristics can affect instrument performance,
therefore it is essential to monitor beads throughout the course of
an experiment. A simple method used by several investigators is to
add the fluorescent beads to each sample to monitor any instrumental
drift.

Calibration standards are used to apply absolute values to the
relative units of histogram channels. Volume measurements can be
calibrated, incorporating standard volume microspheres, which give
absolute dimensions on the volume axis. One equates the size of a
natural particle to an equivalent sphere. Equally important is the
need for absolute calibration of fluorescence. This is far more
difficult to achieve than volume calibration. Calibrated fluorescein
fluorescent beads are available (Flow Cytometry Standards, P.O. Box
12621, Research Triangle Park, NC 27709). For chlorophyll, bulk
measurements must still be made on extracts from a specific number of
cells obtained by sorting. Both calibrated chlorophyll and phycoerythrin beads are needed for quantitative work, as of this writing they
are under development but have not been reliably produced.

All of the components mentioned above combine to make measurements of individual cells at rates exceeding 10^3 per second. Given
sufficient cell concentrations in a sample ($>10^5$ per ml), tens of
thousands of cells can be analyzed in near real-time.

Viability of Sorted Cells

If cells are to be analyzed for their chemical content, it is
important that they remain intact and viable (Hutter and Eipel,
1978). During cell analysis and sorting, cells are (1) injected into

a laminar flow system, (2) intersect the path of a laser beam, and (3) are contained in an electrically charged droplet during the sorting process. All these events may either breakup cells or introduce stress which may alter physiological processes.

The viability of sorted cells was examined in Dunaliella tertiolecta. Cells having low chl a fluorescence and cells having high fluorescence were sorted from a D. tertiolecta culture. These sorted populations gave distinct high and low fluorescence histograms. The histogram peak of the high chl a sort shortly disappeared with the accompanying reformation of a low fluorescence peak, suggesting that virtually all of the high chl a cells were about to divide and did so to form daughter cells having half the original chl a (Campbell et al., submitted).

The low fluorescence sort advanced to higher fluorescence levels with time and after reaching a high chl a level, it too disappeared to reform as low fluorescence daughter cells. These observations demonstrate that sorted cells retain their capacity for chl a synthesis and cell division after cell sorting procedures. Subsequent efforts with other phytoplankton cultures have also shown that sorted algal cells remain intact and viable (D. Phinney, pers. comm.). Yet, physiological damage can result (R. Rivkin, pers. comm.; J. Cullen, pers. comm.).

Increased Cell Sorting Rates

Sample flow needs to be increased as much as possible in order to sort the often large numbers of cells needed for chemical analysis, eg. C:H:N (Table II). Sample flow rates are controlled by sample pressure which is read on a magnehelic differential pressure gauge (EPICS V). Sample flow rates are normally 30-40 $\mu l \cdot min^{-1}$ using the 76 μm aperature flow cell. While it is possible to increase sample flow rates to approximately 140 $\mu l \cdot min^{-1}$ (Fig. 7), particle alignment in the flow cytometer laser beam becomes less precise resulting in a loss of analytical resolution. Using standard 10μm fluorescent latex spheres it was only possible to increase sample pressure levels to magnehelic readings of 1 and 0 (higher flow rates at low magnehelic readings) on forward angle light scatter and fluorescence scales, respectively, before histogram resolution began

Table II. Carbon and nitrogen cell content of phytoplankton with an estimation of cell numbers needed for analysis.

Species	Cell volume (μ^3)	$C \cdot cell^{-1}$ (ng)	$N \cdot cell^{-1}$ (ng)	Cell number for analysis[1] ($\times 10^3$)	
				C	N
Chlorophyceae					
Tetraselmis maculata	310	0.8×10^{-1}	0.19×10^{-1}	63,000	263,000
Dunaliella salina	400	0.59×10^{-1}	0.12×10^{-1}	85,000	417,000
Chrysophyceae					
Monochrysis lutheri	28	1.54×10^{-2}	0.27×10^{-2}	325,000	1,852,000
Syracosphaera carterae	1,760	1.67×10^{-1}	0.43×10^{-1}	30,000	116,000
Bacillariophyceae					
Chaetoceros sp.	650	-	0.13×10^{-1}	-	385,000
Skeletonema costatum	1,390	2.09×10^{-1}	0.39×10^{-1}	24,000	128,000
Phaeodactylum tricornutum	120	1.07×10^{-2}	0.11×10^{-2}	467,000	4,545,000
Dinophyceae					
Amphidinium carteri	740	1.18×10^{-1}	0.11×10^{-1}	42,000	455,000
Exuviella sp.	780	1.25×10^{-1}	0.13×10^{-1}	40,000	385,000
Myxophyceae					
Agmenellum quadrauplicatum	1.5	3.74×10^{-4}	0.48×10^{-4}	13,337,000	104,167,000

[1] assuming at 5 μg analytical target

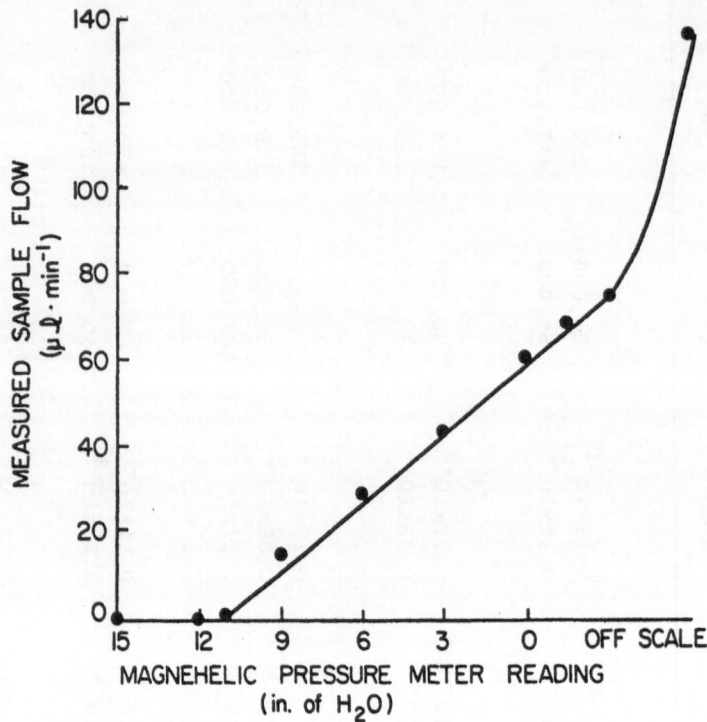

Figure 7. Sample flow rates from the Coulter EPICS V flow cytometer, measured gravimetrically.

to suffer greatly (Fig. 8). These sample pressure levels corresponded to flow rates of approximately 60 μl$\cdot$min^{-1}, significantly increasing normal flow rates. Flow rates higher than this are characterized by expanded histogram ranges, histogram peaks which exhibit greater coefficients of variation (C.V.) and an overall loss of histogram resolution or integrity.

While sample flow rates may be increased by increasing cell sorting speeds, it will also be necessary to concentrate cells, as is discussed later.

Allometric Relationships

Several workers (Sheldon and Parsons, 1967; Sheldon et al., 1972; Sieburth et al., 1979) have subdivided oceanic particles on the basis of size. Allometric relationships are of great interest to

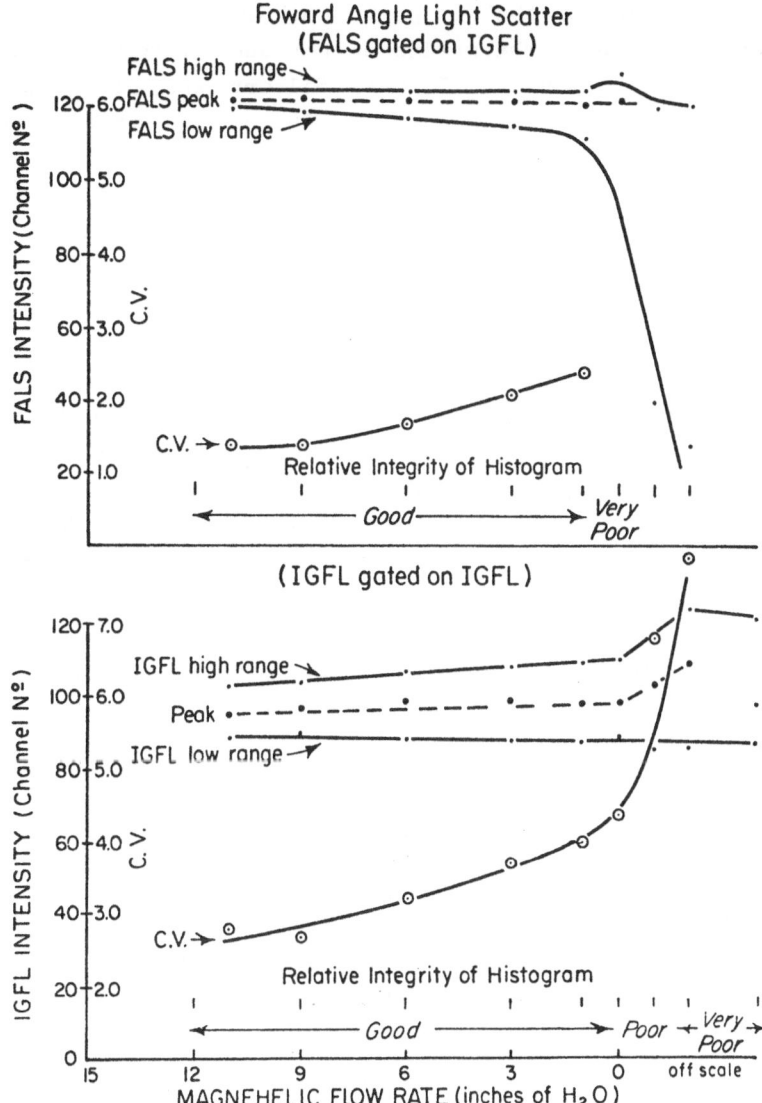

Figure 8. The effects of increased sample flow rates on the quality of forward angle light scatter and fluorescence histograms, using a Coulter EPICS V flow cytometer.

theoretical and experimental researchers, and as previously discussed, electronic particle counters have been used for almost 2 decades in the marine environment.

Recently, due to the recognition of strong ecological implications of body size in the marine environment (Margalef, 1963; Kerr, 1974; Platt and Denman 1977,1978; Silvert and Platt, 1978,1980; Malone, 1980; Turpin and Harrison, 1980; Platt and Silvert, 1981; Thomann, 1981; Blasco et al., 1982; Peters, 1983), the small end of the size/volume spectra has been of particular interest. Unfortunately, this is the region that is at present poorly resolved analytically due to both instrument constraints and the detrital component. We have applied the FACS analyzer to research in this area for two reasons. 1) Impedence volume measured by the FACS is identical to Coulter volume measurements. Standard particles are used to calibrate the instrument and the same algorithm is applied to the data, thus cell volume as measured by the FACS is directly comparable to Coulter volume data. The EPICS V measures forward angle light scatter which is generally related to cell size. While this relationship has been established for mammalian cells (Salzman, 1982; Salzman et al., 1979) no clear demonstration of this relationship in phytoplankton has yet been made. 2) Fluorescence is introduced to eliminate the problem of detritus. Gates are set on pigment fluorescence which require a particle to fluoresce in order to be included in an analysis.

Figure 9 shows the results of a FACS analysis of unialgal cultures (14/10 hr, $15^{\circ}C$, f/2 media) representing four pigment groups listed in Table III. It is clear that we are unable to separate or discriminate amongst several of these clones on the basis of these two parameters alone, yet an envelope is obvious which binds the cells within a seemingly prescribed chlorophyll per unit volume regime.

The question arises, does this chlorophyll/volume relationship hold in nature? The natural populations used to address the question were from a profile collected at sea on 6 September 1984. The station was located in the western Gulf of Maine approximately 20 miles offshore. The station depicts a reasonably strong chlorophyll maximum in a stratified water column during late summer. Dominant species were Gymnodinium sp. and Gyrodinium sp. (naked dinoflagellates), Prorocentrum minima (dinoflagellate), Pyramimonas sp. (prasinophyte), Eutreptia (Euglenophyte), and several species of coccolithophores. Four discrete depths were sampled from the pump profile and run on the FACS Analyzer aboard the R/V Cape Hatteras. Results are shown in

Table III. Cells used in analysis on FACS Analyzer and EPICS V comparison.

cell size	cell type; clonal designation	fluorescence emission
~1 µm	cyanobacteria; DC-2	predominantly phycoerythrin
~3 µm	prasinophyte; unidentified Ω48-23	chl a
~4 µm	diatom; Thalassiosira pseudonana 3H	chl a
~4-6x8-12 µm	green alga; Dunaliella tertiolecta DUN	chl a
~12-24 µm	diatom; Phaeodactylum tricornutum Phaeo	chl a
~8-12 µm	cryptomonad; Chroomonas salinas 3C	chl a + phycoerythrin
~14-24 µm	dinoflagellate; Heterocapsa triquetra HT 984	chl a
~15 µm	dinoflagellate; Prorocentrum sp. EXUV	chl a
~25-35 µm	dinoflagellate; Gonyaulax tamarensis	chl a

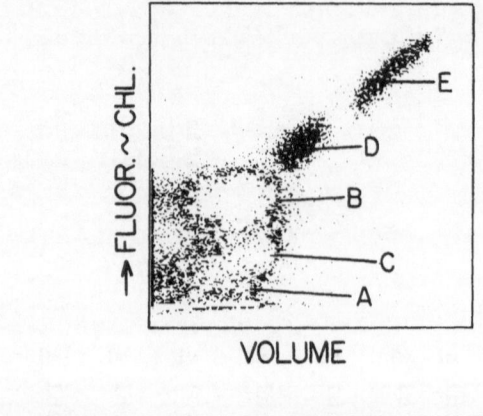

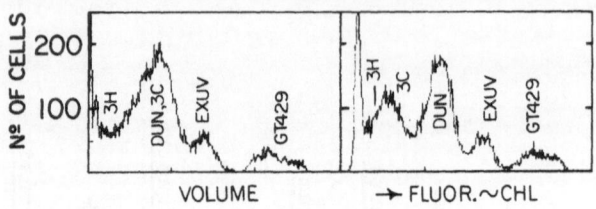

Figure 9. Top plate. Bivariate plot showing the analysis of a mixed
sample of phytoplankton cultures analyzed with the FACS
Analyzer. The X and Y axes are volume (3 decade log scale) a
fluorescence approximating chlorophyll (3 decade log scale),
respectively, with total number of events (cells measured)
along the Z axis represented by individual dots. (The denser
the dots, the more cells occupy that channel.) Cultures
represented by B (Dun), D (Exuv), and E (GT429) are easily
discernible from each other. However, cultures A (3H) and C
(3C) have the similar fluorescence intensity and volumes and
cannot be easily separated by these two measured parameters.
Bottom plate. The same data, but presented as volume and
chlorophyll fluorescence distribution histograms.

Fig. 10. In each case, 25,000 events were analyzed, gated on chlorophyll fluorescence. While these data sets look remarkably similar
there are a few details to note. 1) The slope (1.25) of the chlorophyll per cell/volume per cell relationship appears constant with
depth. The size and shape of an envelope around these populations
differs only slightly. 2) Abundance is weighted heavily toward
smaller cells at all depths. 3) The relative chlorophyll per cell at
any particular volume is similar in 1.8 m and 5 m samples, decreased
at SCM, 10.5 m, and increased below the SCM, at 15.8 m.

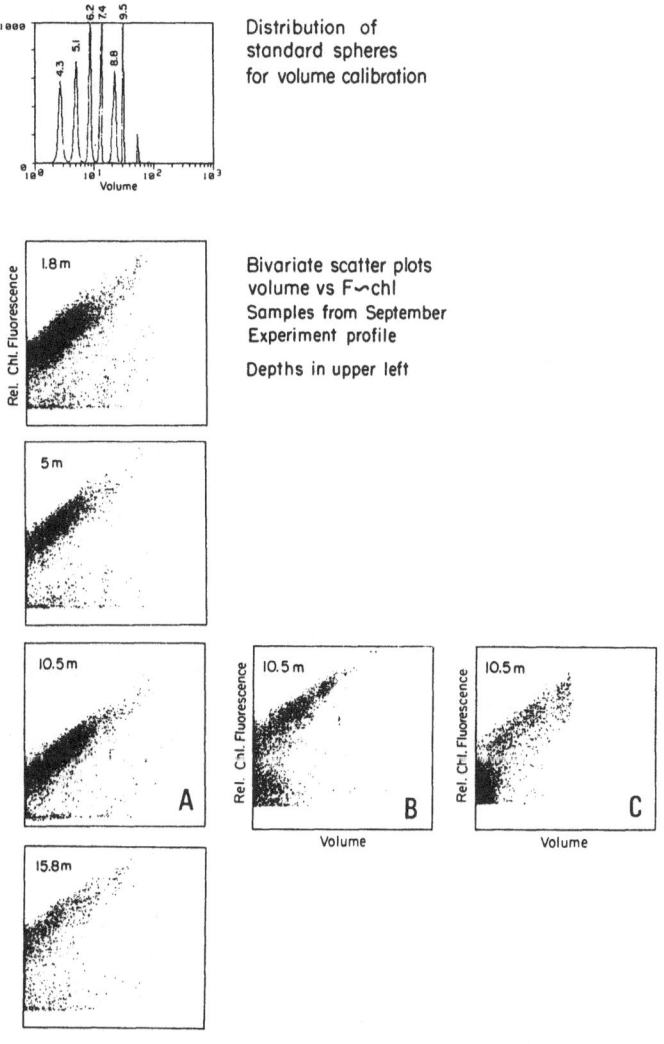

Distribution of
standard spheres
for volume calibration

Bivariate scatter plots
volume vs F~chl
Samples from September
Experiment profile

Depths in upper left

Figure 10. Bivariate plots of volume (three decade log scale) vs.
relative fluorescence approximating chlorophyll (three
decade log scale) of natural populations sampled at the
four depths during the September Experiment (1.8 m, 5 m,
10.5 m (A), and 15.8 m). The FACS Analyzer was used for
the analyses with 25,000 events counted for each sample,
except the 15.8m sample which had a count of only 2500
events. All analyses were gated on chlorophyll. (B) The
same as A, sampled from the holding tank during the
September Experiment 9 hours after the initial sampling
of the water column. (C) The same as A, sampled from the
holding tank during the September Experiment 72 hours
after the initial sampling of the water column. Comparing
A, B, and C, note the increase in the number of less
fluorescent cells, that is, the loss of the rather tight
envelop. The top histogram shows the calibration of the
volume scale using 6 different standard microspheres.

An important concern is, to what extent will a population distribution be altered in time during transport of sample after collection? This is of special concern because currently few flow cytometers accessible to biological oceanographers are seagoing on research vessels. The SCM, 10.5 m, water sample was run in the laboratory 9+ hours after collection (Fig. 10b), during which time the FACS Analyzer was disassembled on the ship, transported back to the laboratory, reassembled, and restandardized with fluorescent microspheres. Some changes have occurred in this late summer SCM Gulf of Maine population over the 9+ h time frame. However, even when the populations are incubated in a 1000 liter tank for 72 hr ($\sim$300 $\mu Ein \cdot cm^{-2} \cdot sec^{-1}$; 24:0 L/D; $12^{o}C$; no nutrients added), many similarities about the FACS discernible population were observed (Fig. 10c).

Ataxonomic Relationships

Yentsch and Phinney (1984, 1985) have subdivided phytoplankton on the basis of accessory pigmentation in an attempt to describe functional groups combined in natural populations. This ataxonomic approach based on fluorescent spectral signatures (Yentsch and Yentsch, 1979), utilizes the differences in light absorption by phytoplankton for photosynthetic processes. Figure 11 illustrates different methods used to characterize water masses ataxonomically. We assess the presence of chlorophyll and/or phycoerythrin fluorescence simultaneously on individual cells using the 514 nm argon laser line of the EPICS V. An ability to discriminate different functional/pigment groups permits detection of different water masses.

The FACS is not suitable for this research as the mercury line at 488 nm lacks the power to efficiently excite phycoerythrin in cyanobacteria while the 546 line is too far in the green to excite chlorophyll. The result is that both pigments cannot be assessed simultaneously. A new mercury lamp source, enriched in cadmium, is now being developed by Becton-Dickinson which may allow simultaneous measurements of chlorophyll and phycoerythrin with the FACS Analyzer.

Referring back to Table II, the cyanobacteria and cryptomonads, which both contain phycoerythrin, can each be distinguished easily from other chlorophyll containing cells when analyzed on a two parameter bivariate plot with relative phycoerythrin fluorescence on

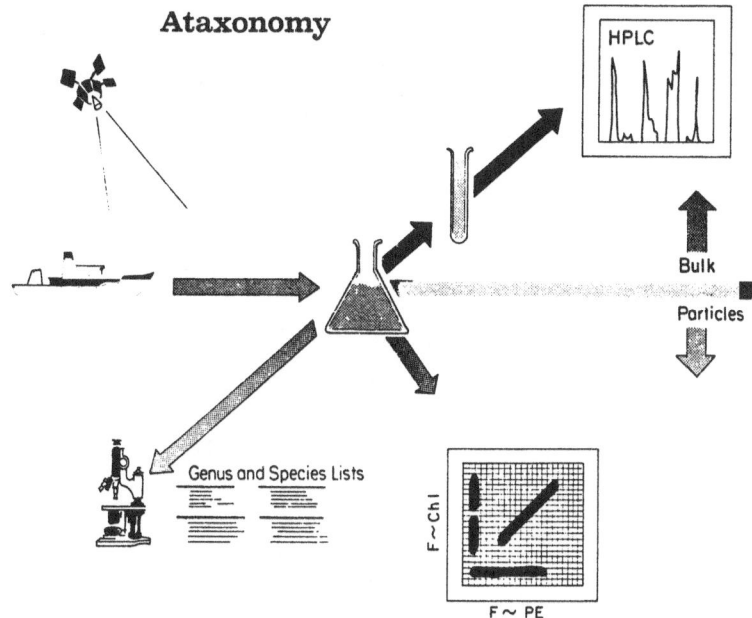

Ataxonomy

Figure 11. Schematic diagram of ataxonomic analysis. Bulk analysis
(upper) results in extracts and sophistication of HPLC
(High Pressure Liquid Chromatography). Individual particle analysis (bottom) results in distributions as either
genus and species lists or automated characterization
into pigment/functional groups. It is clear that one
water mass can be identified from another via FCM
characterization.

the x-axis and relative chlorophyll fluorescence on the y-axis. Gain
and power settings have been adjusted in our natural population
scheme (Yentsch, et al., 1984) to optimize this. The cultures of
Table III are shown in Fig. 12. Note the facility of detection of
cryptomonads by the EPICS V with a natural population reproduced in
Fig. 13. Compare this to the data set collected from a natural
population on the FACS Analyzer which was unable to discriminate
between cryptomonads and chlorophytes or dinoflagellates (Fig. 10).

How are data sets of natural populations verified? First, let
us acknowledge that we would have no indication had it not been for
the collaborative efforts in cell enumeration (E. Haugen, pers.
comm.). Second, we have no way of confirming what cell type falls

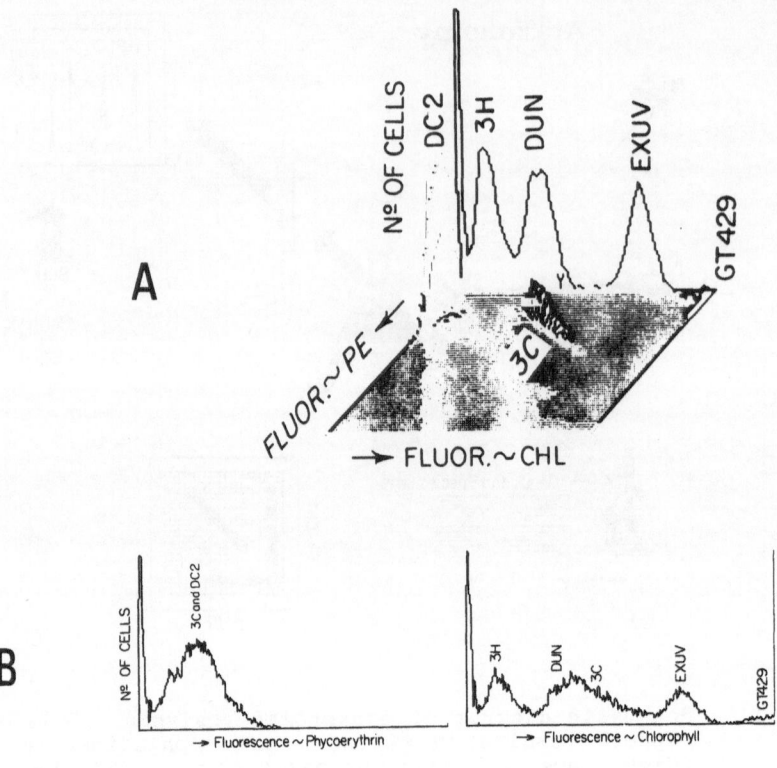

Figure 12.   (A).   Culture mixtures segregated by pigment composition
             using a Coulter EPICS V flow cytometer.   Clones are
             GT429, <u>Gonyaulax tamarensis</u>, a dinoflagellate; Exuv,
             <u>Prorocentrum</u> sp., a dinoflagellate; Dun, <u>Dunaliella</u>
             <u>tertiolecta</u>, a chlorophyte; 3C, <u>Chroomonas salinas</u>, a
             cryptomonad; 3H, <u>Thalassiosira pseudonana</u>, a diatom; and
             DC-2, <u>Synechococcus</u> sp., a cyanobacterium.   The X-axis
             represents relative phycoerythrin fluorescence;   the
             Y-axis represents relative chlorophyll fluorescence; and
             the Z-axis is the total number of cells.   Both X and Y
             axes as three decade log scale.
             (B).   The same date, but presented as phycoerythrin
             fluorescence and chlorophyll fluorescence distribution
             histograms.

into what fluorescence region without additive and/or subtractive
experiments.   In the additive experiments, natural populations were
spiked with clones of specific size and pigment characteristics.   To
insure that the 1 μm diameter low chlorophyll fluorescent particles
(e.g. prasinophytes) were above the non-fluorescent debris channels,

laser power was increased. In the subtractive experiments, specific peaks were selected, sort windows chosen and cells from that region physically sorted into a small vial or onto a microscope slide, for microscopic identification as previously described by Yentsch et al. (1983b). <u>Confirmation in this way is mandatory for any definitive statement about the analyzed population.</u>

DISCUSSION

We have accumulated data which give either an allometric (Fig. 10) or ataxonomic (Fig. 13) characterization of various oceanic regimes by following a few stations locally in the Gulf of Maine. While these are only snapshots of a highly dynamic region, it is indicative of the degree of resolution achievable by profiles in a mixed or stratified water column.

Discrimination of cryptomonads is only possible from unpreserved natural populations (as in Figure 13). Thus, the issue is raised, why have cryptomonads not been acclaimed as an important component in the phytoplankton community in the Gulf of Maine until this time? Our speculation is based on the fact that they do not preserve well, and most of the samples analyzed up to this time have been preserved. A recent exception is the research of Murphy and Haugen (1985).

A question we are often asked is: How should cells be preserved for FCM analysis? We do not advocate preservation; the biases introduced preclude its use. Incomplete attempts to document results from preserved samples are summarized in Table IV. Therefore, we feel that only fresh, unpreserved samples should be considered for flow cytometric analysis.

Sample preparation is extremely important when analyzing size fractions with low cell concentrations. While raw water samples can be run through the instruments (we prefilter through Nitex netting to eliminate clogging of the orifice), dinoflagellates, or other organisms naturally occurring at concentrations up to 10^3 cells$\cdot$ml^{-1}, may not be sampled adequately when analyzing only a few thousand cells in a water sample containing cyanobacteria at 10^4-10^5 cells$\cdot$ml^{-1}.

Table IV.

	A greens, diatoms dinoflagellates	B cryptomonads	C cyanobacteria
Lugols with subsequent bleaching	NG	NG	NG
Gluteraldehyde + cacodylic acid	OK	NG	NG
Gluteraldehyde ala Haas et al. (1983)	OK	?	NG
Liquid N_2 (K. Davis, pers. comm.)	OK	NG	OK

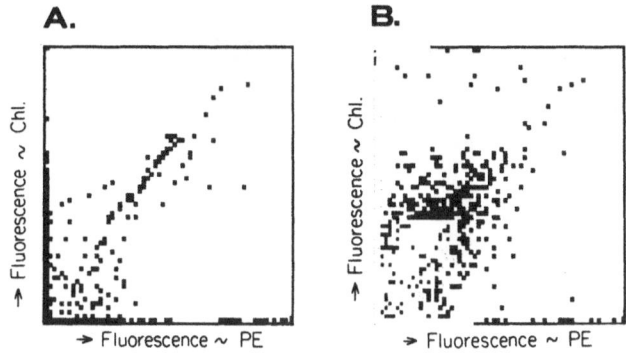

Figure 13. Natural population showing the occurrence of cryptomonads. Analyzed by the Coulter EPICS V, the x and y-axes are three decade log scale represented by relative phycoerythrin and chlorophyll fluorescence, z axis is the total number of events.
(A). Raw water sample from dock study November 29, 1983.
(B). 20-53 μm size fraction from the same water sample.

Conventional methods of obtaining a desired size fraction, i.e., concentration using a series of filters or netting, can damage cells and/or alter their physiological state. Centrifugal elutriation is a new technique for separating particles on the basis of their sedimentation velocity (dependent on size and shape), using centrifugal force plus a highly regulated counter flow (Fig. 14) (Pomponi, pers. comm.; Keng et al., 1981). Cells from large volumes of seawater can be concentrated and individual size fractions sampled with no apparent effects on the cell. After concentrating a natural population at a rate of 60 $\mu l \cdot min^{-1}$ normal phytoplankton and zooplankton movement is noted, however dinoflagellate cells appear healthy but were not swimming.

Neither the FACS Analyzer nor the EPICS V give a direct readout of particles per unit volume. Flow rates are so low (<0.1 ml$\cdot$min^{-1}) that even the most precise mini flowmeters offer little value. Some investigators have chosen a side-by-side Coulter counter to accomplish this end. Others (Olson et al., submitted) have initiated the practice suggested in the biomedical literature of adding standard beads at a known density per unit volume and running the instrument

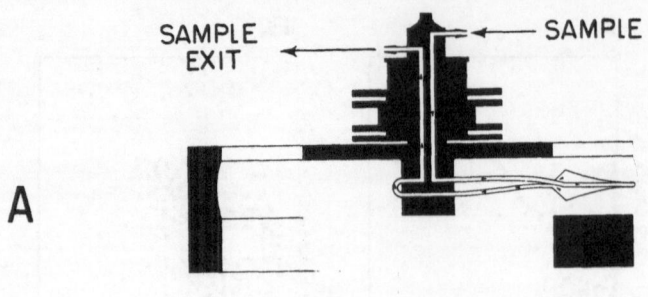

A

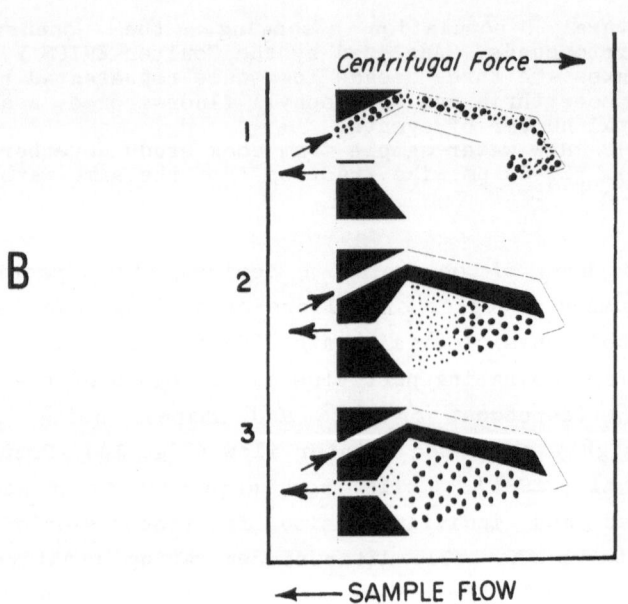

B

Figure 14.   (A). Schematic diagram of the elutriator rotor by Beckman
             Instruments.
             (B). Schematic representation of particles sedimenting in
             a centrifugal elutriator.   Chamber design by Beckman
             Instruments.

until a specific number of beads have been logged. Figure 15 depicts
the analysis of a natural seawater sample with the addition of 9.4μm
beads, at a concentration of 5.2×10^4 per ml.   The total volume
analyzed was calculated to be 20.9ml.   Another method to obtain

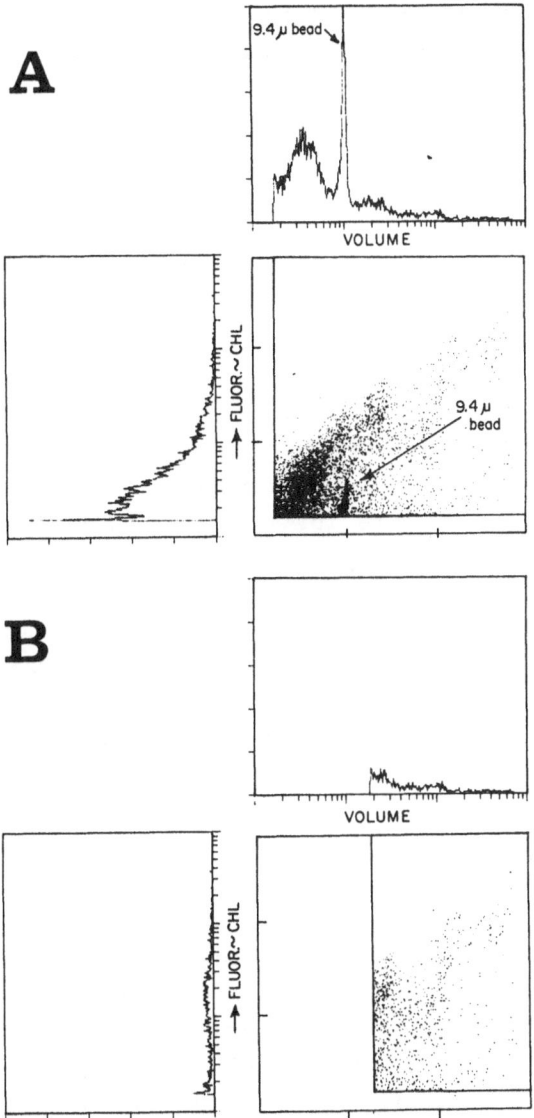

Figure 15. (A). Bivariate plot of a natural population "spiked" with a known concentration of 9.4 μm diameter microspheres (5.2 x 10⁴ per ml) and analyzed with a FACS Analyzer flow cytometer. The x-axis is volume (3 decade log scale); the y-axis is fluorescence approximating chlorophyll; and the z-axis is the number of cells. Gates were set up to analyze >5 μm diameter particles, which were calculated at a concentration of 2.49 x 10³ cells·ml⁻¹. Total volume samples was 20.9 ml.
(B). Same water sample as in (A) but gates were reset to analyze particles >10μm in diameter. A concentration of 45.8 cells·ml⁻¹ was calculated.

particles per unit volume would be gravimetric measurements (Cucci, unpublished). Sample weight is measured before and after analysis to determine volume actually analyzed. When samples are analyzed over minutes vs. seconds, the time can be useful to calculate numbers of cells provided flow rates are kept constant (Shumway et al., 1985; Cucci et al., 1985). Experiments designed to determine changes in known phytoplankton densities due to grazing by bivalve molluscs, the time to analyze 2000 cells after 1 hr. of grazing ranged from 200 to 400 sec longer than the same samples analyzed initially. Cell densities per unit volume and total number of cells grazed can be calculated.

Light scatter, a parameter mentioned only briefly, can be a useful indicator of cell size and/or texture. Some flow cytometers have the ability to measure forward angle light scatter (FALS) and 90° light scatter, both of which are indices of the combined effects of size, shape and refractive index. FALS may, in fact, be a better indicator of size while 90° light scatter is affected more by the shape and refractive index of the cell. When analyzing light scatter data, studies have shown that optical characteristics vary within a population as a function of physiological and/or nutritional state of the population (Spinrad and Yentsch, submitted).

While instrument standards exist and are routinely used to monitor instrument performance, better experimental standards and controls, particularly for fluorescence, are needed. They should be biologically similar to the cell system analyzed and exhibit small variations in the parameter to be measured. In the case of a bivariate analysis of phycoerythrin and chlorophyll, two experimental controls are necessary. 1) A cell containing only chlorophyll (easily available as uniform cells - a chlorophyte) and 2) a cell containing only phycoerythrin (not available in nature). Instrument settings are chosen such that these cells lie on their respective fluorescence axis, namely some intensity of one pigment and none of the other. In this way, unknown cells falling in the middle are relative to these controls. However, we have found that cells from natural populations fluoresce more than cultured cells, the instrument settings chosen using these controls are too sensitive causing non-fluorescent debris found in a natural sample to appear as a fluorescing population. Considerable attention should be addressed to this aspect of oceanographic flow cytometry.

SUMMARY

In recent years, advances in technology have allowed biologists to collect and analyze data far faster than previously possible not only in the laboratory but also at sea. Laborious techniques previously used in oceanography to obtain information on cell numbers, cell size/volume, chemical composition and the physiological state of plankton are gradually being replaced with the advent of new instrumentation. One such group of instruments are flow cytometers which permit simultaneous measurements of cell volume, two colors of fluorescence and light scatter on each individual particle. Never before have such measurements been possible on the smaller picoplankton (~1 μm). Biological data can now be collected at sea and quickly correlated onboard ship with physical (CTD measurements) and chemical (nutrient analyses - autoanalyzer) data.

The determination, enumeration, and separation of bacteria, microheterotrophs and detritus using flow cytometry will be major steps forward in analyzing both biomass and turnover rates of the entire plankton community. These small organisms can now only be detected by light scattering parameters. However, development of methods of fluorescent stains that are useful for bacteria and heterotroph characterization should permit their separation from cells of the autotrophic community, detrital material, and suspended sediment particles.

ACKNOWLEDGEMENTS

This research was made possible by support from NSF OCE81-21331; OCE82-13567; OCE83-19070, BSR-8307081, EXP80-11448, NASA NAS5-27742 (to C.S. Yentsch) and ONR N00014-81-C-0043. Our thanks go to E. Haugen for cell enumeration, R. Selvin at CCMP for culture data, P. Colby for preparing the manuscript and J. Rollins and K. Knowlton for preparing the illustrations. We appreciate constructive criticisms on a previous draft from two reviewers.

REFERENCES

Arvola, L. 1984. A comparison of electronic particle counting with microscopic determination of phytoplankton and chlorophyll a concentrations in three Finnish lakes. Ann. Bot. Fennici 21: 117-142.

Blasco, D., T.T. Packard and P.C. Garfield. 1982. Size dependence of growth rate, respiratory electron transport system activity and chemical composition in marine diatoms in the laboratory. J. Phycol. 18: 56-63.

Campbell, J.W., C.M. Yentsch, J.A. Topinka and T. Cucci. 1985. Flow cytometric observations of chlorophyll synthesis and cell division patterns in clonal cultures of Dunaliella tertiolecta. Submitted.

Cucci, T.L., S.E. Shumway, R.C. Newell, R. Selvin, R.R.L. Guillard, and C.M. Yentsch. 1985. Flow cytometry: a new method for characterization of differential ingestion, digestion and egestion by suspension feeders. Mar. Ecol. Prog. Ser. (in press).

Horan, P.K. and L.L. Wheeless, Jr. 1977. Quantitative single cell analysis and sorting. Rapid analysis and sorting of cells is emerging as an important new tecnology in research and medicine. Science 198: 149-157.

Hutter, J.-J. and H.E. Eipel. 1978. Advances in determination of cell viability. J. Gen. Microbiol. 107: 165-167.

Keng, P.C. , C.K.N. Li, and K.T. Wheeler. 1981. Characterization of the separation properties of the Beckman elutriator system. Cell Biophysics 3: 41-56.

Kerr, S.R. 1974. Theory of size distributions in ecological communities. J. Fish. Res. Bd. Can. 31: 1859-1862.

Kruth, H.S. 1982. Flow cytometry: Rapid biochemical analysis of single cells. Anal. Biochem. 125: 225-242.

Lorenzen, C.J. 1966. A method for the continuous measurement of in vivo chlorophyll concentration. Deep-Sea Res. 13: 223-227.

Malone, T.C. 1980. Algal size. In The Physiological Ecology of Phytoplankton. I. Morris (ed.) pp. 433-463. London: Blackwell Publishers.

Maloney, T.E., E.J. Donovan and E.L. Robinson, Jr. 1962. Determination of numbers and sizes of algal cells with an electronic particle counter. Phycologia 2: 1-8.

Margalef, R. 1963. Successions of populations. Advancing Frontiers of Science 2: 137-188.

Melamed, M.R., P.F. Mullaney and M.L. Mendelson. 1979. Flow Cytometry and Sorting. Wiley, N.Y.

Murphy, L.S. and E. Haugen. 1985. The distribution and abundance of phototrophic ultraplankton in the North Atlantic. Limnol. Oceanogr. 30: 47-58.

Olson, R.J. S.L. Frankel, S.W. Chisholm and H.M. Shapiro. 1983. An inexpensive flow cytometer for the analysis of fluorescence signals in phytoplankton: Chlorophyll and DNA distributions. J. Exp. Mar. Biol. Ecol. 68: 129-144.

Olson, R.J., D. Vaulot, S.W. Chisholm. Marine phytoplankton distributions measured using shipboard flow cytometry. Submitted to Deep-Sea Res.

Patrick, C.W., T.J. Nilson, P.W. McFadden and R.H. Keller. 1984. Flow Cytometry and Cell Sorting. Laboratory Medicine 15: 740-745.

Paau, A.S., J. Oro and J.R. Cowles. 1978. Application of flow cytometry to the study of algal cells and isolated chloroplasts. J Exp. Bot. 29: 1011-1020.

Parsons, T.R., K. Stephens and J.D.H. Strickland. 1961. On the chemical composition of eleven species of marine phytoplankters. J. Fish. Res. Bd. Canada 18: 1001-1016.

Peters, R.H. 1983. Ecological Implications of Body Size. Cambridge Univ. Press, Cambridge 329 pp.

Platt. T. and K. L. Denman. 1977. Organization in the pelagic ecosystem. Helgol. Wiss. Meeresunter. 30: 575-581.

Platt, T. and K. L. Denman. 1978. The structure of pelagic marine ecosystems. Rapp. P.-V. Reun. Cons. Int. Explor. Mer 173: 60-65.

Platt, T. and W. Silvert. 1981. Ecology, physiology, allometry and dimensionality. J. Theoret. Biol. 93: 855-860.

Salzman, G.C., P.F. Mullaney, B.J. Price. 1979. Light scattering approaches to cell characterization. In Flow Cytometry and Cell Sorting, M.R. Melamed, P.F. Mullaney and M.L. Mendelson (eds.) John Wiley and Sons, N.Y. pp. 105-124.

Salzman, G.C. 1982. Light scattering analysis of single cells. In N. Catsimpoolas (ed.), Cell Analysis, Vol. 1. Plenum Publ. Corp.

Shapiro, H.M. 1985 Practical Flow Cytometry. Alan R. Liss, Inc., N.Y. 300 pp.

Sheldon, R.W 1978. Sensing zone counters in the laboratory. In A Phytoplankton Manual, A. Sournia (ed.), Monogr. Ocean Methodol. 3. UNESCO.

Sheldon, R.W. and T.R. Parsons. 1967. A continuous size spectrum for particulate matter in the sea. J. Fish. Res. Bd. Can. 24: 909-915.

Sheldon, R.W., A. Prakash and W.H. Sutcliffe, Jr. 1972. The size distribution of particles in the ocean. Limnol. Oceanogr. 17: 327-340.

Shumway, S.E., T.L. Cucci, R.C. Newell, and C.M. Yentsch. 1985. Particle selection, ingestion and absorption in filter feeding bivalves. J. Exp. Mar. Biol. Ecol. (in press).

Sieburth, J. McN, V. Smetacek and J. Lenz. 1979. Pelagic ecosystem structure: Heterotrophic compartments of the plankton and their relationship to plankton size fractions. Limnol. Oceanogr. 23: 1256-1263.

Silvert, W. and T. Platt. 1978. Energy flux in the pelagic ecosystem: A time-dependent equation. Limnol. and Oceanogr. 23: 1248-1255.

Silvert, W. and T. Platt. 1980. Dynamic energy flow model of the particle size distribution in pelagic ecosystem. In Evolution and Ecology of Zooplankton Communities. W.C. Kerfoot (ed.), pp. 754-763. Univ. Press of New England, Hanover, N.H.

Spinrad, R.S. and C.M. Yentsch. 1984. The optical state of marine phytoplankton. Submitted Limnol. Oceanogr.

Thomann, R.V. 1981. Equilibrium model of fate of micro-contaminants in diverse aquatic food chains. Can. J. Fish. Aquat. Sci. 38: 280-296.

Trask, B.J., G.J. Van den Engh and J.H.B.W. Elgershuizen. 1982. Analysis of phytoplankton by flow cytometry. Cytometry 2: 258-267.

Turpin, D.H. and P.J. Harrison. 1980. Cell size manipulation in natural marine planktonic diatom communities. Can. J. Fish. Aquat. Sci. 37: 1193-1195.

Van Dilla, M.A. and M.L. Mendelsohn. 1979. Introduction and resumé of flow cytometry and sorting. In Flow Cytometry and Sorting, M.R. Melamed, P.F. Mullaney and M.L. Mendelson (eds.) Wiley and Sons, N.Y. pp. 11-37.

Waterbury, J.B., S.W. Watson, R.R.L. Guillard and L.E. Brand. 1979. Wide-spread occurrence of unicellular, marine, planktonic cyano-bacterium. Nature, 277: 293-294.

Yentsch, C.M. and C.S. Yentsch. 1979. Fluorescence spectral signatures: The characterization of phytoplankton populations by the use of excitation and emission spectra. J. Mar. Res. 37: 471-483.

Yentsch, C.M., F.M. Mague, P.K. Horan and K. Muirhead. 1983a. Flow cytometric DNA determinations on individual cells of the dino-flagellate Gonyaulax tamarensis var. excavata. J Exp. Mar. Biol. Ecol. 67: 175-183.

Yentsch, C.M., P.K. Horan, K. Muirhead, Q. Dortch, E. Haugen, L. Legendre, L.S. Murphy, M.J. Perry, D.A. Phinney, S.A. Pomponi, R.W. Spinrad, M. Wood, C.S. Yentsch and B.J. Zahuranec. 1983b. Flow cytometry and cell sorting: A technique for analysis and sorting of aquatic particles. Limnol. Oceanogr. 28: 1275-1280.

Yentsch, C.M., T.L. Cucci and D.A. Phinney. 1984. Flow cytometry and cell sorting: Problems and promises for biological ocean science research. In: O. Holm-Hansen, L. Bolis and R. Giles (eds.), Marine Phytoplankton and Productivity. Springer-Verlag, Berlin, pp. 141-155.

Yentsch, C.S. and D.A. Phinney. 1984. Observed changes in spectral signatures of natural phytoplankton populations: the influence of nutrient availability. In: O. Holm-Hansen, L. Bolis and R. Giles (eds.), Marine Phytoplankton and Productivity. Springer-Verlag, Berlin, pp. 129-140.

Yentsch, C.S. and D.A. Phinney. 1985. The use of fluorescence spectral signatures for studies of marine phytoplankton. In: A. Zirino (ed.), Mapping Strategies in Chemical Oceanography, Advances in Chemistry Series, No. 209. American Chemical Society, Washington, D.C., pp. 259-274.

DEVELOPMENT OF A MOORED IN SITU FLUOROMETER
FOR PHYTOPLANKTON STUDIES

T.E. Whitledge
Oceanographic Sciences Division
Brookhaven National Laboratory
Upton, NY 11973

C.D. Wirick
Department of Marine Science
University of South Florida
St. Petersburg, FL 33701

INTRODUCTION

Marine ecology is a complex science because of the many interactions between physics, chemistry and biology that occur simultaneously. In addition, these physical, chemical and biological processes are quite variable in time and space as a result of secondary
processes, and produce an almost unlimited combination of possibilities for ocean variability. At present, one of the most critical
areas in biological oceanography is to observe and understand the
variability of biological parameters through the range of time scales
of minutes to years. Special instrumentation is needed which will
enhance our ability to collect many data points over a relatively
long period of time. In both the technological sense and biological
importance, a phytoplankton biomass indicator, chlorophyll, is widely
considered to be the "best" initial in situ Eulerian measurement.
Hence an in situ fluorometer to estimate chlorophyll a concentration
was developed at the Division of Oceanographic Sciences at Brookhaven
National Laboratory to be moored in the ocean at depths up to 300 m
for as long as three months. During a given three month data gather-
ing experiment, with data being collected at rates up to 75 times
each day, approximately 7000 data points could be available to show
the range and scales of variability of chlorophyll at a fixed point
in the ocean. Correlation of the estimated chlorophyll a with other
simultaneous measurements such as current speed and direction, salinity, temperature and pressure would allow a comprehensive analysis to
be made of the fluorofield, the hydrographic structure of the water
column, and possible interactions with the phytoplankton. Unfortun-

ately, other biological parameters associated with phytoplankton dynamics such as growth rate, losses to zooplankton grazing or taxonomic composition are not presently capable of being measured remotely, so simultaneous shipboard measurements of these processes are still required.

Design Philosophy

The prototype in situ instrument was constructed by placing a standard bench fluorometer (Turner Designs Model 10-005 RU) in a waterproof housing and using the battery power package as an anchor. Several trial deployments were successful (Whitledge and Wirick, 1983) and new plans were formulated for the construction of a fluorometer which was designed specifically for in situ moorings. Some of the design criteria for the new instrument included: low power consumption, relatively small, light in weight, a long time deployment capability, reduced biological fouling of sensor system and a reasonable construction cost.

A new instrument that has the preceding attributes has been constructed, field tested, and is now being deployed in field experiments. The resulting instrument when mounted in its protective cage, has dimensions of 1.2 x 0.3 x 0.3 m and is versatile enough to be added to a mooring with current meters and other sensors (Fig. 1) to allow a physical-biological integration of data that was heretofore impossible.

Basic Components

The basic in situ fluorometer design is similar to a fluorescence sensor that was developed by Aiken (1981). The excitation light is produced by a xenon flash tube and a photodiode is used to detect the emitted fluorescence. The energy of each xenon flash is also detected and recorded for reference. The sample cell is an open-ended tube (Fig. 2), which is periodically scrubbed with a brush to prevent fouling.

A pack of "D" cell batteries powers the instrument (Fig. 3) and its life (~2 to 3 months) is determined by the sampling and cell cleaning frequency. Each instrument has a four channel data acquisition system with a crystal time base. The analog channels are burst

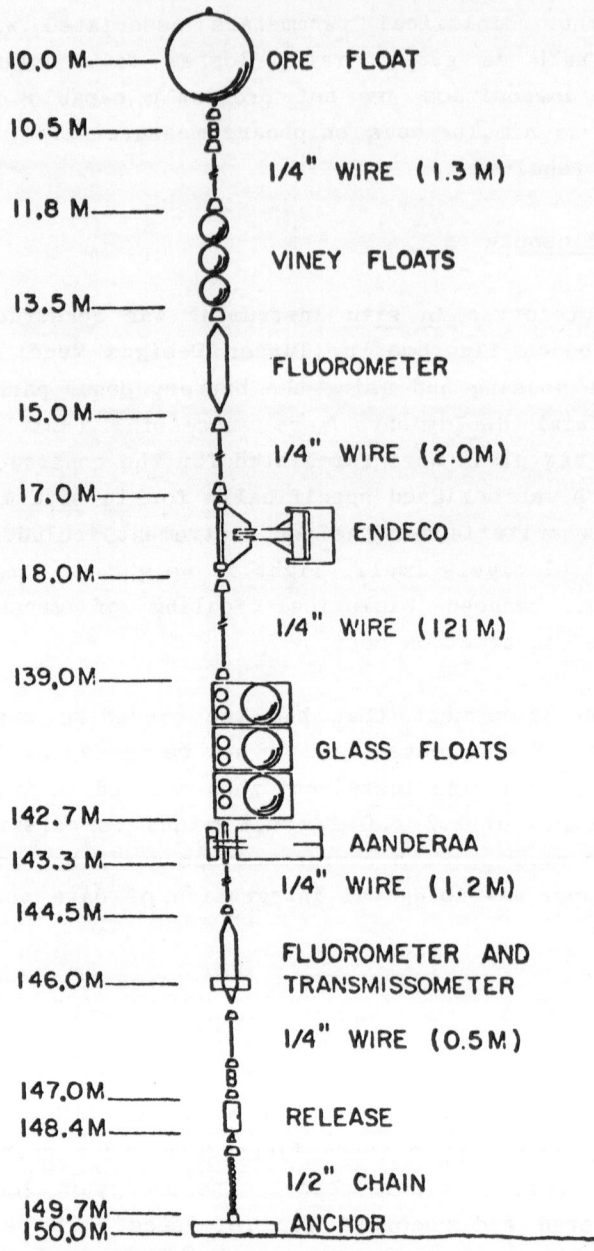

Figure 1. Typical array of Shelf Edge Exchange Processes (SEEP) mooring that contains two fluorometers, one transmissometer, one Endeco current meter and one Aanderas current meter.

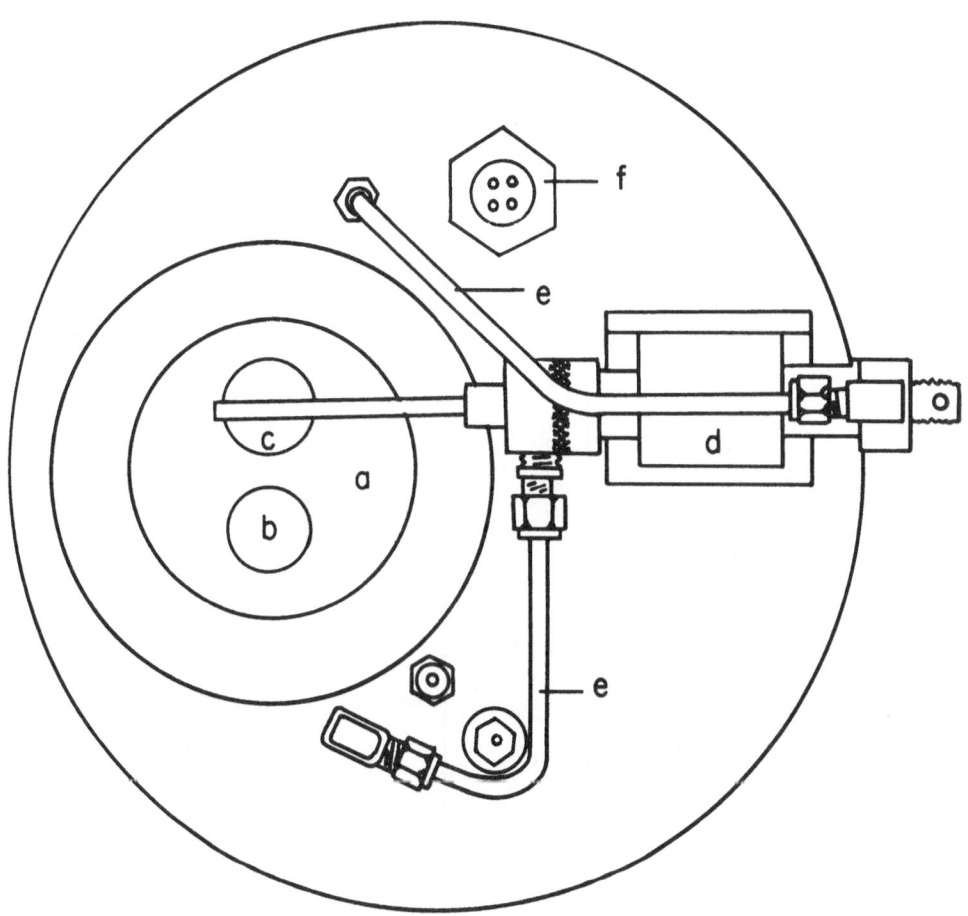

Figure 2. End view of fluorometer showing "O" ring sealed glass window (a) with light excitation path (b) and light emission path (c). Now shown are glass prism, mirrored cuvet or light shield. Window scrubber hydrolic ram (d) minus abrasive is driven by two internal micropumps via pressure tubing (e). Waterproof plug (f) is used for power and data transmission of auxiliary sensors such as transmissometers and pressure.

sampled at fixed intervals of time and the digital values are recorded on magnetic tape (Table 1). Each tape has a capacity of 50,000-16 bit words. Two data channels are used for the fluorescence sensor with the others committed to temperature and transmissometer or pressure sensors. The temperature sensor is either a YSI or a

Table I. Fluorometer characteristics.

Size (cm)

Uncaged 23 (diameter) x 91
Caged   30 x 30 x 120

Weight (Kg)	Air	Water
PVC Case	31.8	4.5
Aluminum Case	40.9	13.6
Stainless Steel Case	68.1	40.9

Detector Range ± precision (μg Chl ℓ^{-1})

High sensitivity   0.1 to 4 ± 0.05
Low sensitivity   25 to 100 ± 1

Excitation wavelength   470 nm
Emission wavelength   570 nm

Bursts/sample

6 at 8 second intervals

Sample Rate

Every 8, 16, 32 or 64 minutes

Duration

110 days at 32 min sampling rate

Power

39 "D" cell alkaline batteries

Auxiliary sensors

Temperature YSI (±0.05°C)
Transmissometer Sea Tech (25 cm)
Pressure (± 0.01 decibars)

454

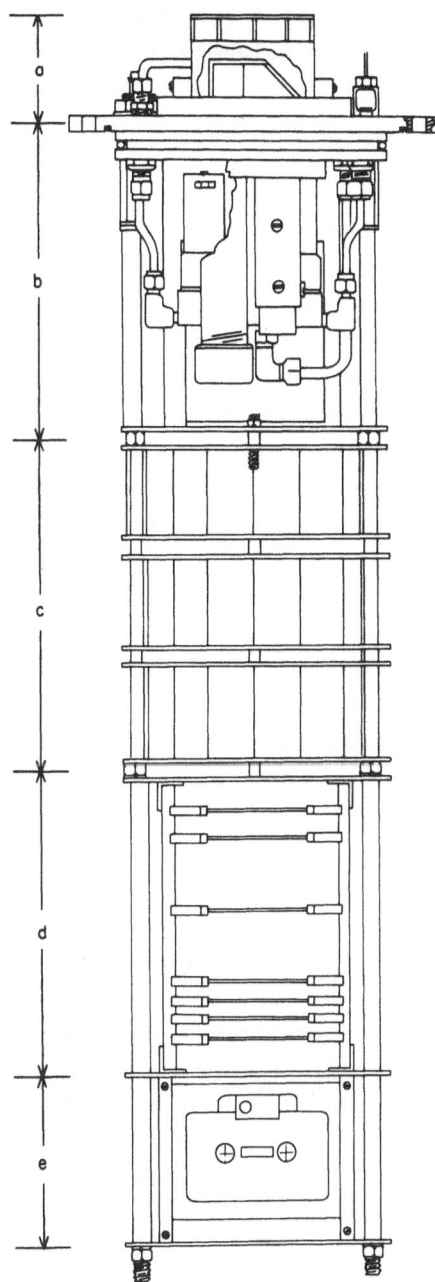

Figure 3. Side view of fluorometer showing (a) external fluorescence cuvet and prisms in cut-away light shield, (b) internal xenon strobe, photodiode detector and pumps for cuvet scrubber, (c) three-layered battery pack, (d) electric control and logic cards, and (e) data recording digital cassette.

Fenwal thermistor, while light transmission is sensed by a Sea Tech transmissometer, with a path length of 25 cm, to estimate total particles for comparison with the fluorescence records.

Calibration

The estimation of chlorophyll a concentration in the ocean from in vivo fluorescence requires some shipboard calibration when the fluorometers are deployed and recovered. Bench testing of the fluorometers against a laboratory fluorometer for both in vivo and in situ determinations insures that the instruments are operating properly, but really does not produce a fully calibrated instrument. Experience from deployment shows that in situ calibration with (a) profiling fluormeter, (b) discrete samples from a rosette attached to the profiling fluorometer for both in vivo and in situ analysis, and (c) in situ pump samples for shipboard in vivo and in situ determinations are much preferred to bench testing. All three methods have shortcomings but the phytoplankton cells used are representative of the ambient population in terms of light and nutritional history which is not the case with bench calibration procedures. Once these calibrations are made, the accuracy of the in situ fluorescence estimate of chlorophyll is within the range of half or double the true value (Whitledge and Wirick, 1983).

Fouling Considerations

Biological fouling of the sensor system was a key problem that had to be solved before deployments of more than a few hours were considered practicable. Algaecides were impracticable for an open sensor system so a mechanical means was chosen. After the design of the cuvet cell and mechanical scrubber was completed, a trial deployment of scrubbed and unscrubbed cuvets was placed in Port Jefferson Harbor, New York. The difference between cleaned and uncleaned cells was obvious, with no accumulation after two weeks in the scrubbed set but plainly visible buildup of material in the uncleaned set.

Nearshore and Mid Shelf Experimental Tests

A field test of the self-contained in situ fluorometer was initiated in shallow water with large chlorophyll concentrations in order to test the dynamic range of the sensors and the antifouling scrubber system. As a result of nutrient discharge within Port Jefferson Harbor on the north shore of Long Island, a large horizontal chlorophyll gradient exists between the Harbor and adjacent Long Island Sound. A test of the fluorometer was made in the mouth of the harbor in a water depth of 10 m from 9 to 15 July 1982. Advection by the tidal currents, high chlorophyll and warm temperature harbor water past the moored fluorometer at 7 m were quite evident (Fig. 4). In this experiment, a burst of 9 xenon flashes occurred every 8 minutes to generate 1320 data sets of in vivo fluorescence, with calibration samples taken during deployment and recovery of the mooring. The data were low pass filtered (frequency centered at 1.0 cycles hr^{-1}) to yield the 1982 time series, clearly depicting (Fig. 4) an ability to resolve high and low chlorophyll features in the sea with in situ sensors. No evidence of fouling was observed during this experiment.

We subsequently deployed two of the new fluorometers at the 60 m isobath on the mid-Atlantic shelf, east of Maryland (Fig. 5), during 8-15 April 1983. At this time of year, nutrient depletion of surface waters at mid-shelf leads to a surface minimum of chlorophyll in this region between wind events (Walsh et al., 1978), with the riverine and shelf-break sources of nitrogen clearly demarcated by the 2 µg ℓ^{-1} isopleths of chlorophyll at each boundary (Fig. 5).

Before a wind resuspension event ($\sim$10 m sec^{-1}) at mid-day on 9 April, the chlorophyll concentration seen by the fluorometer at a depth of 20 m (Fig. 6a) was less than that at 57 m (Fig. 6b), i.e. the phytoplankton had sunk out of the water column. Between 10-13 April, winds were downwelling favorable so nutrients were not introduced into the euphotic zone and a chlorophyll increase did not occur. Another significant event during the experiment was the appearance of Warm Core Ring 25 off shelf, whose signature was clearly visible in the temperature records.

A diurnal variation is also apparent in the fluorescence signal, especially in the upper layer at 20 m depth (Fig. 6a). This daily variation was also observed in the prototype deployments off Long Island (Whitledge and Wirick, 1983) and was speculated to be diel phytoplankton growth and sinking or zooplankton grazing (Owens et al., 1980).

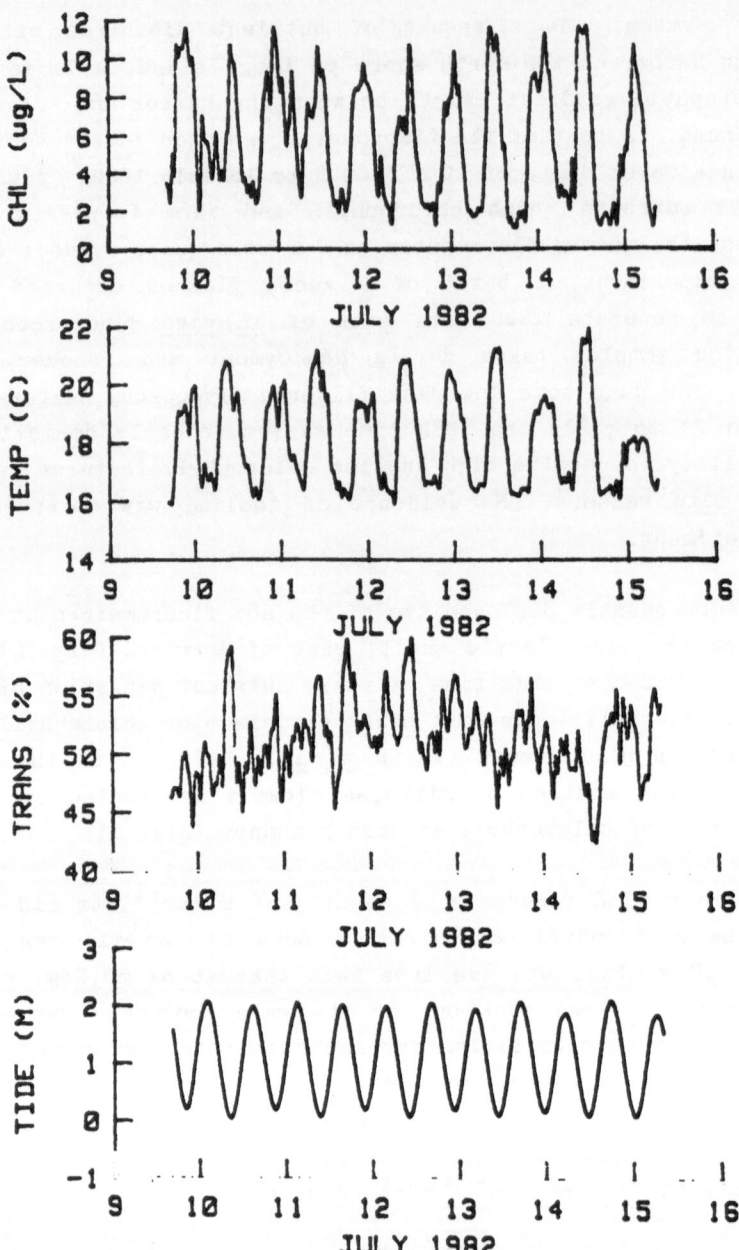

Figure 4. Time series of chlorophyll, temperature, and light transmission with respect to the local semi-diurnal tide at 7.0
m during 9-15 July 1982 within Long Island Sound.

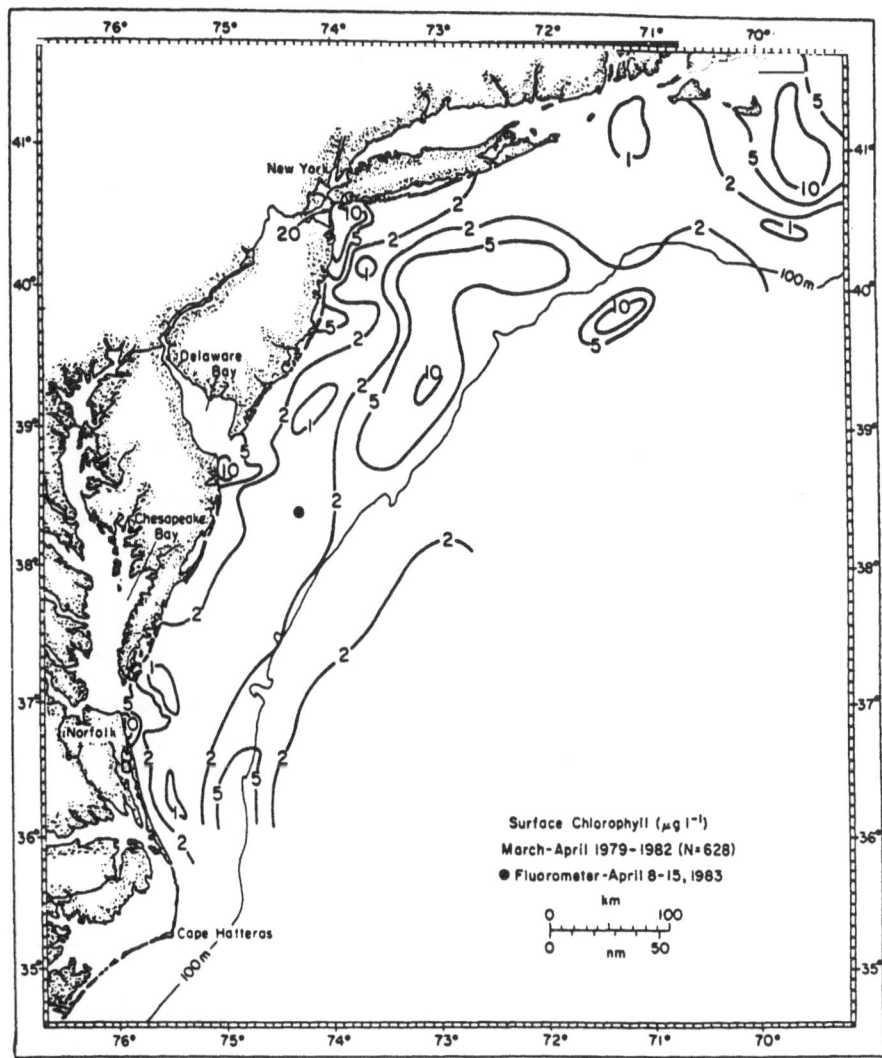

Figure 5. Location of a moored fluorometer array in April 1983 with respect to shipboard surface chlorophyll (µg ℓ$^{-1}$) distributions of a composite spring bloom (1979-1982) of the Mid-Atlantic Bight.

The higher frequency variations with periods of minutes to several hours are real because both fluorescence and light transmission (Figs. 6b and 6f) responded simultaneously. These changes may have resulted from the advection of small phytoplankton patches with slightly different chlorophyll and particle concentrations.

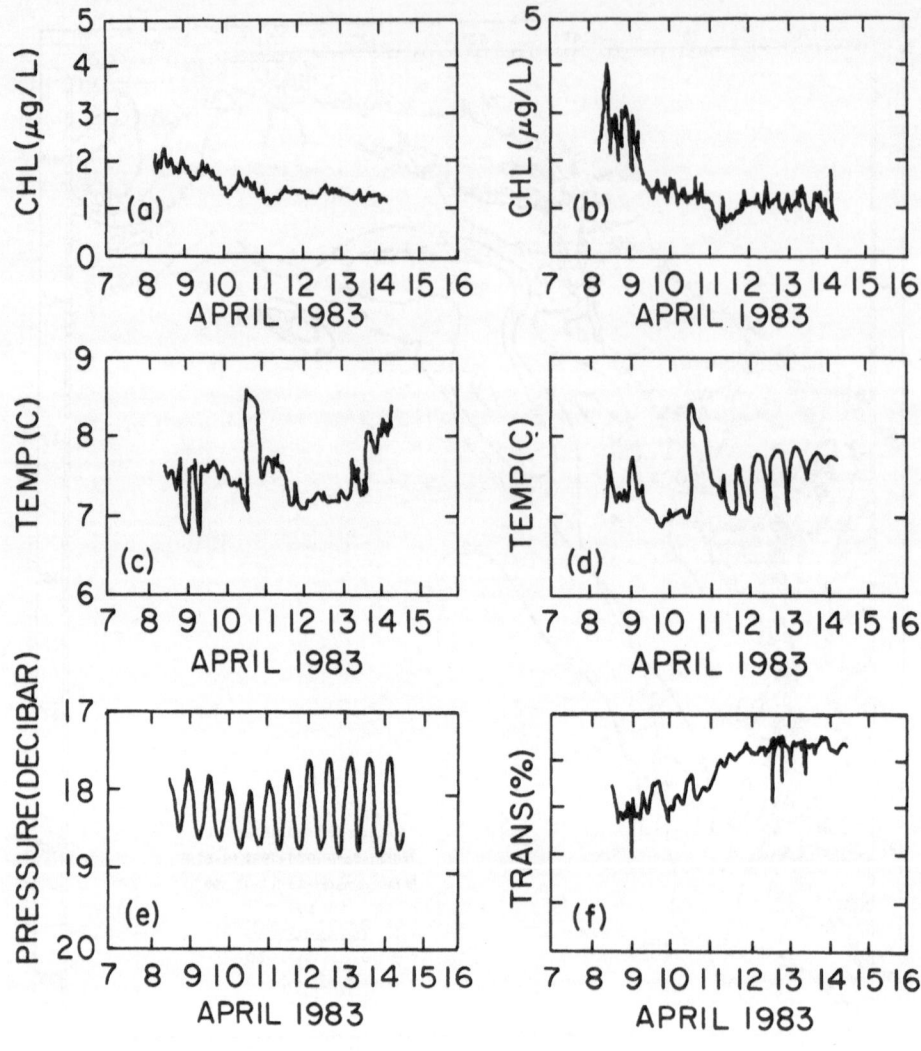

Figure 6. Time series of chlorophyll at (a) 20 m and (b) 57 m, temperature at (c) 20 m and (d) 57 m, (e) pressure at 20 m, and
(f) light transmission at 57 m during the 1983 spring bloom
at the 60 m isobath off Delaware Bay.

An Example of Use on the Shelf

One contemporary area of research into the biological oceanography of the New York Bight is the apparent imbalance between the
increase of phytoplankton biomass from primary production and losses

due to sinking and/or zooplankton grazing (Walsh et al., 1978). Based on the collective data of many investigators, it is widely believed that the sudden increase of phytoplankton biomass during the spring bloom is consumed partially by zooplankton grazing and the remaining balance sinks from the euphotic zone while being carried along by the shelf currents. Since no measurable chlorophyll or phytoplankton population accumulates in the shelf sediments, it is thought that much of the uneaten organic production must eventually be lost from the shelf and deposited on the upper slope where the sediment contains relatively larger amounts of organic carbon.

To test the above hypothesis, a Shelf Edge Exchange Process (SEEP) program was designed and executed in which the main biological objective was to study the fate of phytoplankton production on the East Coast continental shelf. A field experiment to test the hypothesis was devised utilizing our newly tested moored in situ fluorometers in order to observe mesoscale variations in chlorophyll concentrations in both space and time. Balancing cost versus the need for a dense set of measurements in space and time, two across-shelf transects were instrumented with 15 fluorometers, 45 current meters, 7 transmissometers, several sediment traps and numerous temperature and conductivity sensors (Fig. 7). Primary production, zooplankton grazing and sediment-water column interaction measurements were collected aboard ship while synoptic distributions of temperature and chlorophyll were estimated with both active and passive aircraft sensors and the Coastal Zone Color Scanner (CZCS) on the NIMBUS-7 satellite (Fig. 7).

The SEEP-I experiment collectively contains enough measurements to make some first order estimates on the shelf export hypothesis. Central to the measurement program was the use of successfully moored in situ fluorometers and transmissometers to gather local chlorophyll and total particle time series. Contemporaneous current measurements and wind data at the 80 and 120 m isobaths were used to estimate a 0.35-0.47 gC m^{-2} day^{-1} algal export during February 1984. Such fluxes constitute 78-100% of the March 1984 mean primary production, 0.45 gC m^{-2} day^{-1} and 23-30% of that in April 1984, 1.55 gC m^{-2} day^{-1} (P.G. Falkowski, pers. comm.). Other losses of algal carbon due to sinking, zooplankton grazing and benthic consumption can possibly remove an additional 25% of the April primary productivity not exported from the shelf (Walsh, 1983).

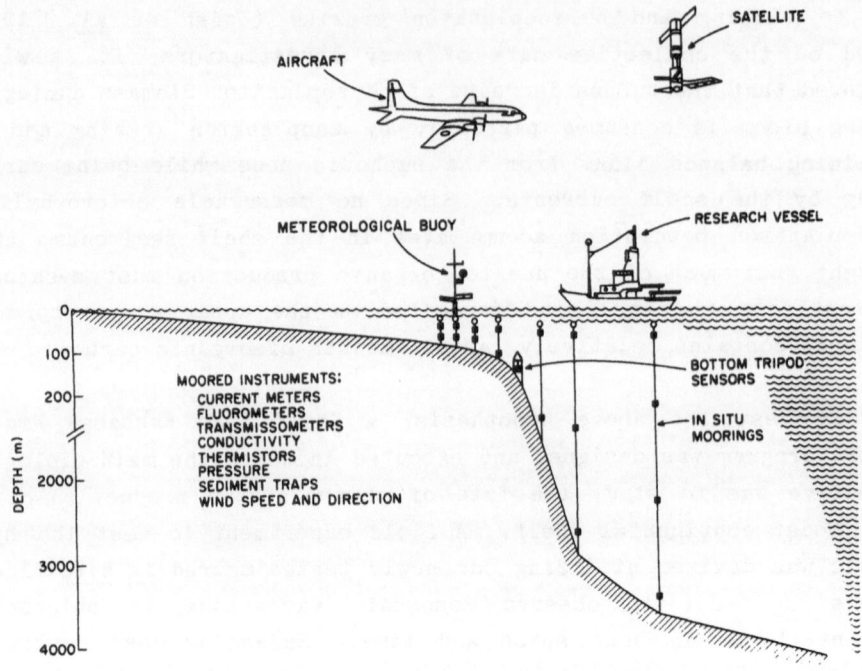

Figure 7. Shelf Edge Exchange Processes (SEEP-I) sampling instruments used to investigate the production and fate of phytoplankton in the Mid-Atlantic Bight south of Nantucket Island.

DISCUSSION

The use of moored <u>in situ</u> fluorometers is clearly beneficial in many phytoplankton studies in the marine environment. Our understanding of marine ecology has progressed steadily over the past few years as better equipment and sampling strategy has been developed and tested. The development of a moored <u>in situ</u> fluorometer is another additon to our overall capabilities in modern oceanographic data collection that is allowing us a more complete understanding of the biomasses and fluxes in the marine ecosystem.

The limitations of an <u>in situ</u> fluorometer are somewhat obvious but they should be mentioned. The data from an <u>in situ</u> fluorometer will show variations in chlorophyll fluorescence, but the causes of the variations may not be easily evaluated. It is not advisable to emplace an <u>in situ</u> instrument at a discontinuity such as in a horizontal frontal region or at a depth near and within the pycnocline. The populations across a front or pycnocline may be radically different, hence the chlorophyll to fluorescence ratio may show considerable variation (perhaps even an order of magnitude). On the other hand, the fluorescence variation will be small which may lead to an underestimate of phytoplankton patchiness.

The future development of <u>in situ</u> fluorometers should bring a more compact instrument that will be based on microprocessor electronics to control all logic functions and with non-volatile memory to store the data. Additional design changes will increase the maximum period of deployment (see Table I).

ACKNOWLEDGEMENTS

This research was performed under the auspices of U.S. Department of Energy Contract DE-AC02-76CH00016. The authors wish to thank Dr. J.J. Walsh for his encouragement and E. Divis, R. Sick, J. Yelk and K. von Bock for their help in construction and deployments.

REFERENCES

Aiken, J. 1981. A chlorophyll sensor for automatic, remote operation in the marine environment. Mar. Ecol. Prog. Ser. 4: 255-239.

Owens, T.G., P.G. Falkowski and T.E. Whitledge. 1980. Diel periodicity in cellular chlorophyll content in marine diatoms. Mar. Biol. 59: 71-77.

Walsh, J.J. 1983. Death in the sea: enigmatic phytoplankton losses. Progress in Oceanography 12: 1-86.

Walsh, J.J., T.E. Whitledge, F.W. Bervenik, C.D. Wirick, S.O. Howe, W.E. Esaias and J.T. Scott. 1978. Wind events and food chain dynamics within the New York Bight. Limnol. Oceanogr. 23: 659-683.

Whitledge, T.E. and C.D. Wirick. 1983. Observations of chlorophyll concentrations off Long Island from a moored <u>in situ</u> fluorometer. Deep-Sea Res. 30: 297-309.

preservative for the variables of interest. Sampling intervals can be varied over the ~1 hr–~10 day time scale.

With some companion hardware, the samplers can function as their own time release mechanism, and presently a mooring system that can be used for repeated deployments is being tested. An array of these sampling devices could provide truly synoptic "pictures" of chemical and biological fields while single moorings would enhance the ability to collect time series data. Another important attribute of the moored sampler is that it can collect samples during conditions that exceed the sea-keeping ability of many coastal research vessels.

Our interest in moored water sampling devices developed during studies of coastal upwelling when it became evident that it would be desirable to have detailed time series and truly synoptic sections and maps of chemical variability. The continuous current meter and temperature records gathered by physical oceanographers revealed physical features that could have important effects on chemical distributions, but an inadequate data base frustrated some of our attempts to relate measured or inferred currents and mixing to the distribution of nutrients. The relative paucity of chemical observations was largely due to high ship rentals that made it prohibitively expensive to take time series observations at a fixed station on a routine basis. Ships are also not suited for investigating shorter term variability because their limited speeds make it impossible to obtain truly synoptic maps and sections. In addition, they frequently cannot sample during stormy periods when important mixing and growth events occur. As well as limiting our understanding of upwelling systems, these sampling problems have clearly inhibited the pace of research into the understanding of tidal mixing and plankton dynamics.

Initial ambitious thoughts about _in situ_ water chemistry analysis (e.g., an _in situ_ autoanalyzer) as a solution to some of these sampling problems were soon discarded since the cost of providing enough samplers to adequately define the spatial nutrient variability in many coastal seas would clearly be prohibitive. In addition, the inherent complexity of such instrumentation would make serious failures very likely. Instead, we designed a simpler instrument to collect and preserve discrete water samples for future analysis. As well as low cost (vs. in situ analysis), this type of instrument would not be restricted to a few analyses; it could collect samples for any analysis that did not require a large sample volume and for which a suitable preservative (if needed) could be found. The instrument shown in Figure 1 evolved through a number of iterations.

Although the water column cannot be sampled continuously by the moored sampler, our previous studies of chemical variability in coastal upwelling systems (e.g. Codispoti, 1981; Friederich and Codispoti, 1981) suggested a relatively tight coupling between nutrient concentrations, temperature and density over short time scales. Thus, our water sampling instruments in combination with, for example, a thermistor chain, would go a long way toward resolving small scale chemical variability in many coastal regions.

By presenting a description of our moored water sampler and some results of preliminary field tests that were conducted during the spring and summer of 1983, we hope to stimulate interest in designing innovative instruments for the study of the biology and chemistry of the sea.

RELATED DEVICES DESIGNED BY OTHERS

What follows is a brief description of some sampling devices that employ novel techniques to obtain oceanic water samples for a variety of analyses. This list is not intended to be exhaustive but merely enumerates some of the recent advances in collecting samples for special observations that depart from conventional bottle cast methods.

Broenkow (1969) and Blakar (1979) designed closely spaced, multilevel syringe sampling devices to study microstratification that employed rubber tube (messenger-tripped) or pneumatic (hand-activated) "springs", respectively. Clasby et al. (1972) modified Broenkow's (1969) mechanism by using a manual system that allowed in situ flushing of the syringes before sampling. A rosette-mounted glass syringe device powered by compressed springs was constructed by Cline et al. (1982). Hargrave and Connolly (1978) invented a clock-driven, spring-activated syringe sampler to collect time series samples (a total of 8 at 3 hour intervals) from a bell jar for the study of dissolved compound fluxes across the sediment-water interface. These authors point out that their instrument may also be used for water column sampling. An instrument designed by Sholkovitz (1970) for the collection of closely spaced water samples above the seabed, employs a dissolving magnesium rod link which activates a mechanism similar to Broenkow's (1969). Upon sampling, the instrument is released from its deployment weight and floats to the surface for retrieval.

Each of these instruments is similar in some respects to our system. However, with the exception of Hargrave and Connolly's (1978) sampler (which is limited in sample number and sampling frequency), none of these instruments may be left unattended to collect a series of samples. Our instrument is specifically designed to allow studies of nutrient dynamics over a period of days to months, and in this respect is unique.

DESCRIPTION OF THE INSTRUMENT

The major components of the moored water sampler can be seen in Figs. 1-4. It consists of a pressure case that houses the electronics and two removable PVC cylinders that serve as mounting brackets for the syringes. The pressure case also serves as the strain member in a mooring and provides the attachment point for the trawl floats that provide bouyancy and act as bumpers during deployment and retrieval.

Each PVC cylinder (Figs. 2-3) provides mounts for 20 syringes that are held in their starting (empty) positions by stainless steel wires. The wires are corroded by electrolytic action upon command from the electronics unit. In the present configuration, half of the syringes are equipped with one-way valves, and the other half are equipped with a filter sandwiched between two one-way valves. The arrangement on the latter syringes prevents contact of the filter with the ambient environment until sampling begins and prevents contact between the filtered sample and the filter itself.

Figures 2 and 3 illustrate the mechanical components. In the step by step "arming" sequence that is described below, all letters refer to those shown in the figures:

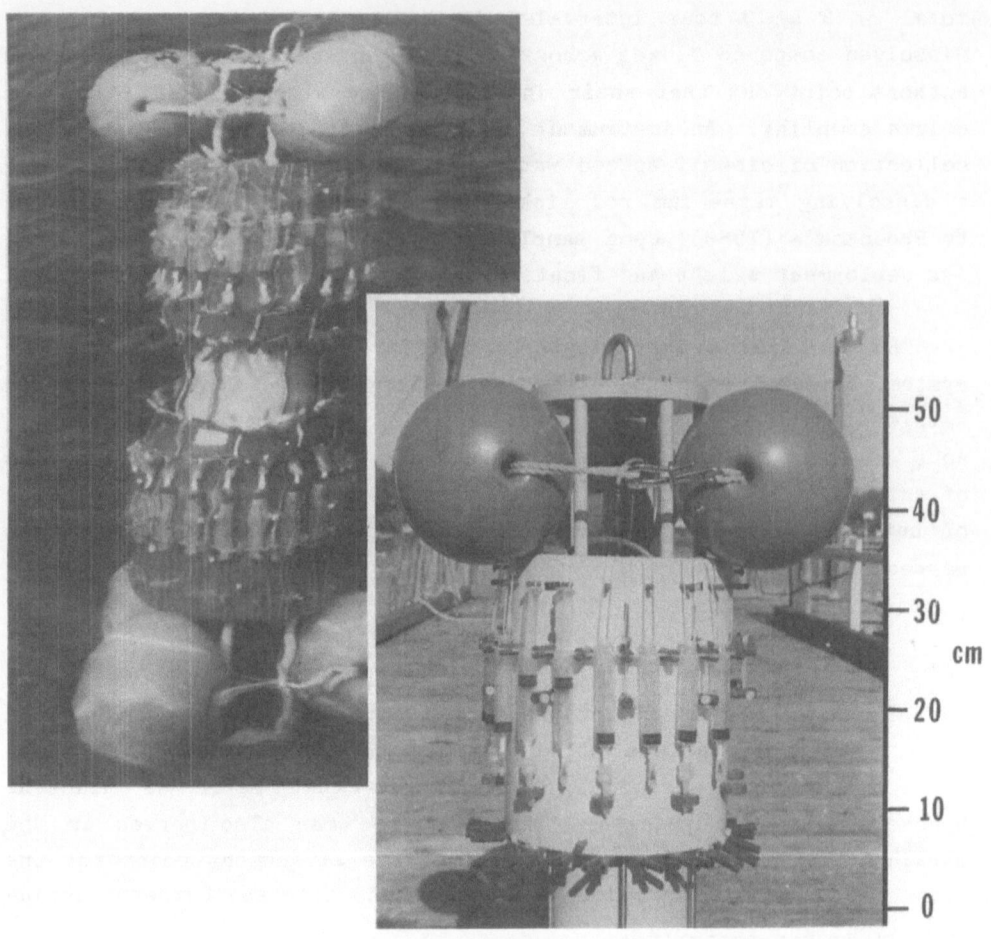

Figure 1. (A) A completely assembled sampler being readied for deployment. This is an earlier version than the one shown in the following photographs. (B) A sampler cylinder mounted on its pressure case. An identical unit is mounted on the other end of the case (total length = 115 cm.)

1) Ten lengths of nylon-coated stainless wire are spliced to a ten conductor cable leading to each underwater connector (one for each PVC cylinder). This splice typically has a useful life of about 10 deployments since a small portion of the stainless leader will be lost during each operation. Each of the leaders (a) is fed through its appropriate stub (b) and about 1/2" of insulation is removed from

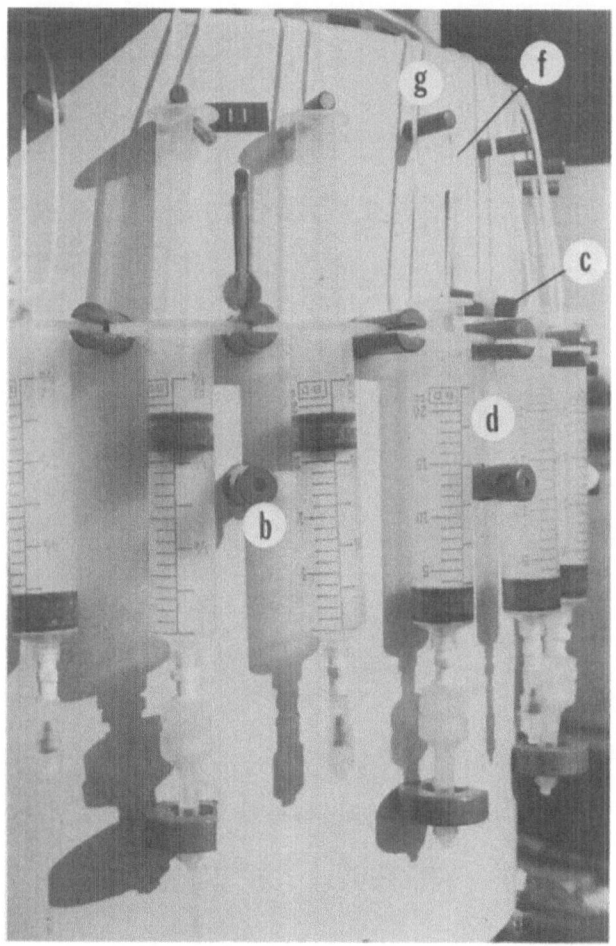

Figure 2. Partially assembled cylinder illustrating the mechanical components. Labeled parts are referred to in the text. Syringe pair #11 is released.

the end of the leaders before they are clamped in place between two nylon screws. The release lever (c) is secured in the down position with a cable tie (d) that is also attached to the release wire.

2) The rubber "springs" (e) on the inside of the PVC cylinder are attached to the release lever (c) via a monofilament lanyard (f).

3) The syringes can now be put into position and their lanyards (g) attached to the springs at position (h).

Figure 3. Partially assembled cylinder illustrating the mechanical component. Labeled parts are discussed in the text.

4) The entire ring is then mounted on a pressure case and the instrument is ready for deployment.

The syringes themselves also need some preparation. The rubber syringe plungers contain a small air space which must be filled with silicone grease to reduce deformation under pressure and aid in lubricating the barrel. All dead spaces in the syringe, valves and filter assembly must also be filled with an appropriate fluid. In the case of salinity samples, all dead spaces are filled with seawater of a salinity close to that expected from the samples. For nutrient chemistry samples, a mercuric chloride solution is used since it acts as a preservative.

The electronics unit that controls the sampling process (Fig. 4) consists of a timer and 20 release drivers. The timing interval may be altered in one hour increments over a range of 1 to 256 hours. When a release is triggered, power is supplied to the stainless wire. A zinc electrode mounted on the pressure case acts as an electrical ground, and the wire is corroded enough to part in approximately 90 seconds. This in turn releases the rubber tubing "springs" and initiates the sampling process that can be inferred from Figures 2 and 3 (the reverse of the arming sequence described above). A total of approximately 0.05 amp-hours is required for all twenty releases, and the power for this is easily provided by six alkaline "D-cells" (flashlight batteries). Six additional batteries operate the quartz crystal timing mechanism for up to five months. To enable easy removal of the PVC cylinders for servicing, the release wire leads are fed through 10 part connectors at each end of the pressure case.

The cost of construction (materials plus labor) for these instruments when built in small quantities is approximately $5,000 per instrument.

SOME PRACTICAL CONSIDERATIONS

Sample Treatment and Analysis

Three major obstacles that are encountered when analyzing water collected by this device are: (i) small sample size; (ii) preservation; and (iii) contamination. Solutions to these complications have been obtained for nutrient chemistry and salinity samples. When developing methods for other analyses, the following conditions must be accommodated:

(i) approximately 16 ml of sample are collected per syringe;

(ii) there is a dead space of about 0.2 ml in syringes with a single one-way valve and 1 ml when a filter holder and two one-way

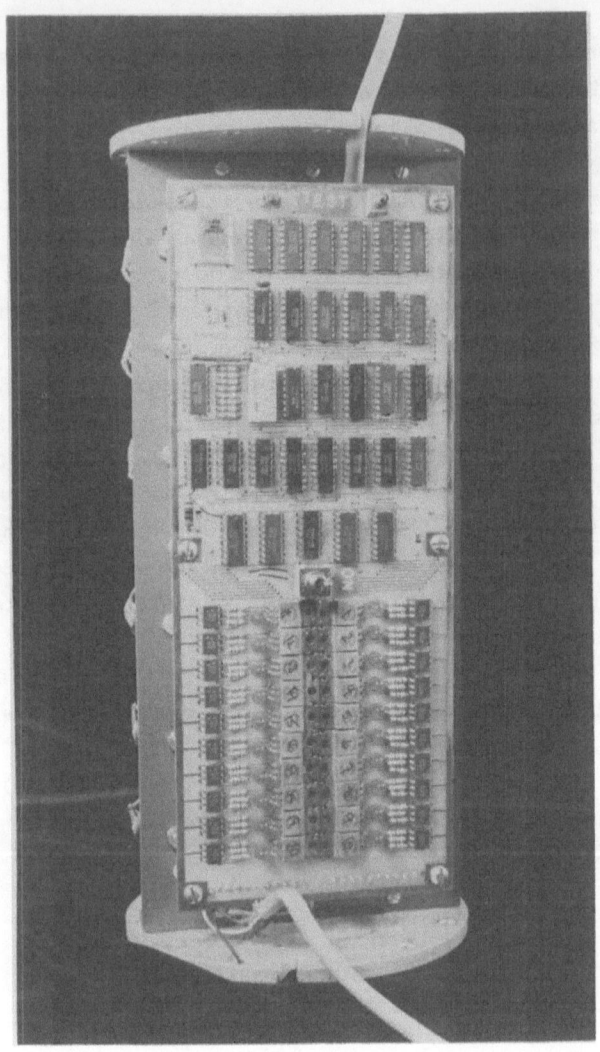

Figure 4. The electronics unit (twelve 9-volt batteries are mounted on the reverse side).

valves are used. These volumes must be filled with a neutral fluid or preservative; and

(iii) material in contact with the sample (polypropylene, nylon, ethylene propylene rubber) must be compatible with the preservation method and not affect the final analyses.

Nutrient Measurements

Nutrient analyses were performed on a Technicon AAII [R] Auto-Analyzer system, using slight modifications to the methods of Whitledge et al. (1981). After removing the valves and filter, the syringes were attached directly to the inlet system, thus eliminating any waste of sample due to rinsing and minimizing the chance of contamination. The dead-air space in the nutrient syringe filter holder and one-way valve was filled with a mercuric chloride solution (0.3 g/ℓ; 1.1 mM) which acted as a preservative. This solution interferes with the ammonium chemistry, hence, the nutrients that we have successfully measured are reactive phosphorus, dissolved silicon, nitrate and nitrite.

To determine the effect of the preservative on these nutrient analyses, we periodically compared standard curves run in the normal way against curves obtained from standards passed through a syringe set up for sample collection (i.e., with mercuric chloride in the valve portion). Ratios of the normal response to the response when mercuric chloride was present were then determined. The results of five such comparisons are presented in Table 1. Since dilution of the sample by the volume of mercuric chloride solution in the valve results in a ratio (sample/total volume) of approximately 0.94, it can be concluded that the mercuric chloride solution enhances color development in the dissolved silicon analysis and reduces it for nitrate and reactive phosphorus. The ratio for nitrite appears to show only the effect of dilution. The factors used in calculating the nutrient concentrations reported here are the product of normal factors times the average of the ratios in Table 1 for each nutrient. Correction factors should be determined for each new batch of preservative.

In initial standardizations, with and without preservative, we noted a dissolved silicon contamination due to the addition of mercuric chloride. When comparing low-nutrient seawater, used in the determination of factors, to low-nutrient seawater plus mercuric chloride, the amount of this contamination may be determined. Contamination in all nutrients being analyzed was examined and dissolved silicon was the nutrient most significantly affected by the mercuric chloride solution (average contamination = 1.42 μM, n = 4), the others being contaminated by less than 0.2 μM on average.

Table 1.   Ratios of calibration factors for normal standards to
factors of standards which include mercuric chloride
(Normal Factor/Mercuric Chloride Factor).

	NUTRIENT			
Date	$PO_4^\equiv$	$Si(OH)_4$	NO_3^-	NO_2^-
7/8/83	–	1.01	0.87	–
7/19/83	0.92	0.98	0.89	0.94
7/27/83	0.88	1.03	0.91	0.94
1/25/84	–	1.01	0.91	0.95
11/26/84	0.90	–	–	–
average	0.90	1.01	0.90	0.94

Contamination effects should be determined for each batch of mercuric chloride. In more recent experiments using new batches of mercuric chloride, no contamination was found.

Since the filters in the syringes set up for collecting nutrient samples were immersed in the mercuric chloride preservative until the sample was collected, any contaminants in or on the filter had a good chance of being dissolved. Due to the occurrence of extraordinary spikes of nitrite in our sampler field tests (as compared to bottle cast samples), we tested for contamination from the filters. A quick comparison of filtered (0.45 µ Millipore HA filters) and unfiltered seawater showed appreciable contamination in one run, but we later found small effects during a more rigorous test. The results of the latter test are presented in Table 2.

In this test we examined two filter types and the effect of acid-washing (10% HCl) in anticipation of substantial contamination from the untreated Millipore filters. The results indicate only minor contamination for both filter types regardless of acid treatment. Acid washed Nuclepore filters appeared to be best (Table 2), and discussions with colleagues suggest that they are less likely to introduce contamination than untreated Millipore filters.

Salinity Measurement

Salinity samples were analyzed in a Guildline Autosal salinometer. The conductivity cell overflow tube was replaced with a small diameter tube, reducing the fill volume to ∿5 ml. Samples were injected directly from the sample syringe into the salinometer inlet tube. Since the temperature equilibration coil in this instrument has a volume of ∿5 ml, it was found necessary to "chase" the last of the three subsamples from a 15 ml syringe into the conductivity cell. This was usually done with the first subsample from the next syringe. Tests conducted with samples that had a large salinity range have shown that this method does not introduce measurable carryover effects.

The dead-air space in the salinity syringe valves is filled with seawater of salinity similar to that being sampled. In order to determine the salinity of the sampled water, the volume of "fill" water must be known. This is determined by filling the valve with

Table 2. Comparison of filtered and unfiltered seawater for new and acid-soaked (4h in 10% HCl) .45 µ Millipore and Nuclepore filters. The filters sat in filter holders with mercuric chloride solution for six days prior to sampling. Values reported are averages for triplicate samples. Δ denotes the change due to filtration.

Treatment	$PO_4^=$	$Si(OH)_4$ (µg-atoms/ℓ)	NO_3^-	NO_2^-
Unfiltered	0.14	3.42	-0.06	0.02
New Millipore	0.29	3.52	0.29	0.08
Δ	0.15	0.10	0.35	0.06
New Nuclepore	0.23	3.51	0.05	0.07
Δ	0.09	0.09	0.11	0.05
Soaked Millipore	0.17	3.49	0.07	0.08
Δ	0.03	0.07	0.13	0.06
Soaked Nuclepore	0.19	3.36	0.05	0.06
Δ	0.05	-0.06	0.11	0.04

de-ionized water, sampling a salinity sub-standard, and then calculating the dead air space based on dilution of the sub-standard. For
the syringes used in this study, the variability in the determination
of the dead-air space ($\pm$ 0.01 ml, n = 10) translates into a salinity
uncertainty of $\sim \pm$ 0.001 ^{O}/oo (for a 1.5 ^{O}/oo difference in sample
and fill water).

In comparing sampler and bottle cast salinity data, a discrepancy was found; the sampler data were consistently higher. On
testing the syringes for evaporation, we discovered changes of $\sim$ 0.3
^{O}/oo, resulting from improper sealing of the original ball valve.
New bunsen valves have been tested successfully. They did not leak
when submerged, even during severe temperature cycling tests.
Evaporation in air was avoided by removing the valve and covering the
syringe opening with a septum upon sampler recovery.

Mooring Method

The mooring system that was used during the field tests conducted in 1983 is illustrated in Figure 5. A 300 kg anchor was lowered
on a 1/2" polypropylene line that was approximately 5 meters longer
than the water depth at high tide. The upper 5 meters were equipped
with several 8" plastic trawl floats and a float with 150 kg of net
buoyancy was attached to the end of this line. The instruments were
then suspended from the surface buoy via a 1/4" steel cable. The
lower instrument was equipped with a release mechanisms that dropped
a disposable weight when the last sample was taken. After the weight
is released, the string of instruments remains submerged, but can be
brought to the surface easily. This system functioned well, but it
would not be suitable for use in open sea conditions where more
conventional mooring techniques should be employed.

FIELD TESTS AND OBSERVATIONS

In 1982, we conducted short term tests of the sampler from the
pier at Bigelow Laboratory. These tests showed that the sampler
system worked satisfactorily. During the spring and summer of 1983,
we conducted extensive field tests using the mooring system described
above at a location near Monhegan Island, Maine (Fig. 6). During
this testing period some further material and procedural problems

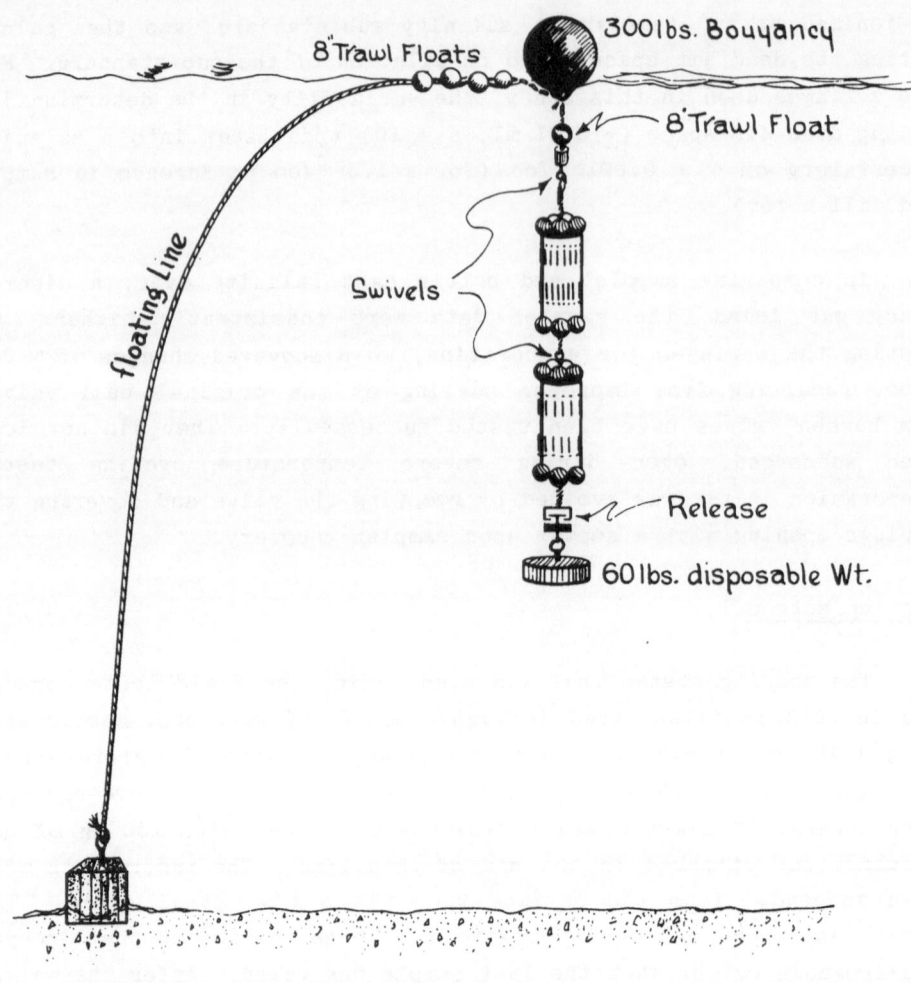

Figure 5. Schematic diagram of the mooring system used in the field
test of 1983.

were discovered and eliminated. For example, we used a brand of
stainless wire that was much more resistant to electrolysis than the
material previously employed. This new corrosion resistant wire led
to the failure of a number of deployments until the cause was
isolated.

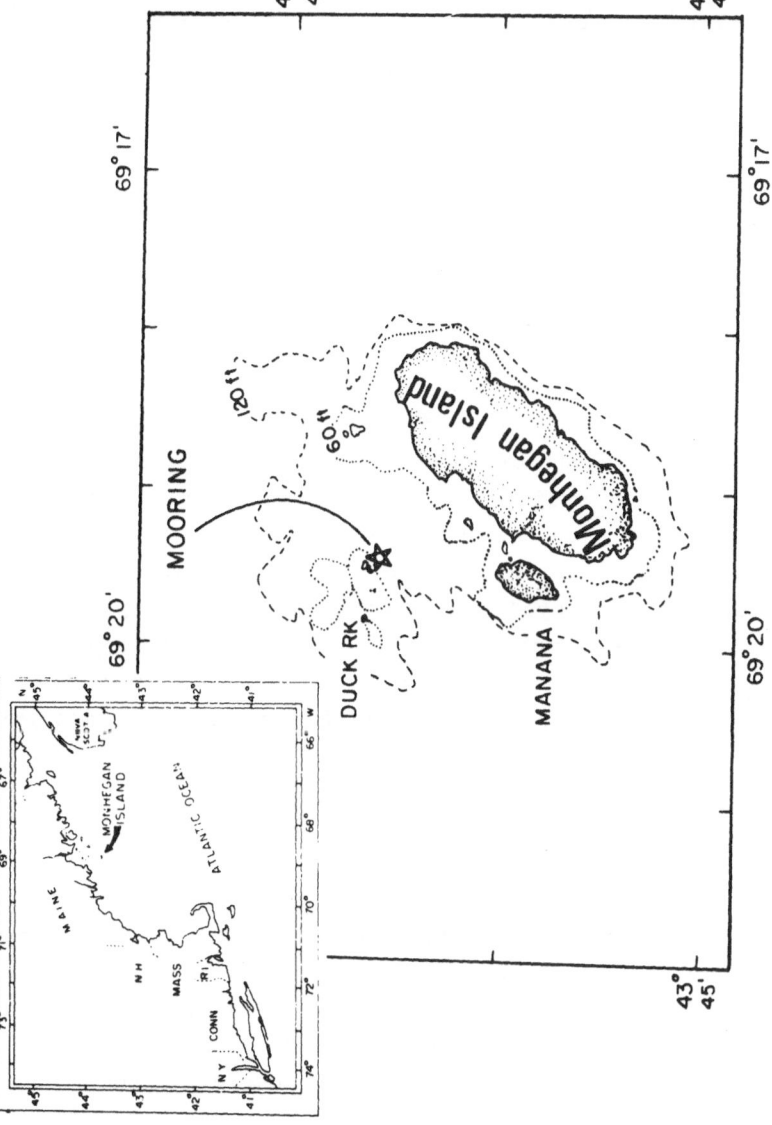

Figure 6. Location of the mooring used in the 1983 field test.

Despite the problems noted above, some of the test data appear to be realistic. For example, the data presented in Figure 7 demonstrate a correlation between the semidiurnal tide and nutrient concentration. Figure 8 shows a decreasing trend in salinity that was observed at 0 and 11 m during a several day period and was in agreement with Niskin bottle data taken at the beginning and end of the deployment.

DISCUSSION OF RESULTS

From the standpoint of instrument development, these tests have helped us to develop adequate mooring systems, and to discover how to improve some of our procedures for preparing the samplers for deployment. As a result of these tests, we estimate that more than 90% of all deployed syringes will yield acceptable samples if the samplers are set up with reasonable care. During a recent investigation of the processes on the shelf of the southeastern United States, five samplers were deployed from four taut wire moorings for a period of two months. The samplers were attached at depths of 7 and 67 meters. In spite of extreme fouling on the shallow moorings, 98 out of 100 syringe pairs collected samples for nutrient analysis and phytoplankton identification.

FUTURE DEVELOPMENTS

Since it is important to obtain simultaneous physical measurements such as temperature and pressure along with the water samples, a data logging system is being included in one of the instruments that we have been developing. The data acquisition and recording system consists of a Hewlett-Packard HP41CV calculator and HP-IL/GPIO interface. The interface and sensors are powered by the same circuit that operates the releases. The calculator runs on its own internal batteries. This unit is small in size and can be accommodated in a standard pressure case along with the standard syringe release circuit.

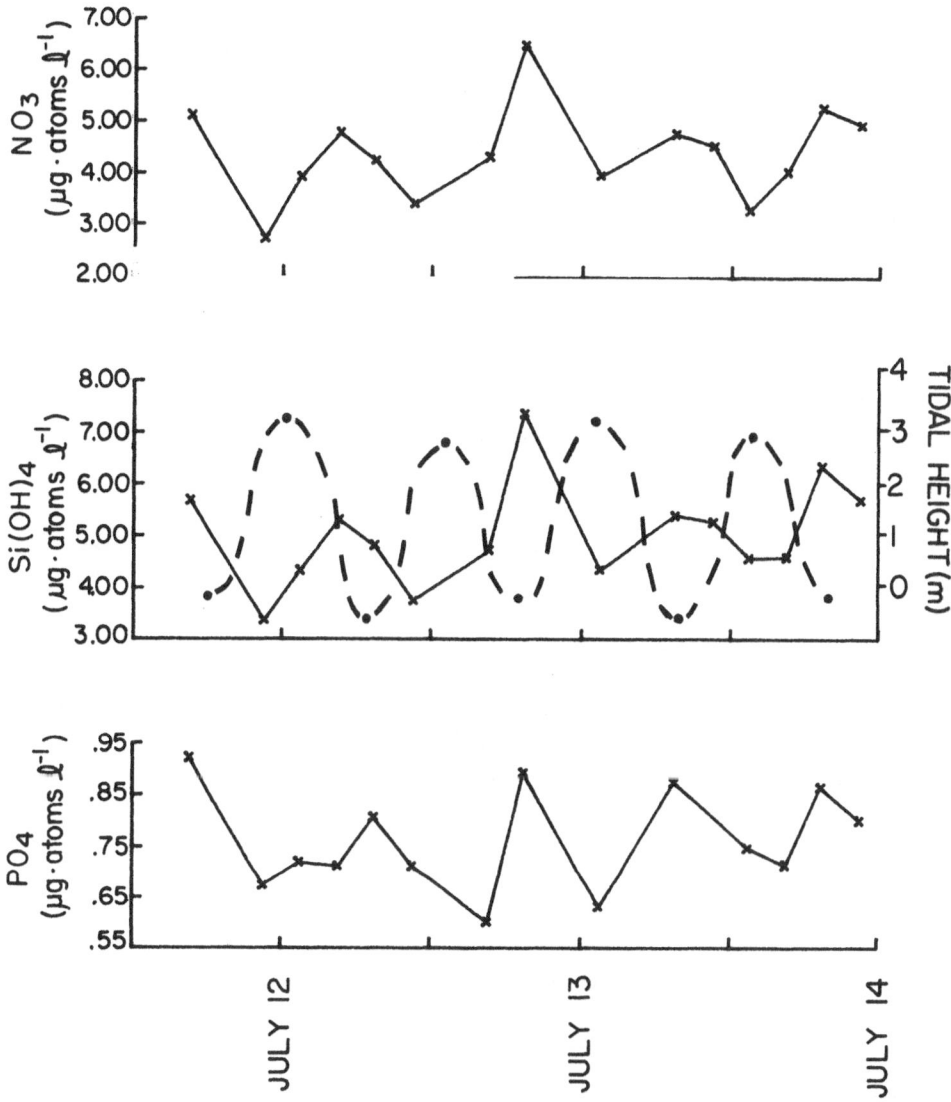

Figure 7. Time series plots of selected nutrient data from an instrument deployed at 27 m from 11–14 July 1983. The sampling
interval was 3 hours and an inverse correlation with tidal
height is evident. Nutrient concentrations are in μg-
atom/ℓ. The dashed line on the dissolved silicon plot indicates tidal height.

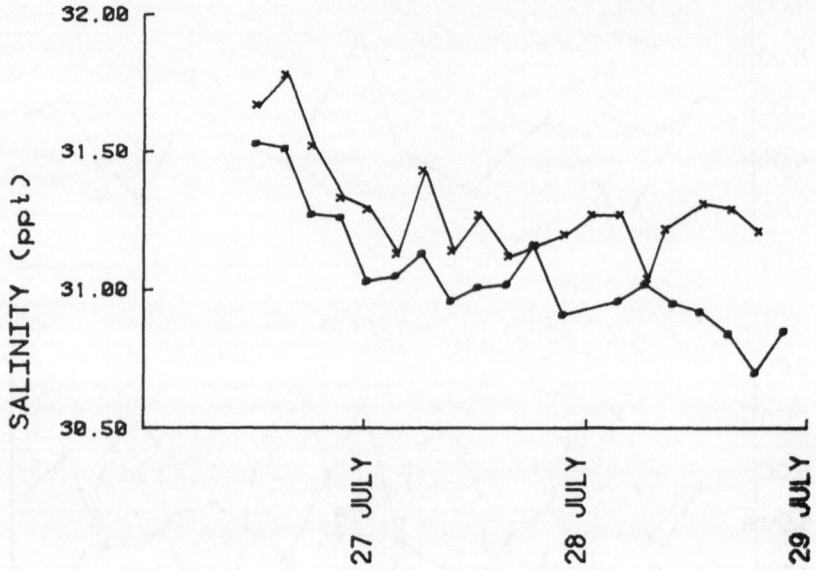

Figure 8. Time series plot of salinity at the surface (o's) and 11 m (x's) from 26-29 July 1983. The sampling interval was 8 hours. These data have been corrected for an average evaporation error of 0.3%. The generally decreasing trend was in accord with Niskin bottle data taken at the beginning and end of the deployment.

ACKNOWLEDGEMENTS

We thank R. Johnson, P. Sherman, J. Rollins, P. Colby and P. Boisvert for their technical assistance. We are also grateful for the cooperation extended by the fishermen of Monhegan Island, especially L. Krause who recovered one of our instruments after a storm.

Financial support was provided by the Office of Naval Research (contract N00014-81-C-0043). This is Bigelow Laboratory Contribution #84016.

REFERENCES

Anderson, J.J. 1982. The nitrite-oxygen interface at the top of the oxygen minimum zone in the eastern tropical North Pacific. Deep-Sea Res. 29: 1193-1201.

Blakar, I.A. 1979. A close-interval water sampler with minimal disturbance properties. Limnol. Oceanogr. 24: 983-988.

Broenkow, W.W. 1969. An interface sampler using spring-activated syringes. Limnol. Oceanogr. 14: 288-291.

Clasby, R.C., W.S. Reeburgh and V. Alexander. 1972. A close-interval syringe sampler. Limnol. Oceanogr. 17: 632-633.

Cline, J.D., H.B. Milburn and D.P. Wisegarver. 1982. A simple rosette-mounted syringe sampler for the collection of dissolved gases. Deep-Sea Res. 29: 1245-1250.

Codispoti, L.A. 1981. Temporal nutrient variability in three different upwelling regions. In: Coastal Upwelling. F.A. Richards (ed.), Am. Geophys. Union, Washington, D.C. 529 pp.

Friederich, G.E. and L.A. Codispoti. 1981. The effects of mixing and regeneration on the nutrient content of the upwelling waters off Peru. In: Coastal Upwelling. F.A. Richards (ed.), Am. Geophys. Union, Washington, D.C. 529 pp.

Hargrave, B.T. and G.F. Connolly. 1978. A device to collect super-natant water for measurement of the flux of dissolved compounds across sediment surfaces. Limnol. Oceanogr. 23: 1005-1010.

Packard, T.T. In press. The use of enzyme analysis to measure and map oxygen consumption in the ocean. In: Chemical Oceanography: Analytics of Mesoscale and Macroscale Processes. A. Zirino (ed.); Advances in Chemistry Series No. 208, American Chemical Society, Washington, D.C.

Schink, D.R., P.J. Setser, S.T. Sweet, and N.L. Guinasso, Jr. In press. Sampling the upper 100 m of a warm core ring with a towed pumping system. In: Chemical Oceanography: Analytics of Mesoscale and Macroscale Processes. A. Zirino (ed.); ADVANCES IN CHEMISTRY SERIES No. 208, American Chemical Society, Washington, D.C.

Sholkovitz, E.R. 1970. A free vehicle bottom-water sampler. Limnol. Oceanogr. 15: 641-644.

Smith, R.L. 1978. Poleward propagating perturbations in currents and sea level along the Peru coast. J. Geophys. Res. 83: 6083-6092.

Whitledge, T.E., S.C. Malloy, C.J. Patton and C.D. Wirick. 1981. Automated Nutrient Analyses in Seawater. Brookhaven National Lab. report #51398. Upton, NY. 216 pp.

Yentsch, C.M. and C.S. Yentsch. 1984. Emergence of optical instru-mentation for measuring biological properties. Oceanogr. Mar. Biol. Ann. Rev. 22: 55-98.

AN INDUCED MIXING EXPERIMENT?

J.H. Simpson
Marine Science Laboratories
Menai Bridge
Gwynedd LL59 5EY
United Kingdom

P.B. Tett
Scottish Marine Biological Association
Dunstaffnage
Oban, Argyll PA23 4ADH
United Kingdom

INTRODUCTION

As the diverse contributions to this volume show, there is now widespread recognition of the importance of tidal stirring as one of the fundamental influences on the shelf sea environment. Not only does it control the intensity of vertical mixing thus partitioning the shelf into mixed and stratified regions separated by fronts, but tidal stirring is also responsible for the resuspension of sediments which strongly influence the absorption and scattering of light in the water column. The impact of these physical processes on the biology has been the focus of considerable interdisciplinary effort in the last decade. During this period, in spite of rather slow progress in understanding the detail of vertical mixing processes and frontal dynamics, a substantial body of evidence has been assembled which suggests that the productivity and standing crop in the water column may be related to the level of tidal stirring. In particular there is some suggestive evidence that frontal zones are regions of generally higher primary production because they provide phytoplankton with a favourable balance of light and nutrients.

It must, however, be acknowledged that the available evidence for enhancement of production at fronts is limited in several important respects. Most of the data are concerned with the distribution of standing crop rather than the rate of carbon fixation and derive mainly from single quasi-synoptic surveys of different frontal regions. In these circumstances there may be a tendency for the literature to accumulate positive results showing significant

Lecture Notes on Coastal and Estuarine Studies, Vol. 17
Tidal Mixing and Plankton Dynamics. Edited by J. Bowman, M. Yentsch and W. T. Peterson
© Springer-Verlag Berlin Heidelberg 1986

enhancement, while ambiguous or even negative results (showing depletion at fronts) would go unpublished. Only in a very few cases has
there been a systematic repetition of surveys over the seasonal cycle
with a commitment to publish all the results. One such study
(Richardson et al., 1985) does demonstrate a significant enhancement
but it also clearly shows the high variability and short time scales
of biological processes which greatly limit our ability to draw firm
conclusions from isolated surveys. Clearly there is a need to develop
instruments for continuous observation of biological variables and
thus remove the 'aliasing' inherent in existing survey strategies.

The related problem of assessing island enhancement of biomass
should be simpler, if only because we are dealing with what is
effectively a point source of nutrients geographically tied to the
island. By contrast, fronts approximate to line sources which show
considerable physical variability due to advection and eddy formation
(see Fig. 1). It should be clear, however, from the discussion of
the available data in Simpson and Tett (this volume) that, even in
the island case, there are difficulties in establishing unambiguously
the causal connection between increased biomass and the extra tidal
stirring induced by the island. The island may be insufficiently
isolated from land masses or other islands and the island region may
experience significant mean advection which will sweep away the
nutrient rich water into a wake or even expose the island region to
other nutrient sources.

Perhaps the most that can be said at this stage is that a strong
prima facie case for the existence of some significant enhancement at
fronts and around islands has been made. To make further progress in
clarifying this connection and establishing it in a quantitative
form, we may need a new strategy which involves measuring the
response to controlled mixing inputs. It is the purpose of this note
to make a first exploration of the feasibility and usefulness of such
"experiments."

Experimental Requirements

The island mixing problem serves as a useful starting point for
our thinking. Insteady of accepting the various difficulties associ-

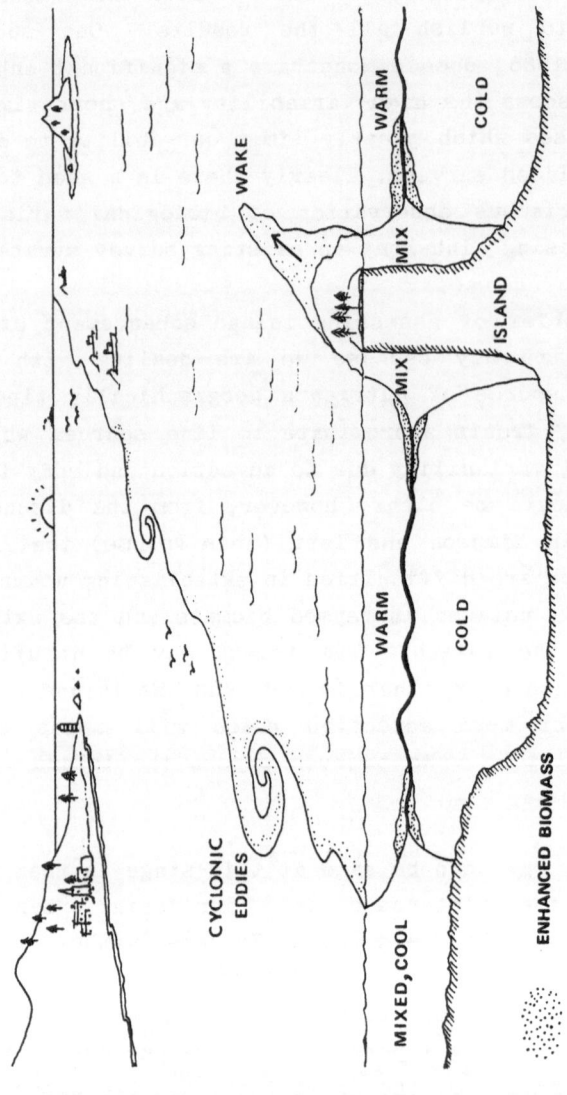

Figure 1. Zones of potential biomass enhancement in tidally energetic shelf seas.

ated with real islands, we might think in terms of a mixer approximating to an idealized island.

The mixer would be positioned in a laterally uniform region of the stratified shelf seas where the mean advection is weak. During the mature summer regime, when nutrients are depleted in the surface layer, the mixer would be used to break down the thermocline locally and provide a quantifiable input of nutrient rich water into the surface layer (Fig. 2). This injection would need to stimulate a biological response that clearly exceeded natural variability in the area.

Ideally the mixer should be capable of being switched on and off to permit the direct investigation of response times and the effects of periodic mixing episodes. Experiments of this kind would represent an important improvement over observations of natural islands where, apart from the spring-neap cycle, we are forced to accept a fixed mixer output.

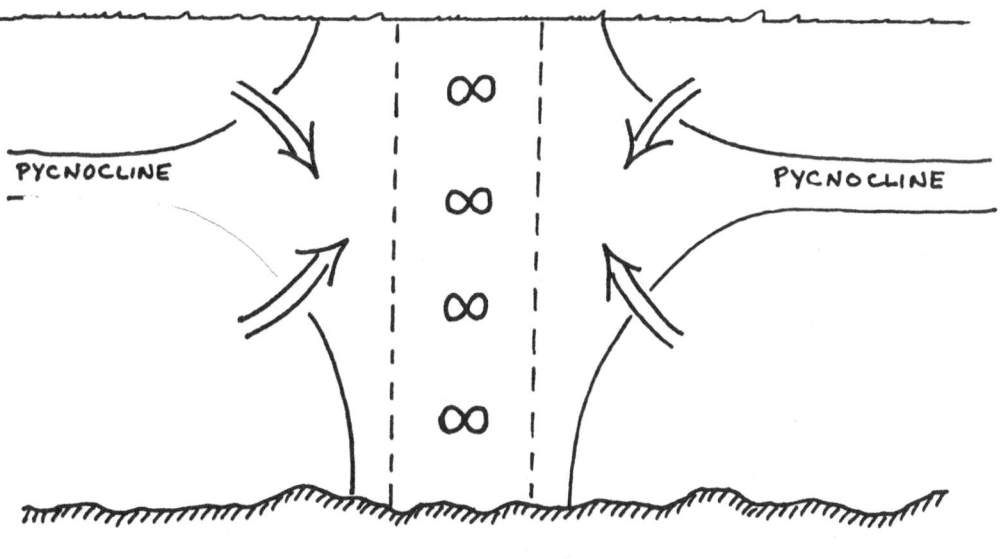

∞ = STIRRING ELEMENT

Figure 2. Breakdown of the pycnocline by forced mixing producing a nutrient flux into the surface layers.

It is also desirable that the artificial mixing should be similar to the natural process of tidal stirring and not involve other processes that may complicate the assessment of stirring effects. It may be prudent therefore to think in terms of utilizing energy, which is available almost everywhere in the shelf seas, to drive the mixing process.

In nature, the efficiency of tidal stirring is very low with only 0.5% of turbulent kinetic energy produced by bottom stresses being used to work against the buoyancy forces. This is largely because the turbulence is generated so far from the buoyancy input by surface heating. A system generating turbulence in the pycnocline may be expected to have a higher efficiency.

At the same time there is the option of extracting more power per square meter from the tidal flow. Frictional dissipation by the bottom boundary is limited to ~1 watt m^{-2}. By using high-drag elements located in midwater much more power can be extracted. This, together with improved efficiency, permits the production of significant mixing rates from a relatively modest structure - certainly very much smaller than the equivalent natural island.

Estimates of Mixing Rates and Power Demand

To clarify these general points, we will consider the output of a suitable mixing source in more detail.

Because of the tidal flow, the mixing source is moving relative to the water and creates a central zone of more or less uniform nutrient concentration C_o with a scale r_o which is effectively the radius of the tidal ellipse (Fig. 3). As a first approximation we may treat the subsequent spreading as the purely radial diffusion of a conservative scalar: the concentration C thus represents all dissolved and particulate forms of the nutrient. The diffusion equation for such radial spreading is just

$$\frac{\partial}{\partial r} \left(rK \frac{\partial C}{\partial r} \right) = 0$$

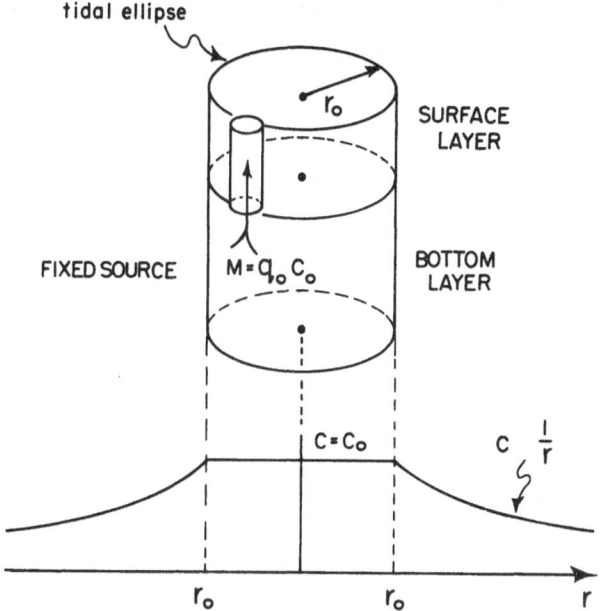

Figure 3. Schematic of nutrient supply to the surface layer by a mixer source which moves (relative to the water) around the tidal ellipse.

r_o = scale of the tidal ellipse

C_o = surface nutrient concentration enhancement within the tidal ellipse

q_o = rate of production of mixed water

C_b = nutrient concentration of mixed water injected into the photic zone

h_1 = depth of the mixed layer

$\dot{M}$ = mass flux of nutrients from mixing zone into the euphotic zone

For the case of a scale dependent diffusivity of the form $K = \alpha r$ the solution is

$$C = \frac{\dot{M}}{2\pi\alpha hr}$$

where $\dot{M}$ is rate of nutrient supply to the surface layer of depth h. This solution implies a steady flux $\dot{M}$ outwards across each circular section surrounding the source due to radial spreading at the "diffusion velocity" α. The concentration in the source region will be

$$C_o = \frac{\dot{M}}{2\pi\alpha h r_o} = \frac{q_o C_b}{2\pi\alpha h\, r_o} \qquad (1)$$

where q_o is the rate of supply of mixed water at concentration C_b.

C_o is an excess over 'natural' or 'background' photic zone nutrient concentration C'; a proportion of the excess will appear as a biomass enhancement X_o above the 'normal' concentration X'; we assume that for small C_o the yield y is unchanged from its normal value. That is

$$X' = y.C' \quad \text{and} \quad X_o = y.C_o$$

Standardized enhancements are defined as

$$C* = C_o \,/\, C' \;=\; X_o \,/\, X' \;= X*$$

In order to be detectable, $X*$ must exceed normal variation. The natural coefficient of variation of biomass is given by

$$V* = \sigma_X' \,/\, X'$$

and thus the detectable minimum is defined by

$$X*_{min} \;=\; C*_{min} \;=\; 2V*$$

Hence the minimum volume production rate for the mixer will be from (1)

$$q_o(min) \;=\; 2V* \; \frac{.C'}{C_b} \;.\; 2\pi\alpha h r_o$$

Figure 10 in Simpson and Tett (this volume) suggests that for measured forms of nitrogen in stratified water in the approaches to the Scilly Isles $C'/C_b \simeq 0.2$.

The value of V* is critical and must be determined on spatial and temporal scales, and under stratification conditions, equivalent to those of the proposed experiments. We estimate it to be about 0.10 on the basis of data in Figure 9, Simpson and Tett (this volume). Thus $q_o(min)$ must be $\simeq 160$ m^3 s^{-1} with h = 25 m, α = 5 x 10^{-3} ms^{-1} and r_o = 5 x 10^3 m.

To provide mixed water at this rate it will be necessary to work against the buoyancy forces associated with the density difference $\Delta\rho$ between the surface and bottom layers of depths h_1 and h_2, respectively. Assuming a two-layer structure of this kind, the necessary rate of working will be

$$P_m = \frac{q_o \, g\Delta\rho h_1 h_2}{2 \, (h_1 + h_2)}$$

For $\Delta\rho$ = 1 kg m^{-3}, h_1 = 20 m, h_2 = 60 m we have $P_m \simeq 12$ kW, a relatively modest power demand, which could in principle be met in several ways.

REALIZATION OF A PRACTICAL MIXER

One obvious option would be to use a moored grid of horizontal bars to generate turbulence in the pycnocline as the tidal flow streams past the mooring (Fig. 4). The use of a single point mooring enables the grid to be aligned perpendicular to the flow for all phases of the tidal ellipse.

The power dissipated by such a grid will be given by

$$p = \tfrac{1}{2}C_D A\rho \overline{u^3}$$

where C_D is the drag coefficient based on the total aspect area A of the grid and $\overline{u^3}$ is the mean cube of the tidal velocity. For a grid in which the ratio of bar area to total area is ~ 0.35 (optimum for mixing), we may take C_D = 1.6 (Naudascher and Farell, 1970).

The fraction of energy available for mixing is the flux Richardson number ε and we may assume a conservative value of ε = 0.05 based on the laboratory experiments of Linden (1980). Setting the mixing power $P_m = \varepsilon P$ we have for the required grid area

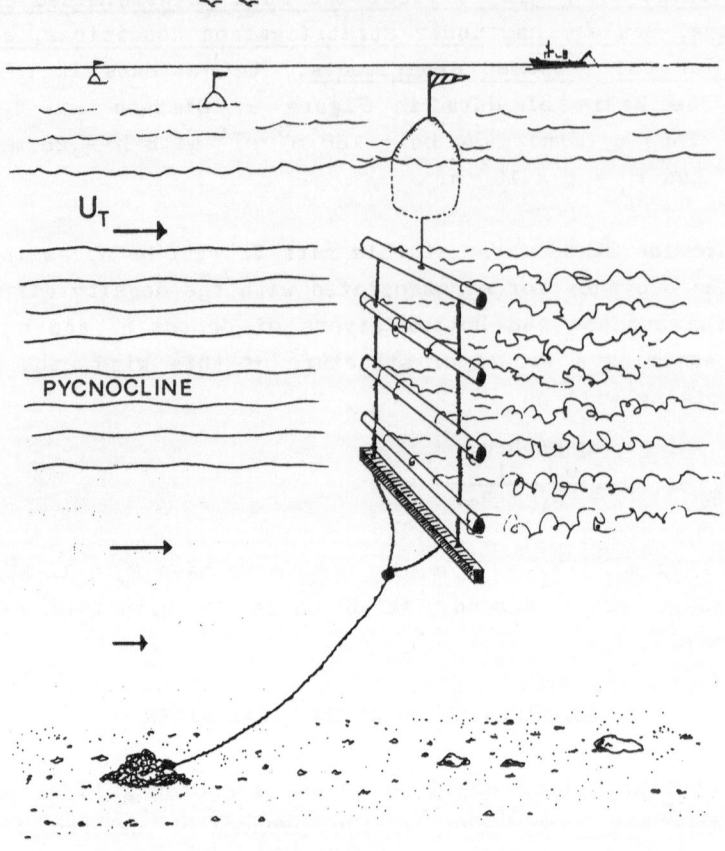

Figure 4. A moored grid of horizontal bars using tidal energy to force vertical mixing. Significant mixing could be produced by a number of such units each with a grid area of 20 x 25 m^2. U_T is the tidal stream velocity.

$$A = \frac{2 P_m}{C_D \rho \overline{u^3} \varepsilon}$$

Assuming a tidal stream amplitude of 0.5 ms^{-1}, $\overline{u^3} \simeq (4/3\pi)x(0.5)^3 = 0.053\ m^3\ s^{-3}$, so that $A \simeq 5500\ m^2$. This may be realized as a number of smaller units moored close together, say twelve grids each of 450 m^2. The maximum drag force on each unit would be

$$F_D = \tfrac{1}{2}C_D A\rho u^3 = 92.5 \text{ kN} \simeq 9.4 \text{ tonnes wt.}$$

With careful design the grid efficiency may be improved up to ∿10% with consequent reductions in the scale of the engineering involved.

LIMITATIONS AND ALTERNATIVES

The extraction of power from the tidal flow which mimics the natural process is an attractive, but not an essential, aspect of an experimental arrangement. Mixing could equally well be forced by an air-lift bubbler system in which air would be pumped from an array of nozzles at the sea bed. This approach would allow ready control of the mixer and may also lead to a better distribution of nutrients into the surface layer with some of the input being carried right up to the surface. By contrast, mixing by a grid would tend to take the form of a thickening of the pycnocline with most of the nutrient input into the lowest layers of the euphotic zone.

A bubbler system would require power input of ∿100 kW so that it would only be practical in areas like the northern North Sea where there are existing oil platform installations. Bubblers may also induce other biologically significant effects apart from vertical mixing.

A further alternative would be to use thermal convection to bring bottom water up to the surface layers. The power input here is, however, very large (∿2 x 10^6 kW for q_o = 160 m^3 s^{-1} and a surface to bottom temperature difference of ΔT = 4°C) so that it would not be practical unless a prolific source of waste heat was available.

FURTHER IMPLICATIONS

The above estimates of the mixer requirements are necessarily crude and they should be regarded only as a starting point for more informed thought experiments which would be a necessary prelude to any serious experimental commitment. In order to undertake a full modelling study we need to know more about some of the key processes and the numerical values of certain parameters. A valuable inter-mediate step in understanding the dispersion by turbulent diffusion and advection would be the release of a discrete dump of one of the new purposeful tracers (e.g. SF_6) which do not occur in nature and have a very low detection threshold ($\sim 10^{-16}$). Such a tracer may be used to determine the rate of input of nutrient to the surface layer from mixed water introduced into the pycnocline. Water pumped from the pycnocline would be labelled with the tracer and discharged back into the pycnocline. Subsequent monitoring of tracer concentration in the surface layer and pycnocline should permit estimates of both the entrainment rate into the surface layer and the subsequent lateral diffusion in the surface layer.

On the biological side we need improved models of the nutrient response in a real experiment; such models may be developed from existing algal nutrient-growth theory (e.g. Tett and Droop, in press) and could be used to refine estimates of the minimum source strength required. As well as developing models to predict overall enhancement of nitrogen (or other nutrients) in the photic zone, and the aggregate yield of plant, animal and bacterial plankton, there are interesting questions to answer concerning how the species composition of the phytoplankton might be influenced by an extra source of inorganic nutrients, and by enhanced mixing (see e.g. Margalef, 1978), and how the microzooplankton, mesozooplankton, and bacterioplankton would respond to these changes in mixing and food quality and quantity.

The prospect of a major experimental commitment in this area may seem daunting in the present funding climate but we should recognize the weakness of persisting with more of the same descriptive survey work which is unlikely even to provide the basis for properly testing our models.

We should also not lose site of the fact that a clear and unambiguous demonstration of enhancement by such an experiment would open up the possibility of improving the fertility of the shelf seas. If our basic ideas about nutrient limitation are correct, then production could be substantially improved by using an array of tidal stirrers to recruit a much higher proportion of the available nutrient into the photic zone. By a suitable choice of stirrer spacing we would effectively turn large areas of the shelf into frontal zones in which the balance of stratification, light and nutrients would be optimal (Fig. 5).

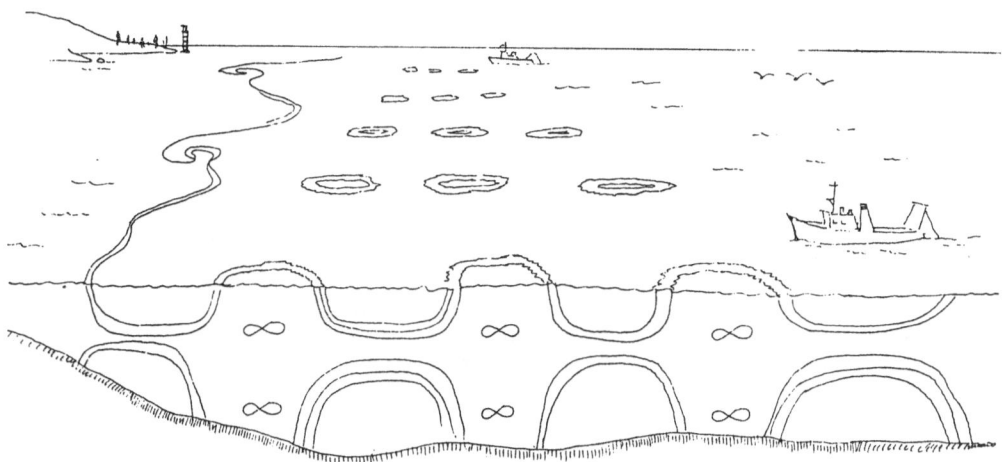

Figure 5. Increasing production of the shelf seas by an array of mixer units spaced to provide alternating regions of strong stratification and mixing.

A significant fraction of the energy expended by developed economies is used in the manufacture of fertilizers and related agriculture activities aimed at the production of high protein foods. It is possible that harnessing tidal energy to the direct enhancement of organic production in the shelf seas, and thus to an increased harvest of fish protein, might make a more effective contribution to these economies than using tidally generated electrical power to manufacture fertilizers for terrestrial applications.

ACKNOWLEDGEMENTS

The ideas presented here were stimulated by a workshop on tidal mixing held in San Francisco in December 1984 and organized by Dr. C.M. Yentsch. Paul Linden provided helpful advice on the efficiency of stirring grids and Ed Hill kindly assisted with the diagrams.

REFERENCES

Linden, P.F. 1980. Mixing across a density interface produced by grid turbulence. J. Fluid Mechanics 100: 691-703.

Margalef, R. 1978. Life forms of phytoplankton as survival alternatives in an unstable environment. Oceanol. Acta 1: 493-509.

Naudascher, E. and C. Farell. 1970. Grid turbulence. Proc. A.S.C.E. Journal of Engineering Mechanical Division 96: 121.

Richardson, K., M.F. Lavin-Peregrina, E.G. Mitchelson and J.H. Simpson. 1985. Seasonal distribution of chlorophyll a in relation to physical structure in the western Irish Sea. Oceanol. Acta 8: 77-86.

Tett, P. and M.R. Droop. In press. Cell quota models and planktonic primary production. In: Handbook of Laboratory Model Systems for Microbial Ecosystem Research. J.W.T. Wimpenny (ed.). CRC Press.

SUBJECT INDEX*

* In most cases, the page numbers listed in the subject index
  indicate only the earliest references to that topic within each
  chapter. Reading further in the chapter(s) will generally yield
  more information on the subject desired.